Springer Proceedings in Physics **416**

Indexed by Scopus

The series Springer Proceedings in Physics, founded in 1984, is devoted to timely reports of state-of-the-art developments in physics and related sciences. Typically based on material presented at conferences, workshops and similar scientific meetings, volumes published in this series will constitute a comprehensive up to date source of reference on a field or subfield of relevance in contemporary physics. Proposals must include the following:

- Name, place and date of the scientific meeting
- A link to the committees (local organization, international advisors etc.)
- Scientific description of the meeting
- List of invited/plenary speakers
- An estimate of the planned proceedings book parameters (number of pages/articles, requested number of bulk copies, submission deadline).

Please contact:

For Americas and Europe: Dr. Zachary Evenson; zachary.evenson@springer.com
For Asia, Australia and New Zealand: Dr. Loyola DSilva; loyola.dsilva@springer.com

Ery Fitrianingsih · Johan Muhamad ·
Yazdi Ibrahim Jenie · Joko Widodo
Editors

Proceedings of the 10th International Seminar on Aerospace Science and Technology; ISAST 2024; 17 September, Bali, Indonesia

Integrating Aviation, Aerospace Science and Technology for Climate Solution

Springer

Editors
Ery Fitrianingsih
Research Organization for Aeronautics and Space
National Research and Innovation Agency of Indonesia (BRIN)
Jakarta, Indonesia

Johan Muhamad
Research Organization for Aeronautics and Space
National Research and Innovation Agency of Indonesia (BRIN)
Jakarta, Indonesia

Yazdi Ibrahim Jenie
Faculty of Mechanical and Aerospace Engineering
Bandung Institute of Technology (ITB)
Bandung, Indonesia

Joko Widodo
Research Organization for Electronics and Informatics
National Research and Innovation Agency of Indonesia (BRIN)
Jakarta, Indonesia

ISSN 0930-8989 ISSN 1867-4941 (electronic)
Springer Proceedings in Physics
ISBN 978-981-96-1343-4 ISBN 978-981-96-1344-1 (eBook)
https://doi.org/10.1007/978-981-96-1344-1

This Springer imprint is published by the registered company Springer Nature Singapore Pte Ltd.
The registered company address is: 152 Beach Road, #21-01/04 Gateway East, Singapore 189721, Singapore

Preface

The 10th International Seminar of Aerospace Science and Technology 2024 (ISAST 2024) serves as a global scientific platform advancing the theme, "**Bridging Asian Aerospace Research and Innovation to the Global Stage**," focusing on integrating aviation and aerospace technologies to address climate challenges. This year, ISAST 2024 was co-organized by the Research Organization for Aeronautics and Space under Indonesia's National Research and Innovation Agency (*Badan Riset dan Inovasi Nasional*, BRIN), Institut Teknologi Bandung (ITB), and the Indonesian Researcher Union (*Perhimpunan Periset Indonesia,* PPI). Since its inception in 2013, ISAST, originally under the Indonesian National Institute of Aeronautics and Space (*Lembaga Penerbangan dan Antariksa Nasional*, LAPAN), has been a flagship conference fostering interdisciplinary collaboration and knowledge sharing for Indonesian researchers, academia, and professionals in the aerospace and aviation sectors, reaching both regional and international communities. In 2021, LAPAN transitioned into the Research Organization for Aeronautics and Space under BRIN, further enhancing its role in national and international research initiatives.

This year's ISAST 2024 was held in Bali, Indonesia, in conjunction with the Bali International Air Show 2024, organized by the Coordinating Ministry for Maritime Affairs and Investment of the Republic of Indonesia (Kemenkomarves RI), the Indonesian Airforce (TNI AU), and the Ministry of Transportation (Kemenhub RI). The event marks a decade of ISAST, making it particularly significant. This milestone gathering brought together a wide range of stakeholders, aiming to foster synergy and collaboration among researchers, academia, Government bodies, and industry, further reinforcing the conference theme of bridging Asian aerospace research with the global stage.

This conference proceeding compiles peer-reviewed papers accepted and presented at ISAST 2024. Reflecting the theme, these contributions are categorized into four key topics:

1. Climate Change Detection and Monitoring Using Aeronautics and Space Technology.
2. Space Environment and Climate Change Interactions.
3. Innovative Aviation and Aeronautics Technology Toward a Green and Sustainable Environment.
4. Advanced Aerospace Technology in Combating Climate Change.

The first topic highlights the role of aeronautics and space technology, including vehicles, sensors, and algorithms, as critical tools in addressing global climate change. The second topic delves into the interactions between space environment dynamics and Earth's climate. The third topic showcases innovations in unmanned systems, aerodynamics, and structural design aimed at promoting a green and sustainable aerospace industry. The fourth topic presents further applications of aerospace technology in combating climate change.

ISAST 2024 also featured an opening speech followed by keynote addresses from notable figures, including Ir. Christianus R. Dewanto, M.Eng (BRIN), Mr. Atsushi

Watanabe (National Space Policy Secretariat, Cabinet Office, Japan), Dr. Robertus Heru Trihardjanto, MSc. (BRIN), and Prof. Dr. Erna Sri Adiningsih (INASA). The event also featured invited talks from distinguished speakers such as Dr. Gan Kai Poh (NASA Goddard Space Flight Center, USA), Prof. Taufiq Hidayat (ITB), Dr. Yazdi I. Jenie (ITB), Dr. Rezy Pradipta (Boston University, USA), Prof. Jian Jiang (University of Science and Technology, China), Dr. Mikio Kurita (Kyoto University, Japan), Prof. Koji Shimoyama (Kyushu University, Japan), and Dr. Hiroyo Ohya (Chiba University, Japan).

The successful organization of ISAST 2024 was made possible through the support of numerous individuals and organizations. On behalf of the ISAST 2024 Organizing Committee, we would like to express our sincere gratitude to all contributors, including BRIN, ITB, PPI, the Kemenkomarves RI, TNI AU, Kemenhub RI, the Climate Promise of the United Nations Development Programme (UNDP), the Archipelagic and Island States (AIS) Forum, Dassault Systèmes, and our esteemed sponsors: PT. Nusantara Secom IntoTech, PT. Dahana, and Ocean Solution Technology Co., Ltd.

Finally, we extend our heartfelt thanks to the authors, speakers, and delegates for their participation. Your contributions and presence made this conference a success.

Oktober 2024 The Editors of ISAST 2024

Organization

ISAST 2024 Committee

General Chair

Peberlin Sitompul

Co-chairman

Anton Timur Jaelani
Yazdi Ibrahim Jenie

Technical Program Committee

Chairs

Ery Fitrianingsih
Johan Muhamad
Yazdi I. Jenie
Joko Widodo

Editorial Board Members and Reviewers

Heri Budi Wibowo
Thomas Djamaluddin
Heri Budi Wibowo
Farohaji Kurniawan
Cahya Edi Santosa
Muhammad Ilham Adhynugraha
Mario Batubara
Dandy Aditya Novresiandi
Matza Guszo Andika
Hadyan Luthfan Prihadi
Akhmad Farid Widodo
Novi Andria
Ibnu Nurul Huda
Harry Septanto
Abdul Rahman
Rhorom Priyatikanto
Rianto Adhy Sasongko
Pramudita Satria Palar
Ridanto Eko Putro
Leonardo Gunawan
Babag Purbantoro
Mario Batubara
Edy Trihatmoko
Clara Y. Yatini

Eriko Nasemudin Nasser
La Ode Muhammad Musafar
Afni Restasari
Idris Eko Poetro
Tiar Dani
Udhi C. Nugroho

Advisory Board

I Gede Wenten
Robertus Heru Triharjanto
Tatacipta Dirgantara
Wahyu Srigutoma
Lavi Rizki Zuhal
Yuyu Wahyu
Taufiq Hidayat
Dhani Herdiwijaya
Erna Sri Adiningsih
Thomas Djamaluddin
Heri Budi Wibowo
Arif Nur Hakim
Fadillah Hasim
Wahyudi Hasbi
Emanuel Sungging Mumpuni

Secretary

Elvina Ayu Ratnasari

Organizing Committee

Timbul Manik
Dewi Anggraeni
Gerhana Puannandra Putri
Ari Nuryadi
Estu Kriswati
Christine Widianingrum

IT, Multimedia and Website

Rizal Suryana
Christine Widianingrum

Publications

Noviastuti Putri Indrasari
Indah Susanti

Sponsored by

Contents

Space Environment and Climate Change Interactions

Climate Change Detection and Monitoring Using Aeronautics and Space Technology

Remote Sensing-Based Analysis of the Relation Between Land Surface Temperature and Major Air Pollutants in Western Part of Java Island

Agil Akbar Fahrezi(✉)

Remote Sensing Study Program, Universitas Gadjah Mada, Yogyakarta 55281, Indonesia
agilakbarfahrezi@mail.ugm.ac.id

Abstract. Highly populated urban areas in the Western Part of Java Island are the center point of many anthropogenic activities that lead to high pollution levels. Increasing pollution is associated with various public health risks, decreased environmental quality, and increased land surface temperature (LST). LST and air pollution are unclear, and studies are often limited to city-based analysis. This study will analyze the relationship between several Major Air Pollutants (MAP) and LST via Ordinary Least Square (OLS) and correlation analysis. Three major indicators of MAP are observed: CO, NO_2, SO_2, and O_3 gasses. The MAP is derived from Sentinel-5P Tropomi measurement, while the LST data is derived from the MODIS sensor onboard the Terra satellite; both satellites have matching spatial resolution, which results in the same scale of analysis. This study used Google Earth Engine (GEE) to derive the respected satellite's average concentration and temperature in 2023. OLS analysis resulted in multiple R^2 values of 0.647751 and adjusted R^2 values of 0.645827. Based on these results, MAP explains approximately 64% of the LST in the western part of Java Island. The correlation analysis between MAP and LST shows a strong correlation (CO, NO_2, and O_3) and a weak correlation between SO_2 and LST.

Keywords: LST · MAP · Pollution · OLS · Correlation

1 Introduction

Urban centers, characterized by dense anthropogenic activity and industrial development, are often associated with sources of various air pollutants [1, 2]. One of the regions with such conditions is the western part of Java Island, especially in Jabodetabek (Jakarta, Bogor, Depok, Tangerang, and Bekasi) metropolitan areas. High population mobility, urbanization, industrial development, and poor mitigation efforts culminated in the region experiencing its worst air quality in 2023, attracting substantial media attention.

A study by the World Health Organization (WHO) concluded that air pollution is a significant threat to human health [3]. Another study indicates that anthropogenic air pollution emissions are responsible for approximately 70% of global health problems [4]. Beyond its impact on human health, air pollution also adversely affects the environment

E. Fitrianingsih et al. (Eds.): ISAST 2024, SPPHY 416, pp. 3–13, 2025.
https://doi.org/10.1007/978-981-96-1344-1_1

and is characterized by declining environmental quality. One consequence of declining ecological quality is increased Land Surface Temperature (LST) [1, 5, 6].

Numerous studies concluded that LST is often influenced by land use or land cover change [7, 8]. This change is frequently associated with increased LST, leading to high air pollution [9]. On the other hand, the relationship between LST and air pollution is not clearly understood, and studies are often limited to city-based analysis [6]. Such studies conducted by Lou et al. [10] state that temperature changes significantly impact air quality and its subsequent health effects. Fuladlu and Altan [6] also found a relationship between LST and air pollution in Tehran, Iran. This study also highlighted the use of Major Air Pollutants as an indicator for studying the relationship.

Major Air Pollutants (MAP) are a series of gasses that consist of Nitrogen Dioxide (NO_2), Particulate Matter (PM_x), Carbon Monoxide (CO), Sulphur Dioxide (SO_2), Ozone (O_3), and Lead. MAP surface concentrations can increase public health risks and ultimately contribute to increased rates of respiratory disease and death [11]. Therefore, monitoring and mapping air pollutants is essential in public health mitigation efforts [12]. Remote sensing is known for providing the primary data sources of many biophysical indicators, including MAP and LST. LST become a standard product of many satellite-based sensors. One is the Moderate Resolution Spectroradiometer (MODIS) onboard Terra and Aqua satellites, capable of recording LST twice daily with 1 km spatial resolution [13]. This capability allows daily and large-scale LST studies to comprehensively monitor LST's global and regional variations.

The Sentinel-5P Tropospheric Monitoring Instrument (TROPOMI) is one of the newest satellite missions to monitor air quality [12]. The sensor is designed for mapping and tracking major atmospheric constituents, including ozone (O_3), nitrogen dioxide (NO_2), carbon monoxide (CO), sulfur dioxide (SO_2), methane (CH_4), formaldehyde (CH_2O), aerosols, and clouds [14]. Several studies have implemented this satellite for mapping and monitoring air pollution, such as those by Zheng et al. [15] and Halder et al. [16]. These studies concluded that Sentinel-5P can monitor variations of air pollution over large areas with higher resolution than its previous sensors. The development of cloud-based computing also improved the application of this sensor, enabling a more comprehensive spatiotemporal analysis of air pollution [17].

Based on the abovementioned background, this research analyzes the relationship between LST derived from MODIS and MAP from Sentinel-5P. Referring to works conducted by Fuladlu and Altan [6] and Wang et al. [18], the MAP used in this research is a direct product of Sentinel-5P imagery. This research will be carried out in the western part of Java Island throughout 2023. This location was chosen because the west part of Java Island experienced fluctuating air pollution conditions during 2023. Regression analysis between MAP and LST gas concentrations will be used to quantify the magnitude of the influence of MAP gas on LST. Correlation analysis will also be applied to assess how MAP gas concentrations relate to LST.

2 Materials and Methods

2.1 Location

This research was conducted in the western part of Java Island, including Banten, DKI Jakarta, West Java, parts of Lampung, and parts of Central Java Province. The location's extent is chosen because air pollution does not tend to be limited within an administrative border. To further clarify the location, the area of interest in this study is shown in the orange box in Fig. 1.

Fig. 1. Map of the research location shown in the orange box.

2.2 Data Processing

Data Collection and Processing. This study used two satellite products, namely LST from MODIS and MAP gas concentrations from Sentinel-5P Tropomi. MODIS daily LST recorded by Terra and Aqua satellites are derived using two algorithms, namely the generalized split window and the day/night algorithm [19]. For this study, daily daytime LST was used to match the recording time of Sentinel-5P. The MAP included in this study were CO, SO_2, and O_3 vertical column concentrations. The vertical column concentration included measurement from the ground to 6 km above sea level. For NO_2 gas concentration, the total tropospheric layer concentration is used in this study.

The data collection process in this study refers to the 'calling' of MODIS and Sentinel-5P image datasets on the GEE platform. This dataset, also known as the image collection, is a collection of images on GEE stored in a sizeable cloud-based data structure [20]. Two types of Sentinel-5P data are available in GEE, offline and Near Real Time (NRT). This study used offline data due to broader coverage compared to NRT [14]. The mean value reducer command summarizes the LST and MAP from January 2023 to December 2023.

Processing of the MAP gas concentrations includes cloud masking for NO_2 gas and cropping the image to the study area boundary shown in Fig. 1. Cloud masking for NO_2 gas was performed due to the availability of cloud fraction data, which is not available in other Sentinel-5P products [14]. Processing MODIS LST required the calculation of scale factors provided by NASA (Eq. 1). Therefore, users only need to apply the equation to convert the DN value into temperature in Kelvin [21].

$$LST_{celcius} = (MODIS_{LST} * 0.02) - 273.15 \quad (1)$$

Data Analysis. The Ordinary Least Square (OLS) global regression model is used to analyze the relationship between a dependent variable (LST) and independent variables (MAP). Compared to much more complex regression analysis, such as Geographically Weighted Regression (GWR), OLS is more general and simple, commonly used in multivariate data analysis [22]. OLS could also handle multicollinearity, which GWR sometimes causes problems in GWR analysis. Pearson correlation analysis is also used to model the strength and type of the relationship between LST and MAP. This analysis was conducted based on the values of the specified sample locations. The Pearson correlation values range from –1 to 1, with higher values corresponding to a stronger correlation. Related to interpreting the correlation value, work by Schober et al. [23] will be used as a reference for this study.

Due to the extent of the research location, 750 random samples were generated over land in the area of interest. The samples were distributed in the western part of Java Island, encompassing Banten, Jakarta, and West Java. This decision was made to focus the analysis on the western part of Java Island. In addition, the area of interest only included parts of Lampung and Central Java; therefore, those provinces were excluded from the analysis. The number of samples was determined based on Congalton and Green's [24] rule, which states that the minimum number of samples in an area of 1000000 acres is 50 for each category. Since the study area is 243688 km^2, the minimum number of samples is 300. To better represent the area of interest, the number of samples was increased to 750.

3 Result and Discussion

3.1 Land Surface Temperature in the Western Part of Java Island

363 MODIS LST imageries were used to create an image showing the average LST value from January 1, 2023, to December 31, 2023. The average temperature during this period fluctuated considerably, with the highest temperature recorded on May 24, 2023, while the lowest was recorded on July 14, 2023. The temperature trend also shows an increase in July - December, which is in line with the peak summer season in the Java Island region.

The distribution of the average LST value in 2023 is shown in Fig. 2. Based on the map, the Jabodetabek (Jakarta, Bogor, Depok, Tangerang, and Bekasi) metropolitan region has the highest temperature, indicated by the red color gradation. The Bodetabek region (Bogor, Depok, Tangerang, and Bekasi) is considered a supporting region (hinterland) for Jakarta, making it the center of various anthropogenic activities. High population mobility was shared in these regions, especially among commuters working in Jakarta [25].

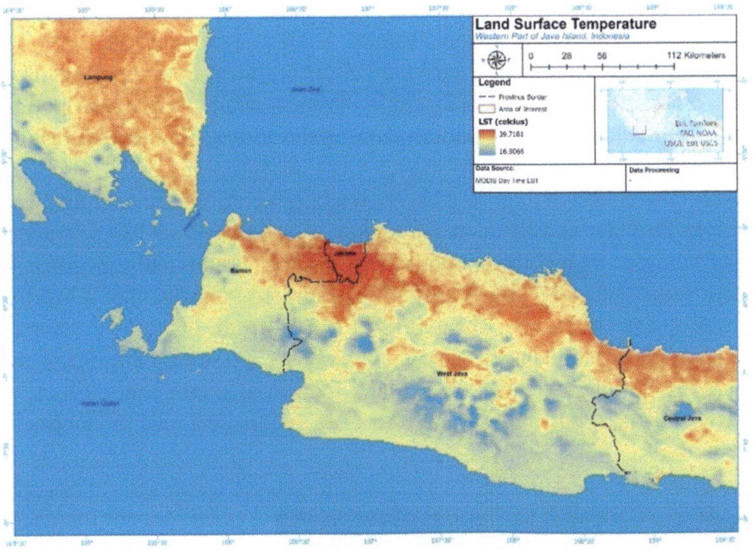

Fig. 2. LST Map derive from MODIS Terra satellite.

The statistical values for each mapped province are shown in Table 1. It can be seen that DKI Jakarta has a high average temperature value, followed by Banten and West Java. The average value of West Java, which is relatively low compared to other regions, larger areas compared to other provinces, and mountainous topography, especially in the southern part of West Java, could result in a lower LST value in the province.

Table 1. Zonal statistics of LST based on province administrative levels

No	Province	Pixel	Mean	Median	Std. Dev	Min	Max
1	West Java	36960	29.331	29.183	3.433	17.141	39.621
2	Jakarta	649	37.560	37.800	1.225	32.258	39.718
3	Banten	9262	30.078	29.646	3.121	20.430	39.592

Figure 2 also shows that the northern region of Java Island recorded higher temperatures compared to the southern region. This temperature difference is caused by the topography of this region, in which the northern region consists of lowlands as opposed to the mountainous southern region. In addition, the Trans Java highway developed in the northern part of the island, connecting the west and eastern end, which resulted in higher anthropogenic development compared to the southern part. This region has been used for transporting goods since British and Dutch colonialization.

3.2 MAP Concentration in the Western Part of Java Island

The MAP distributions in Fig. 3, 4, 5 and Fig. 6 show the average concentration value obtained from image processing from January 1, 2023 - December 31, 2023. Data processing was done one by one at each Sentinel-5P pollutant concentration. The results of this data processing are then visualized in the form of an average concentration map (mol/m^2) with a color gradation pattern from green to red. Red tends to show higher concentrations, while green tends to show lower concentrations.

The NO_2 concentration map (Fig. 3) shows a high concentration along the northern part of Banten Province and Jabodetabek region. NO_2 results from NO conversion in the presence of Volatile Organic Compounds (VOCs) [26]. Various anthropogenic activities contributed to this gas's creation, especially motorized vehicles, the industrial sector, power plants, and air conditioners/heaters [27]. Therefore, the highest NO_2 concentration value is shown in the Jabodetabek region, which hosts various anthropogenic activities. In addition, the Suralaya Steam Power Plant, located in Pulomerak District, northern Banten Province, also contributes to high-level emissions recorded by Sentinel-5P.

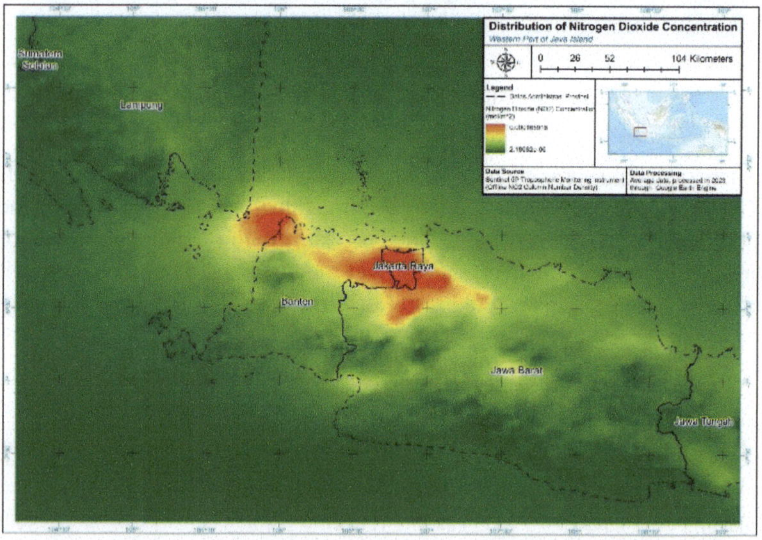

Fig. 3. The distribution of a major air pollutants: Nitrogen Dioxide (NO_2).

The CO concentration map (Fig. 4) shows a similar distribution with NO_2 concentration. However, a clear difference can be seen where areas with orange-yellow gradations tend to be more numerous. These color gradations indicate higher CO concentrations than light green-green color gradations. The southern region of West Java shows less CO gas concentrations than the other regions because the region is more mountainous. CO gas comes from the incomplete combustion process of fuels and organic materials, steam engine processes, industry, solid waste disposal, and various other sources [28]. Based on Fig. 4, the concentration of CO gas in the northern region of the western part of Java Island can be caused by the high population mobility and numerous population centers in the region. The northern region is also traversed by the Trans Java highway, which resulted in even higher mobilization. Therefore, the high mobilization could contribute to the increase of this gas.

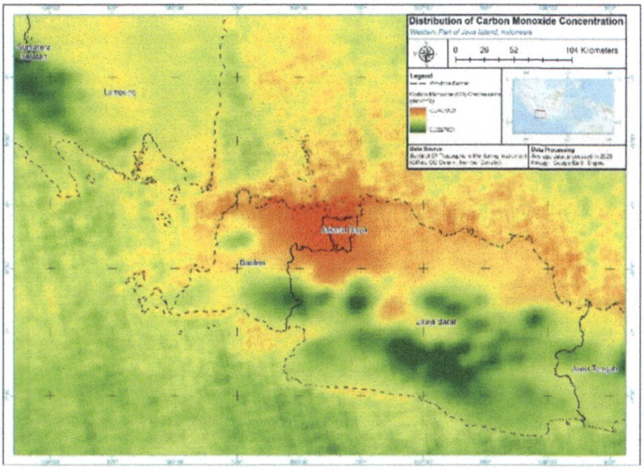

Fig. 4. The distribution of several major air pollutants: Carbon Monoxide (CO).

Surface ozone (O_3) is often called 'bad' ozone due to potential damage to cells and organs at high concentrations. This gas primarily originates from ozone in the stratosphere, which descends through processes involving VOCs and oxides of nitrogen (NOx), which mainly consist of NO_2 and nitric oxide (NO) [29]. VOCs are also the primary source of CO gas, resulting from the complex photochemical reaction between the two. Therefore, it can be seen that there is a similar distribution pattern between the distribution of O_3 (Fig. 5) and CO gas (Fig. 4).

The distribution of SO_2 concentration (Fig. 6) differs slightly from other maps due to the nature of SO_2 gas, which is primarily caused by volcanic eruptions [30]. Additionally, this gas is a byproduct of coal and oil combustion and copper mining activities. Coal mining activities were located in Sukabumi regency, West Java; some parts of Sukabumi show elevated SO_2 concentrations. According to Fig. 6, this area's primary source of SO_2 can be linked to Mount Anak Krakatau's volcanic activity in the Sunda Strait between Banten and Lampung Province.

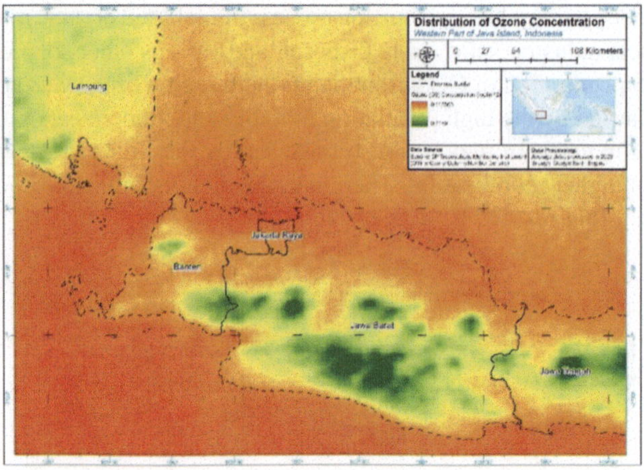

Fig. 5. The distribution of several major air pollutants: Ozone (O3).

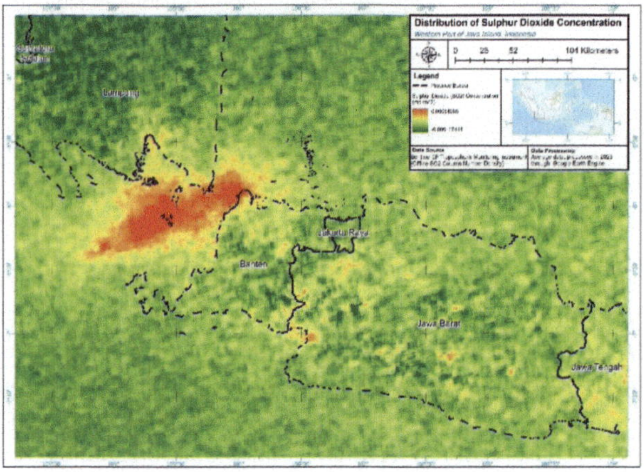

Fig. 6. The distribution of several major air pollutants: Sulphur Dioxide (SO_2).

3.3 Regression and Correlation Analysis

Ordinary Least Square Analysis. OLS analysis resulted in multiple R^2 values of 0.647751 and an adjusted R^2 value of 0.645827. This value indicates that MAP explains approximately 64% of the LST in the western part of Java Island. This statement is derived from multiple R^2 and adjusted R^2 values, which measure the strength of the relationship between the dependent variable and all independent variables. Adjusted R^2 values are lower than numerous R^2 values because they reflect model complexity concerning the data and produce a more accurate value. Table 2 shows the OLS summary report that shows several statistical indicators.

The degree of relationship of the independent variable can be sorted based on the t-statistic value. Table 2 indicates that NO_2, CO, and O_3 were the most influential pollutants

Table 2. OLS report of the analysis between LST and MAP.

Variable	Coefficient [a]	StdError	t-Statistic	Probability [b]	Robust_Pr [b]	VIF [c]
Intercept	−187.608604	41.4540	−4.5257	0.000009*	0.000002*	-
O_3	1713.793732	369.4386	4.6389	0.000006*	0.000001*	6.43846
SO_2	−6561.787368	1850.832	3.5453	0.000431*	0.000123*	1.00692
CO	469.068847	74.7984	6.2711	0.000000*	0.000000*	11.8154
NO_2	47986.779904	5714.819	8.3969	0.000000*	0.000000*	3.73696

in the Western Part of Java; SO_2 gas was the least influential pollutant in the area of interest. Based on the coefficient value, O_3, CO, and NO_2 have a positive relationship with LST, indicating that higher concentrations of these gases will lead to higher temperatures. On the other hand, SO_2 has a negative relationship with LST, which suggests that lower SO_2 concentration will lead to higher temperatures. This finding aligns with the study by Fuladlu and Altan [8] and Suthar et al. [31], which concluded that SO_2 has minimal impact on the LST. Other interesting results are observed from the VIF (Variance Inflation Factor) value, indicating the multicollinearity severity. Variance Inflation Factor value above 7.5 indicates severe multicollinearity. Table 2 shows that CO concentration has a Variance Inflation Factor value above 7.5; therefore, the concentration indicates a redundancy. Removing the variable from the analysis could improve the OLS result. Still, the variable will be included in this study, considering that CO gas is one of the main contributors to air pollution in the area of interest.

Correlation Analysis Table 3 shows that all MAP gases except SO_2 strongly correlate with surface temperature, with values ranging from 0.7 to 0.8. As mentioned earlier, the primary source of SO_2 gas is emissions from volcanic eruptions, of which there were no significant volcanic eruptions in western Java during 2023. Several small eruptions of the Anak Krakatau volcano were recorded in 2023, which resulted in higher SO_2 emissions in the Sunda Strait. These eruptions are not strong enough to create a high concentration of SO_2 in mainland Java. In the case of other MAP gases, anthropogenic activities are primarily carried out in areas that are population centers, which also produce high temperatures.

A strong correlation was also observed between CO and O_3 gas and NO_2 and CO gas. The strong correlation is also supported by the spatial distribution of these gases in Fig. 3, Fig. 4 and Fig. 5. Conversely, SO_2 gas has neglected the correlation between other gasses and even a negative correlation between LST. The different sources of SO_2 emission could result in no meaningful correlation between SO_2 and other MAP gases. Based on OLS and correlation analysis, it can be concluded that SO_2 has a negligible impact on the temperature increase compared to other MAP gases in the area of interest. This result could be different depending on the geographical condition of the region and the scale of analysis conducted.

Table 3. Correlation analysis of each MAP gas and LST

	O_3	SO_2	CO	NO_2	LST
O_3	1	0.021241	0.861152	0.427038	0.713469
SO_2	0.021241	1	0.055814	0.080006	-0.01546
CO	0.861152	0.055814	1	0.744669	0.836086
NO_2	0.427038	0.080006	0.744669	1	0.710327
LST	0.713469	-0.01546	0.836086	0.710327	1

4 Conclusion

Land Surface Temperature (LST) distribution reveals higher temperatures in urban areas, particularly within the Jabodetabek metropolitan region. Zonal statistics of LST indicate that DKI Jakarta exhibits higher temperatures than other provinces in this study. This increase in temperature can be attributed to intense anthropogenic activities in the region. In addition, these activities also resulted in high concentrations of Major Air Pollutants (MAP), namely, NO_2, CO, and O_3. Conversely, the average concentration of SO_2 gas remains relatively stable throughout 2023. The SO_2 concentration map indicates that the highest concentration of this gas was recorded in the Sunda Strait. In contrast, volcanic activity from Anak Krakatau volcano could result from a high concentration of SO_2.

OLS analysis resulted in multiple R^2 values of 0.647751 and an adjusted R^2 value of 0.645827. Based on these results, MAP explains approximately 64% of the LST in the western part of Java Island. The OLS analysis also revealed two different relationships of MAP and LST. The NO_2, CO, and O_3 gas concentrations have a positive relationship, while SO_2 gas has a negative relationship. Higher NO2, CO, and O3 gas concentrations will result in a higher LST value.

Meanwhile, a higher concentration of SO_2 gas will lower the LST value. OLS regression also highlights NO_2, CO, and O_3 as the primary pollutants significantly affecting the LST. Meanwhile, SO_2 was the least influential pollutant that affected the LST. Pearson correlation analysis also reveals that the three primary contaminants in the study area strongly correlate against LST. On the other hand, SO_2 has neglected correlation with LST.

References

1. Weng, Q., Yang, S.: Environ. Monit. Assess. **117**, 463 (2006)
2. Tian, Y., Yao, X., Chen, L.: Comput. Environ. Urban Syst. **75**, 35 (2019)
3. Yang, J., Shi, B., Shi, Y., Marvin, S., Zheng, Y., Xia, G.: Sustain. Cities Soc. **54**, 101941 (2020)
4. Chowdhury, S., et al.: Environ. Int. **159**, 107020 (2022)
5. Fugiel, A., Burchart-Korol, D., Czaplicka-Kolarz, K., Smoliński, A.: J. Clean. Prod. **143**, 159 (2017)
6. Fuladlu, K., Altan, H.: Urban Clim. **39**, 100958 (2021)

7. Zheng, S., Zhou, X., Singh, R., Wu, Y., Ye, Y., Wu, C.: Atmosphere (Basel). **8**, 110 (2017)
8. Naim, M.N.H., Kafy, A.A.: Environ. Challenges **4**, 100107 (2021)
9. Tran, D.X., Pla, F., Latorre-Carmona, P., Myint, S.W., Caetano, M., Kieu, H.V.: ISPRS J. Photogramm. Remote Sens. **124**, 119 (2017)
10. Lou, J., Wu, Y., Liu, P., Kota, S.H., Huang, L.: Curr. Pollut. Reports **5**, 144 (2019)
11. Kaplan, G., Avdan, Z.Y.: Int. J. Eng. Geosci. **3**, 130 (2020)
12. Viatte, C., et al.: Surv. Geophys. **41**, 1583 (2020)
13. Phan, T.N., Kappas, M.: J. Appl. Remote. Sens. **12**, 1 (2018)
14. Venter, Z.S., Chowdhury, S.: Air Pollution and Population Exposure. In: Cardille, J.A., Crowley, M.A., Saah, D., Clinton, N.E. (eds.) Cloud-Based Remote Sensing with Google Earth Engine, pp. 725–744. Springer, Cham (2024)
15. Yang, Z., Marinello, W.: Remote Sens. **11**, 1939 (2019)
16. Halder, B., et al.: Sci. Rep. **13**, 7968 (2023)
17. Gandhi, U., Cloud-based remote sensing with google earth engine. In: Cardille, J., Crowley, M., Saah, D., Clinton, N. (eds.), pp. 3–18. Springer, Cham (2024)
18. Wang, Y., Guo, Z., Han, J.: Ecol. Indic. **129**, 107976 (2021)
19. Wan, Z.: MODIS Land Surface Temperature Products Users' Guide. University of California, Santa Barbara, USA, Institute for Computational Earth System Science (2013)
20. Nicolau, A.P., Dyson, K., Saah, D.: N. Clinton. In: Cardille, J., Crowley, M., Saah, D., Clinton, N. (eds.) Cloud-Based Remote Sens. with Google Earth Engine, pp. 41–66. Springer International Publishing, Cham (2024)
21. Chakraborty, T.: Cloud-based remote sensing with google earth engine. In: Cardille, J., Crowley, M., Saah, D., Clinton, N. (eds.), pp. 745–771. Springer, Cham (2024)
22. Fahrezi, A., Mawari, S.: Eighth Geoinf. Science Symposium 2023 Geoinf. Sci. Sustain. Planet, edited by C. M. Roelfsema, A. Besse Rimba, S. Arjasakusuma, and A. Blanco (SPIE, 2024), p. 10
23. Schober, P., Boer, C., Schwarte, L.A.: Anesth. Analg. **126**, 1763 (2018)
24. Congalton, R.G., Green, K.: Assessing the Accuracy of Remotely Sensed Data, 2nd edn. CRC Press, Boca Raton (2008)
25. Hasibuan, H.S., Soemardi, T.P., Koestoer, R., Moersidik, S.: Procedia Environ. Sci.Environ. Sci. **20**, 622 (2014)
26. Vîrghileanu, M., Săvulescu, I., Mihai, B.-A., Nistor, C., Dobre, R.: Remote Sens. **12**, 3575 (2020)
27. Baldasano, J.M.: Sci. Total. Environ. **741**, 140353 (2020)
28. Levy, R.J.: Neurotoxicol. Teratol. **49**, 31 (2015)
29. Zhang, J., Wei, Y., Fang, Z.: Front. Immunol. 10 (2019)
30. B. M. Sukojo, N. A. Q. Fadhilah, and H. Sanjaya, in 2023 IEEE Int. Conf. Aerosp. Electron. Remote Sens. Technol. (IEEE, Bali, 2023), pp. 1–7
31. Suthar, G., Singhal, R.P., Khandelwal, S., Kaul, N.: Remote Sens. Appl. Soc. Environ. **32**, 101011 (2023)

Utilization of Himawari and GEMS for Forest Fire Aerosol Detection in Sumatera and Kalimantan

Asri Indrawati[1](✉), Amalia Nurlatifah[1], Risyanto[1], Dyah Aries Tanti[1], Novita Ambarsari[1], Muhammad Gilang Eka Hermawan[2], and Moch Syarif Romadhon[1]

[1] Reserach Center for Climate and Atmosphere, BRIN, Bandung, Indonesia
asri.004@brin.go.id

[2] Faculty of Mathematics and Natural Science, Universitas Negeri Semarang, Semarang, Indonesia

Abstract. Indonesia frequently experiences forest fires, posing significant environmental, public health, and economic challenges. Effectively monitoring these events is crucial for early warning systems and disaster management. Aerosol Optical Thickness (AOT) measurements provide critical data for assessing aerosol concentrations from such fires. This study compares two satellite-based AOT products: AOT-GEMS (Geostationary Environmental Monitoring Spectrometer) and AOT-Himawari (Himawari-8). Both offer real-time, high-temporal-resolution data essential for tracking aerosol dispersal. Data from Sumatra and Kalimantan during forest fire events in August and September 2023 were analyzed. Results indicate higher AOT values in September, correlating with increased hotspot activity. HIMAWARI consistently recorded higher AOT values than GEMS, despite a weak correlation ($r = 0.14$), suggesting differences in sensitivity and measurement techniques. GEMS showed semi-diurnal peaks that were not observed in HIMAWARI data. These findings highlight the strengths and limitations of each system, emphasizing the need for integrated approaches to enhance monitoring accuracy during forest fires.

Keywords: Forest Fire · Aerosol · HIMAWARI · GEMS

1 Introduction

Forest fires are a recurrent and significant environmental challenge in Southeast Asia, particularly on the islands of Sumatra and Kalimantan in Indonesia [1]. These fires, often triggered by natural causes and anthropogenic activities such as land clearing for agriculture, release substantial amounts of aerosols into the atmosphere [2]. The resulting haze profoundly impacts air quality, public health, regional climate, and global atmospheric circulation [3]. Effective monitoring and detection of these aerosol emissions are critical for mitigating their adverse effects and informing policy-making and firefighting efforts.

E. Fitrianingsih et al. (Eds.): ISAST 2024, SPPHY 416, pp. 14–23, 2025.
https://doi.org/10.1007/978-981-96-1344-1_2

Indonesia, with its sprawling archipelago and tropical climate, frequently experiences forest fires that pose significant threats to the environment, public health, and economic stability. The need for robust monitoring systems is underscored by the importance of early warning, disaster management, and mitigation strategies. Traditional ground-based monitoring systems, while valuable, are often limited by geographic coverage and accessibility, especially in remote or densely forested regions. As a result, satellite remote sensing has become an indispensable tool for monitoring forest fires and associated aerosol emissions on a large scale.

Among the array of available satellites, the Himawari series, operated by the Japan Meteorological Agency (JMA), and the Geostationary Environmental Monitoring Spectrometer (GEMS), aboard South Korea's GEO-KOMPSAT-2B satellite, are particularly notable for their high temporal resolution and ability to provide near-real-time data. With its advanced multi-band imaging capabilities, Himawari offers detailed observations of cloud cover, atmospheric phenomena, and surface conditions across the Asia-Pacific region. Its frequent imaging, every 10 min, enables the detection of rapidly evolving fire events and the tracking aerosol plumes [4]. Meanwhile, as the first geostationary air quality monitoring instrument, GEMS provides continuous, high-resolution data on atmospheric pollutants, including aerosols, across Asia [5].

Aerosol Optical Thickness (AOT) measurements from these satellite systems have become invaluable for assessing the presence and concentration of aerosols resulting from forest fires. AOT is a key parameter in monitoring atmospheric conditions, offering a quantifiable measure of the extinction of solar radiation due to aerosol particles [6]. Integrating data from both satellite systems provides a comprehensive perspective on the dynamics of forest fire aerosols, offering essential insights into their spatial and temporal patterns.

The utilization of satellite-based AOT data has become increasingly crucial for monitoring atmospheric aerosols and understanding their impact on air quality and climate. Instruments like the Advanced Himawari Imager (AHI) aboard Himawari-8 and GEMS provide high spatial and temporal resolution data, enabling the tracking of aerosol distribution over large regions in near real-time. Studies have demonstrated the potential of Himawari-8 for aerosol retrieval, particularly over East Asia and Japan, with good agreement to ground-based observations like AERONET. However, challenges remain in areas with cloud contamination and high humidity [7, 8]. Similarly, GEMS has shown promise in monitoring aerosols across complex terrains, though issues such as calibration and the resolution of small-scale variations in AOT remain [9]. Comparative studies of Himawari-8 and GEMS have highlighted both strengths and limitations in their algorithms, underscoring the need for improved retrieval techniques to enhance the accuracy of AOT data for air quality and climate applications [10, 11].

This study aims to explore the synergistic utilization of Himawari and GEMS data for detecting and analyzing aerosols resulting from forest fires in Sumatra and Kalimantan. By leveraging the strengths of both satellite platforms, this research seeks to enhance the accuracy and timeliness of aerosol detection, contributing to improved early warning systems and more effective strategies for managing and mitigating the impacts of forest fires in Indonesia [12]. The comprehensive comparison of AOT products from GEMS and

Himawari will further illuminate their respective capabilities and potential applications in monitoring forest fire dispersal in the region [13].

2 Method

2.1 Location

This study focuses on the regions of Sumatera and Kalimantan, Indonesia, which are highly prone to forest fires and land-clearing activities that lead to significant aerosol emissions. Sumatera is situated between latitudes –4° N and 2° N and longitudes 95° E and 105.5° E, while Kalimantan is located between latitudes –4.5° N and 2.5° S and longitudes 109° E and 116° E. In addition to the broader study of aerosol detection across Sumatera and Kalimantan, this research also investigates the relationship between Aerosol Optical Thickness (AOT) and PM2.5 concentrations in the city of Pontianak, located at coordinates 109.393° E, 0.0897° S.

2.2 Data

This study utilizes data from two key geostationary satellites, Himawari-8 and the Geostationary Environment Monitoring Spectrometer (GEMS), to analyze Aerosol Optical Thickness (AOT) over Sumatera and Kalimantan. Additionally, ground-based PM2.5 concentration data for Pontianak are used to explore the relationship between satellite-derived AOT and surface-level particulate matter concentrations.

Himawari-8 AOT Data. Himawari-8, operated by the Japan Meteorological Agency (JMA), provides Level 3 AOT data with a temporal resolution of 1 h and a spatial resolution of 5 km. The AOT data used in this study were obtained from the Himawari-8 V3.0 dataset, available for download at JAXA's P-Tree System (https://www.eorc.jaxa.jp/ptree/index.html) [9]. The analysis focuses on two time periods: from August 10th to 25th, 2023, and from September 15th to 30th, 2023, covering critical months during the dry season when forest fires are most prevalent.

GEMS AOT Data. The Geostationary Environment Monitoring Spectrometer (GEMS) onboard South Korea's GEO-KOMPSAT-2B satellite provides high-resolution, continuous observations of atmospheric pollutants, including AOT. The comparison between GEMS and Himawari-8 AOT data is conducted specifically for September 2023, allowing for an assessment of the consistency and complementarity between the two satellite datasets.

PM2.5 Concentration Data. For ground-level validation and to explore the correlation between AOT and PM2.5, daily average PM2.5 concentration data for Pontianak were sourced from the Air Quality Monitoring System (AQMS) managed by the Indonesian Meteorological, Climatological, and Geophysical Agency (BMKG). These data provide insights into the impact of aerosol emissions detected by the satellites on surface-level air quality.

Hotspot data indicating the locations of active fires were downloaded from the National Research and Innovation Agency/Badan Riset dan Inovasi Nasional (BRIN) [14]. This data is crucial for identifying the sources of aerosol emissions and understanding their spatial distribution across Sumatera and Kalimantan during the study periods.

2.3 Data Processing

HIMAWARI AOT and GEMS AOT utilize satellite-based observations to estimate aerosol optical thickness (AOT). HIMAWARI employs a look-up table approach, correlating reflected solar radiation with AOT values. Conversely, GEMS leverages a radiative transfer model to calculate AOT based on observed radiation and atmospheric conditions. While GEMS AOT offers higher spatial resolution and accuracy, HIMAWARI AOT benefits from more frequent measurements. These complementary datasets provide valuable insights into aerosol distribution and variations across different regions and time scales. The AOT data from GEMS and Himawari, provided in NetCDF format, are processed using Python to extract and analyze aerosol concentrations during forest fire events. The processed data are compared to assess the accuracy and reliability of both satellites in detecting aerosols using statistical methods such as correlation analysis and basic descriptive statistics. Additionally, a simple linear correlation is performed between the AOT data from both satellites and ground-based PM2.5 concentrations in Pontianak to explore their relationship.

3 Result and Discussion

3.1 AOT HIMAWARI

Forest fires in Indonesia are often exacerbated by slash-and-burn agriculture, particularly on the islands of Sumatra and Kalimantan. These fires contribute to significant haze episodes, affecting air quality and visibility and posing health risks to millions of people [15, 16]. Based on the results obtained in Fig. 1, the AOT values were significantly higher in September than August 2023. More hotspots were observed in September than in August. The average AOT values for August were 0.45 for Kalimantan and 0.52 for Sumatra. In contrast, the average AOT values in September were 0.58 for Kalimantan and 0.60 for Sumatra. The increase in AOT and hotspots in Kalimantan and Sumatra from the 27th to the 30th is likely attributed to various factors, including dry weather conditions, human activities, and atmospheric circulation patterns.

Based on the semi-diurnal data presented in Fig. 2, the Aerosol Optical Thickness (AOT) exhibits distinct peak patterns between the regions of Sumatra and Kalimantan. In Kalimantan, AOT shows two significant peaks at 9:00 AM and 1:00 PM. Conversely, Sumatra displays a single peak in AOT, which occurs later in the day, between 3:00 PM and 4:00 PM. It is important to note that no nighttime peaks are recorded, as the Himawari satellite data, which provides this information, is only available from 7:00 AM to 5:00 PM.

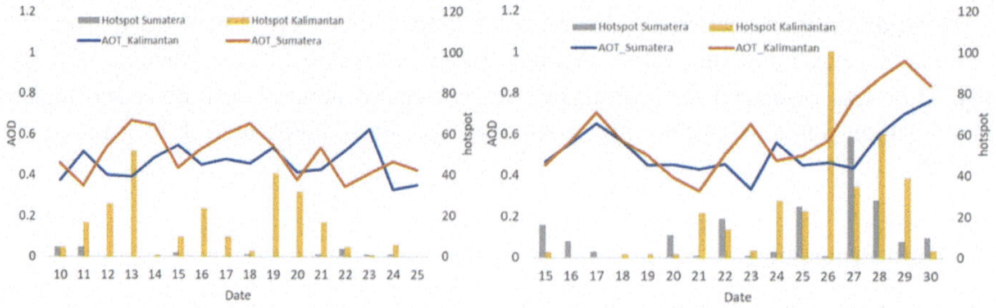

Fig. 1. Daily AOT from Himawari in (A) Agustus and (B) September 2023 for Sumatera and Kalimantan.

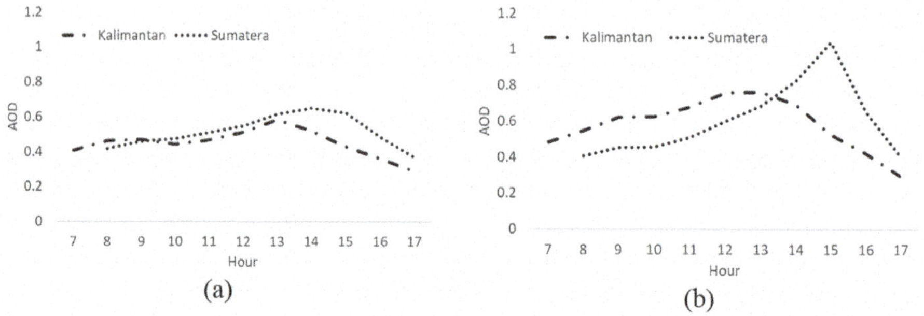

Fig. 2. Semi-diurnal time series of AOT for Kalimantan and Sumatera in (A) August and (B) September 2023 based on HIMAWARI.

3.2 GEMS and HIMAWARI

Based on the Aerosol Optical Thickness (AOT) values from HIMAWARI and hotspot data, it is evident that in September 2023, mainly from September 15th to 30th, the AOT values were higher than in August 2023. Therefore, the HIMAWARI and GEMS was compared for September 2023 in the regions of Sumatera and Kalimantan. As shown in Fig. 3a, for Sumatera, the AOT values from HIMAWARI were higher than those from GEMS, with GEMS values ranging from 0.27 to 0.33, while HIMAWARI values ranged from 0.33 to 0.77. The overall patterns between GEMS and HIMAWARI were similar, though some inverse trends were observed.

In Fig. 3b, the data for Kalimantan shows a better correlation, where increases in GEMS AOT values were accompanied by increases in HIMAWARI AOT values. The AOT values from GEMS were generally higher, with a range of 0.70 to 0.81, compared to HIMAWARI's range of 0.33 to 0.96. The correlation coefficient between GEMS and HIMAWARI AOT data was $r = 0.14$ for Sumatera and $r = 0.47$ for Kalimantan, indicating a stronger correlation in Kalimantan.

The differences in AOT values between GEMS and HIMAWARI suggest potential variations in sensitivity, calibration, or measurement techniques between the two satellite instruments. The GEMS data exhibited a relatively stable pattern throughout the period, while the HIMAWARI data showed a significant peak on September 29, 2023. This discrepancy highlights that, although both satellites monitor the same regions, the data they collect can differ significantly, underscoring the importance of understanding the limitations and calibration differences of each satellite system.

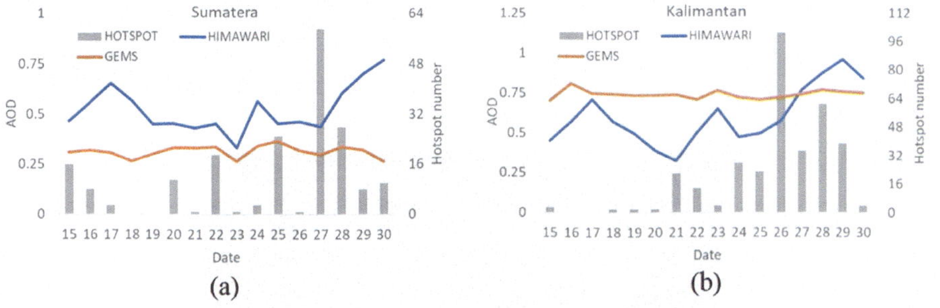

Fig. 3. Comparison AOT between HIMAWARI and GEMS in (A) Sumatera (B) Kalimantan

The distribution of Aerosol Optical Thickness (AOT) values from GEMS and HIMAWARI on September 27, 2023, as depicted in Figs. 4a and 4b, shows that both satellites present similar AOT distributions for Kalimantan. However, in Sumatera, the AOT distribution differs between the two satellites: GEMS detects higher AOT concentrations in the southern part of the island, extending towards Bandar Lampung, while HIMAWARI identifies higher concentrations in the northern regions, particularly in Riau, Jambi, and Palembang. Despite these differences, the pattern of low AOT values (0.2–0.4) observed by GEMS (Fig. 4a) aligns well with the distribution of hotspots (Fig. 4c). This indicates that GEMS is more sensitive in detecting increases in AOT, even at lower levels above the threshold of 0.1, compared to HIMAWARI.

The hotspot distribution in Sumatera and Kalimantan, as shown in Fig. 4c, further supports these findings. In Sumatera, the hotspot distribution closely matches the AOT pattern detected by GEMS, while in Kalimantan, the hotspot distribution aligns more closely with the AOT values observed by HIMAWARI. These results suggest that GEMS is more effective at detecting lower AOT values, whereas HIMAWARI shows greater sensitivity to higher AOT levels. This complementary sensitivity highlights the importance of using both satellite systems for a more comprehensive understanding of aerosol distributions across these regions.

The semi-diurnal patterns depicted in Fig. 5a for HIMAWARI and GEMS in Sumatra exhibit significant discrepancies. GEMS data shows pronounced peaks at 10:00 AM and 6:00 PM, with the lowest points occurring at 9:00 AM and 1:00 PM. In contrast, the HIMAWARI data does not follow this pattern, indicating substantial differences between the two datasets. These variations likely stem from differences in the temporal resolution and measurement methodologies employed by the two satellite systems. Conversely, in Kalimantan, the semi-diurnal patterns are more consistent between the two satellites, as illustrated in Fig. 5b. HIMAWARI data shows a peak in Aerosol Optical Thickness (AOT) at 1:00 PM, while GEMS exhibits a more stable, flat pattern throughout the day. This consistency in Kalimantan, compared to the discrepancies observed in Sumatra, suggests that regional factors may influence the satellite data, or that the two systems may respond differently to varying environmental conditions.

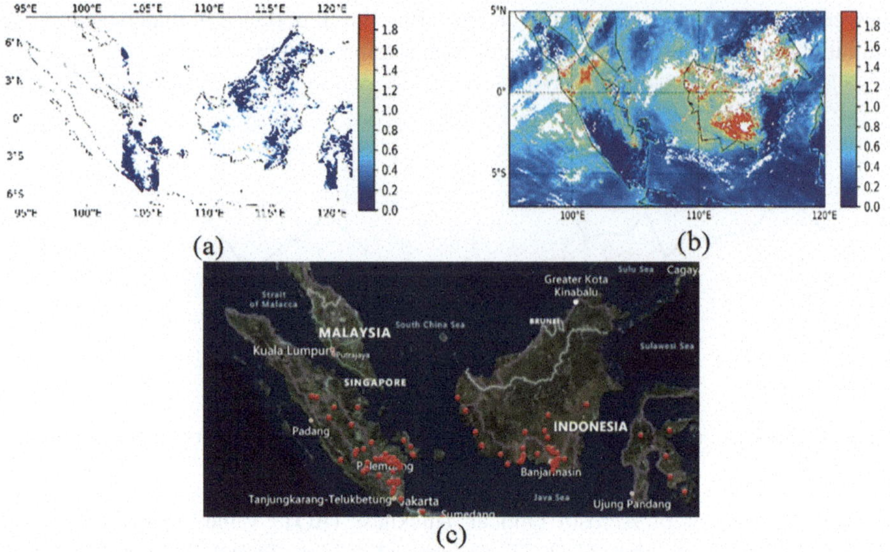

Fig. 4. The spatial distribution of AOT for Kalimantan and Sumatera in September, 27 2023 based on (a) HIMAWARI, (b) GEMS dan (c) Hotspot

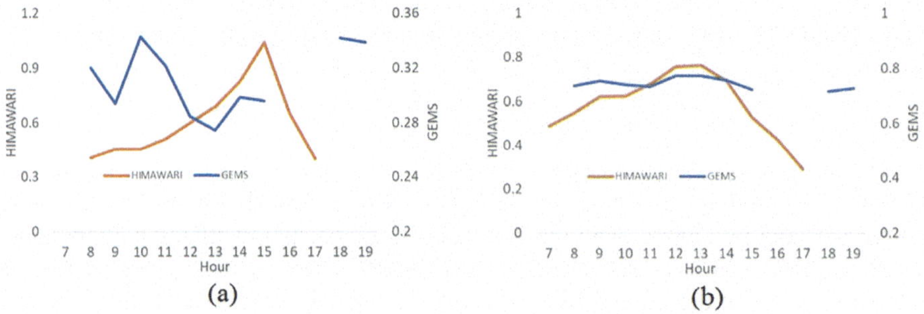

Fig. 5. Semi-diurnal pattern of HIMAWARI and GEMS in (a) Sumatera (b) Kalimantan

3.3 AOT and PM2.5

The correlation between Aerosol Optical Thickness (AOT) from GEMS and HIMAWARI with PM2.5 concentrations was analyzed in the Supadio, Pontianak area, using daily average PM2.5 concentrations obtained from the AQMS instrument. As shown in Fig. 6a, a similar pattern is observed between PM2.5 concentrations and AOT values from both GEMS and HIMAWARI. The correlation between PM2.5 and AOT, presented in Fig. 6b, indicates that the correlation for HIMAWARI is stronger than that for GEMS, with correlation coefficients of $r = 0.60$ and $r = 0.41$ for HIMAWARI and GEMS, respectively.

The regression equation derived from the correlation between AOT and PM2.5 for GEMS is $y = 59.828x + 16.139$, while for HIMAWARI it is $y = 49.2x + 27.39$. Several factors influence this correlation, particularly meteorological conditions and atmospheric corrections, as AOT is measured at higher altitudes, while PM2.5 concentrations are measured at the surface level.

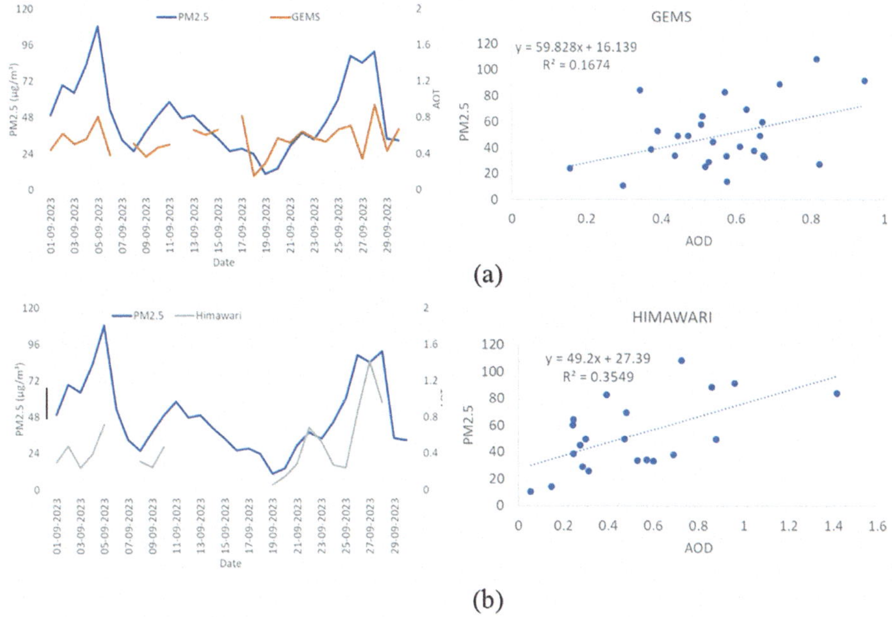

Fig. 6. Correlation between PM2.5 concentration at Supadio, Pontianak and AOT from (A) GEMS and (B) HIMAWARI

Based on those explanation, the differences in Aerosol Optical Thickness (AOT) values between GEMS and Himawari can be attributed to variations in their retrieval algorithms and sensor characteristics. Himawari-8 typically uses a Dark Target Algorithm (DTA), which estimates aerosol concentrations by contrasting the dark surface of the Earth with the brighter atmosphere. This method works well over dark surfaces such as oceans or vegetated areas but can face challenges over brighter surfaces, leading to potential inaccuracies. On the other hand, GEMS employs a more complex Radiative Transfer Model (RTM), which simulates the passage of radiation through the atmosphere, accounting for factors such as aerosol scattering and absorption, as well as gas interactions. These fundamental differences in retrieval approaches result in variations in how each satellite derives AOT, and consequently, their respective AOT values may differ even when observing the same region at the same time.

In addition to the algorithms, the sensor characteristics of GEMS and Himawari also contribute to discrepancies in their AOT measurements. Himawari primarily uses visible and near-infrared (NIR) bands, which limit its ability to detect aerosols under certain conditions, such as when cloud cover is present or during high humidity. GEMS, however, incorporates additional shortwave infrared (SWIR) bands, providing more comprehensive information on aerosol properties, especially in complex atmospheric conditions. Furthermore, GEMS offers higher spatial resolution compared to Himawari, which allows for more detailed observations of small-scale variations in aerosol distribution. These differences in spectral bands, spatial resolution, and overall sensitivity lead to variations in the AOT data reported by each satellite, reflecting their distinct capabilities and limitations. Himawari is better for detecting the size of aerosol, while GEMS is more suited for identifiying absorbing properties.

4 Conclusion

This study has demonstrated the utility of satellite-based monitoring systems, specifically the Himawari series and the Geostationary Environment Monitoring Spectrometer (GEMS), in tracking aerosol emissions from forest fires. By comparing Aerosol Optical Thickness (AOT) data from both satellites, we have identified key differences in their sensitivity, calibration, and measurement methodologies. While Himawari data exhibited greater sensitivity to higher AOT levels, GEMS provided more stable and consistent measurements, particularly at lower AOT values. The analysis revealed that both satellite systems have distinct strengths that can complement each other, offering a more comprehensive understanding of aerosol distributions across Sumatra and Kalimantan. Furthermore, the correlation analysis between AOT and ground-based PM2.5 concentrations in Pontianak highlighted that Himawari data had a stronger correlation with PM2.5 levels compared to GEMS, indicating its potential for more accurate surface air quality assessments. Overall, the integration of data from Himawari and GEMS presents a powerful approach for enhancing the accuracy and timeliness of aerosol detection in regions prone to forest fires. This combined dataset offers valuable insights into the spatial and temporal dynamics of forest fire emissions, which are crucial for informing early warning systems, disaster management strategies, and mitigation efforts in Indonesia.

References

1. Field, R.D., van der Werf, G.R., Shen, S.S.P.: Human amplification of drought-induced biomass burning in Indonesia since 1960. Nat. Geosci. **2**(3), 185–188 (2009). https://doi.org/10.1038/ngeo443
2. Tacconi, L.: Preventing fires and haze in Southeast Asia. Nat. Clim. Chang. **6**(7), 640–643 (2016). https://doi.org/10.1038/nclimate3008
3. Wooster, M.J., Perry, G.L.W., Zoumas, A.: Fire, drought and El Niño relationships on Borneo (Southeast Asia) in the pre-MODIS era (1980–2000). Biogeosciences **9**(1), 317–340 (2012). https://doi.org/10.5194/bg-9-317-2012
4. Zhang, W., Xu, H., & Zhang, L. (2019). Assessment of Himawari-8 AHI Aerosol Optical Depth Over Land. Remote Sensing, 11(9). https://doi.org/10.3390/rs11091108
5. Kim, J., et al.: Monitoring of air pollution episodes using satellite observations. Remote Sens. Environ. **250**, 112045 (2020)
6. Wang, L., Yu, C., Cai, K., Zheng, F., Li, S.: Retrieval of aerosol optical depth from the Himawari-8 advanced himawari imager data: application over Beijing in the summer of 2016. Atmos. Environ. **241**, 117788 (2020). https://doi.org/10.1016/j.atmosenv.2020.117788
7. Zhang, Z., et al.: Validation of Himawari-8 aerosol optical depth retrievals over China. Atmos. Environ. **199**, 32–44 (2019). https://doi.org/10.1016/j.atmosenv.2018.11.024

8. Cho, Y., et al.: First atmospheric aerosol-monitoring results from the geostationary environment monitoring spectrometer (GEMS) over Asia. Atmos. Meas. Tech. **17**(14), 4369–4390 (2024). https://doi.org/10.5194/amt-17-4369-2024
9. Yamamoto, Y., Ichii, K., Higuchi, A., Takenaka, H.: Geolocation accuracy assessment of Himawari-8/AHI imagery for application to terrestrial monitoring. Remote Sens. **12**(9) (2020). https://doi.org/10.3390/rs12091372
10. Tan, Y., Wang, Q., Zhang, Z.: Improved Himawari-8 10-minute scale aerosol optical depth product using deep neural network over Japan. Atmos. Pollut. Res. **15**(3), 102005 (2024). https://doi.org/10.1016/j.apr.2023.102005
11. Seinfeld, J.H., Pandis, S.N.: Atmospheric Chemistry and Physics: From Air Pollution to Climate Change. Wiley, New York (2016)
12. Choi, W.J., Moon, K.-J., Kim, G., Lee, D.: Reliability analysis based on air quality characteristics in East Asia using primary data from the test operation of geostationary environment monitoring spectrometer (GEMS). Atmosphere, 14(9) (2023). https://doi.org/10.3390/atmos14091458
13. Bessho, K., et al.: An introduction to Himawari-8/9— Japan's new-generation geostationary meteorological satellites. J. Meteorol. Soc. Japan. Ser. II, **94**(2), 151–183 (2016). https://doi.org/10.2151/jmsj.2016-009
14. Holben, B.N., et al.: AERONET – a federated instrument network and data archive for aerosol characterization. Remote Sens. Environ. **66**(1), 1–16 (1998)
15. BRIN Hotspot Data. https://hotspot.brin.go.id/. Accessed 19 June 2024
16. Balasubramanian, R., See, S.W.: Impact of forest fires on regional air quality in Southeast Asia. Chin. J. Geochem. **25**(1), 221–222 (2006). https://doi.org/10.1007/BF02840172

The Use of Landsat-8 and ASTER Satellite Imagery to Analyze the Environmental Impact of Artisanal Mining in Santong Bay

Rakhmat Senoaji[1], Muhammad Rayhan Syahrir[1(✉)], Ichsan Fachri Siroj[1], Anugrah Arya Muhajir[1], and Bintang Yanuar Putra Taranggana[2]

[1] Faculty of Earth Sciences and Technology, Bandung Institute of Technology, Bandung, Indonesia
rayhansyahrir03030@gmail.com
[2] Faculty of Mining and Petroleum Engineering, Bandung Institute of Technology, Bandung, Indonesia

Abstract. From all the minerals obtained from mining operations in Indonesia, it is reported that up to 20% of them came from artisanal or informal mining. Unlike large-scale industrial mining, artisanal mining brings various uncontrollable and undetected ecological and natural hazards, mainly land use conversion and contamination of nearby water bodies by mercury metal, which is the byproduct of gold extraction in artisanal mines. This can lead to severe health risks for the local population and wildlife. To assess this, we confirmed the Indonesian Body of Geological Resources research findings about the locations of potential mineral areas and artisanal mines in Santong Bay, West Nusa Tenggara, Indonesia, which has mineralization and gold potential. This is done by analyzing the mineralization area detected by imagery band ratios, and the results are compared to the findings of the Indonesian Body of Geological Resources. The imagery is then analyzed to assess the impacts of land use change and mercury precipitation in nearby water bodies.

Keywords: Environmental Remote Sensing · Hazard Assessment · Artisanal Mining

1 Introduction

With its geological condition, Indonesia has abundant mining products, such as gold, copper, and iron, especially in Sumbawa, West Nusa Tenggara [1]. Various mining products can be found through exploration activities. With the development of aerial mapping technology, minerals contained in the soil can be found and mapped using satellite imagery [2].

Extracting mineral resources from Earth will impact the ecological condition of the mines' surrounding areas, such as land conversion and contamination of natural water springs with mercury [3]. These are prominent issues in artisanal mining because mercury is used in an amalgamation process to extract the gold from the ore—the mercury and gold

E. Fitrianingsih et al. (Eds.): ISAST 2024, SPPHY 416, pp. 24–33, 2025.
https://doi.org/10.1007/978-981-96-1344-1_3

precipitate form a mercury-gold amalgam, which is then heated at high temperatures. Finally, the gold is extracted via evaporation of the mercury [4]. Furthermore, local miners can further process the precipitation using the cyanide method to extract the remaining gold from the initial processing waste. This paper analyzes the ecological impacts of artisanal gold mining in the Santong Bay Area, West Sumbawa Regency.

The Indonesian Center of Geology Resources had prospected Sumbawa Island for an abundance of hydrothermally altered rocks [5], indicating a high concentration of gold and copper in the area. Imagery from Landsat 8 and ASTER was then used to estimate the total area of minerals in the prospected zones proposed by the Indonesian Center of Geology Resources and subsequently, the potential mining area [6].

We assessed the impact on the ecology around gold mining sites by comparing them to the effects in other areas with gold mining sites, including Indonesia. At Batu Hijau Mine, West Sumbawa Regency, Sumbawa, Indonesia, land use changed from 2007 to 2020. Land previously covered by vegetation became an urban complex supporting mining activities at Batu Hijau Mine [7]. Land use in this area distinctly features a mix of agroforestry, primarily dominated by coconut palm trees, along with some intensive farming activities and human settlements.

2 Material and Methods

2.1 Satellite Imagery

Two satellite sensors are used to obtain imagery data of Santong Bay, Sumbawa Island imagery data: ASTER and Landsat 8. These two sensors are well-established as reliable sources for mineral analysis [8]. The Landsat 8 imagery is obtained from USGS' Earth Explorer and consists of one low cloud cover level 1 Landsat 8 product. The Landsat 8 imagery was taken on December 10, 2023. The images were pre-georeferenced to the UTM zone 50 south projection using the WGS-84 datum. After obtaining the imagery, it was radiometrically calibrated, and then atmospheric correction was applied using the Fast Line-of-sight Atmospheric Analysis of Spectral Hypercubes (FLAASH) algorithm. During the atmospheric correction, raw radiance data from the imaging spectrometer is re-scaled to reflectance data [9]. The false color result is then overlayed with boundaries of the potential mineral area according to the Indonesia Center of Geology Resources. Finally, the actual area of the potential mineral field is estimated.

The ASTER imagery used in this analysis was selected before 2008 when the sensor operated optimally before the SWIR (Shortwave Infrared) malfunctioned due to a cryocooler failure [10]. This imagery, also sourced from USGS' Earth Explorer (https://earthexplorer.usgs.gov), consists of 14 spectral bands covering VNIR (Visible Near Infrared), SWIR, and TIR (Thermal Infrared) [11]. Like the Landsat 8 data, the ASTER imagery required preprocessing, including radiometric calibration and atmospheric correction using the Quick Atmospheric Correction (QUAC) method. To facilitate the preprocessing, the nine VNIR and SWIR bands were layer-stacked [11].

2.2 Environmental Impacts of Mining

To assess the potential environmental impact, Landsat-8 imagery is used to map land use over the past eight years (2014 and 2023) and the spread of mercury during the same period as a parameter of environmental impact from gold mining activities in Santong Bay.

The Normalized Different Turbidity Index (NDTI) and Normalized Different Suspended Sediment Index (NDSSI) are utilized to detect the presence of mercury in water. These indices are vital tools in water quality monitoring using satellite imagery, specifically Landsat eight imagery, in this study. NDSSI is a remote sensing index developed to estimate and map suspended sediment concentrations in water bodies, calculated using specific spectral bands from satellite imagery [12]. On the other hand, NDTI is a remote sensing index used to estimate and map turbidity in water bodies [13]. Turbidity refers to the cloudiness or opacity of water caused by suspended particles such as silt and clay.

Mercury is strongly associated with fine particles, especially organic matter and clays. NDTI utilizes the reflectance of red and green bands, understanding that the red band increases reflectance as turbidity rises [14]. When mercury is released into the environment, it often attaches to suspended sediment particles. High turbidity levels can indicate increased sediment loads in the water, which may also suggest higher levels of mercury contamination.

Table 1 outlines the specific wavelength ranges of the relevant bands to understand better the spectral characteristics of the Landsat-8 imagery used in this study. This information is crucial for interpreting the indices and their sensitivity to various environmental factors.

Table 1. Landsat-8 Band wavelength ranges.

Bands	Wavelength (μm)	Resolution (m)
Band 1 – COASTAL AEROSOL	0.43–0.45	30
Band 2 – BLUE	0.45–0.51	30
Band 3 – GREEN	0.53–0.59	30
Band 4 – RED	0.64–0.67	30
Band 6 – SWIR 1	1.57–1.65	30
Band 7 – SWIR 2	2.11–2.29	30
Band 8 – PAN	0.50–0.68	15
Band 9 – CIRRUS	1.36–1.38	30
Band 10 – TIR 1	10.60–11.19	100
Band 11 – TIR 2	11.50–12.51	100

The NDSSI (Normalized Difference Suspended Sediment Index) is calculated using the Near-Infrared (NIR) and Blue (BLUE) bands' reflectance values from the imagery, as shown in Eq. 1:

$$NDSSI = \frac{BLUE - NIR}{BLUE + NIR} \quad (1)$$

The NDTI (Normalized Difference Turbidity Index) is calculated using the Red (RED) and Green (GREEN) bands' reflectance values from the imagery, as shown in

Eq. 2:

$$NDTI = \frac{RED - GREEN}{RED + GREEN} \quad (2)$$

Both indices produce values between –1 and + 1. Those are the indices' output values for every pixel in the imagery. Higher values of NDSSI signify a more significant presence of suspended sediments, while lower values indicate more transparent water with fewer sediments. Similarly, higher NDTI values represent more turbid water, and lower values indicate more transparent water with lower turbidity levels.

By analyzing NDSSI and NDTI data from 2014 and 2023, we can identify areas experiencing increased sedimentation and turbidity due to mining activities. This is very important in evaluating the impact of mercury spread from the gold mining process.

Another technique used in this research is the Normalized Difference Vegetation Index (NDVI), a unique type of band ratio between band 5 and band 4, which are Near Infrared and Red, respectively, that measures the amount of vegetation present in the subject area. It is widely used to differentiate between vegetation, barren soil, and bodies of water alike [12].

Equation 3 shows that the NDVI (Normalized Difference Vegetation Index) calculation yields values ranging from –1 to + 1. Higher NDVI values indicate more abundant green vegetation, while lower values suggest vegetation under stress or areas with bare soil or concrete [12]. A lower NDVI value indicates minimal vegetation, which signifies differences in vegetation coverage. These differences can then be used to determine the pattern of changes in vegetation and its surroundings [12].

$$NDVI = \frac{NIR - RED}{NIR + RED} \quad (3)$$

NDVI is used here to detect land conversion by comparing the value of NDVI in each distribution map: one taken from 2014 and the other from 2023. By comparing and analyzing data from 2014 and 2023, we can identify whether or not the subject area experienced a mass land conversion. In this study, Landsat 8 imagery is used to determine the NDVI value.

3 Result and Discussion

3.1 Mining Prospects in Santong Bay

False color imagery using band ratios derived from image spectra (4/2, 6/7, 5, and 10 in RGB) in Fig. 1 allows the identification of altered rocks, lithological units, and vegetation in Landsat 8 imageries [14]. In this area, the hydrothermally altered rocks are purple-yellowish, with the boundaries between the confirmed areas with alteration minerals and volcanic rocks visually delimited. Dark purple areas indicate vegetation. The prospected areas and the confirmed altered mineral zones are each marked with yellow and cyan dashed lines. The estimated total area of the prospective zone is 13.56 km^2, with 1.89 km^2 or 13.9% of the area being altered minerals.

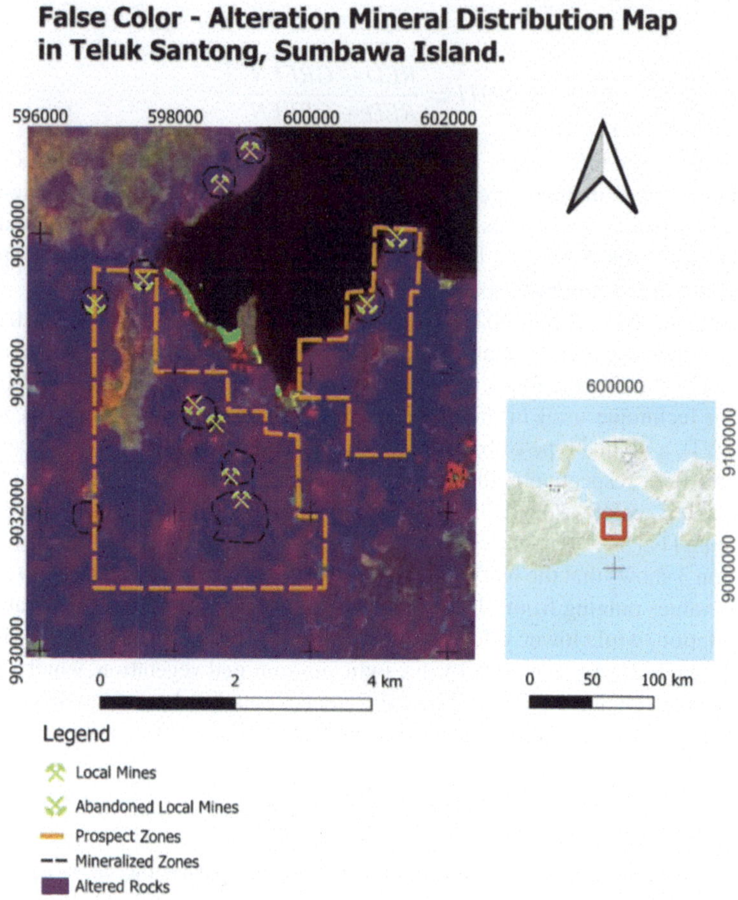

Fig. 1. Alteration mineral distribution using Landsat 8 imagery.

ASTER band ratio image using bands 4/8, 4/2, and 8/9 in RGB highlights the mineralized sections of the alteration zones, with bluish-white areas within pink regions, indicating mineralized surface weathering (see Fig. 2). This image was created to enhance the distinction of mineralized parts within the alteration zones. The 4/8 ratio was selected to amplify the response of iron oxides in the altered mineralized rocks. The 4/2 ratio helps distinguish altered mineralized rocks from other background materials, including altered non-mineralized and unaltered ones. Finally, the 8/9 ratio generated an image with better contrast by representing the averaged values of all rock units in the area [15]. This shows that the area is an abundant source of minerals and gold, evidenced by the existing and abandoned mines identified by the Indonesian Body of Geological Resources.

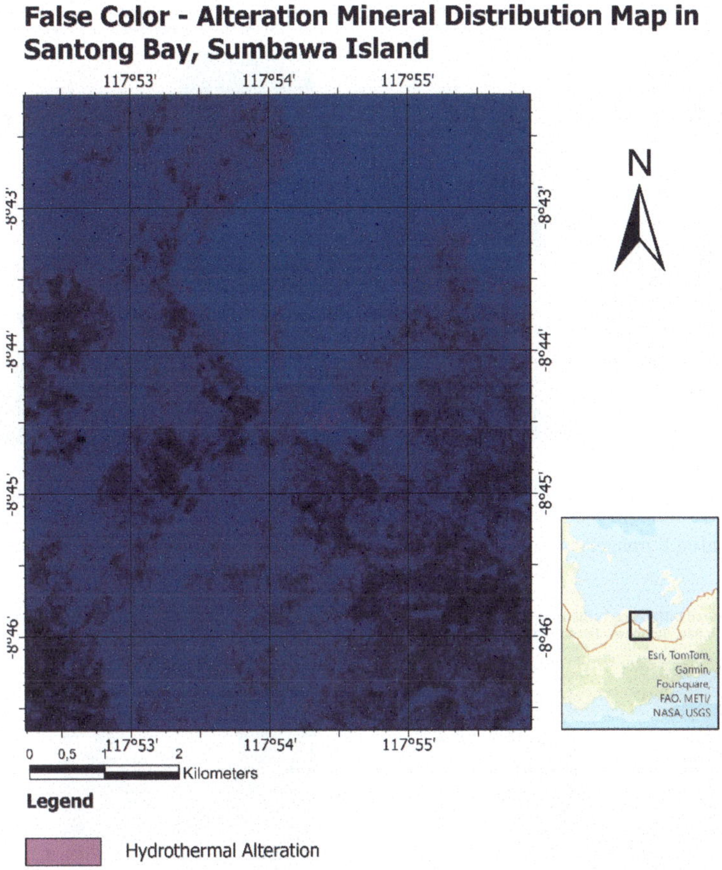

Fig. 2. Distribution of gold mineral potential using RGB Band ratios 4/8, 4/2, and 8/9 with ASTER imagery.

3.2 Environmental Impacts in Sumbawa Island

Figure 3 and Fig. 4 illustrate the distribution of NDSSI values in Santong Bay. In 2014, as shown in Fig. 3, the NDSSI index ranged from a minimum of –0.4056 to a maximum of 0.1147, with an average of 0.0034. These values suggest that Santong Bay had relatively clear water with moderate levels of suspended sediments during that period. By 2023, as depicted in Fig. 4., there was a slight decrease in the NDSSI index, with values ranging from –0.4288 to 0.1189 and an average of 0.0022. Despite the decrease, the water remained relatively clear, indicating minimal sedimentation over the nine years.

Figure 5 and Fig. 6 present the distribution of NDTI values in Santong Bay. In 2014, as shown in Fig. 5., the NDTI index ranged from a minimum of –0.0928 to a maximum of 0.0445, with an average of –0.0364. These results indicate that the water was generally clear with no significant turbidity issues. By 2023, as shown in Fig. 6., the NDTI values increased slightly, ranging from –0.0922 to 0.0406 and an average of –0.0354. This

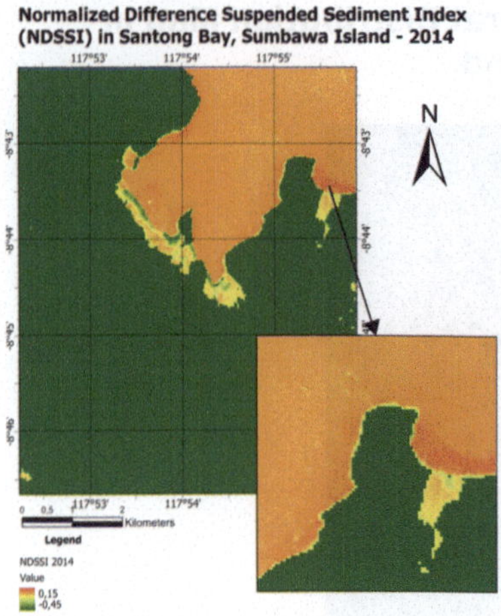

Fig. 3. Distribution of Normalized Difference Suspended Sediment Index (NDSSI) in 2014 using Landsat 8 imagery.

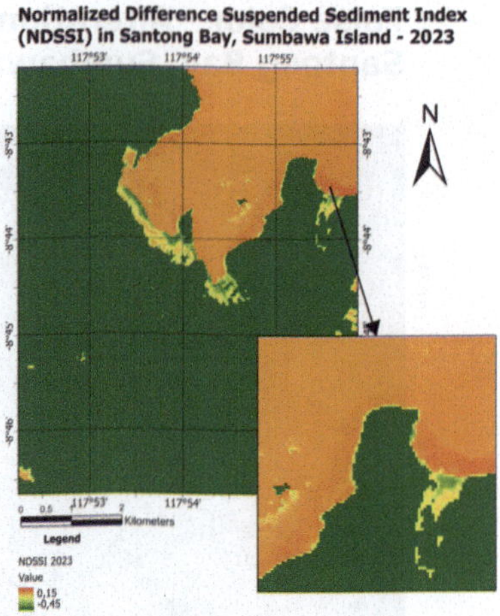

Fig. 4. Distribution of Normalized Difference Suspended Sediment Index (NDSSI) in 2023 using Landsat 8 imagery.

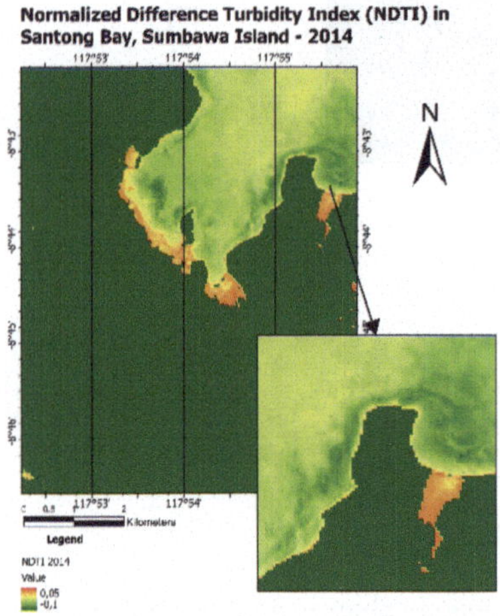

Fig. 5. Distribution of Normalized Difference Turbidity Index (NDTI) in 2014 using Landsat 8 imagery.

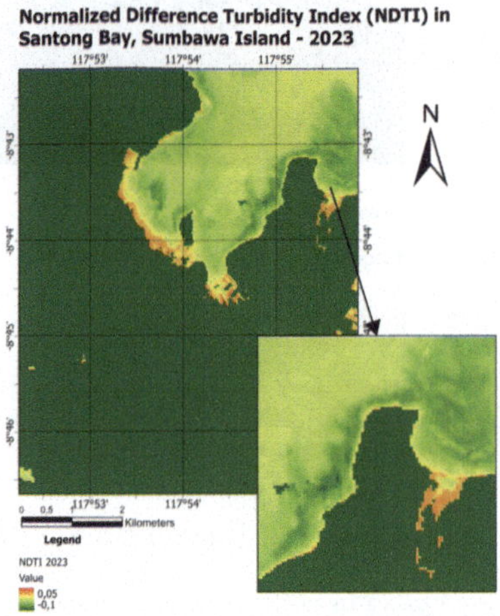

Fig. 6. Distribution of Normalized Difference Turbidity Index (NDTI) in 2023 using Landsat 8 imagery.

marginal increase in NDTI suggests a slight deterioration in water quality, particularly in turbidity levels.

Overall, the comparison of NDSSI and NDTI values from 2014 to 2023 indicates that Santong Bay has not experienced significant changes in water quality. The region does not show a substantial potential for mercury contamination, implying that artisanal mining activities in Santong Bay may not negatively impact water quality at a macro scale. However, given the established practice of mercury dumping in the area, further research using higher-resolution imagery is necessary to assess the micro-scale effects of mercury contamination.

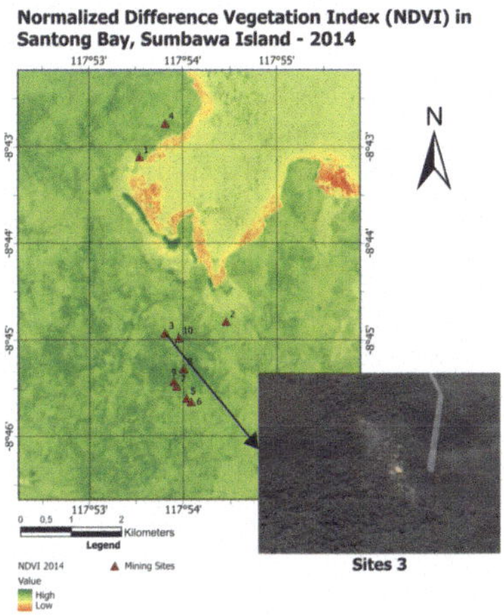

Fig. 7. Distribution of Normalized Difference Vegetation Index (NDVI) in 2014 using Landsat 8 imagery.

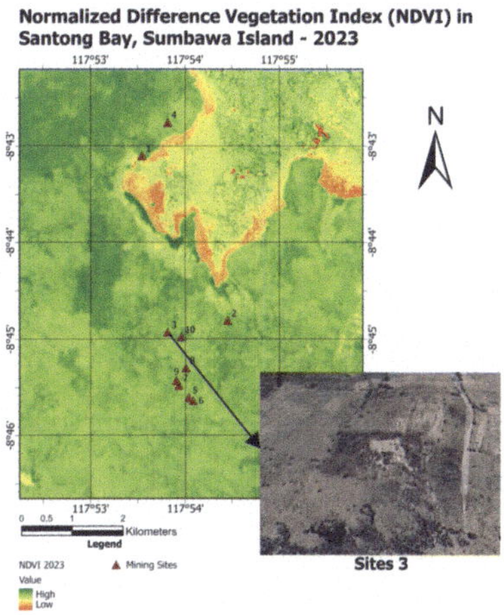

Fig. 8. Distribution of Normalized Difference Vegetation Index (NDVI) in 2023 using Landsat 8 imagery.

Figure 7 and Fig. 8 are the results of the NDVI to detect land conversion. From Fig. 7, we can notice that the lands are greener than in Fig. 8; this indicates that a wide range of vegetation is still available. Meanwhile, the yellowish color outside the islands is the water surrounding the island. Figure 8 shows that the land in Santong Bay is also green but not as clear as in Fig. 7 because of the slight discoloration, which might suggest a denser urban area or urban settlements. Figure 7 shows that the NDVI has a minimum and maximum range of –0.9954 and 0.8329, respectively, with an average value of 0.3425; meanwhile, Fig. 8 has a minimum and maximum range of –0.9721 and 0.9961, with an average value of 0.3713. Notice that the average value of the NDVI slightly increases; the increase in average value might suggest a land conversion from mining activities on a large scale, but no evidence supports this. Thus, more research is needed.

From a closer look at one of the mining sites, one can see that the artisanal mine negatively impacts the vegetation quantity on a micro-scale. Other sites show the same impacts on the vegetation. This negative effect might not show on a macro scale because of several factors, such as occasional forest fires, urbanization, and seasonal effects on the vegetation. The spatial resolution of Landsat 8 imagery also may miss finer-scale changes. Future studies should integrate higher-resolution imagery, advanced atmospheric corrections, and in-situ verification data of the reflectance To improve accuracy.

4 Conclusion

The alteration of mineral zones and an analysis of artisanal mining effects in Santong Bay have been conducted using remote sensing technology. About 1.89 km^2 the land is confirmed to be altered mineral zones with high amounts of gold in its soil. This tends to strongly attract artisanal mining practices, which have been confirmed in the area. From those alteration zones, the NDVI data from 2014 to 2023 are compared to extract the difference in land cover area. The results show that the land cover area has not changed significantly. This is due to the small scale of artisanal mining operations in Santong Bay, resulting in minimal changes in land use from 2014 to 2023 if examined from a macro scale. A closer look at individual sites yields a consistent decrease in vegetation in the mining area over time. An analysis of mercury precipitation in water is also done using NDTI and NDSSI. The results show that the distribution of mercury in Santong Bay's water does not exceed the value of 0.25. This indicates that mercury dumping is not easily examined on the macro-scale. Therefore, additional research with higher-resolution imagery is needed to yield better information.

References

1. Lloyd, B.: Indonesia's mineral resources: Performance and prospects. Resour. Policy **1**(6), 326–342 (1975)
2. Rajesh, H.M.: Application of remote sensing and GIS in mineral resource mapping-an overview. J. Mineral. Petrol. Sci. **99**(3), 83–103 (2004)
3. Worlanyo, A.S., Jiangfeng, L.: Evaluating the environmental and economic impact of mining for post-mined land restoration and land-use: a review. J. Environ. Manage. **279**, 111623 (2021)
4. Veiga, M.M., Angeloci-Santos, G., Meech, J.A.: Review of barriers to reduce mercury use in artisanal gold mining. Extr. Ind. Soc. **1**(2), 351–361 (2014)
5. Moe'tamar, E.: Penyelidikan Logam Emas Kabupaten Sumbawa, Provinsi Nusa Tenggara Barat. In: *Prosiding Hasil Kegiatan Pusat Sumberdaya Geologi Tahun 2011*. https://www.scribd.com/document/387499554/Proseding-Sumbawa
6. Pramumijoyo, S., Setijadji, L. D.: Alteration mineral mapping to identify primary tin potential using Landsat 8 images and geographic information system in Rimba Kulit Area, Southern of Bangka Island. In: Franto, Pramumijoyo, S., Setijadji, L. D. IOP Conference Series: Earth and Environmental Science, vol. 212. IOP Publishing (2018)
7. Krisnayanti, B.D.: ASGM status in West Nusa Tenggara Province, Indonesia. *Malang* (2017). https://garuda.kemdikbud.go.id/documents/detail/614059. Accessed 17 June 2024
8. Pour, A., Hashim, M.: Hydrothermal alteration mapping from Landsat-8 data, Sar Chesmeh copper mining district, south-eastern Islamic Republic of Iran (2015). https://www.sciencedirect.com/science/article/pii/S1658365514001290. Accessed 17 June 2024
9. Sabins, F.: Remote sensing mineral exploration (1999). https://www.sciencedirect.com/science/article/pii/S0169136899000074. Accessed 17 June 2024
10. Shimoda, H., Kimura, T.: Japanese Space Program. In: Liang, S. Comprehensive Remote Sensing, vol. 1, p 246–279. Elsevier (2018) https://www.sciencedirect.com/science/article/pii/B9780124095489103203. Accessed 18 June 2024
11. Hajaj, S., et al.: ASTER data processing and fusion for alteration minerals and silicification detection: Implications for cupriferous mineralization exploration in the western Anti-Atlas, Morocco., Artificial Intelligence in Geosciences, Morocco (2024). https://www.sciencedirect.com/science/article/pii/S2666544124000182. Accessed 18 June 2024
12. Hossain, A.K.M., Jia, Y., Chao, X.: Development of remote sensing based index for estimating/mapping suspended sediment concentration in river and lake environments (2010)
13. Bid, S., Siddique, G.: Identification of seasonal variation of water turbidity using NDTI method in Panchet Hill Dam, India. Model. Earth Syst. Environ. **5** (2019). https://doi.org/10.1007/s40808-019-00609-8

14. Mas, J.F.: Monitoring land-cover changes: a comparison of change detection techniques. Int. J. Remote Sens. **20**(1), 139–152 (1999)
15. Gabr, S., Ghulam, A., Kusky, T.: Detecting areas of high-potential gold mineralization using ASTER data. Ore Geol. Rev. **38**, 59–69 (2010)

Confined Aquifer Damage Analysis in Jakarta from Disaster Monitoring of Land Subsidence and Water Table Decline Correlation Data Inversion

Redho Surya Perdana[1](✉), Heri Andreas[2], Ongky Anggara[1], Aulia Try Atmojo[1], Satrio Muhammad Alif[1], and Adam Irwansyah Fauzi[1]

[1] Geomatics Engineering, Institut Teknologi Sumatera, Lampung, Indonesia
redho.perdana@gt.itera.ac.id

[2] Geodesy and Geomatics Engineering, Institut Teknologi, Bandung, Indonesia

Abstract. Jakarta is an industrial and biggest Government city in Indonesia, making Jakarta one of the provinces in Indonesia that uses groundwater a lot in various activities. Excessive groundwater extraction will cause a vacuum in soil pores so that hydrostatic pressure in the confined aquifer layer will be reduced, and rock compaction will occur in the confined aquifer layer, which will cause damage to the confined aquifer layer. Many studies show that water table decline data can describe the damage factor to the confined aquifer layer. However, the lack of groundwater observation wells causes a decline in water table data in Jakarta. Many GNSS observation points are spread in Jakarta, which can help observe the land subsidence phenomenon. Various studies also state a correlation between land subsidence data and water table decline. This study aims to change GNSS measurement data that can describe the phenomenon of land subsidence into water table decline data through simple average equations from various correlation graphs between land subsidence data and water tables in various countries. The analysis shows that Jakarta's overall confined aquifer situation is low in southern Jakarta and high in northern Jakarta.

Keywords: Confined Aquifer Damage · Land Subsidence · Water Table Decline · Jakarta

1 Introduction

The decrease in groundwater availability has occurred in various cities in Indonesia, including Jakarta [1, 2]. This is reinforced by data from the Meteorology, Climatology and Geophysics Agency in Indonesia (BMKG) which explains spatially that the level of groundwater availability in Indonesia, especially Jakarta, has been declining in a relatively short period. This phenomenon has caused impacts such as groundwater crises, material losses, and natural disasters such as drought and land subsidence [1, 3, 4]. The decrease in groundwater availability is caused by large volumes of groundwater

E. Fitrianingsih et al. (Eds.): ISAST 2024, SPPHY 416, pp. 34–43, 2025.
https://doi.org/10.1007/978-981-96-1344-1_4

extraction, which causes a reduction in the volume of water in the rocky layer below the surface of the soil that contains water and can be water permeated or often called the aquifer layer [5–8]. Groundwater levels, or aquifers, are classified into free and confined aquifers [9]. Excessive groundwater extraction will cause a vacuum in the soil pores so that the hydrostatic pressure in the confined aquifer layer will be reduced and rock compaction will occur in confined aquifer layer which causes damage to confined aquifer layer [6, 10].

As a disaster mitigation effort, previous studies have been carried out to identify the damage to confined aquifers through a model, such as making groundwater flow models and aquifer compaction models [10–12]. However, to identify confined aquifer damage spatially, data from groundwater wells is needed which is evenly distributed in study area to know water table decline phenomenon. However, Jakarta only has 9 groundwater wells that can observe the water table decline. Meanwhile, Jakarta has many GNSS observation points for monitoring land subsidence.

In previous studies, confined aquifer damage models that integrate land subsidence data, water table decline, and supporting spatial data have never been made. Water table decline data is essential as the main parameter of confined aquifer damage because if an area stops carrying out groundwater extraction, the phenomenon of land subsidence will stop causing the confined aquifer layer to not suffer damage [13, 14]. Groundwater monitoring wells systems located at nine 9 points in Jakarta are not as extensive as GNSS observation for land subsidence monitoring distributed in 45 points (see Fig. 1).

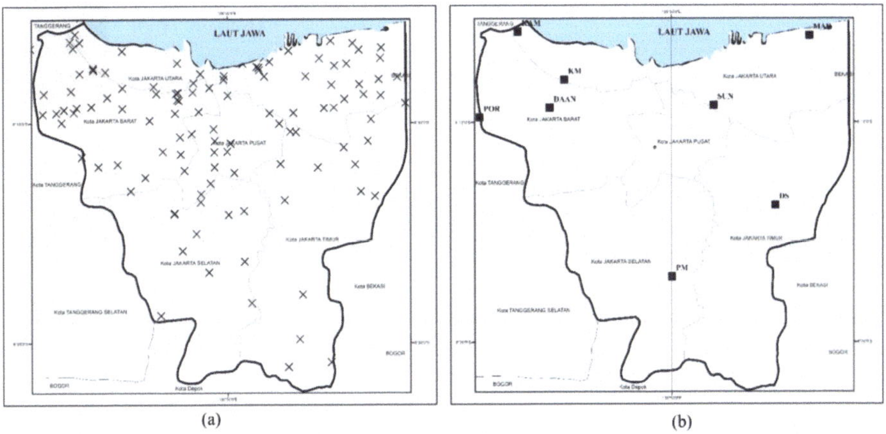

(a) (b)

Fig. 1. Map of GNSS observation point land subsidence (a) and groundwater wells point (b) in Jakarta.

Therefore, for the analysis of confined aquifer damage spatially, we need a model that can integrate land subsidence data, water table decline data and spatial data to analyze the phenomenon of confined aquifer damage in an area, especially Jakarta. The result can be used as a reference for decision-makers in implementing various needs, such as spatial-based groundwater extraction regulations, environmental conservation, infrastructure design planning, and disaster mitigation [10, 15, 16].

2 Material and Methods

2.1 Confined Aquifer

Todd explains that an aquifer can also mean a geological formation that contains water and is significantly capable of flowing water through its natural conditions. Aquifers have two important functions, namely as a store of water and as a water distributor, both of these functions are assigned to pores or cavities in aquifer rock [6]. Aquifers are grouped into free aquifers and confined aquifers. Groundwater from free aquifers is generally found at relatively shallow depths, which is less than 40 m. Whereas in confined aquifers, groundwater comes from a deeper depth and has a pressure greater than air pressure. At the top of confined aquifer layer there is impermeable rock, so the pressure exceeds atmospheric pressure and in this confined aquifer layer more groundwater will be found than in the free aquifer layer [12]. A confined aquifer is one whose entire water body is limited by an upper and lower waterproof layer and has a saturation pressure greater than atmospheric pressure [17] (Table 1).

Table 1. Categories of confined aquifers damaged from cumulative water table decline, rates and coloring [6, 17–19].

Cumulative Water Table Decline (m)	Rate Water Table Decline (m/year)	Classification	Color
> 45.72	> 27	Extremely High	
30.48 - 45.72	14 - 27	Very High	
15.24 - 30.48	9 - 14	High	
7.62 - 15.24	5 – 9	Medium	
3.048 - 7.62	2 - 5	Low	
< 3.048	< 2	Very Low	

2.2 Methodology

The methods used in this study is continuous GNSS measurement at GNSS observation points to obtain land subsidence in DKI Jakarta. Furthermore, the process of data acquisition as input from the confined aquifer damage modeling process. The data used are periodic GNSS observation and INSAR data which are then processed to see phenomenon of land subsidence, water table decline data, and supporting spatial data. The data was collected in the form of primary data (data taken directly in field and historical data of events. Land subsidence data is transformed into water table decline using a simple equation from the combined correlation between land subsidence data and water table decline data. The results of this aquifer damage modeling are used to carry out mitigation of groundwater storage programs in aquifer layers in the Jakarta area.

3 Theory

3.1 Equation of Correlation Between Land Subsidence and Water Table Decline in the World

Jakarta has nine groundwater observation wells which aim to monitor the state of groundwater and soil layers. In addition, benchmarks for GNSS observations are used to describe phenomena in Jakarta. The graphs in Fig. 2 show the comparison between land subsidence and water decline table at the location of groundwater observation wells and adjacent GNSS benchmark points.

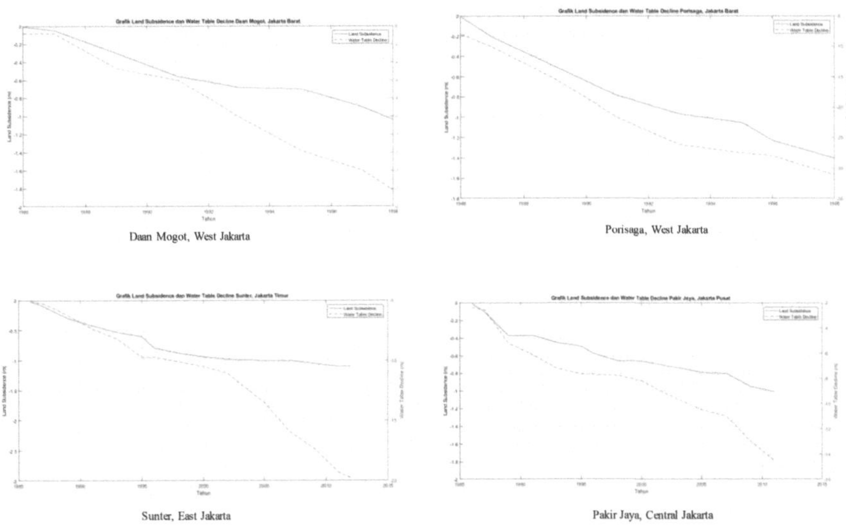

Fig. 2. Sample of graph land subsidence and water table decline in Jakarta.

From various sources (journal) in many countries, data were obtained and combined to form a correlation graph between land subsidence and water decline table which aims to find average correlation value between these two phenomena. This value aims to convert land subsidence data in the study area into water table decline data which will be categorized to determine the phenomenon of confined aquifer damage. [1–10, 14, 19, 20].

Figure 3 illustrates the trend cross-correlation of land subsidence and water table decline is

$$y = 18.755x + 0.4591 \tag{1}$$

This graph explains that, on average, all over the world experience land subsidence and water table decline each year with a ratio of around 1:18.7. This graph data was

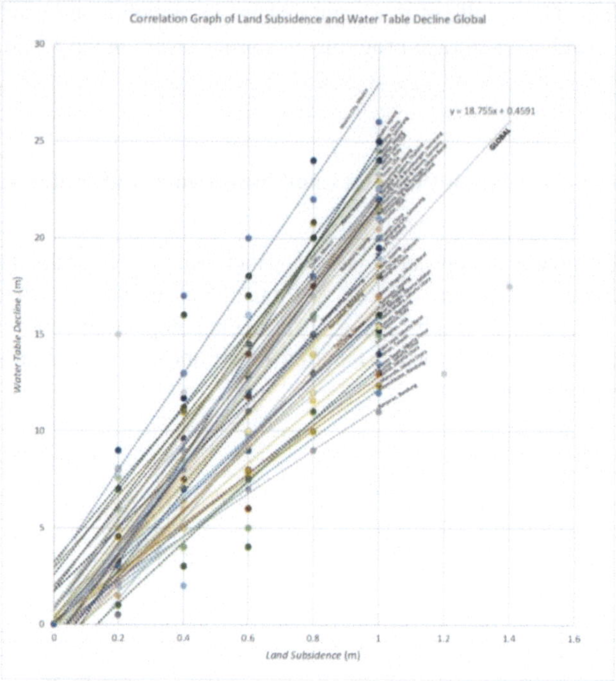

Fig. 3. Global cross-correlation of land subsidence and water table decline [20–28].

used as the basis for creating a data inversion from land subsidence data to water table decline data and classifying to create a confined aquifer damage map in DKI Jakarta.

4 Results

The result was obtained from GNSS survey and InSAR data in more than 100 observation points in DKI Jakarta. The GNSS data survey is processed using interpolation and classification to produce results that shows the land subsidence phenomena.

4.1 Land Subsidence Map of DKI Jakarta in 1925–2020

Based on the topographic and GNSS surveys from 1925–2020 presented in Fig. 4, we found that the cumulative land subsidence in DKI Jakarta ranges from 0.5–4.6 m, and the cumulative rate of land subsidence is 1–28 cm. The worst affected area for land subsidence is the northern area of DKI Jakarta.

4.2 Map of Confined Aquifer Damage Year 1925–2020 and Predict in 2025

The data inversion process is carried out to create a water table decline model, defined as a confined aquifer damage parameter. The map of confined aquifer damage of DKI

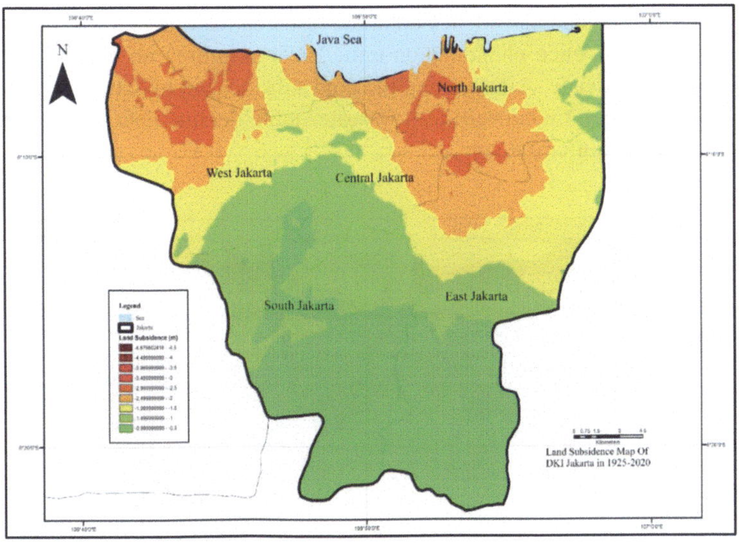

Fig. 4. Cumulative land subsidence map of DKI Jakarta in 1925–2020.

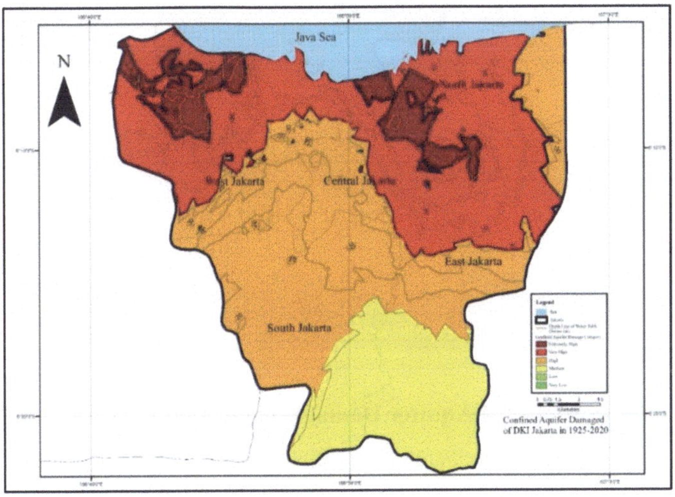

Fig. 5. Confined aquifer damaged of dki Jakarta in 1925–2020.

Jakarta in 1925–2020 (see Fig. 6) results in an extremely high damage area, which are Pluit, Cengkareng Regional Hospital, Palm Bay Water Park, Ancol, Kemayoran Station, and Cempaka Mas with water table decline reaching more than 46 m. In comparison, those categorized as very high are Pantai Indah Kapuk, Daan Mogot, WTC Mangga Dua, Ancol, Tanjung Priok Port, Mal Kelapa Gading, Pegangsaan Dua Street, and Jakarta State University.

Based on existing land subsidence data, the research team uses linear regression to predict the land subsidence phenomenon until 2025, assuming no disaster mitigation efforts for land subsidence. After getting the land subsidence prediction, the prediction data is processed again by converting it to water table decline data to make a spatial model for confined aquifer damage.

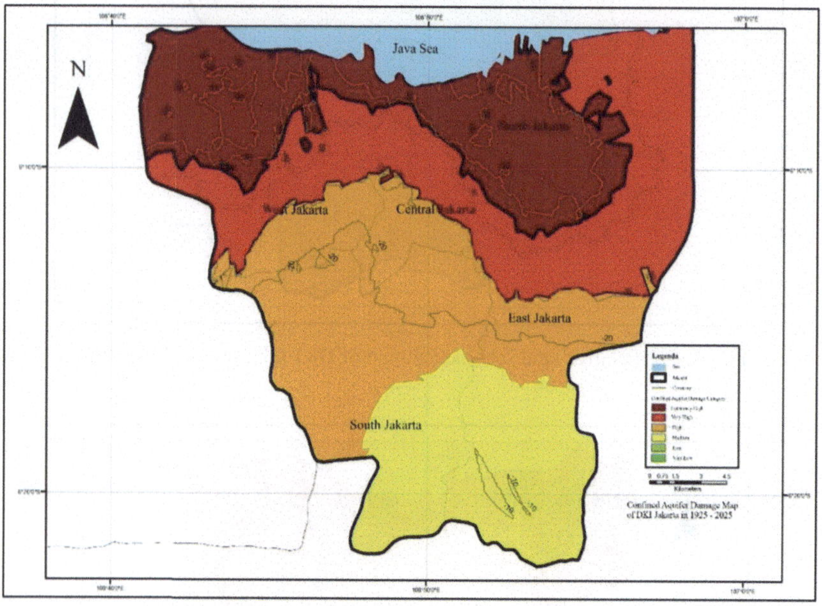

Fig. 6. Confined aquifer damaged of DKI Jakarta in 1925–2025.

5 Discussion

5.1 The Factor of Confined Aquifer Damage in Jakarta

North Jakarta, categorized as extremely high and very high in the confined aquifer damage, is a densely populated residential and industrial area. Meanwhile, in the South Jakarta area, categorized as low, it is a green open area. The result confirms that the densely populated area, which has a high use of groundwater, resulting a water void in the aquifer layer, which causes the soil surface to decrease and the water table layer to decrease.

Another influencing factor is the type of soil. It was known that the North Jakarta area is marine clay soil, the northern part of West Jakarta is clay, the southern part of West Jakarta is alluvial soil, the East Jakarta area is cemented silty sand, and the South Jakarta area is alluvial soil.

When viewed from the type of soil, marine clay and/or clay are types of soil often found in coastal areas with clay/sticky properties that are difficult to absorb water. This soil type is very susceptible to land subsidence and water table decline. Meanwhile, the North Jakarta area, whose area is marine clay soil type, is used as a residential and industrial area that often carries out large-scale groundwater extraction. Those factors make the northern area of DKI Jakarta very vulnerable to aquifer damage and land subsidence because theoretically, the clay soil type area has poor hydraulic conductivity and permeability values.

5.2 Limitation of the Study

This study uses GNSS data obtained periodically, and groundwater wells are used to obtain water table decline data. The construction of a confined aquifer damage spatial model requires groundwater wells spread over many locations in DKI Jakarta. Currently, DKI Jakarta only has nine groundwater wells. A confined spatial model for aquifer damage needs to be complemented by a model based on correlation data to describe better the phenomenon of confined aquifer damage in DKI Jakarta.

6 Conclusion

Jakarta is an area predicted to sink in the next few years due to the land subsidence that has occurred in most areas of Jakarta. It has become a global concern because Jakarta is the capital of Indonesia. By inversion of land subsidence data to water table decline data, the confined aquifer damage phenomenon can be observed using the correlation of land subsidence and water table decline data obtained from various data in the world (from journals). Land subsidence data can be inverted with this equation into water table decline data and further classified confined aquifer damage data.

Based on the results of data processing and modeling, it can be concluded from the confined aquifer damage of DKI Jakarta in 1925–2020, the area that fell into the extremely high category are Pluit, Cengkareng Regional Hospital, Palm Bay Water Park, Ancol, Kemayoran Station, and Cempaka Mas with water table decline reaching more than 46 m. At the same time, those categorized as very high are Pantai Indah Kapuk, Daan Mogot, WTC Mangga Dua, Ancol, Tanjung Priok Port, Mal Kelapa Gading, Pegangsaan Dua Street, and Jakarta State University.

The decision-makers can use the result to establish policies related to regulations on groundwater extraction in areas classified as high and very high to stop land subsidence and groundwater level decline, so that the aquifer is not further damaged and groundwater capacity is maintained.

References

1. Andreas, H., Zainal Abidin, H., Anggreni Sarsito, D., Meilano, I., Susilo, S.: Investigating the tectonic influence to the anthropogenic subsidence along northern coast of Java Island Indonesia using GNSS data sets. In: E3S Web of Conferences, vol. 94 (2019). https://doi.org/10.1051/e3sconf/20199404005
2. Chaussard, E., Amelung, F., Abidin, H., Hong, S.H.: Sinking cities in Indonesia: ALOS PALSAR detects rapid subsidence due to groundwater and gas extraction. Remote Sens. Environ. **128** (2013). https://doi.org/10.1016/j.rse.2012.10.015

3. Abidin, H.Z., Andreas, H., Gumilar, I., Fukuda, Y., Pohan, Y.E., Deguchi, T.: Land subsidence of Jakarta (Indonesia) and its relation with urban development. Nat. Hazards, **59**(3) (2011). https://doi.org/10.1007/s11069-011-9866-9
4. Béjar-Pizarro, M., et al.: Mapping groundwater level and aquifer storage variations from InSAR measurements in the Madrid aquifer, Central Spain. J. Hydrol. **547** (2017). https://doi.org/10.1016/j.jhydrol.2017.02.011
5. Chai, J.C., Shen, S.L., Zhu, H.H., Zhang, X.L.: Land subsidence due to groundwater drawdown in Shanghai. Geotechnique, **54**(2) (2004). https://doi.org/10.1680/geot.2004.54.2.143
6. Todd, D.K., Mays, L.W.: Groundwater Hydrology Third edn. (2005)
7. Döll, P., et al.: Impact of water withdrawals from groundwater and surface water on continental water storage variations. J. Geodyn. **59–60** (2012). https://doi.org/10.1016/j.jog.2011.05.001
8. Phi, T.H., Strokova, L.A.: Prediction maps of land subsidence caused by groundwater exploitation in Hanoi, Vietnam. Resour. Technol. 1(2) (2015). https://doi.org/10.1016/j.reffit.2015.09.001
9. Zorn, M.R.: Fundamentals of hydrology. Southeast. Geographer, 44(1) (2004). https://doi.org/10.1201/b14926-8
10. Hung, W.C., Hwang, C., Liou, J.C., Lin, Y.S., Yang, H.L.: Modeling aquifer-system compaction and predicting land subsidence in central Taiwan. Eng. Geol. **147–148** (2012). https://doi.org/10.1016/j.enggeo.2012.07.018
11. Leake, S.A., Galloway, D.L.: Use of the SUB-WT package for MODFLOW to simulate aquifer-system compaction in Antelope Valley, California, vol. 339. USA (2010). in IAHS-AISH Publication
12. Galloway, D.L., et al.: Detection of aquifer system compaction and land subsidence using interferometric synthetic aperture radar, Antelope Valley, Mojave Desert, California. Water Resour. Res. **34**(10) (1998). https://doi.org/10.1029/98WR01285
13. Andreas, H., Zainal Abidin, H., Pradipta, D., Anggreni Sarsito, D., Gumilar, I.: Insight look the subsidence impact to infrastructures in Jakarta and Semarang area; Key for adaptation and mitigation. In: MATEC Web of Conferences, vol. 147 (2018). https://doi.org/10.1051/matecconf/201814708001
14. Sato, C., Haga, M., Nishino, J.: Land subsidence and groundwater management in Tokyo. Int. Rev. Environ. Strateg. **6**(2) (2006)
15. Abidin, H.Z., et al.: Land subsidence of Jakarta (Indonesia) and its geodetic monitoring system. Nat. Hazards, **23**(2–3) (2001). https://doi.org/10.1023/A:1011144602064
16. Wang, H., Gao, J.E., Jie Zhang, M., Hua Li, X., Long Zhang, S., Zhi Jia, L.: Effects of rainfall intensity on groundwater recharge based on simulated rainfall experiments and a groundwater flow model. Catena, **127** (2015). https://doi.org/10.1016/j.catena.2014.12.014
17. Sen, Z.: Practical and Applied Hydrogeology (2014).https://doi.org/10.1016/C2013-0-14020-2
18. Nadiri, A.A., Taheri, Z., Khatibi, R., Barzegari, G., Dideban, K.: Introducing a new framework for mapping subsidence vulnerability indices (SVIs): ALPRIFT. Sci. Total Environ. **628–629** (2018). https://doi.org/10.1016/j.scitotenv.2018.02.031
19. McGuire, V.L.: Water-level changes and change in water in storage in the High Plains aquifer, predevelopment to 2013 and 2011–13, U.S. Geol. Surv. Sci. Investig. Rep. no. 5218 (2014)
20. Andreas, H., Zainal Abidin, H., Gumilar, I., Purnama Sidiq, T., Anggreni Sarsito, D., Pradipta, D.: On the acceleration of land subsidence rate in Semarang city as detected from GPS surveys. In: E3S Web of Conferences, vol. 94 (2019). https://doi.org/10.1051/e3sconf/20199404002
21. Phien-wej, N., Giao, P.H., Nutalaya, P.: Land subsidence in Bangkok, Thailand. Eng. Geol. **82**(4) (2006). https://doi.org/10.1016/j.enggeo.2005.10.004

22. Ho Tong Minh, D., Van Trung, L., Le Toan, T.: Mapping ground subsidence phenomena in Ho Chi Minh City through the radar interferometry technique using ALOS PALSAR data. Remote Sens. 7(7) (2015). https://doi.org/10.3390/rs70708543
23. Andreas, H., Abidin, H.Z., Gumilar, I., Sidiq, T.P., Sarsito, D.A., Pradipta, D.: Insight into the correlation between land subsidence and the floods in regions of Indonesia. Nat. Haz. Risk Assess. Vulnerability Reduction (2018). https://doi.org/10.5772/intechopen.80263
24. Carbognin, L., Gatto, P., Mozzi, G., Gambolati, G.: LAND SUBSIDENCE OF RAVENNA AND ITS SIMILARITIES WITH THE VENICE CASE (1978)
25. Teatini, P., Ferronato, M., Gambolati, G., Bertoni, W., Gonella, M.: A century of land subsidence in Ravenna, Italy. Environ. Geol. 47(6) (2005). https://doi.org/10.1007/s00254-004-1215-9
26. Wolkersdorfer, C., Thiem, G.: Ground water withdrawal and land subsidence in northeastern Saxony (Germany). Mine Water Environ. **18**(1) (1999). https://doi.org/10.1007/bf02687252
27. Ortega-Guerrero, A., Rudolph, D.L., Cherry, J.A.: Analysis of long-term land subsidence near Mexico City: field investigations and predictive modeling. Water Resour. Res. **35**(11) (1999). https://doi.org/10.1029/1999WR900148
28. Calderhead, A.I., Therrien, R., Rivera, A., Martel, R., Garfias, J.: Simulating pumping-induced regional land subsidence with the use of InSAR and field data in the Toluca Valley, Mexico. Adv. Water Resour. 34(1) (2011). https://doi.org/10.1016/j.advwatres.2010.09.017

The Use of Remote Sensing to Assess Economic Damage from Return Period Flooding in Bojonegoro City, Indonesia

Lisa Agustina[1(✉)] and Andrew Lowe[2]

[1] Department of Chemical and Environmental Engineering, University of Nottingham, University Park, Nottingham NG7 2RD, UK
lisa.agustina@mespom.eu

[2] Department of Earth and Environmental Sciences, Oxford Road, Manchester M13 9PL, UK

Abstract. The Bengawan Solo River, the longest river on Java, Indonesia, frequently causes severe flooding, particularly affecting Bojonegoro. This study assesses flood risk and economic damage in Bojonegoro's urban areas using remote sensing and flood modelling techniques. A 2D Flood Model was created using Flood Modeller software by Jacobs, with rainfall-runoff analysis of six sub-catchments via the SCS-CN method to generate input for the model. Economic damage and flood-prone village mapping were conducted using ArcGIS Pro and R Programming. Results show that a 25-year return period flood will inundate urban areas, increasing economic damage from $6.988 million to $16.219 million compared to a 10-year flood. Several previously unaffected villages, including Sumbang and Sukorejo in Bojonegoro District and Ngampel and Tikusan in Kapas District, will experience significant flooding.

Keywords: Remote Sensing · Economic Damage · Flood Risk Assessment

1 Introduction

Flooding is a recurrent hydrometeorological hazard causing economic losses and fatalities worldwide. Indonesia, a tropical country with high-intensity rainfall, is significantly affected by river floods in lowland areas. Indonesia has 5,590 main rivers, with 600 posing flood risks [1]. Between 2014 and 2019, Indonesia experienced 21,239 disasters, with flooding being the most frequent, accounting for 7,768 events [2]. A World Bank study estimates that one in four Indonesians lives in high-risk flood zones [3], leading to significant casualties and economic losses. In early 2019, Indonesia faced 1,107 flood events resulting in 375 deaths, 1,340 injuries, and the displacement of 850,772 people [4]. This impact hampers optimal development, affecting its citizens' physical infrastructure and socio-economic welfare.

The Bengawan Solo River, Indonesia's longest river on Java island, frequently causes floods. Bojonegoro Regency, located along a low-altitude floodplain and traversed by a high-capacity river, has experienced 147 flood events from 2008 to 2022, making it highly

E. Fitrianingsih et al. (Eds.): ISAST 2024, SPPHY 416, pp. 44–54, 2025.
https://doi.org/10.1007/978-981-96-1344-1_5

flood-prone due to increased river flow in the rainy season [5]. Bojonegoro, located in the flood-prone Bengawan Njero area of north-central East Java, is highly susceptible to flooding due to the overflowing Bengawan Solo river [6]. Bojonegoro has a population of around 1.3 million [7], with many residents vulnerable to flooding. The city itself, with a population of 87,563, is situated in the floodplain of the Bengawan Solo River, making it highly vulnerable to flooding. Recent data from 2018 to 2022 shows that nearly 12,000 houses were submerged, and 76 public facilities were damaged due to flooding [8]. Frequent floods cause material losses and disrupt various sectors, including politics, education, and the economy [9]. This persistent issue hinders the region's development. Forecasting floods and effective flood management are crucial to mitigate these impacts.

Flood modelling is essential for predicting floods and formulating management strategies. While Jakarta has detailed flood models, Bojonegoro requires similar attention. Microscale models are crucial for accurate flood risk assessment and management, including calculating economic losses and flood depth. This study focuses on assessing flood risk and economic damage in Bojonegoro due to the Bengawan Solo River. Objectives include reviewing literature, gathering data for a 2D hydraulic flood model, analysing sub-catchments, forecasting flood risk, calculating economic losses, and identifying flood-prone areas and crops.

Climate change is expected to worsen flooding in Indonesia, with more frequent heavy rainfall and increased hydrometeorological disasters [10]. The AQUEDUCT Floods Analyzer predicts a nearly two-fold increase in annual affected people by 2080 compared to 2010 [11]. Effective flood management strategies are essential to ensure the safety and well-being of people in flood-prone areas.

Bojonegoro's flood management requires attention at all disaster stages. However, recent studies indicate a lack of synergy among stakeholders, hindering effective flood management and mitigation [6]. Encouraging stakeholder collaboration is crucial for enhancing flood management strategies. Flood modelling, especially the 2D hydraulic model with rainfall-runoff approaches, is widely used to simulate flood events and forecast risks. Detailed flood modelling and economic damage assessment at a microscale are necessary for Bojonegoro. Using the microscale approach provides more precise outcomes, aiding better flood risk management and decision-making.

2 Methods

This study investigates the fluvial flood risk of the Bengawan Solo River in Bojonegoro City and its surroundings, located on the island of Java. The river flows from the mountains of southern Central Java into northern East Java, with Bojonegoro Regency situated along the border of Central Java at coordinates 6°59′ to 7°37′ South and 112°25′ to 112°09′ East. Roads and railroads in Bojonegoro connect the two provinces on Java Island, with the city of Bojonegoro centrally located within the regency, as depicted in Fig. 1.

The design flooding model was built using Flood Modeller software from Jacobs, employing the SCS-CN method. Data collected for building the model is listed in Table 1. The sub-catchments used in this study, derived from watershed delineation with ArcGIS Pro, are shown in Fig. 2. Various parameter values from each sub-catchment are listed in

Table 2 and serve as inputs for the hydrological boundary unit, or Generic Rainfall-Runoff Boundary (GERRBDY).

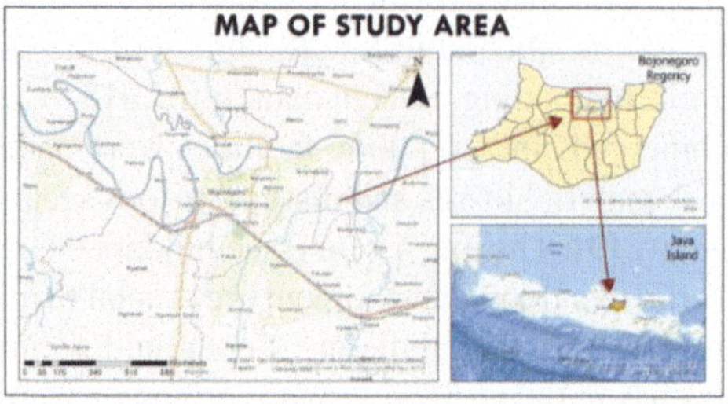

Fig. 1. Map of study area and active area of the flood model, showing the region where simulations are focused.

Table 1. Data collected during study.

Data	Source
Digital Elevation Model (DEM)	Ina-Geoportal (BIG Indonesia, 2017)
Intensity Duration Frequency (IDF)	This information was obtained from Fauziyah et al. (2013)
Curve Number (CN)	CN values of each sub-catchment were obtained from Jaafar et al. (2019)
Land Cover	Esri Global Land Cover 2022

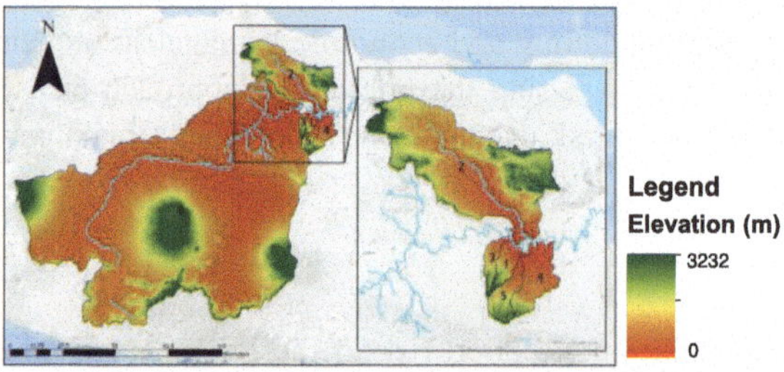

Fig. 2. Sub-catchments map derived from watershed delineation techniques.

In Table 2, area is measured in km2 using the Area (geodesic) function in ArcGIS Pro with the WGS 1984 UTM Zone 49S coordinate system. Rainfall profiles include storm duration, depth, and Area Reduction Factor (ARF), with a 24-h storm for design rainfall. The ARF equation, based on Fiddes [12], adjusts for spatial variability in rainfall. Curve Number (CN) values are used to estimate total inflow in the SCS-CN model. The initial abstraction ratio, set to a default of 0.2 but ranging from 0.05 to 0.20 per Moglen et al. [13], affects runoff volume and hydrograph peaks. Time of Concentration (Tc),

Table 2. Parameter Values of Each Sub-Catchment.

Sub-Cathment	CN	Length	Area	Slope	Tc	ARF
1	78.25	13	50	0.004538	6.27	0.9
2	79.33	53	844	0.006566	17.91	0.6
3	78.83	17	40	0.003588	8.79	0.91
4	78.8	20	129	0.0055	8.45	0.84
5	79.65	18	71	0.006667	8.79	0.88
6	79.72	180	11597	0.000828	70.82	1

calculated using the Bransby William equation from Wanielista [14], considers sub-catchment length, area, and slope, with measurements obtained from ArcGIS Pro and DEM data.

The model in this study discusses the flood risk associated with the Bengawan Solo River, excluding the tributaries in different return period years including 1 in 2 years, 1 in 5 years, 1 in 10 years, 1 in 25 years, 1 in 50 years, and 1 in 100 years. The model provides flood extent, depth, and velocity data for each flooding scenario. The return period is an approach that is widely used for flood risk assessment based on exceedance probabilities [15].

The study utilised Flood Modeller software to construct the flood model, specifically employing the 2D ADI Solver for simulating fluvial and overland flooding. This solver uses shallow water equations and a method for calculating rainfall in sub-catchments. A 2D model was created after pre-processing the DEM on ArcGIS Pro, using the WGS 1984 UTM Zone 49S spatial reference. Input data for the Domain Details are provided in Table 3.

Table 3. Input in Domain Details

Topography	Computational Area and Time Step	Roughness Data
DEM derived from the National DEM of Indonesia with a 7.5 m spatial resolution	**Grid size** is 14 m and **time step** is 1 s	**Roughness Law** used is Manning's
Road lowered by 0.1 m to ensure it is accurately represented in the model compared to the DEM	**Number of cols** is 1225 and **number of rows** is 925	
Building heights increased by 0.2 m to account for typical elevation above flood levels, ensuring accurate representation in the model	**Rotation** is 0	

Manning's roughness is vital in hydraulic modelling as it influences water flow over the ground, affecting flood area, velocity, and water level. The Manning's roughness coefficient used in this study was sourced from Papaioannou [16], who estimated it using CORINE land cover data and typical values from Chow's study [17].

The modelling results include flood depths, extents, and velocities in shapefiles (.shp) and ASCII (.asc) formats, which are exported and visualized using ArcGIS Pro. This software maps the flood extents and assesses impacted communities. The flood depths in raster format are also used as input data for economic damage modelling.

In economic damage modelling, input data include depth-damage curves, land use and flood depth maps in raster format, and land use values. These inputs are used to estimate losses and convert them into economic damage, as outlined in Fig. 3. The depth-damage curves link flood depth to damage for various building and material types, with estimates based on the research of [18], which divides buildings into seven categories with five material types. Depth-damage curves for non-building land uses, such as forestry and agriculture, are derived from Budiyono et al. [19]. Both building and non-building land uses are assessed for valuation using the same study, as shown in Table 4. The damage assessment uses raster data in R to calculate each cell's damage based on land use and flood depth, building on Pinos et al. (2020)'s microscale economic flood loss methodology in Ecuador [18].

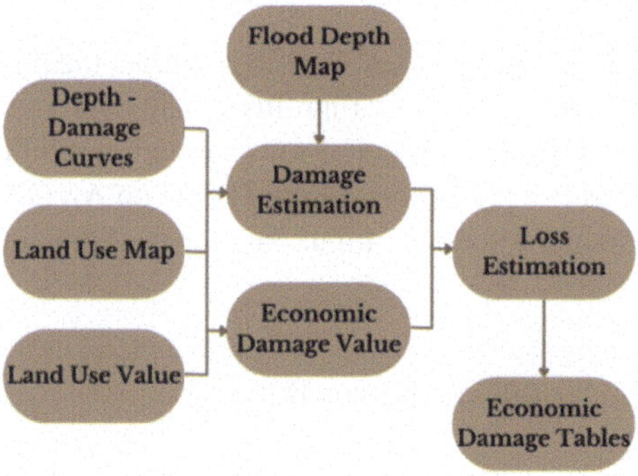

Fig. 3. Economic Damage Modelling Workflow.

Flood-prone areas are identified by overlaying administrative maps with flood forecast maps to determine the extent of impact in each village. A trend slope calculation assesses changes in affected areas over different flood return periods, helping stakeholders prepare for potential risks.

Table 4. Land Use Value of Each Category

Land Use Class Name	New Maximum Economic Exposure Value (thousand USD per hectare)
Government Facility	301
Forestry	10.4
Industry and Warehouse	517.9
Residential	150.6
Agriculture	1.6
Agriculture and Open Space (Rangeland)	3.1

3 Results

Design flooding for various return periods was analysed using Log-Pearson Type III rainfall to generate inflow hydrographs from each sub-catchment, with results displayed in Fig. 4(a–f). A 100-year flood affects nearly half of the active area, almost doubling the inundation compared to a 25-year return period. Flow components for each return period are checked, with peak flow occurring between 65 and 70 h, as shown in Fig. 4(g). Figure 4(h) shows that the active area for flood modelling has an extent of 14,991.4 hectares.

Figure 4 illustrates the extent of flooding for different return periods. For a 1-in-2-year event, the flood extent is 942.7 hectares, which covers 6.3% of the active area. This extent increases to 1,096.8 hectares (7.3%) for a 1-in-5-year flood and to 1,747.3 hectares (11.7%) for a 1-in-10-year event. In the case of a 1-in-25-year flood, the extent grows to 3,914.3 hectares, representing 26.1% of the active area. For a 1-in-50-year flood, the extent reaches 5,992.9 hectares (40.0%), and for a 1-in-100-year flood, it covers 7,089.1 hectares, or 47.3% of the active area.

After calculating the economic damage assessment using R, the output data shows building losses in Fig. 5(a) and crop losses in Fig. 5(b). The graphs indicate a significant increase in economic damage for 10-year and 25-year return periods. Building losses more than doubled, with residential damage nearly doubling and religious facilities damage quadrupling. Crop losses saw nearly a three-fold increase in economic damage to crop fields.

The flood extent maps from design flooding identify areas prone to inundation over several return periods. The percentage of inundated areas is calculated relative to each village's total area, and the slope measures the rate of increase in inundation from 2-year to 100-year return periods. This analysis helps local authorities prepare for future floods. For instance, in Tikusan village, Kapas district, the highest slope value of 22.5 indicates no inundation during the 2–10 year period but over 50% inundation at the 25-year period. Sumbang Village, Bojonegoro District, shows a similar pattern with a slope value of 21.5. Other affected areas include Ngampel Village in Kapas District and Sumbang Village in Bojonegoro District, as shown in Fig. 6.

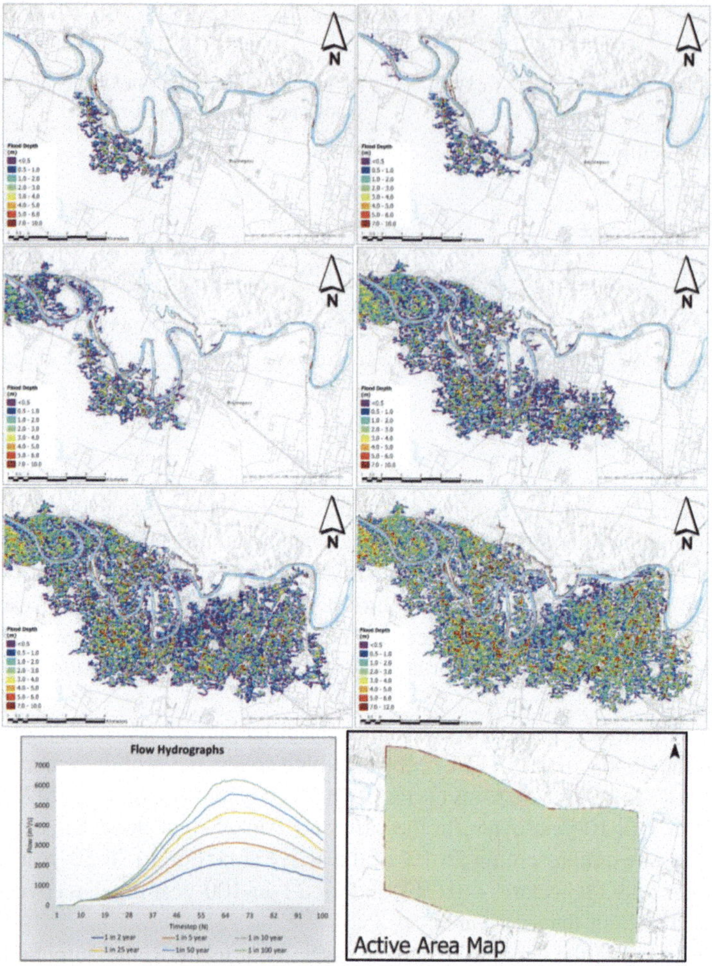

Fig. 4. Return period flood extent simulation maps; (a) 1 in 2 year, (b) 1 in 5 year, (c) 1 in 10 year, (d) 1 in 25 year, (e) 1 in 50 year, (f) 1 in 100 year, (g) flow hydrographs for design flood simulation, (h) active area of flood modelling.

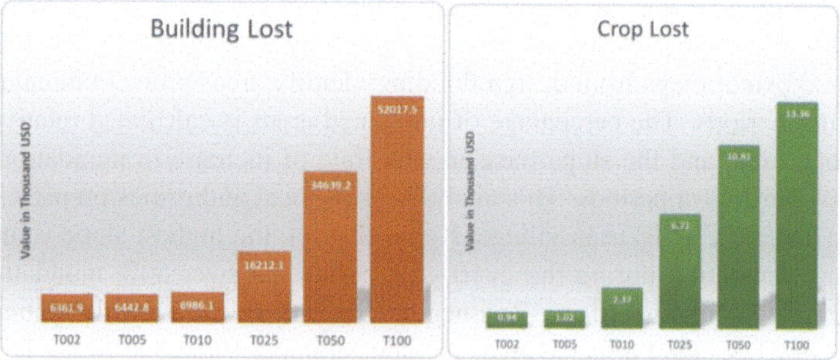

Fig. 5. Graph of economic damage from; (a) building areas, (b) crop areas.

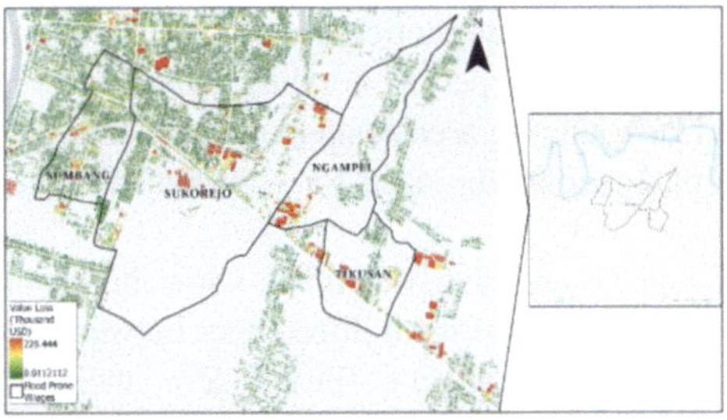

Fig. 6. Flood prone villages from 1 in 25 year return period flooding.

4 Discussion

4.1 Design Flooding

Designing flooding using rainfall return periods helps estimate the recurrence of flood events, ranging from low to extreme, aiding in extreme flood disaster planning. Figure 4(g) shows that peak flow values increase significantly for return periods of 2, 5, 10, 25, 50, and 100 years, with values of 2156m^3/s, 3193m^3/s, 3797m^3/s, 4740 m^3/s, 5611 m^3/s, and 6329 m^3/s, respectively. Prastica [9] indicates that the Bengawan Solo River in Bojonegoro City has a maximum capacity of 3605.14 m^3/s, which is exceeded by over 1000 m^3/s during a 25-year return period flood (Fig. 4), leading to severe flooding. Estimating peak flow is crucial for designing flood protection structures like dams and bridges [20]. Rainwater harvesting through structures like infiltration wells can also help reduce peak flow [21]. In urban areas, managing flood risks and avoiding settlements in high-risk floodplain areas is essential. This is vital for Indonesia, where understanding flood extent aids government and stakeholders in decisions about relocation or restoring floodplain functions [22].

The SCS-CN method has shown good results in modelling return period flooding. Meresa [20] found that the SCS-CN method outperformed other techniques like Artificial Neural Networks (ANN) and HEC-HMS, with lower peak flow errors and better hydrograph agreement. Zin et al. [23] also reported high validation for the SCS-CN method, showing a good correlation between observed and simulated flows in the Bago River Basin, Myanmar.

4.2 Assessment of Microscale Economic Damage

Return period flooding analysis helps create flood depth and extent maps for micro-scale economic damage assessment. These assessments serve as benchmarks for potential impacts if flood mitigation measures are not implemented, guiding land use planning and protection. They also aid in vulnerability analysis of elements-at-risk and encourage cooperation among local governments and stakeholders [24].

In the results section, a significant increase in economic damage to buildings and crops was noted during the 25-year return period, especially in urban areas. The total economic damage rose from approximately 7 million USD for the 10-year return period to 16 million USD for the 25-year return period, indicating a more than twofold increase.

This highlights the need for adequate anticipation and management of 25-year return period floods. The 100-year return period flooding is projected to inundate the entire city of Bojonegoro and surrounding areas, causing significant economic damage up to 52 million USD. This underscores the need for government action in flood protection and management.

The economic damage assessment outcomes show high values, consistent with Wijayanti et al. [25], who found significant differences between economic damage estimates from surveys and damage scanners. These discrepancies are due to factors like vulnerability curves, depth-damage analysis, and asset values [26].

4.3 Communities at Risk of Return Period Flooding

The flood extent map from design flooding helps identify administrative regions at high risk of flooding due to different return periods. In the 25-year return period, several villages like Ngampel and Tikusan in Kapas District, and Sumbang and Sukorejo in Bojonegoro District, experience new inundations. These areas, shown in Fig. 6, include high-value urban buildings, making flood mitigation measures crucial.

This assessment provides local governments with a reference for implementing flood management strategies at the village level. Measures could include developing flood early warning systems and enhancing community preparedness through initiatives like the Disaster Resilient Village (DESTANA). The assessment also helps prioritize areas for intervention and identifies the most vulnerable regions. DESTANA has proven effective in raising disaster awareness and reducing risks by leveraging local resources and wisdom for a more sustainable community [2, 6].

5 Conclusion

The primary goal of this study was to spatially assess the flood risk in Bojonegoro from the Bengawan Solo River. Flood modelling was conducted using the SCS-CN method to analyse the rainfall-runoff of six sub-catchments, generating inflow hydrographs as input for a 2D hydraulic flood model. The model's results were compared with observed floods recorded by BBWS. Although the model had some limitations and the validation was not fully adequate, it performed well for fluvial flood modelling.

The analysis of return period flooding, particularly the 25-year return period, revealed that urban areas in Bojonegoro City are at significant risk. The economic damage assessment showed that buildings and crops experienced substantial increases in damage compared to a typical 10-year flood event, with building damage doubling and crop damage tripling. Further analysis identified that several villages, previously unaffected by 2, 5, and 10-year floods, experienced over half of their area flooded during the 25-year event. Notably, Tikusan and Ngampel Villages in the Kapas District and Sumbang and Sukorejo Villages in the Bojonegoro District, which are urban areas with high-value assets, were significantly impacted.

References

1. Rahadianto, H., Fariza, A., Akhmad, J., Hasim, N.: Risk-level assessment of Bengawan Solo River basin floods using analytic hierarchy process. J. Water Resour. Manage. **29**(6), 2325–2340 (2015)
2. Ningtyas, T., Larasati, E., Warsono, H., Purnaweni, H.: Optimization of Destana policies for flood disaster preparedness in Bojonegoro District. Manag. Entrep. Trends Dev. **1**(15), 113–126 (2021)
3. Rentschler, J., Klaiber, C., Vun, J.: Mapping poverty and flood risk in Indonesian cities. East Asia & Pac. on the Rise (2021). https://blogs.worldbank.org/eastasiapacific/floods-neighborhood-mapping-poverty-and-flood-risk-indonesian-cities. Accessed 8th May 2023
4. Sholihah, Q., Kuncoro, W., Wahyuni, S., Puni Suwandi, S., Dwi Feditasari, E.: Analysis of flood disaster causes and impacts from an environmental law perspective. IOP Conf. Ser.: Earth Environ. Sci. **437**(1), 012056 (2020)
5. Bengawan Solo River Regional Office: Sejarah Singkat | BBWS Bengawan Solo. https://sda.pu.go.id/balai/bbwsbengawansolo/portal/index.php/sejarah-singkat/. Accessed 2 May 2023
6. Widyarta, A., Hidayat, E., Kriswibowo, A.: Collaborative governance in flood mitigation in Bojonegoro regency. Int. J. Educ. Res. Soc. Sci. **3**(5), 1873–1878 (2022)
7. Bojonegoro Regency in Figures 2021. BPS Bojonegoro, Bojonegoro (2021)
8. Data Management and Information Systems of BNPB: Data Informasi Bencana Indonesia. DIBI BNPB. [Online] [Accessed 2nd May 2023] https://dibi.bnpb.go.id/xdibi (2023)
9. Prastica, R.M.S., Maitri, C., Hermawan, A., Nugroho, P.C., Sutjiningsih, D., Anggraheni, E.: Design flood estimation and HEC-RAS modeling in Bojonegoro city. IOP Conf. Ser. Mater. Sci. Eng. **316**(1), 012042 (2018)
10. BMKG: Proyeksi Perubahan Iklim: Peta Proyeksi Iklim Indonesia. https://www.bmkg.go.id/iklim/?p=proyeksi-perubahan-iklim. Accessed 11 May 2023
11. World Resources Institute: AQUEDUCT Floods Risk: Indonesia Country Profile (2020). https://www.wri.org/aqueduct. Accessed 11 May 2023
12. Fiddes, D.: The TRRL East African Flood Model. TRRL Laboratory Report 706. TRRL, Berkshire (1976)
13. Moglen, G.E., McCuen, R.H., Moglen, R.L.: Consequences of changes to the NRCS rainfall-runoff relations on hydrologic design. J. Hydrol. Eng. **23**(8), 04018032 (2018)
14. Wanielista, M.P., Kersten, R., Eaglin, R.: Hydrology: Water Quantity and Quality Control. Wiley, New York (1996)
15. Ward, P.J., De Moel, H., Aerts, J.C.J.H.: Flood risk estimates and return-periods. Nat. Hazards Earth Syst. Sci. **11**(12), 3181–3195 (2011)
16. Papaioannou, G., et al.: Flood directive implementation in ungauged urban areas. Hydrology **5**(2), 24 (2018)
17. Chow, W.: Open-Channel Hydraulics, 1st edn. McGraw-Hill, New York (1959)
18. Pinos, J., Orellana, D., Timbe, L.: Microscale economic flood losses: Santa Bárbara River. Ecuador. Nat. Hazards **103**, 2323–2337 (2020)
19. Budiyono, Y., Aerts, J.C.J.H., Tollenaar, D., Ward, P.J.: River flood risk in Jakarta under future scenarios. Nat. Hazards Earth Syst. Sci. **16**(3), 757–774 (2016)
20. Meresa, H.: Modelling river flow in ungauged catchments using remote sensing data. Model. Earth Syst. Environ. **5**(1), 257–273 (2019)
21. Suprayogi, S., Rifai, I., Latifah, R.: HEC-HMS model for urban flood analysis in Belik River. Yogyakarta. ASEAN J. Sci. Technol. Dev. **38**(1), 15–20 (2021)
22. Kusratmoko, E., Marko, K., Elfeki, A.M.M.: Spatial modeling of flood inundation in Pesanggrahan floodplain. J. Geogr. Environ. Earth Sci. Int. **5**(3), 1–10 (2016)

23. Zin, W.W., Kawasaki, A., Win, S.: River flood inundation mapping in the Bago river basin. Hydrol. Res. Lett. **9**(4), 97–102 (2015)
24. Țîncu, R., Zêzere, J.L., Crăciun, I., Lazăr, G., Lazăr, I.: Micro-scale flood risk assessment in the Trotuş River. Land Use Policy **95**, 103881 (2020)
25. Wijayanti, P., Zhu, X., Hellegers, P., Budiyono, Y., van Ierland, E.C.: Estimation of river flood damages in Jakarta. Nat. Hazards **86**(3), 1059–1079 (2017)
26. Jongman, B., et al.: Comparative flood damage model assessment. Nat. Hazards Earth Syst. Sci. **12**(12), 3733–3752 (2012)

Recent Analysis of Land Subsidence Rates in Medan, Indonesia, Using Timeseries InSAR from 2015–2023

Ongky Anggara[1], Putri Sonya[1], Redho Surya Perdana[1](✉), Dita Mulia Pangestika[2], Aulia Try Atmojo[1], Argo Galih Suhadha[3], Satrio Muhammad Alif[1], and Agung Mahadi Putra Perdana[1]

[1] Geomatics Engineering, Institut Teknologi Sumatera, South Lampung, Indonesia
redho.perdana@gt.itera.ac.id
[2] Geodesy and Geomatics Engineering, Institut Teknologi Bandung, Bandung, Indonesia
[3] National Research and Innovation Agency (BRIN), Bogor, Indonesia

Abstract. The increasing population demand for water resources in urban areas has forced excessive groundwater extraction and increased excessive building loads, resulting in many urban areas experiencing land subsidence. The land subsidence phenomenon has occurred in many big cities in Indonesia, including Medan. This research used Sentinel-1A satellite imagery from LiCSBAS time series analysis, which is integrated with the automated Sentinel-1 InSAR processor (COMET-LiCSAR) and adopted to obtain the land subsidence rate in 2015–2023. The InSAR processing process considers atmospheric effects using the Generic Atmospheric Correction Online Service for InSAR (GACOS). Time series InSAR indicates Medan City experienced an average land subsidence rate ranging from ~−2 mm/year to ~−40 mm/year. Land subsidence spread throughout all sub-districts in Medan, with sinking occurring predominantly in heavily populated regions in the central part of Medan City. Excessive building loads and excessive groundwater use due to the increasing population growth rate are suspected to cause land subsidence in Medan.

Keywords: Medan · Land Subsidence · InSAR · SBAS Method

1 Introduction

Subsidence is primarily the downward movement of the Earth's surface, typically due to a lack of support from below, an added load, or both. Subsidence is triggered by natural processes, human activities, or often by human actions that destabilize natural systems [1]. Natural subsidence occurs from sediment consolidation and loss of fluids due to lithostatic loading, compaction, and tectonic activities [2–4]. Human-induced subsidence mainly arises from fluid extraction such as hydrocarbons and groundwater [5], land-use changes [6], urban-induced loading [7], geothermal exploitation, and mining [8]. Subsidence results from a wide variety of circumstances and processes. This poses significant challenges, leading to increased costs and delays in new developments, damage to existing structures, infrastructure, and land, and occasionally causing injuries or fatalities.

E. Fitrianingsih et al. (Eds.): ISAST 2024, SPPHY 416, pp. 55–65, 2025.
https://doi.org/10.1007/978-981-96-1344-1_6

Land subsidence occurs in many areas of the world, especially in densely populated cities [9], irrigated areas around the coast [10, 11], and alluvial land [12]. In Indonesia, land subsidence occurs in many urban areas [1, 13]. One of the cities that indicated land subsidence is Medan. Medan, the capital of North Sumatra Province, is a strategically located city with a population of 2,494,166 in 2023 [14].

In the last two decades, monitoring land subsidence uses various techniques such as the Global Navigation Satellite System (GNSS) [15–17], levelling [18], geology and geotechnics [19], which is costly and relatively small coverage. In the last decade, research on monitoring land subsidence using Interferometric Synthetic Aperture Radar (InSAR) data [20, 21] has proven successful in monitoring spatially in large areas such as Mexico City [22], Zhuhai City, China [23], Karapinar City, Turki [24], and Surabaya City [25]. In large cities in Indonesia, land subsidence research using GNSS from 1982–2010 showed a significant decrease with a value of 20–28 cm/year due to groundwater extraction, construction loads, and related to city development [1]. The consequences of land subsidence in Jakarta include cracking and damage to buildings, damage to infrastructure, expansion of flood areas, and changes to the drainage system, resulting in infrastructure, environmental, economic, and social impacts [26].

InSAR has various data processing methods; one of the methods used in this study is the SBAS method [27]. SBAS InSAR (Small Baseline Subset InSAR) minimizes speckle interference problems and improves measurement accuracy. It can also be used to estimate the height of the earth's surface with high accuracy [28]. This technology utilizes phase differences in SAR imagery to detect small changes in the ground surface, indicating land subsidence. This research uses Sentinel-1 with the InSAR Small Baseline Subset (SBAS) method from 2015 to 2023 to monitor land subsidence in Medan City. This research suggests references in implementing regional development policies as a disaster mitigation effort by satellite monitoring in Medan City.

2 Methodology

This research area covers the entire city of Medan City, which has an area of 265.1 km^2 and consists of 21 sub-districts, as shown in Fig. 1. This research uses Sentinel-1 imagery from 2015 to 2023 in ascending and descending imagery. This study employs interferogram data processing to map ground surface deformation. The interferogram data utilized in this research were downloaded from the COMET-LiCS Sentinel-1 InSAR web portal at https://comet.nerc.ac.uk/COMET-LiCS-portal/. The coverage area of the data spans approximately 250 $\times$ 250 km with an Interferometric Wide (IW) mode. The interferogram data processing was conducted using the LiCSBAS software [29]. LiCSBAS was used to process 1368 ascending and 1752 descending interferogram data to obtain the Line of Sight (LOS) deformation values.

The processing steps include tropospheric noise correction using data from the Generic Atmospheric Correction Online Service (GACOS) [30]. Integration of GACOS correction maps with the interferogram data to remove tropospheric noise effects. The analysis results in LOS deformation data, which are further processed to obtain time series deformation with vertical projection [23, 29]. Conversion of LOS deformation measurements to vertical displacement using geometric relationships based on satellite acquisition geometry and look angles. A 2.5 D decomposition is also applied to derive the vertical velocity rate from the LOS deformation [31]. This method enables more accurate identification of vertical surface changes in the study area.

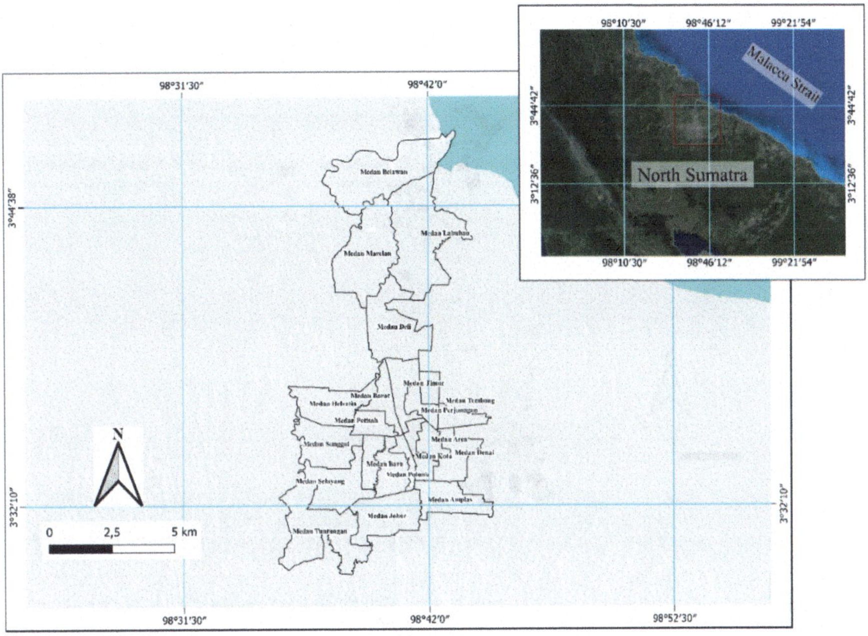

Fig. 1. Study area for land subsidence in Medan City.

3 Results and Discussion

Coherence is the correlation coefficient of pairs of SAR images. This correlation shows the similarity of each pixel between the master image and the slave image. Land cover dominantly influences the coherence value due to scattering from the phase when the object reflects. The coherence value ranges from 0 to 1 [32]. A coherence value close to 0 indicates little information (high noise level) contained in the image. This is because the radar sensor receives a small amount of backscattered signal.

Meanwhile, a coherence value of 1 means that the image pair is identical, namely, the strength of the signal reflected to the radar sensor. Ascending coherence corrected for the atmosphere in the Medan City area from 2015–2023 is presented in Table 1. The ascending coherence value in Medan City is 0.023 to 0.958, with an average of 0.536. The sub-district that has the highest average coherence value in ascending order is the Medan area Sub-district at 0.766. The average coherence values in Subdistricts resulting from ascending and descending image processing using atmospheric correction in the Medan City area for 2015–2023 are presented in Table 1. The coherence values in Medan City are 0.028 to 0.973, with an average of 0.555. The subdistrict with the highest average coherence value in descending order is Medan area Subdistrict, which is 0.839. High coherence values dominate the central part of Medan City, as shown in Fig. 2.

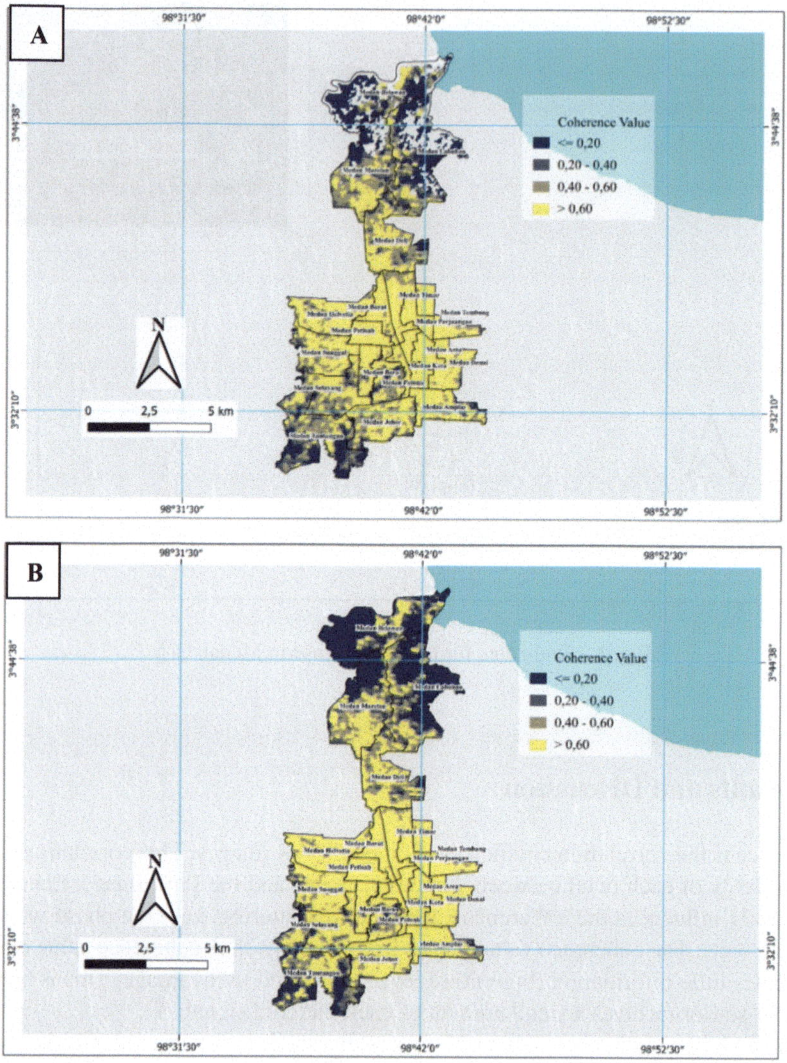

Fig. 2. Coherence in image pair from 2015–2023. (a) satellite ascending and (b) satellite descending.

The results of the InSAR analysis provided Line of Sight (LOS) velocity values for both ascending and descending images. This correction is essential to reduce atmospheric effects, ensuring more accurate ground deformation measurements. The LOS velocity

Table 1. The average coherence image in Medan City.

No	Sub-Districts	Ascending	Descending
1	Medan Barat	0.699	0.776
2	Medan Timur	0.686	0.748
3	Medan Amplas	0.626	0.675
4	Medan Area	0.766	0.839
5	Medan Baru	0.671	0.736
6	Medan Belawan	0.266	0.244
7	Medan Deli	0.563	0.586
8	Medan Denai	0.698	0.748
9	Medan Helvetia	0.659	0.734
10	Medan Johor	0.583	0.681
11	Medan Kota	0.755	0.814
12	Medan Labuhan	0.307	0.309
13	Medan Maimun	0.690	0.755
14	Medan Marelan	0.467	0.469
15	Medan Perjuangan	0.746	0.813
16	Medan Petisah	0.736	0.817
17	Medan Polonia	0.548	0.584
18	Medan Selayang	0.564	0.625
19	Medan Sunggal	0.616	0.681
20	Medan Tembung	0.690	0.749
21	Medan Tuntungan	0.388	0.444

value in the atmospheric correction ascending image in the Medan City area from 2015–2023 is presented in Table 2. The LoS velocity value in Medan City ranges from −64.9 mm/year to +62.5 mm/year with an average value LOS velocity of −5.5 mm/year. The subdistrict with the highest land subsidence in LOS velocity is Medan Tembung Subdistrict, with an average value of −36.5 mm/year. Negative LOS deformation values dominate the central part of Medan City. The deformation value of LOS velocity in Medan City is −69.6 mm/year up to +66.5 mm/year with an average LOS velocity value of −0.5 mm/year. The subdistrict that experienced the highest land subsidence in terms of LOS velocity was Medan Tembung Subdistrict, with an average value of −34.8 mm/year. Negative LOS deformation values dominate the central part of Medan City. Visualization of LOS velocity deformation in Medan City is shown in Fig. 3.

Vertical displacement was used to analyze the land surface that occurs through uplift and subsidence. Vertical displacement is obtained by transforming the LoS velocity value into a vertical (upward) displacement value using data from ascending and descending image processing. Based on the results obtained in Fig. 4, a positive value means an

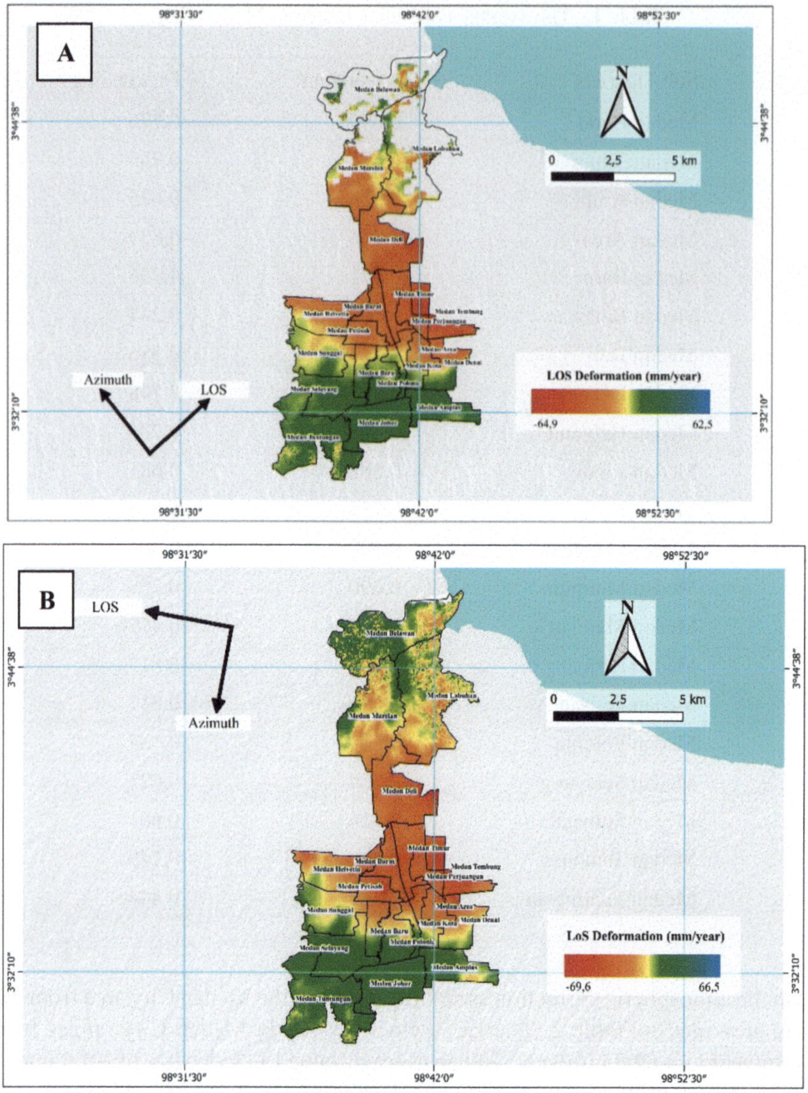

Fig. 3. Line of Sight (LOS) Displacement obtain from 2015–2023. (a) Satellite Ascending and (b) Satellite Descending.

uplift, and a negative value indicates a subsidence. Vertical displacement values with atmospheric correction in Medan City range from −48.5 mm/year to +48.8 mm/year, thus averaging by −3.9 mm/year. Apart from that, the average subsidence value for Medan City is −19.2 mm/year. The subdistrict that has the highest average subsidence value for vertical displacement in Medan City is the Medan Tembung Subdistrict, which is −38.9 mm/year. Subsidence is dominant in the central part of Medan City.

Moreover, Medan has experienced a significant population increase from 2005 to 2023 [14], with a growth rate of 11.9%, as shown in Table 2. This population growth has likely contributed to increased land use and subsurface exploitation, which are important factors in understanding ground deformation patterns in the area. This demographic data provides a critical context for interpreting the deformation results and assessing the implications for urban planning and disaster mitigation strategies. The significant rates of subsidence observed in Medan, particularly in the Medan Tembung Subdistrict, are impacted by various factors such as excessive groundwater extraction, soil compaction, and the weight of buildings and infrastructure. These findings are consistent with previous studies on land subsidence in urban areas of Indonesia, which highlight the impact of anthropogenic activities on ground stability [16, 26]. In Medan City, the central area dominance in subsidence is linked to dense urban development and concentrated groundwater extraction. Similar patterns have been observed in other major cities in Indonesia, such as Jakarta and Bandung, where rapid urbanization and groundwater extraction have led to significant subsidence issues [33, 34].

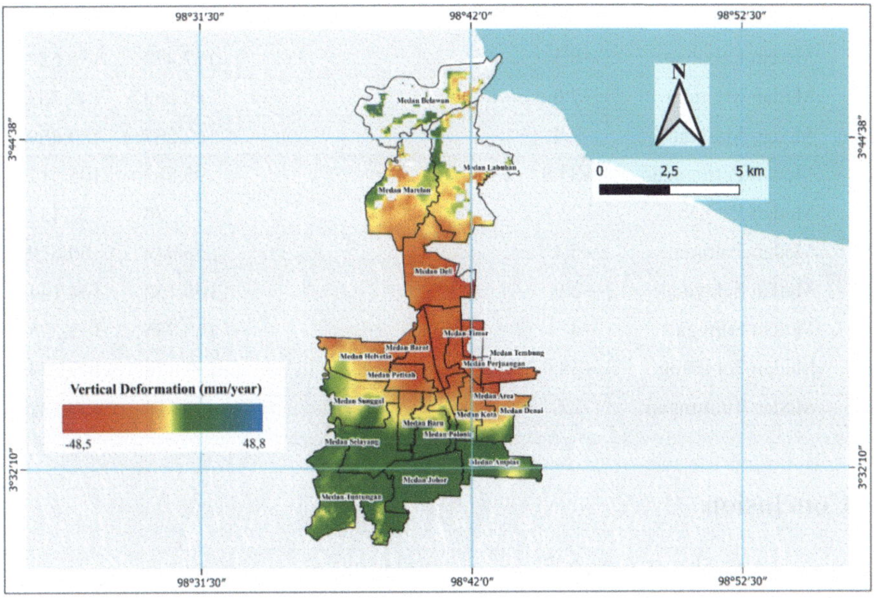

Fig. 4. Map of vertical displacement in Medan City.

Table 2. Population increase in Medan City from 2005 to 2023

No	Sub-Districts	Average Subsidence Rate (mm/year)	Population	
			2015	2023
1	Medan Barat	−29.9	72.683	89.248
2	Medan Timur	−31.1	114.720	117.035
3	Medan Amplas	−0.6	123.850	131.770
4	Medan Area	−17.9	98.992	118.057
5	Medan Baru	−11.4	40.540	36.191
6	Medan Belawan	−5.9	98.113	110.238
7	Medan Deli	−26.1	181.460	191.743
8	Medan Denai	−16.5	146.061	171.896
9	Medan Helvetia	−15.1	150.721	168.292
10	Medan Johor	–	132.012	154.868
11	Medan Kota	−15.6	74.439	84.778
12	Medan Labuhan	−8.3	117.472	135.622
13	Medan Maimun	−7.6	40.663	49.708
14	Medan Marelan	−9.4	162.267	189.469
15	Medan Perjuangan	−33.8	95.882	105.317
16	Medan Petisah	−23.8	63.374	72.432
17	Medan Polonia	−5.4	55.949	60.679
18	Medan Selayang	−2.9	106.150	104.144
19	Medan Sunggal	−4	115.785	133.273
20	Medan Tembung	−38.9	137.178	149.274
21	Medan Tuntungan	−2.6	85.613	100.132

4 Conclusion

Based on the results of InSAR processing, Medan City experienced an average land subsidence ranging from −2.0 mm/year to −38.5 mm/year from 2015–2023. Land subsidence is almost spread throughout all sub-districts in Medan City, but the one with the highest is in Medan Tembung District. Excessive building loads and excessive use of groundwater cause land subsidence in Medan.

Acknowledgment. The authors thank Institut Teknologi Sumatera for providing the research grant (No. 1539r/IT9.2.1/PT.01.03/2024) through "Hibah Penelitian ITERA 2024". We thank the COMET-LiCSBAS for providing the Sentinel-1 interferogram data used in this study.

References

1. Abidin, H.Z., Andreas, H., Gumilar, I., Fukuda, Y., Pohan, Y.E., Deguchi, T.: Land subsidence of Jakarta (Indonesia) and its relation with urban development. Nat. Hazards **59**(3), 1753–1771 (2011). https://doi.org/10.1007/s11069-011-9866-9
2. Higgins, S.A., Overeem, I., Steckler, M.S., Syvitski, J.P., Seeber, L., Akhter, S.H.: InSAR measurements of compaction and subsidence in the Ganges-Brahmaputra Delta, Bangladesh. J. Geophys. Res. Earth Surf. **119**(8), 1768–1781 (2014). https://doi.org/10.1002/2014jf003117
3. Wöppelmann, G., et al.: Is land subsidence increasing the exposure to sea level rise in Alexandria, Egypt? Geophys. Res. Lett. **40**(12), 2953–2957 (2013). https://doi.org/10.1002/grl.50568
4. Becker, R.H., Sultan, M.: Land subsidence in the Nile Delta: inferences from radar interferometry. The Holocene **19**(6), 949–954 (2009). https://doi.org/10.1177/0959683609336558
5. Hung, W.C., et al.: Surface deformation from persistent scatterers SAR interferometry and fusion with leveling data: a case study over the Choushui River Alluvial Fan, Taiwan. Remote Sens. Environ. **115**(4), 957–967 (2011). https://doi.org/10.1016/j.rse.2010.11.007
6. Minderhoud, P.S.J., Coumou, L., Erban, L.E., Middelkoop, H., Stouthamer, E., Addink, E.A.: The relation between land use and subsidence in the Vietnamese Mekong delta. Sci. Total. Environ. **634**, 715–726 (2018). https://doi.org/10.1016/j.scitotenv.2018.03.372
7. Solari, L., et al.: Combined use of C-and X-Band SAR data for subsidence monitoring in an urban area. Geosciences **7**(2), 21 (2017). https://doi.org/10.3390/geosciences7020021
8. Dong, S., Samsonov, S., Yin, H., Yao, S., Xu, C.: Spatio-temporal analysis of ground subsidence due to underground coal mining in Huainan coalfield, China. Environ. Earth Sci. **73**, 5523–5534 (2015). https://doi.org/10.1007/s12665-014-3806-4
9. Ng, A.H.M., Ge, L., Li, X., Abidin, H.Z., Andreas, H., Zhang, K.: Mapping land subsidence in Jakarta, Indonesia using persistent scatterer interferometry (PSI) technique with ALOS PALSAR. Int. J. Appl. Earth Obs. Geoinf. **18**, 232–242 (2012). https://doi.org/10.1016/j.jag.2012.01.018
10. Ciampalini, A., Solari, L., Giannecchini, R., Galanti, Y., Moretti, S.: Evaluation of subsidence induced by long-lasting buildings load using InSAR technique and geotechnical data: the case study of a Freight Terminal (Tuscany, Italy). Int. J. Appl. Earth Obs. Geoinf. **82**, 101925 (2019). https://doi.org/10.1016/j.jag.2019.101925
11. Chen, F., Lin, H., Zhang, Y., Lu, Z.: Ground subsidence geo-hazards induced by rapid urbanization: implications from InSAR observation and geological analysis. Nat. Hazards Earth Syst. Sci. **12**(4), 935–942 (2012). https://doi.org/10.5194/nhess-12-935-2012
12. Fryksten, J., Nilfouroushan, F.: Analysis of clay-induced land subsidence in Uppsala City using Sentinel-1 SAR data and precise leveling. Remote Sens. **11**(23), 2764 (2019)
13. Yastika, P.E., Shimizu, N., Abidin, H.Z.: Monitoring of long-term land subsidence from 2003 to 2017 in coastal area of Semarang, Indonesia by SBAS DInSAR analyses using Envisat-ASAR, ALOS-PALSAR, and Sentinel-1A SAR data. Adv. Space Res. **63**(5), 1719–1736 (2019). https://doi.org/10.1016/j.asr.2018.11.008
14. Statistik, B.P.: Kota Medan Dalam Angka 2023. Medan (2023)
15. Alif, S.M., Fattah, E.I., Kholil, M., Anggara, O.: Source of the 2019 Mw6. 9 Banten Intraslab earthquake modelled with GPS data inversion. Geodesy Geodyn. **12**(4), 308–314 (2021). https://doi.org/10.1016/j.geog.2021.06.001
16. Gumilar, I., Abidin, H.Z., Hutasoit, L.M., Hakim, D.M., Sidiq, T.P., Andreas, H.: Land subsidence in Bandung Basin and its possible caused factors. Procedia Earth Planet. Sci. **12**, 47–62 (2015). https://doi.org/10.1016/j.proeps.2015.03.026

17. Susilo, S., Salman, R., Hermawan, W., Widyaningrum, R.: GNSS land subsidence observations along the northern coastline of Java, Indonesia, 1–9 (2023). https://doi.org/10.1038/s41597-023-02274-0
18. Amighpey, M., Arabi, S.: Studying land subsidence in Yazd province, Iran, by integration of InSAR and levelling measurements. Remote Sens. Appl. Soc. Environ. **4** (2016). https://doi.org/10.1016/j.rsase.2016.04.001
19. Tzampoglou, P., Loupasakis, C.: Evaluating geological and geotechnical data for the study of land subsidence phenomena at the perimeter of the Amyntaio coalmine. Int. J. Min. Sci. Technol. (2018). https://doi.org/10.1016/j.ijmst.2017.11.002
20. Anggara, O., et al.: Monitoring ground deformation of Sinabung volcano eruption 2018–2019 using DInSAR technique and GPS data. In: AIP Conference Proceedings, vol. 2654 (2023). https://doi.org/10.1063/5.0114428
21. Natadikara, R., et al.: Monitoring deformation of Anak Krakatoa Volcano using differential interferometry synthetic aperture radar (DInSAR) method. In: AIP Conference Proceedings, vol. 2941, no. 1 (2023). https://doi.org/10.1063/5.0181540
22. Cigna, F., Tapete, D.: Present-day land subsidence rates, surface faulting hazard and risk in Mexico City with 2014–2020 Sentinel-1 IW InSAR. Remote Sens. Environ. **253**, 112161 (2021). https://doi.org/10.1016/j.rse.2020.112161
23. Sun, H., et al.: Land subsidence in a coastal city based on SBAS-InSAR monitoring: a case study of Zhuhai, China. Remote Sens. **15**, 2424 (2023). https://doi.org/10.3390/rs15092424
24. Orhan, O.: Monitoring of land subsidence due to excessive groundwater extraction using small baseline subset technique in Konya, Turkey. Environ. Monit. Assess. **193**(4), 10661 (2021). https://doi.org/10.1007/s10661-021-08962-x
25. Hakim, W., Fadhillah, F., Park, S., Pradhan, B., Won, J.-S., Lee, C.-W.: InSAR time-series analysis and susceptibility mapping for land subsidence in Semarang, Indonesia using convolutional neural network and support vector regression. Remote Sens. Environ. **287**, 113453 (2023). https://doi.org/10.1016/j.rse.2023.113453
26. Abidin, H.Z., Andreas, H., Gumilar, I., Brinkman, J.J.: Study on the risk and impacts of land subsidence in Jakarta. Proc. Int. Assoc. Hydrol. Sci. **372**(372), 115–120 (2015). https://doi.org/10.5194/piahs-372-115-2015
27. Raspini, F., Caleca, F., Del Soldato, M., Festa, D., Confuorto, P., Bianchini, S.: Review of satellite radar interferometry for subsidence analysis. Earth Sci. Rev. **235**, 104239 (2022). https://doi.org/10.1016/j.earscirev.2022.104239
28. Dong, J., Niu, R., Li, B., Xu, H., Wang, S.: Potential landslides identification based on temporal and spatial filtering of SBAS-InSAR results. Geomat. Nat. Hazards Risk **14**(1), 52–75 (2023). https://doi.org/10.1080/19475705.2022.2154574
29. Morishita, Y., Lazecky, M., Wright, T.J., Weiss, J.R., Elliott, J.R., Hooper, A.: LiCSBAS: an open-source InSAR time series analysis package integrated with the LiCSAR automated Sentinel-1 InSAR processor. Remote Sens. **12**(3), 424 (2020). https://doi.org/10.3390/rs12030424
30. Yu, C., Li, Z., Penna, N.T., Crippa, P.: Generic atmospheric correction model for interferometric synthetic aperture radar observations. J. Geophys. Res. Solid Earth **123**(10), 9202–9222 (2018)
31. Fuhrmann, T., Garthwaite, M.C.: Resolving three-dimensional surface motion with InSAR: constraints from multi-geometry data fusion. Remote Sens. **11**(3), 241 (2019). https://doi.org/10.3390/rs11030241
32. Alif, S.M., Anggara, O., Ristiana, V.: Coherence analysis of Sentinel-1A images in various land. Jurnal Geografi Gea **23**(2), 135–143 (2023). https://doi.org/10.17509/gea.v23i2.61258

33. Chaussard, E., Amelung, F., Abidin, H.Z., Hong, S.: Sinking cities in Indonesia: ALOS PALSAR detects rapid subsidence due to groundwater and gas extraction. Remote Sens. Environ. **128**, 150–161 (2013)
34. Alif, S.M., Anggara, O., Perdana, S.P., Hasannah, U., Azizah, F.N.: Analysis of presumed land subsidence in the cities of Lampung Province using InSAR and GNSS data. J. Geosci. Eng. Environ. Technol. **9**(3), 287–293 (2024). https://doi.org/10.25299/jgeet.2024.9.3.14096

Modeling of Tidal Flood Potential in the Coastal Areas of Medan City and Deli Serdang Regency North Sumatera Using Remote Sensing Data

Muhammad Sayyid Manan, Redho Surya Perdana, Aulia Try Atmojo, Ongky Anggara(✉), and Kaleb Kristian Kaban

Geomatics Engineering, ITERA, South Lampung, Indonesia
{redho.perdana,ongky.anggara}@gt.itera.ac.id

Abstract. Medan City and Deli Serdang Regency are directly adjacent to the Malacca Strait, which has high fluctuations that cause sea level rise in these waters. Sea level rise can affect coastal geographical conditions, such as shoreline changes, which can potentially cause tidal flooding. This study estimates the Tidal Flood Potential of six sub-districts located in the coastal areas of Medan City and Deli Serdang Regency. Identification of shoreline changes was carried out using Sentinel-2A satellite image observations from 2019–2022 with the Digital Shoreline Analysis System (DSAS) method to determine the condition of abrasion and sedimentation and modeling the potential for tidal flooding in 2024–2028. The result is abrasion occurs in Percut Sei Tuan and Pantai Labu Subdistricts, Deli Serdang Regency and that concluded the coastal area has the potential to be affected by tidal floods. The tidal flood modeling results show that six sub-districts on the coast of Medan City and Deli Serdang Regency are potentially affected by tidal floods, namely Medan Belawan, Medan Labuhan, Medan Marelan, Hamparan Perak, Percut Sei Tuan, and Pantai Labu. From the modeling results, it is known that 19 existing tidal flood villages are affected by tidal floods when the sea water experiences high tides.

Keywords: tidal flooding · sea level rise · shoreline change · abrasion

1 Introduction

The coastal area is a part of the earth's surface that becomes a meeting between two areas, namely land and sea. Based on topographic conditions, coastal areas are very vulnerable to the impact of sea level rise [1]. The coastal areas are also affected by river flooding, however sea level rise is not the only way that climate change affects coasts [2]. Large concentrations of people, wealth, and industry are concentrated along coastal areas, making susceptible to compound floods brought on by oceanic and hydrological forces [3]. There are various potential threats to coastal communities and ecosystems such as coastal erosion, coastal abrasion, storm surges, and inundation of seawater on land or tidal floods [4]. Tidal flooding is known as an inundation of water in the land area around the coast that has a lower elevation than the sea level during high tidal [5].

E. Fitrianingsih et al. (Eds.): ISAST 2024, SPPHY 416, pp. 66–75, 2025.
https://doi.org/10.1007/978-981-96-1344-1_7

Coastal subsidence increases the risk of floods and contributes to the relative rise in sea level and subsidence rates are caused by both anthropogenic and natural [6]. Tidal flooding is caused by several factors, two of which are sea level rise and land subsidence impact on high population [6, 7].

Coastal areas with high urbanization will experience land subsidence caused by the dynamism of the earth's surface by factors such as excessive groundwater extraction and land subsidence due to building loads [8]. The cities of Medan and Deli Serdang are among the most densely populated cities on the island of Sumatra. According to data recorded until 2023, the population in Deli Serdang Regency reached 2.018.164 people and ranks second with the largest population in North Sumatra Province after Medan City [9]. The rate of land subsidence in Medan City and Deli Serdang Regency reached ~3.9 cm/year and became the region with the highest land subsidence in Sumatra Island [10]. Land subsidence increases the risk of flooding in some cities along the coast or near sea level and significant rivers [11]. The area will eventually be inundated by sea water due to ongoing land subsidence and rising sea levels, particularly during high tide [12].

Medan City and Deli Serdang Regency are two of the cities or regencies located in North Sumatra Province and are directly adjacent to the Malacca Strait. The characteristics of the Malacca Strait waters are high fluctuations and unstable water dynamics. This is one of the factors for sea level rise in the Malacca Strait [13]. Rising sea levels pose an increasingly uncontrollable threat of tidal flooding. The dangers of tidal flooding are unpredictable, and the impacts of coastal change are increasing every year. Regional changes with rapid data acquisition can be done by utilizing remote sensing technology [14]. The use of remote sensing technology can identify areas affected by tidal floods in coastal areas and support in estimating changes in areas that occur [11].

Tidal floods accompanied by the phenomena of sea level rise and land subsidence also have the potential to expand tidal flood inundation [13]. This modeling utilizes the capabilities of geographic information systems in spatial analysis of raster data and vector data, as well as the mapping process. The tidal flood spatial modeling process uses mathematical formulas and operators in a raster calculator to produce a tidal flood inundation model. This research was conducted to model the potential for tidal flooding in the coastal areas of Medan City and Deli Serdang Regency in 2024–2028, as a short-term mitigation effort for the next few years.

2 Methodology

2.1 Research Location

The location of this research focuses only on six sub-districts located in the coastal areas of Medan City and Deli Serdang Regency. As for the six of these sub-districts, three of them are located in Medan City, namely Medan Belawan, Medan Labuhan, and Medan Marelan. While the other three sub-districts are located in Deli Serdang Regency, namely Hamparan Perak, Percut Sei Tuan, and Pantai Labu sub-districts is shown in Fig. 1.

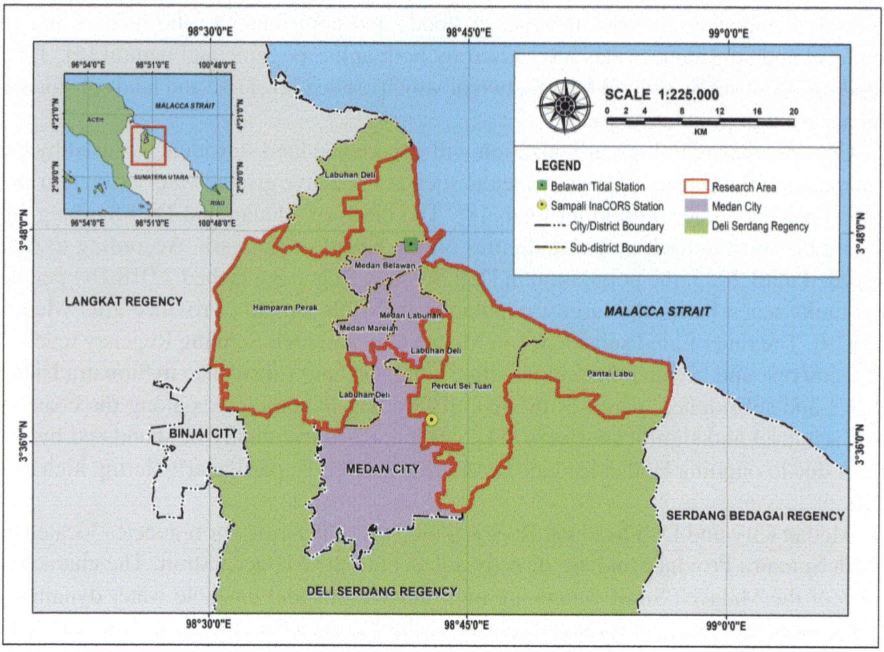

Fig. 1. Research Location Map

2.2 Research Data

The materials used in this study are shown in Table 1. Satellite image data processing was conducted using Google Earth Engine with cloud computing storage and spatial data processing applications. Tidal data was analyzed using the least square method to produce the Highest High-Water Level (HHWL) value of Mean Sea Level (MSL). This modeling also uses DEMNAS data to describe the topography of the earth's surface in the area to be studied. The DEMNAS based elevation model is digitally displayed in the form of raster data to perform spatial modeling of tidal floods in a large area.

2.3 Data Processing Method

Shoreline Change Processing estimates using using the Modified Normalized Difference Water Index (MNDWI) method to separate land and ocean objects in the image [15], with minimize the reflectance of water body in the NIR band [16]. The MNDWI formula is written as follows:

$$\text{MNDWI} = \frac{\text{Green} - \text{SWIR 1}}{\text{Green} - \text{SWIR 2}} \tag{1}$$

The classification results from MNDWI were then digitized to the line separating the land area from the waters of the image in the 2019–2022 time span. Calculation of shoreline changes using Digital Shoreline Analysis System (DSAS) to determine Net Shoreline Movement (NSM) and End Point Rate (EPR).

Table 1. Research Data

No.	Data	Data Source	Year
1	Sentinel-2A	U.S. Geological Survey, Google Earth Engine	2019–2022
2	Administrative Boundary	Ina-Geoportal, BIG	2023
3	Tides	Belawan Tidal Station (BLWN), BIG	2023
4	DEMNAS	Ina-Geoportal, BIG	2018
5	Sea Level Rise	Altimetry Satellite (Topex/Poseidon, Jason-1, Jason-2, Jason-3, and Sentinel-6 MF), NOAA	1992–2023

Modeling of Tidal Flood Potential estimation with the parameters used in calculating the height of the inundation model are the results of tidal data processing, the trend of sea level rise, and the rate of land subsidence. The calculation of the tidal flood inundation model height by the Higher high water level (HHWL) value of Mean Sea Level (MSL) that has been obtained from the tidal data processing results into an equation. The equation for calculating the height of the inundation model using a number processing application is as follows [17]:

$$Y = [(SLR) \times (t_x - t_0) + (LS) \times (t_x - t_0) + (HHWL - MSL)] \tag{2}$$

The results were obtained after calculating the height of the inundation model using various parameters. Then, the spatial analysis of raster data was carried out using the raster calculator tool in the spatial data processing application. A raster calculator is a calculation method that involves raster data as the main input. The Digital Terrain Model (DTM) data in the form of a raster that has been processed before will be used as input in the equation. The equation consists of DTM data and the results of the inundation model height calculation. This modeling using formulas and mathematical operators in the raster calculator that will produce output in the form of a new raster [18, 19]. The equation is written as follows:

$$Tidal\ Flood = CON([RasterDTM] <= [Y], 1, 0) \tag{3}$$

3 Result and Discussion

The identification of potential tidal flooding based on shoreline changes shows the occurrence of shoreline changes dominated by abrasion in Pematang Lalang Village, Sei Tuan Village, Regemuk Village, and Pantai Labu Pekan Village, Deli Serdang Regency. The results show the average sedimentation (+) and abrasion (−) in 4 years of observation. Tidal floods will potentially affect abrasion events that occur continuously every year in several villages located along the coast. The interpretation of abrasion on coastal change is included in Fig. 2.

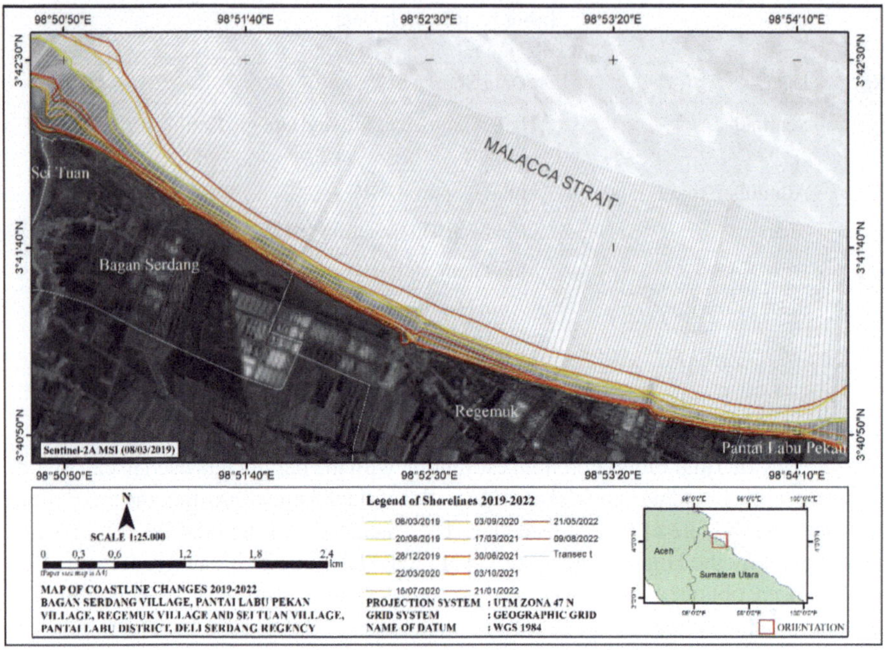

Fig. 2. Shoreline Change in Deli Serdang Regency

Table 2. Calculation Of Shoreline Change

Sub-district	Village	Average Net Shoreline Movement (m)		Average *End Point Rate* (m/year)		Description
		(+)	(−)	(+)	(−)	
Percut Sei Tuan	Pematang Lalang	196.75	−442.30	69.01	−128.56	Abrasion
Pantai Labu	Sei Tuan	243.49	−306.13	84.72	−92.27	Abrasion
Pantai Labu	Regemuk	0.00	−161.36	0.00	−47.19	Abrasion
Pantai Labu	Pantai Labu Pekan	77.13	−181.65	26.84	−53.33	Abrasion

The results of the calculation of shoreline changes for four years using the Net Shoreline Movement (NSM) method and the End Point Rate (EPR) method presented in Table 2 show that the villages are dominated by abrasion. Even in Regemuk Village, Pantai Labu Subdistrict, there is abrasion without sedimentation. The tendency of land surface subsidence due to land cover change, sea level rise, and coastal abrasion makes villages in the Deli Serdang Regency coastal area potentially affected by tidal flooding.

Potential tidal flood areas modelling using the data processing applications. This modeling is based on Digital Terrain Model (DTM) data and the results of inundation model height calculations. The outcomes of the tidal flood potential modeling are shown in Figs. 3 is potential flood in 2024 and Fig. 4 is potential flood in 2028.

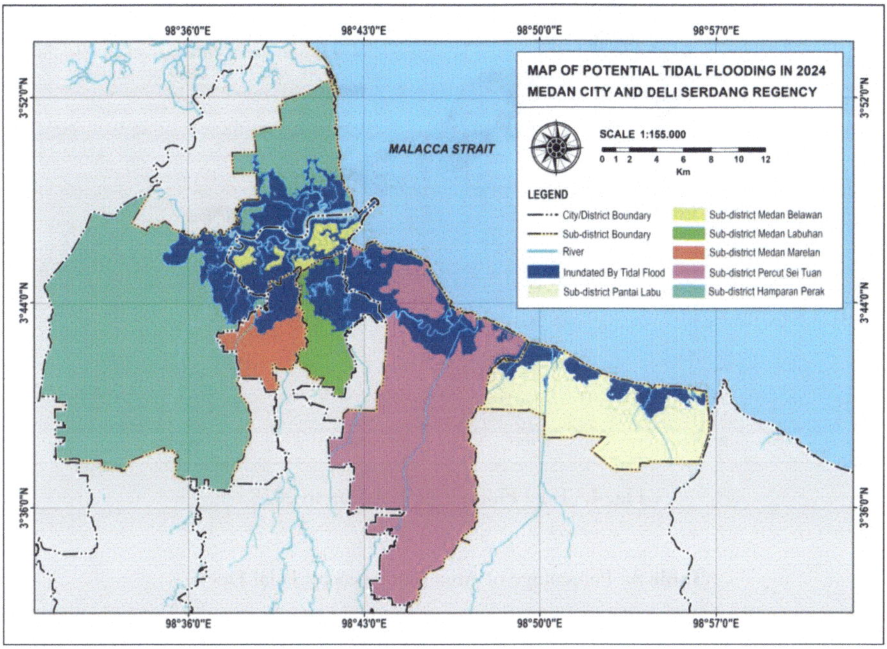

Fig. 3. Tidal Flood Potential Map in 2024

The results of the potential tidal flood maps for 2024 and 2028 show that six sub-districts have the potential to be inundated by tidal floods, namely Medan Belawan, Medan Labuhan, Medan Marelan, Hamparan Perak, Percut Sei Tuan, and Pantai Labu. The increase in tidal flood inundation areas in spatial coverage is due to rising sea level and land subsidence that occur in coastal areas. Land subsidence cause land surface height to be lower than before, thus increasing the risk of widespread tidal flooding. The results are spatially that tidal floods that enter land are spread widely in land areas, not around coastal areas. Some of these areas are areas traversed by rivers. This is caused by high tides that the volume of water in the river also increases and causes tidal flooding. The percentage of inundated areas can be seen in Table 3 below.

Hamparan Perak and Percut Sei Tuan sub-districts have the highest tidal flood inundated areas. However, if calculated based on the area, Medan Belawan Sub-district is the sub-district with the highest percentage of tidal flood inundation area, because its area is only 3,015.22 hectares with a potential inundation area in 2028 of 2,324.39 hectares. The results of the percentage of inundation area based on area show that Medan Belawan Sub-district has a percentage of inundation area of more than 74%. This percentage is

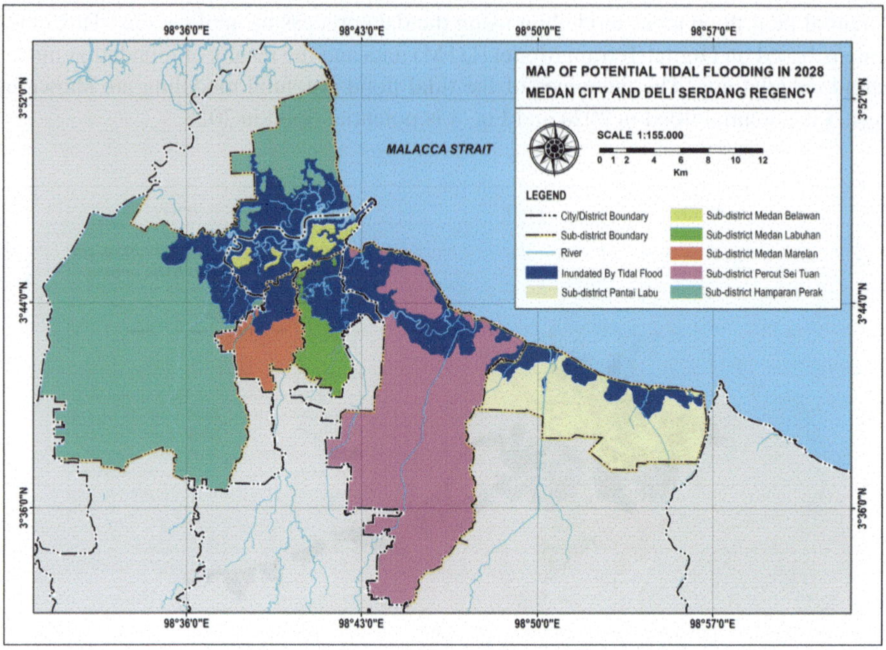

Fig. 4. Tidal Flood Potential Map in 2028

Table 3. Percentage of Area Inundated by Tidal Flood

Sub-district	Sub-district area (ha)	Inundation Area (ha) 2024	Inundation Area (ha) 2028	Percentage of Inundation Area by Area				
				2024	2025	2026	2027	2028
Medan Belawan	3,015.22	2,237.44	2,324.39	74.20%	75.42%	76.42%	76.96%	77.09%
Medan Labuhan	3,532.99	1,731.97	1,778.35	49.02%	49.56%	49.99%	50.26%	50.34%
Medan Marelan	2,956.16	819.12	846.92	27.71%	28.08%	28.40%	28.60%	28.65%
Hamparan Perak	30,257.11	4,276.78	4,506.65	14.13%	14.43%	14.66%	14.83%	14.89%
Percut Sei Tuan	19,028.59	3,119.01	3,267.17	16.39%	16.63%	16.86%	17.07%	17.17%
Pantai Labu	8,583.30	1,640.33	1,763.26	19.11%	19.65%	20.11%	20.45%	20.54%

greater than the percentage in Hamparan Perak and Percut Sei Tuan Sub-districts, which only have a percentage of inundation area of more than 14% and 16%, respectively.

Based on information from the Badan Nasional Penanggulangan Bencana (BNPB), several tidal floods occurred in Medan City and Deli Serdang Regency. Some occurred in Paluh Kurau Village, Paluh Manan Village, Hamparan Perak District, and Medan

Belawan District. The tidal floods in Paluh Kurau and Paluh Manan Villages occurred due to high tides that caused river water to overflow, inundating settlements, roads, and rice fields. The tidal floods in Medan City were also caused by sea tides, with the height of the tidal floods reaching 60 cm [20], such as monitoring using GNSS [21]. Figure 5 below shows the identification map of existing tidal flood villages, which shows that these villages have been affected by tidal floods. The map also includes the results of the inundation model of potential tidal flooding in 2028.

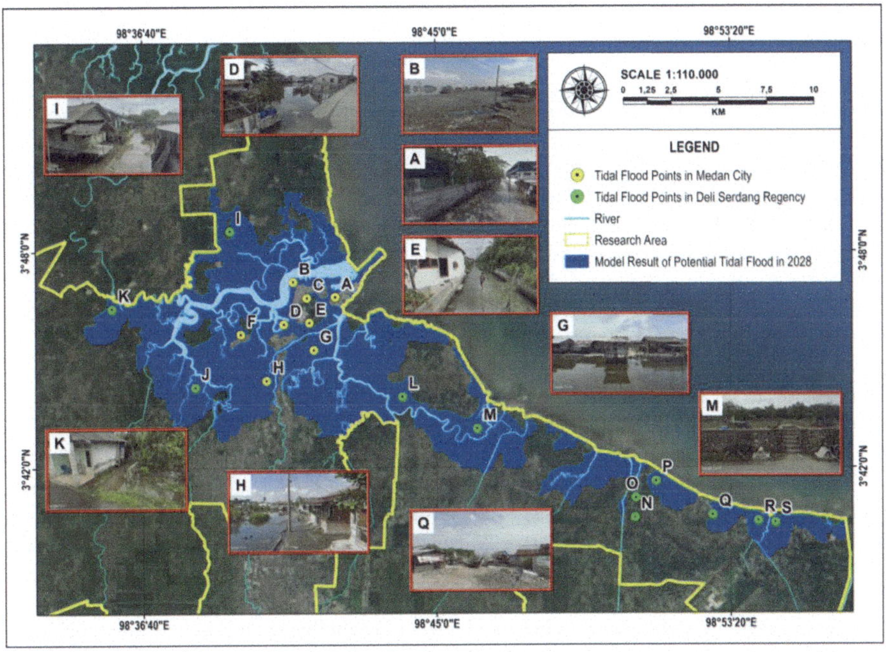

Fig. 5. Map of Existing Tidal Flood Villages

The map results in Fig. 5 above show that 19 villages were identified as experiencing tidal flooding. An area that has experienced tidal flooding in Medan and Deli Serdang areas are marked by inundation of water toward land. The tidal flood that entered the mainland expanded, exceeding the coastal area. Some of these areas are areas that are traversed by rivers. This is caused by high tides, that the volume of water in the river also increases, resulting in tidal flooding.

4 Conclusion

The result of coastline change from 2019–2022 dominated obtain abrasion occurred in the villages of Pematang Lalang, Sei Tuan, Regemuk, and Pantai Labu Pekan, indicating that the village has the potential for tidal flooding in the coastal area. The modeling results of potential tidal floods in 2024–2028 found that six sub-districts in Medan City and Deli

Serdang Regency, namely Medan Belawan, Medan Labuhan, Medan Marelan, Hamparan Perak, Percut Sei Tuan, and Pantai Labu, were potentially affected by tidal floods. Based on the modeling results, it is known that 19 existing tidal flood villages exist, two of which are Regemuk Village and Pantai Labu Pekan. This shows that environmental changes due to sea level rise and land subsidence, such as shoreline changes and tidal flooding, significantly impact the lives of coastal communities. Therefore, villages identified as existing tidal flood villages must seek appropriate mitigation to minimize tidal flooding risk and provide protection to local communities and infrastructure.

References

1. Frederikse, T., et al.: The causes of sea-level rise since 1900. Nature **584**(7821), 393–397 (2020)
2. Zscheischler, J., et al.: A typology of compound weather and climate events. Nat. Rev. Earth Environ. **1**(7), 333–347 (2020). https://doi.org/10.1038/s43017-020-0060-z
3. Xu, K., Wang, C., Bin, L.: Compound flood models in coastal areas: a review of methods and uncertainty analysis. Nat. Hazards **116**(1), 469–496 (2023). https://doi.org/10.1007/s11069-022-05683-3
4. Edwards, T.L., et al.: Projected land ice contributions to twenty-first-century sea level rise. Nature **593**(7857), 74–82 (2021)
5. Dedekorkut-Howes, A., Torabi, E., Howes, M.: When the tide gets high: a review of adaptive responses to sea level rise and coastal flooding. J. Environ. Plann. Manag. **63**(12), 2102–2143 (2020). https://doi.org/10.1080/09640568.2019.1708709
6. Alif, S.M., Anggara, O., Perdana, S.P., Hasannah, U., Azizah, F.N.: Analysis of presumed land subsidence in the cities of Lampung Province using InSAR and GNSS data. J. Geosci. Eng. Environ. Technol. **9**(3), 287–293 (2024). https://doi.org/10.25299/jgeet.2024.9.3.14096
7. Cao, A., et al.: Future of Asian Deltaic Megacities under sea level rise and land subsidence: current adaptation pathways for Tokyo, Jakarta, Manila, and Ho Chi Minh City. Curr. Opin. Environ. Sustain. **50**, 87–97 (2021)
8. Perdana, R.S., Fikri, M., Alif, S.M., Isnaini, E.L.: Determining the location of land subsidence observation points based on lithological data and land cover changes in Lampung Province. GeoEco J. **8**(2), 164–174 (2022)
9. BPS Kabupaten Deli Serdang, Kabupaten Deli Serdang Dalam Angka Deli Serdang Regency in Figures 2024, vol. 21. BPS Kabupaten Deli Serdang, Deli Serdang (2024)
10. Chaussard, E., Amelung, F., Abidin, H.Z., Hong, S.-H.: Sinking cities in Indonesia: ALOS PALSAR detects rapid subsidence due to groundwater and gas extraction. Remote Sens. Environ. **128**, 150–161 (2013)
11. Navarro-Hernández, M.I., Valdes-Abellan, J., Tomás, R., Tessitore, S., Ezquerro, P., Herrera, G.: Analysing the impact of land subsidence on the flooding risk: evaluation through InSAR and modelling. Water Resour. Manag. **37**(11), 4363–4383 (2023)
12. Takagi, H., Esteban, M., Mikami, T., Pratama, M.B., Valenzuela, V.P.B., Avelino, J.E.: People's perception of land subsidence, floods, and their connection: a note based on recent surveys in a sinking coastal community in Jakarta. Ocean Coast. Manag. **211**, 105753 (2021)
13. Koropitan, A.F., Barus, T.A., Cordova, M.R.: Coastal water properties and hydrodynamic processes in the Malacca Strait: case study northeastern coast of Sumatra, Indonesia. J. Ecol. Eng. **22**(11) (2021)
14. Anggara, O., et al.: Monitoring ground deformation of Sinabung volcano eruption 2018–2019 using DInSAR technique and GPS data. In: AIP Conference Proceedings, vol. 2654 (2023). https://doi.org/10.1063/5.0114428

15. Du, Y., Zhang, Y., Ling, F., Wang, Q., Li, W., Li, X.: Water bodies' mapping from Sentinel-2 imagery with modified normalized difference water index at 10-m spatial resolution produced by sharpening the SWIR band. Remote Sens. **8**(4), 354 (2016)
16. Xu, H.: Modification of normalised difference water index (NDWI) to enhance open water features in remotely sensed imagery. Int. J. Remote Sens. **27**(14), 3025–3033 (2006). https://doi.org/10.1080/01431160600589179
17. Trihatmoko, E., Pratiwi, E.S., Sanjoto, T.B., Marfai, M.A.: A numerical study of tidal run up and inundation impact using logical tool-less than equal. In: IOP Conference Series: Earth and Environmental Science, vol. 683, no. 1, p. 012062. IOP Publishing (2021)
18. Rahmanto, M.R., Susetyo, C.: Pemodelan spasial genangan banjir akibat gelombang pasang di wilayah pesisir Kota Mataram. Jurnal Teknik ITS **7**(1), C33–C37 (2018)
19. Seenath, A., Wilson, M., Miller, K.: Hydrodynamic versus GIS modelling for coastal flood vulnerability assessment: which is better for guiding coastal management? Ocean Coast. Manag. **120**, 99–109 (2016)
20. BNPB: [infografis21052023]: Banjir ROB di Kab. Deli Serdang, Prov. Sumatera Utara, Pusat Pengendalian Operasi BNPB (2023)
21. Alif, S.M., Fattah, E.I., Kholil, M., Anggara, O.: Source of the 2019 Mw6. 9 Banten Intraslab earthquake modelled with GPS data inversion. Geodesy Geodyn. **12**(4), 308–314 (2021). https://doi.org/10.1016/j.geog.2021.06.001

Mapping Shallot in Lowland Areas Through Sentinel-2 Imagery and Machine Learning Approach

Robby Arifandri[1(✉)], Vicca Karolinoerita[2(✉)], Fadhlullah Ramadhani[2], Dede Dirgahayu[2], Lukita Devy[3], Muhammad Iqbal Habibie[4], Yudi Riadi Fanggidae[5], Dharmayanti[6], Robby Kurniawan Harahap[6], Swelandiah Endah Pratiwi[7], and Widiastuti[7]

[1] Research Center for Artificial Intelligence and Cyber Security, BRIN, Bandung, Indonesia
robb005@brin.go.id
[2] Research Center for Geoinformatics, BRIN, Bandung, Indonesia
{vicc001,fadh004,dede004}@brin.go.id
[3] Research Center for Horticulture, BRIN, Cibinong, Indonesia
luki002@brin.go.id
[4] Research Center for Environmental and Clean Technology, BRIN, Tangerang Selatan, Indonesia
muha105@brin.go.id
[5] Research Center for Climate and Atmosphere, BRIN, Bandung, Indonesia
yudi020@brin.go.id
[6] Faculty of Industrial Technology, Gunadarma University, Depok, Indonesia
{dharmayanti,robby_kurniawan}@staff.gunadarma.ac.id
[7] Faculty of Computer Science and Information System, Gunadarma University, Depok, Indonesia
{swelandiah,widiastuti}@staff.gunadarma.ac.id

Abstract. As one of the staple horticultural crops, Shallot requires accurate monitoring to understand its cultivation better and to face environmental challenges in lowland regions. Nevertheless, there are still limited studies on mapping shallot agriculture using remote sensing data. In this study, we aimed to enhance the accuracy and efficiency of mapping shallot cultivation in lowland areas by leveraging high-resolution Sentinel-2 imagery combined with machine-learning techniques, specifically Random Forest, to classify the cultivation areas accurately. Preprocessed Sentinel-2 data using Google Earth Engine and validation data from field surveys and high-resolution satellite imagery inspection are utilized to create training and testing datasets for the machine learning model. Our findings reveal that the Random Forest model resulted in an Overall Accuracy of 88% and a total area of 12,121 hectares in July 2023. The proposed method can be used to estimate the planting and harvested area regularly to support smart-climate agriculture.

Keywords: Shallot · Food Crops · Mapping · Random Forest · Sentinel-2

E. Fitrianingsih et al. (Eds.): ISAST 2024, SPPHY 416, pp. 76–84, 2025.
https://doi.org/10.1007/978-981-96-1344-1_8

1 Introduction

Shallot is a vital agricultural crop that plays a significant role in global cuisine and the economy. Its total market value is 24 billion USD in Asia-Pacific and is predicted to grow [1]. Shallot remains one of the most strategic agricultural commodities, especially in Indonesia, where it is the largest horticultural producer, with Central Java province as the biggest producer nationally [2]. Shallot cultivation is integral to small-scale family farms and large agricultural operations. However, shallot production faces substantial challenges, especially in lowland regions prone to flooding and other environmental stresses. Pathogens during the rainy season and inappropriate practices also played a crucial role in decreasing shallot yield in recent years [3, 4]. These problems significantly impact shallot crops' growth and yield, increasing farmers' loss and creating market value volatility.

Satellite-based agricultural monitoring has been studied in the past decade to create crop maps for various crop types for effective agricultural management. [5, 6]. Several studies have shown the effectiveness of optical and radar satellite imaging applications in creating and monitoring crop maps. [7–9] using publicly and commercially available satellite imagery [10], such as Landsat, Sentinel-1, Sentinel-2, and PlanetScope. Supervised classification has also shown promising results for accurate crop map monitoring from high temporal and spatial resolution imageries. The random forest classifier has shown accurate results of over 80%, showing its advantage over other classification methods, such as the Support Vector Machine (SVM) classifier for crop type mapping [11].

The Sentinel-2 satellite from the European Space Agency (ESA) is equipped with a high-resolution (10 m) multispectral imaging sensor ranging from visible to shortwave infrared bands, which covers the range to detect unique spectral signatures of crops and soil. [12]. In several studies, sentinel satellites have been utilized to accurately map crop types such as maize and rice [13–16]. Yet, there is still limited study on mapping shallot cultivation areas, specifically in lowland regions. The primary objective of our research is to develop and validate an in-depth analysis for accurately mapping shallot cultivation areas in lowland regions, leveraging the potential of Sentinel-2 satellite imagery and machine learning.

2 Material and Methods

Brebes Regency, Central Java, Indonesia, is selected as our study area as it is the largest producer of shallot cultivation in Indonesia. Brebes is a prominent lowland that spans approximately 88,453 ha and is characterized by its flat topography with altitudes not exceeding 100 m above sea level (Fig. 1). The region benefits from a tropical climate, marked by a distinct wet and dry season, providing a conducive environment for shallot growth.

Sample areas are spread across the study area, and six classifications are visually inspected through Google Earth and field survey data. The classes are Water Body, Urban, Mixed Garden, Dryland Agriculture, Paddy Rice, and Shallot. Sentinel-2 images are available publicly and provided through the Google Earth Engine (GEE) catalog.

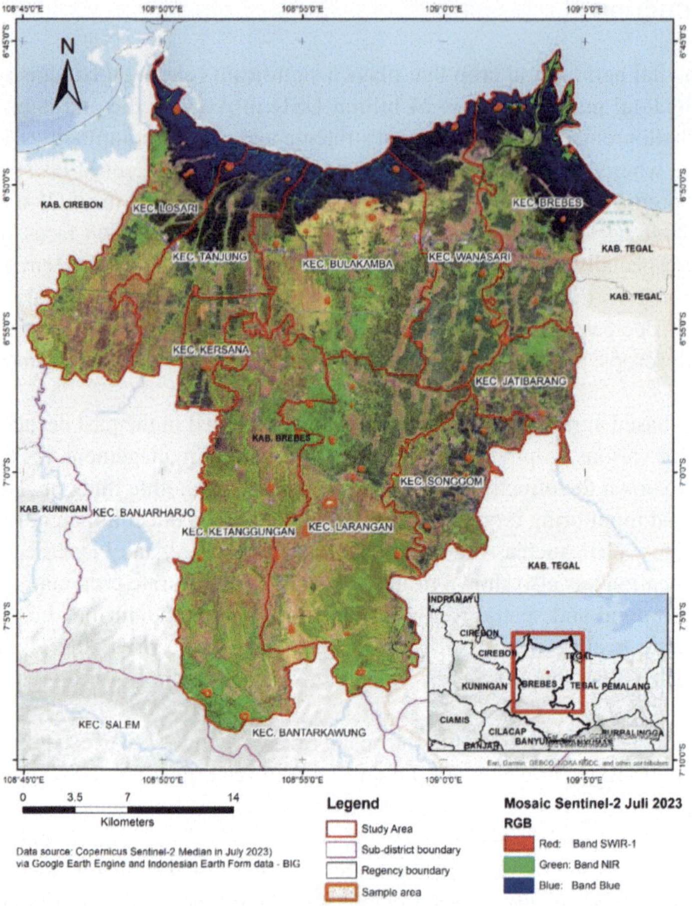

Fig. 1. The study area for shallot cultivation. Small red-bounded areas are the sample areas distributed across the study area, covering multiple classes used to train and validate the classifier.

As shown in Fig. 2, the Sentinel-2 data passes several processes before being used as the input for the machine learning classifier. First, it is preprocessed for orthorectification and atmospheric correction, which is already preprocessed for the Sentinel-2 data from the GEE catalog [17–19]. A cloud-free mosaic of preprocessed Sentinel-2 data is then generated in the GEE platform using the median value of image collections from six scenes in July 2023.

Ten bands of Sentinel-2 with wavelengths ranging from 490nm to 2190 nm are used (Table 1). The reflectance and transmittance within the range of these spectra are known to be related to plant characteristics [12], which is ideal for detecting cultivated areas. Specifically, shallot fields exhibit distinctive spectral signatures due to their unique physiological traits, such as leaf structure, chlorophyll content, and canopy density. These factors influence the absorption and reflectance properties in critical bands, such as the red-edge (705 nm–783 nm) and near-infrared (NIR) regions (841 nm–2190 nm),

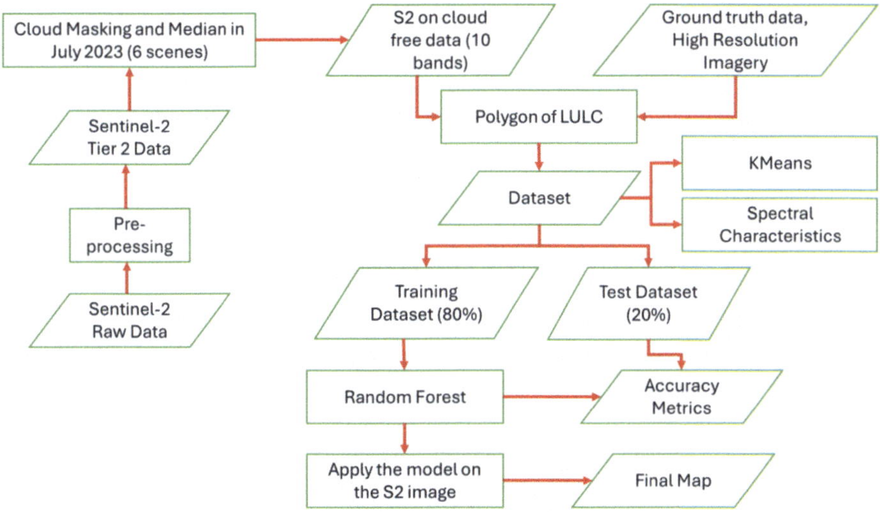

Fig. 2. Processing steps utilized for this study.

which are highly sensitive to vegetation health and biomass [20]. Each crop type also has unique spectral characteristics, which, in this case, differentiate between paddy and Shallot. [21]. The satellite data is combined with the ground truth data, which is then provided for the training and validating stage of the random forest classifier. The data processing and classification steps are being done using GEE, Python, and its supporting packages, such as *scikit-learn* and *rasterio* [22–24].

Table 1. List of Sentinel-2 bands used.

Sentinel-2 Bands	Central Wavelength (μm)	Resolution (m)
Band 2 – Blue	0.490	10
Band 3 – Green	0.560	10
Band 4 – Red	0.665	10
Band 5 – Vegetation Red Edge	0.705	20
Band 6 – Vegetation Red Edge	0.740	20
Band 7 – Vegetation Red Edge	0.783	20
Band 8 – NIR	0.842	10
Band 8A – Vegetation Red Edge	0.862	20
Band 11 – SWIR	1.610	20
Band 12 – SWIR	2.190	20

Random Forest, a machine learning algorithm used in satellite imagery classification, is characterized by its bootstrap and bagging techniques [25], which were employed to classify shallot cultivation areas. The Random Forest classifier offers several advantages in remote sensing applications [13]: it can handle large datasets, run on thousands of input variables, generate feature importance, remain robust to noise and outliers, and has lower computational complexity than other ensemble algorithms.

The validation and testing processes involved quantitative metrics such as Overall Accuracy (OA), Cohen's Kappa coefficient (K), Producer's Accuracy (PA), User's Accuracy (UA), Omission Error (OE), and Commission Error (CE) for each class.

$$OA = \frac{\sum Correct\ Prediction}{Total\ Predictions} \tag{1}$$

The Kappa coefficient metric (K) determines the classifier's accuracy compared to assigning random values. K = 0 means random values, while K = 1 means the classifier is perfect [14].

$$K = \frac{P_{agree} - P_{chance}}{1 - P_{chance}} \tag{2}$$

$$PA = \frac{Number\ of\ correctly\ classified\ \text{samples}\ to\ class}{Total\ samples\ of\ true\ label\ of\ class} \tag{3}$$

$$UA = \frac{Number\ of\ correctly\ classified\ \text{samples}\ to\ class}{Total\ samples\ of\ true\ \text{samples classified}\ as\ class} \tag{4}$$

$$OE = 100\% - PA \tag{5}$$

$$CE = 100\% - UA \tag{6}$$

3 Results and Discussion

The evaluation of the Random Forest (RF) classifier for the shallot cultivation areas using Sentinel-2 imagery showed an overall accuracy of 88%, and Cohen's Kappa coefficient of 0.86 with the shallot class gained a producer's accuracy of 0.78 and user's accuracy of 0.95. The mixed garden class gained the highest producer's accuracy with 0.98, and the water body gained the highest user's accuracy with 1.00.

Moreover, Table 2 shows that the Mixed Garden, with a PA of 0.98 and UA of 0.69, has an OE of 0.02 and a higher CE of 0.31, showing suitable identification but frequent misclassification. Dryland Agriculture has a PA of 0.90, UA of 0.96, OE of 0.10, and CE of 0.04, indicating robust identification with few errors. Paddy Rice, with a PA of 0.82 and UA of 0.96, has an OE of 0.18 and CE of 0.04, indicating some missed instances but few incorrect predictions. Shallot shows a PA of 0.78, UA of 0.95, OE of 0.22, and CE of 0.05, reflecting a higher omission rate but relatively accurate predictions.

As depicted in Fig. 3, The bands B11 and B12 are the most important features, followed closely by B8A. Other bands like B2, B3, B4, B7, B8, B6, and B5 follow in decreasing order of importance. It indicates that bands B11 and B12 contribute the most to this analysis's model predictions.

Table 2. Producer's Accuracy (PA), User's Accuracy (UA), Omission Error, and Commission Error of each class.

Class	PA	UA	OE	CE
Water Body	0.89	1.00	0.11	0.00
Urban	0.81	0.99	0.19	0.01
Mixed Garden	0.98	0.69	0.02	0.31
Dryland Agriculture	0.90	0.96	0.10	0.04
Paddy Rice	0.82	0.96	0.18	0.04
Shallot	0.78	0.95	0.22	0.05

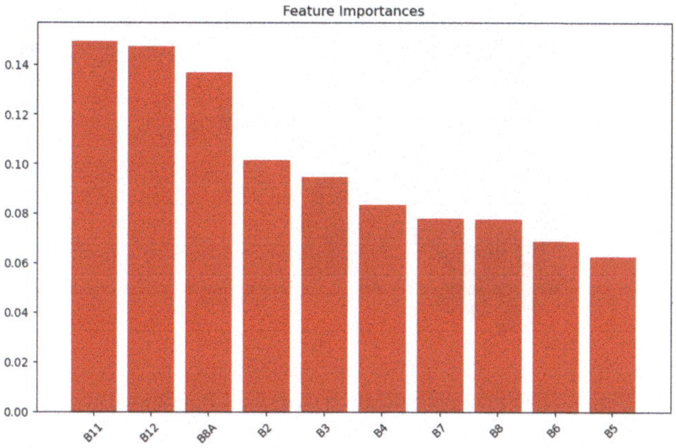

Fig. 3. Feature importance of the input parameter of the classifier.

The areas classified as Shallot (highlighted in pink) are spread throughout the region, demonstrating this crop's spatial extent and cultivation patterns (Fig. 4). The higher omission error (0.22) for Shallot indicates that the model missed some actual areas. In contrast, the low commission error (0.05) suggests that most areas predicted as Shallot were correctly identified. This information is crucial for agricultural planning and monitoring, helping improve land cover classification accuracy in future studies. Shallot cultivation is significant for local economies, making its accurate identification vital for resource management.

Table 3 summarizes the final map's distribution, highlighting the prevalence and distribution of different land cover types within the study area, providing context to the map's classification accuracy and spatial representation. The notable presence of Dryland Agriculture and Mixed Gardens underscores their significant roles in the region's land use. At the same time, the detailed breakdown of classes aids in understanding the landscape composition and planning for targeted interventions. Though comprising 8.33% of the land cover, the shallot areas show significant challenges in classification accuracy, highlighting the need for enhanced techniques in distinguishing this crop.

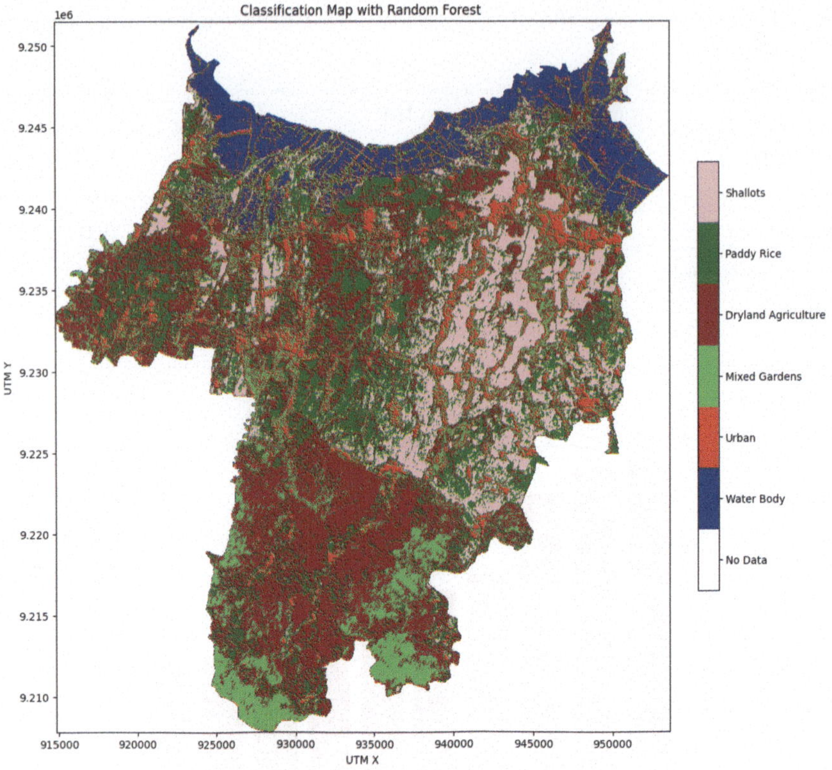

Fig. 4. The final classification map with Random Forest.

Table 3. Summary of study area classes.

Class Label	Class Name	Area in hectares (Ha)	Percentage
1	Water Body	9,430	10.66%
2	Urban	7,881	8.91%
3	Mixed Garden	12,808	14.48%
4	Dryland Agriculture	25,616	28.96%
5	Paddy Rice	20,597	23.29%
6	Shallot	12,121	13.70%

4 Conclusion

Our research has demonstrated the potential of building a shallot cultivation map using Sentinel-2 data and a Random Forest classifier in lowland areas. This approach gained an overall accuracy of 88% in identifying shallot fields with a total area of 12,121 ha. This approach demonstrates the feasibility of remote sensing for mapping high-value crops and establishes a crucial baseline for more advanced studies such as precision farming, yield estimation, crop health monitoring, and cultivation management. Furthermore, this method can be developed into spatiotemporal analysis to monitor shallot cultivation changes over time, allowing for a dynamic understanding of crop development, growth cycles, and environmental interactions.

Acknowledgments. We want to thank The National Research and Innovation Agency (BRIN) for supporting this research through the Satellite-based Decision Support System Prototype Research Program Year 2024 and the support from BRIN's Mahameru High-Performance Computing Facility for providing computation resources. We also would like to thank the reviewers who gave constructive feedback on this paper's writing.

References

1. IndexBox: Overview of the Onion and Shallot Market in Asia-Pacific. https://www.globaltrademag.com/overview-of-the-onion-and-shallot-market-in-asia-pacific/. Accessed 20 June 2024
2. Badan Pusat Statistik, Indonesia: Distribusi Perdagangan Komoditas Bawang Merah di Indonesia 2022. https://www.bps.go.id/id/publication/2022/10/24/958ef61ffcb0e88357bb99d1/distribusi-perdagangan-komoditas-bawang-merah-di-indonesia-2022.html. Accessed 30 July 2024
3. Supyani, Poromarto, S.H., Supriyadi, Hadiwiyono: Moler disease of shallot in the last three years at Brebes Central Java: the intensity and resulting yields losses is increasing. IOP Conf. Ser. Earth Environ. Sci. **810**, 012004 (2021). https://doi.org/10.1088/1755-1315/810/1/012004
4. Widono, S., Poromarto, S.H., Supyani, Noviantoro, W., Hadiwiyono: Relationship of weather factors on the progress of shallot moler disease in Brebes, Central Java in the rainy and dry seasons: intensity increases in humid and warm air. IOP Conf. Ser. Earth Environ. Sci. **1200**, 012062 (2023). https://doi.org/10.1088/1755-1315/1200/1/012062
5. Azzari, G., Jain, S., Jeffries, G., Kilic, T., Murray, S.: Understanding the requirements for surveys to support satellite-based crop type mapping: evidence from sub-Saharan Africa. World Bank Group (2021)
6. Lobell, D.B., et al.: Eyes in the sky, boots on the ground: assessing satellite- and ground-based approaches to crop yield measurement and analysis. Am. J. Agr. Econ. **102**, 202–219 (2020). https://doi.org/10.1093/ajae/aaz051
7. Dakir, A., Bachir Alami, O., Barramou, F.: Crop type mapping using optical and radar images: a review. In: 2020 IEEE International Conference of Moroccan Geomatics (Morgeo), pp. 1–8 (2020). https://doi.org/10.1109/Morgeo49228.2020.9121869
8. Adrian, J., Sagan, V., Maimaitijiang, M.: Sentinel SAR-optical fusion for crop type mapping using deep learning and Google Earth Engine. ISPRS J. Photogramm. Remote Sens. **175**, 215–235 (2021). https://doi.org/10.1016/j.isprsjprs.2021.02.018
9. Triscowati, D.W., Sartono, B., Kurnia, A., Dirgahayu, D., Wijayanto, A.W.: Classification of rice-plant growth phase using supervised random forest method based on landsat-8 multitemporal data. Int. J. Remote Sens. Earth Sci. (IJReSES) **16**, 187–196 (2020)
10. Messina, G., Peña, J.M., Vizzari, M., Modica, G.: A comparison of UAV and satellites multispectral imagery in monitoring onion crop. An application in the 'Cipolla Rossa di Tropea' (Italy). Remote Sens. **12**, 3424 (2020). https://doi.org/10.3390/rs12203424

11. Inglada, J., et al.: Assessment of an operational system for crop type map production using high temporal and spatial resolution satellite optical imagery. Remote Sens. **7**, 12356–12379 (2015). https://doi.org/10.3390/rs70912356
12. Pinter, P.J., Jr., et al.: Remote sensing for crop management. Photogram. Eng. Remote Sens. **69**, 647–664 (2003). https://doi.org/10.14358/PERS.69.6.647
13. Jin, Z., et al.: Smallholder maize area and yield mapping at national scales with Google Earth Engine. Remote Sens. Environ. **228**, 115–128 (2019). https://doi.org/10.1016/j.rse.2019.04.016
14. Ramadhani, F., Pullanagari, R., Kereszturi, G., Procter, J.: Automatic mapping of rice growth stages using the integration of SENTINEL-2, MOD13Q1, and SENTINEL-1. Remote Sens. **12**, 3613 (2020). https://doi.org/10.3390/rs12213613
15. Fatchurrachman, et al.: Automated near-real-time mapping and monitoring of rice growth extent and stages in Selangor Malaysia. Remote Sens. Appl. Soc. Environ. **31**, 100993 (2023). https://doi.org/10.1016/j.rsase.2023.100993
16. Li, M., Shamshiri, R.R., Weltzien, C., Schirrmann, M.: Crop monitoring using Sentinel-2 and UAV multispectral imagery: a comparison case study in Northeastern Germany. Remote Sens. **14**, 4426 (2022). https://doi.org/10.3390/rs14174426
17. Habibie, M.I., et al.: The development land utilization and cover of the Jambi district are examined and forecasted using Google Earth Engine and CNN1D. Remote Sens. Appl. Soc. Environ. **34**, 101175 (2024). https://doi.org/10.1016/j.rsase.2024.101175
18. Sencaki, D.B., et al.: Land cover multiclass classification of wonosobo, Indonesia with time series-based one-dimensional deep learning model. Remote Sens. Appl. Soc. Environ. **32**, 101040 (2023). https://doi.org/10.1016/j.rsase.2023.101040
19. Habibie, M.I., Nurda, N.: Innovative way to support data processing using the geospatial data science. Presented at the International Conference on Engineering and Computer Science (ICECS) 2022: The Use of Innovative Technology in Accelerating Problems Sustainable Development, Bandar Lampung City, Indonesia (2024). https://doi.org/10.1063/5.0204754
20. Ramadhani, F., et al.: Multisource spatiotemporal analysis of cropping patterns on dry upland: a case study in Rubaru Sub-district, Sumenep Regency. Egypt. J. Remote Sens. Space Sci. **27**, 403–415 (2024). https://doi.org/10.1016/j.ejrs.2024.04.008
21. Zhai, Y., Wang, N., Zhang, L., Hao, L., Hao, C.: Automatic crop classification in Northeastern China by improved nonlinear dimensionality reduction for satellite image time series. Remote Sens. **12**, 2726 (2020). https://doi.org/10.3390/rs12172726
22. Gorelick, N., Hancher, M., Dixon, M., Ilyushchenko, S., Thau, D., Moore, R.: Google Earth Engine: planetary-scale geospatial analysis for everyone. Remote Sens. Environ. **202**, 18–27 (2017). https://doi.org/10.1016/j.rse.2017.06.031
23. Pedregosa, F., et al.: Scikit-Learn: machine learning in Python. J. Mach. Learn. Res. **12**, 2825–2830 (2011)
24. Gillies, S., et al.: Rasterio: geospatial raster I/O for Python programmers (2013). https://github.com/rasterio/rasterio
25. Breiman, L.: Random forests. Mach. Learn. **45**, 5–32 (2001). https://doi.org/10.1023/A:1010933404324

SuperDove Sensor to Monitor Illegal Cannabis Plantation in Indonesia

Karunika Diwyacitta[1(✉)], Olivia Maftukhaturrizqoh[1], Fauzan Al Ayyubi[1], Bayu Satya Adhitama[1], Ika Siwi Supriyani[1], Syarif Budhiman[1], Dony Kushardono[1], Dedi Irawadi[2], Dede Dirgahayu Domiri[1], and Muhammad Sulaiman Nur Ubay[1]

[1] Research Center for Geoinformatics, National Research and Innovation Agency, Bogor Regency, Indonesia
karu001@brin.go.id

[2] Research Center for Satellite Technology, National Research and Innovation Agency (BRIN), Bogor Regency, Indonesia

Abstract. The PlanetScope satellite's SuperDove sensor provides daily images with a three-meter spatial resolution. This high-resolution satellite imagery can monitor the Earth's surface, including cannabis plantations, in detail. In Indonesia, cannabis plantations are illegal and classified as Class 1 according to Narcotics Law No. 35 of 2009. The inaccessibility of these area poses significant challenge for local authorities to counter this illegal action. In this paper, SuperDove sensor is used to calculate NDVI; based on RED (650–680 nm) and NIR (845–885 nm) bands; and monitor the land cover changes from natural forest to illegal cannabis plantation land over two years. By analyzing the NDVI from each SuperDove imagery, this study tracked the cannabis growth cycle, including vegetative, flowering and ripening stages. The cannabis NDVI changes of each stage is clearly depicted in the graph. The observation shows how illegal cannabis plantation takes around 6–7 months per cycle before it is either eradicated by authorities or harvested by the suspect. On the other hand, the NDVI of natural forests shows a stable trend over the years. This study presents the success of the extraction of NDVI from SuperDove sensor and displays the signature phenology of illegal cannabis that can be used to differentiate from other nearby objects.

Keywords: PlanetScope Data · Illegal Cannabis · Vegetation Index · Plan Growth Phase

1 Introduction

Cannabis sativa is not native to Indonesia, but historical documents report its use in the country since the 10th century [1]. However, according to Indonesian legislation number 35 of 2009, cannabis is classified as an illegal plant used to produce narcotics and illegal drugs. Illegal cannabis plantation often occurs in remote, sensitive ecosystems, leading to deforestation, water depletion, and soil degradation. These environmental impacts contribute to climate change by reducing carbon sequestration and disrupting local ecosystems. Therefore, identifying and mitigating illegal cannabis farming is

E. Fitrianingsih et al. (Eds.): ISAST 2024, SPPHY 416, pp. 85–94, 2025.
https://doi.org/10.1007/978-981-96-1344-1_9

crucial to addressing both environmental damage and climate change [2]. Cannabis is profitable and grows easily in various geographical conditions with sufficient sunlight [3, 4]. The hilly and densely forested regions of North Sumatra and Aceh are prime areas for cannabis cultivation in Indonesia. Farmers plant cannabis in dense wilderness and hidden hills, making detection difficult.

The psychoactive properties of tetrahydrocannabinol (THC) and cannabidiol (CBD), which can change moods, cause euphoria, and induce hallucinations, are the key reasons why this plant is prohibited in Indonesia [4]. Regardless of this, cannabis is widely used in Indonesia, with around 2 million users reported in 2014 [1]. In Aceh, cannabis is even used as a food.

The process of eradication of cannabis plantations in Indonesia still relies on traditional method, such as gathering information from local residents in areas suspected of illegal cultivation. This approach is neither effective nor efficient, as it is difficult to verify the accuracy of the points located in dense forests and hilly regions. It will take a long time to reach these areas, and they may not necessarily contain cannabis plants. Alternatively, phenological research can improve plant identification by studying the growth and harvesting phases of cannabis, providing more precise and actionable information that is needed to identify plants more easily. Understanding the growth and harvesting phases of cannabis can provide crucial information for plant identification [5].

Since cannabis requires direct sunlight to grow, remote sensing technology can be used as an alternative to monitor the illegal cannabis plantation. Tanasa et al. used a drone to visually inspect cannabis fields in Aceh [6]. The drone's sensing device is Vertical Aerial Photo to capture cannabis plantations. The study only reported the utilization of red, green, and blue channels.

It is also possible to detect the cannabis plantation by manually spotting the cannabis plantation by human crew on a low flying airplane [7]. When the detection is done by a well-trained crew, it tends to give better measurement and analysis of the suspected cannabis fields. The disadvantages of this method are inefficient, costly and requires training for multiple people that will conduct the detection.

Another method to detect cannabis plantations is using remote sensing satellite data. Remote sensing provides wide coverage based on spatial resolution and has been widely used for monitoring the Earth's surface. Sujud et al. processed optical (sentinel-2, landsat) and radar (sentinel-1) imageries to classify cannabis plantations in Lebanon using machine learning, utilizing the NDVI (Normalized Difference Vegetation Index) and other composite bands [8]. Lisita et al. used SPOT5 imagery with the OBIA method (Object-Based Image Analysis) to identify potential cannabis plantation areas in Brazil [9]. The parameters extracted included land-cover classes, slope, distance to water, NDVI and anthropogenic areas. The study demonstrated the effectiveness of OBIA method in locating cannabis plantation areas.

This study retrieved the vegetation index of cannabis plantations using PlanetScope's SuperDove satellite imagery. PlanetScope's second generation PS2.SD sensor offers 4-band imagery [10, 11], which has been successfully used for cannabis classification with machine learning, achieving over 93% accuracy with random forest (RF) method [3]. The latest generation, Planetscope offers SuperDove 8-bands imagery, used in this study [10, 12], offering high-resolution satellite imagery data with daily temporal resolution, and three meters spatial resolution. This allows accurate daily monitoring of the vegetation index. The combination of high spatial resolution and daily temporal data of PlanetScope is essential for this study because it can help minimize the presence of clouds in the region observed. The index of illegal cannabis for two years is cal-culated and the phenology is generated from the PlanetScope imagery. This study demonstrates how PlanetScope imagery can monitor illegal cannabis plantations and visualize the illegal cannabis fields.

2 Methodology

Obtaining data on cannabis in Indonesia is very challenging, especially for field measurements. This difficulty arises because cannabis is classified as an illegal substance under Indonesian law. This study greatly benefited from the support of Indonesian law enforcement agencies, which permitted participation in the eradication of cannabis fields, allowing for the collection of field data. This collaboration enabled the collection of valuable field data during eradication activities in Aceh Besar District, NAD, as shown in Fig. 1. For satellite data, the PlanetScope SuperDove imagery was proposed to monitor cannabis plantations.

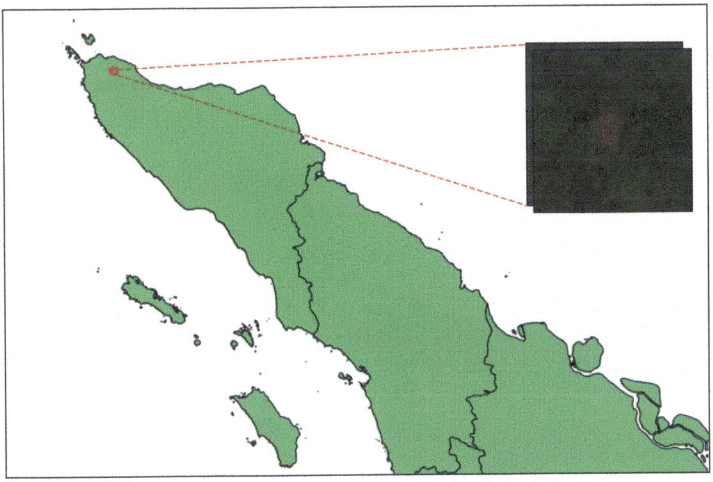

Fig. 1. The study area used in this study. Satellite Imagery © 2024 Planet Labs PBC.

In this study, the PlanetScope satellite constellation with the SuperDove sensor, the latest generation of PlanetScope satellites, was used. The SuperDove sensor provides two types of images based on its bands. The first one is PS2.SD which contains RGBNIR and the second one is PSB.SD which contains RGBNIR, red edge, green, yellow and coastal blue. The SuperDove sensor, launched in early 2020, offered around 3 m spatial resolution and has been providing daily imagery with 8 spectral bands since April 2022 [12]. Details of the bands are shown in Table 1.

The selection of PlanetScope high-resolution satellite imageries was considered crucial because, based on field studies, illegally planted cannabis in the studied area was

Table 1. Spectral bands of SuperDove Imagery.

Instrument	PSB.SD
Spectral Bands	Coastal Blue 431–452 nm
	Blue: 465–515 nm
	Green I: 513–549 nm
	Green: 547–583 nm
	Yellow: 600–620 nm
	Red: 650–680 nm
	Red-Edge: 697–713 nm
	NIR: 845–885 nm

spaced about one meter apart. High spatial resolution images provided more accurate data as each pixel better represented the reflectance value of cannabis plants, minimizing unnecessary elements such as soil, rocks, and other types of weeds.

SuperDove 8-bands imagery was assessed for its radiometric accuracy based on its surface reflectance, compared to surface reflectance from Sentinel-2 and surface reflectance from reference field measurements [13]. Various vegetation indices were also evaluated, and it was found that SuperDove-derived NDVI showed the best performance among all vegetation indices when compared to Sentinel-2 data [13]. The whole process of this study is illustrated in Fig. 2.

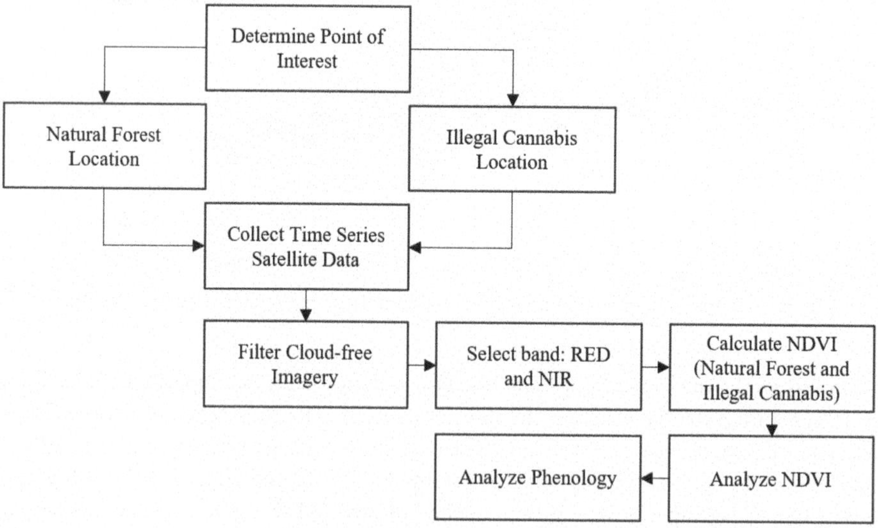

Fig. 2. The process of the study.

The first step was to determine the coordinate of the illegal cannabis plants and natural forest. This step aims to carefully find the difference of illegal cannabis and natural forest characteristic. While the location of the cannabis plants used in this study has been confirmed, the exact coordinates required further processing to accurately point to targeted cannabis plants. This was done using very high-resolution imagery to identify the cannabis plants and selecting the precise coordinates as shown in Fig. 3. Since the illegal cannabis location is surrounded by natural forest, the coordinate of natural forest taken nearby.

The second step was to collect multi-temporal high resolution satellite imagery over time to capture the stages of the cannabis plantation growth cycle and the natural forest behavior. Since the cannabis plant is typically ready to be harvested after 7 months, the satellite data was collected over a two years interval to obtain insights into multiple cannabis plantation growth cycles.

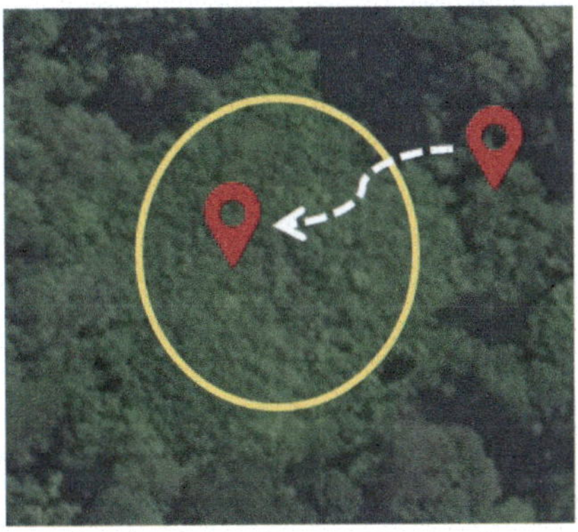

Fig. 3. Determining cannabis plant locations using very high-resolution satellite data.

The multi-temporal data were visually checked to remove any images where the target location was covered by clouds. There are almost 50 free-cloud imageries data collected in two years. All data were automatically processed to extract the NIR and RED band values, forming time-series imagery. The cannabis plant growth cycle was then monitored using the normalized difference vegetation index (NDVI), calculated from the extracted NIR and RED bands value. The NDVI formula was as follows [13]:

$$\mathrm{NDVI} = (\mathrm{R_{NIR}} - \mathrm{R_{Red}})/(\mathrm{R_{NIR}} + \mathrm{R_{Red}}) \tag{1}$$

where R was the reflectance value at the desired wavelength (nm).

The final step involved analysis by creating a phenology graph based on the NDVI values over 2 years. The result, analysis and discussion were explained in the next section.

3 Result and Discussion

The identified illegal cannabis plantations in this study were used to reconstruct the growth chart of cannabis and validate that its growth stages align with those referenced in the literature. The outcome of this work was to build a dataset for developing a machine learning model that was used to predict the presence of illegal cannabis plantations in other areas [14].

Figure 4 shows the NDVI of illegal cannabis at the specified coordinates and the natural forest right beside it. The black line shows that the NDVI fluctuates over time. From May-2022 to January-2024, there are two cycles, each lasting 6–7 months. Each cycle typically features a peak point (the highest point), a trough (the lowest point), and some intermediate points. Compared to the orange line which shows the natural forest area, the NDVI is relatively stable in between 0.8 to 0.9 over two years. There is no significant highest and lowest peak. The initial peak in May-2022, with the highest NDVI index of 0.895, likely represents natural forest as shown in Fig. 5(a). This area was cleared, as indicated by the decrease in the index, suggesting the start of illicit cannabis cultivation as shown in Fig. 5(b). The reflectance index identification was focused solely on cannabis and forest areas. Other objects were not identified in this study, based on the assumption that land use changes from natural forest to other vegetation or bare land in remote areas are primarily for illegal activities (i.e. deforestation and forest degradation), including cannabis plantation [15].

Subsequent cycles reflect the growth patterns of cannabis plantations. To better understand these patterns, they could be compared to cannabis data from legally cultivated cannabis, as detailed in the literature review used in this study [5]. The final cycle shown in Fig. 4 is used to analyzed the illegal cannabis growth (Fig. 6). There are four stages of cannabis plant growth based on the real experience while planting it [5]. The cannabis cultivation typically involves four stages: 1) Seed Germination (3–10 days), it is the procedure of preparing the seed for transfer to a new growing area; 2) Seedling (2–3 weeks), which is characterized by the growth of the distinctive cannabis fan leaves; 3) Vegetative Growth (3–16 weeks), the plants require more space as they grow rapidly; 4) Flowering (8–11 weeks), is the final stage of cannabis cultivation.

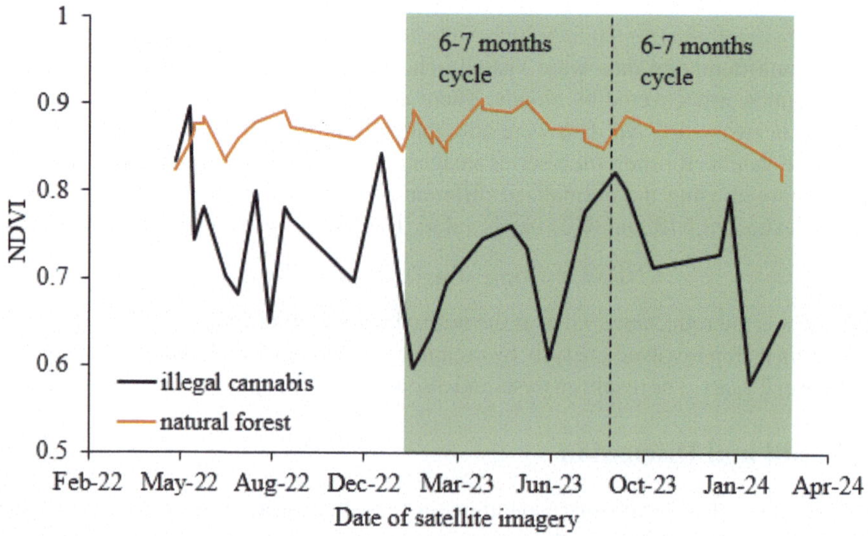

Fig. 4. NDVI of cannabis over several years.

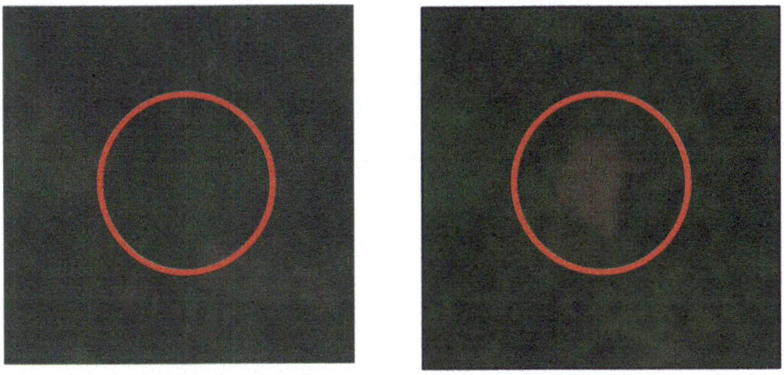

(a) (b)

Fig. 5. (a) Natural forest in May-2022; (b) Illegal cannabis plantation in October-2023. Satellite Imagery © 2024 Planet Labs PBC.

Figure 6 shows that the identifiable stages of cannabis cultivation include only the vegetative and flowering stages, followed by a ripening period of about three months. Seed germination and seedling stages may not be detectable due to insufficient leaf area for NDVI calculation or because these stages occurred outside the cannabis field. According to local authorities, illegal cannabis undergoes germination and seedling stages at separate site before being moved to a larger area which explains why the vegetation index reaches around 0.7 within a month. The vegetative stage lasts 6 weeks, from June 2023 to the end of July 2023, with a significant increase in the NDVI. The flowering lasts less than 5 weeks, during which the illegal cannabis grower harvest only the flowers, causing the index to drop by nearly half. The ripening period usually aligned with cannabis harvesting, but in illegal cultivation, harvesting may be delayed due to "*security concerns*" for the growers. If local authorities discover the presence of cannabis field, the harvest process is halted, and the index vegetation drops significantly due to eradication. The entire cycle is visually summarized in Fig. 7 both in RGB and NDVI composites.

Figure 7 presents visualization of illegal cannabis plantations over time using PlanetScope SuperDove high resolution satellite data. The illegal cannabis area is centrally located in the image with a relatively rectangular shape. On June 20, 2023, the land appeared brown, but by August 30, 2023, it had significantly greened. Figure 7 includes RGB composites and the calculated NDVI composites for better visualization of vegetation changes. In the NDVI composites, purple represents bare land with low vegetation, yellow indicates dense forest, and intermediate greens show varying vegetation levels. The NDVI index rose from 0.641 in June 2023 to 0.824 in August 2023, reflecting peak flowering. In October 2023, the index dropped slightly due to an early harvest, and continued to grow during ripening stage until the local authorities eradicated the plants in January 2024, reducing the index to 0.580, indicating relatively bare land.

PlanetScope SuperDove high resolution satellite data (up to 3 m) is effective for visual interpretation, clearly showing changes in illegal cannabis fields. Although the imagery could not display the detail of cannabis leaves clearly, the data effectively represents the

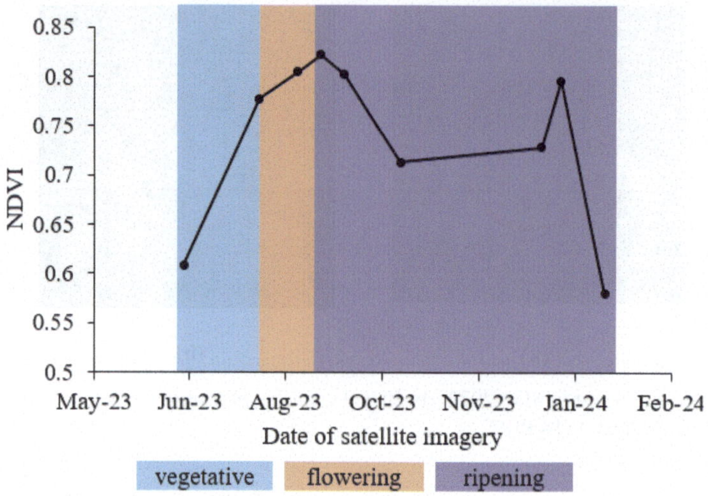

Fig. 6. One cycle of illegal cannabis plantation.

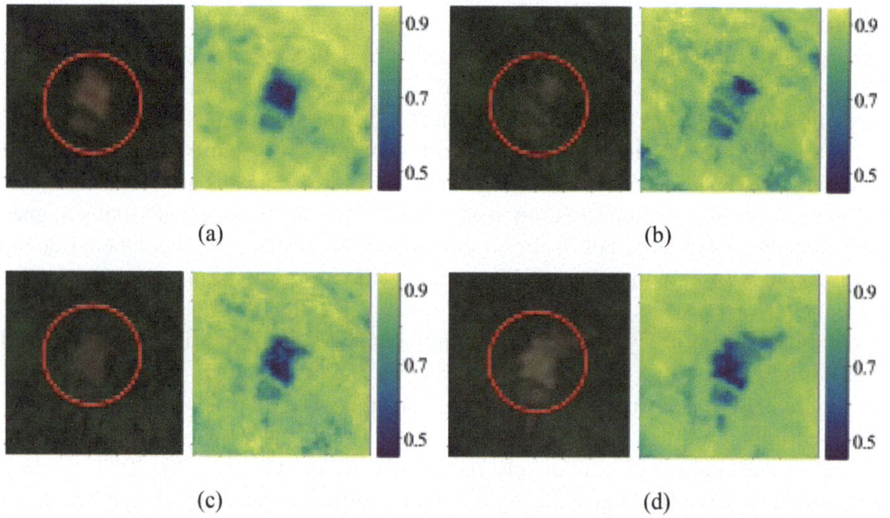

Fig. 7. 7-month cannabis field cycle visualization using PlanetScope SuperDove imagery (a) 20/06/2023; (b) 30/08/2023; (c) 10/10/2023; (d) 24/01/2024. Satellite Imagery © 2024 Planet Labs PBC.

field's changes over months, making it easier to spot within the dense forest areas. These imageries also help to analyze the cannabis field for future research.

4 Conclusion

The extraction of NDVI from PlanetScope imagery using SuperDove sensor has been done successfully. This index, derived from daily multitemporal images, can be used to analyze the phenology of illegal cannabis crops. The phenology graph clearly illustrated the growth stages of illegal cannabis plantations over time. Based on satellite imagery and local authority information, the identifiable stages of cannabis cultivation include only the vegetative and flowering stages, followed by a ripening period. The index fluctuations highlight the unique characteristics of illegal cannabis plantations and may help differentiate them from another nearby object. This data can be used as dataset for illegal cannabis detection with machine learning. SuperDove's high-resolution enhances visual analysis and correlates with the normalized difference vegetation index (NDVI) value. Daily multitemporal images could help researchers in monitoring dynamic changes in illegal cannabis areas.

This study offers preliminary insights into using very high spatial resolution imagery, such as PlanetScope, to track cannabis plantation growth stages in remote areas. The findings highlight the importance of vegetation index values in reconstructing these growth stages. However, the study's limitation lies in its assumption that land use changes from natural forests to other vegetation or bare land in remote areas are mainly for illegal activities, such as converting land into cannabis plantations. Future research should explore reflectance indices for various land types beyond natural forests and cannabis, and analyze multiple samples from different regions. Comparing these results with studies using the NDVI method on other vegetation types will also enhance the analysis.

Acknowledgment. All authors would like to greatly thank Research Organization for Aeronautics and Space, National Research and Innovation Agency (BRIN) for providing all things needed to conduct this research. We also would like to appreciate Research Organization for Electronics and Informatics for the guidance while carrying out this research. Last but not least we are also grateful for National Narcotics Agency (BNN) and National Intelligence Agency (BIN) for providing the materials.

References

1. Putri, D., Blickman, T.: Cannabis in Indonesia. Drug Policy Brief **44**, 1–24 (2016). https://doi.org/10.13140/RG.2.1.4573.8000
2. Zheng, Z., Fiddes, K., Yang, L.: A narrative review on environmental impacts of cannabis cultivation. J. Cannabis Res. **3**(1) (2021). https://doi.org/10.1186/s42238-021-00090-0
3. Bicakli, F., Kaplan, G., Alqasemi, A.S.: Cannabis sativa L. Spectral discrimination and classification using satellite imagery and machine learning. Agriculture **12**(6) (2022). https://doi.org/10.3390/agriculture12060842
4. Irianto, A., et al.: National survey on drug abuse (2021)
5. Kocábková, B.: 4 stages of cannabis plant growth (2022). https://en.cbdsvet.cz/2022/09/19/4-stages-of-cannabis-plant-growth/. Accessed 10 June 2024
6. Tanasa, I.Y., et al.: Segmentation of drone-captured images for illegal cannabis plantation detection. In: AIP Conference Proceedings, vol. 2941, no. 1 (2023). https://doi.org/10.1063/5.0181482

7. Walthall, C.L., et al.: Detection of illegal cannabis cultivation using remote sensing. In: International Symposium on Geoscience and Remote Sensing, pp. 2281–2284 (2006). https://doi.org/10.1109/IGARSS.2006.590
8. Sujud, L., Jaafar, H., Haj Hassan, M.A., Zurayk, R.: Cannabis detection from optical and RADAR data fusion: a comparative analysis of the SMILE machine learning algorithms in Google Earth Engine. Remote Sens. Appl. Soc. Environ. **24**, 100639 (2021). https://doi.org/10.1016/j.rsase.2021.100639
9. Lisita, A., Sano, E.E., Durieux, L.: Identifying potential areas of Cannabis sativa plantations using object-based image analysis of SPOT-5 satellite data. Int. J. Remote Sens. **34**(15), 5409–5428 (2013). https://doi.org/10.1080/01431161.2013.790574
10. Planet Labs PBC: PlanetScope Product Specifications, pp. 1–38 (2023).https://assets.planet.com/docs/Planet_PSScene_Imagery_Product_Spec_letter_screen.pdf
11. Kington, J., Collison, A.: Scene level normalization and harmonization of planet dove imagery, pp. 1–29. Planet Labs Inc. (2022)
12. Planet Labs PBC: Planet Imagery Product Specifications, pp. 1–100. Planet Labs Inc. (2023). https://www.planet.com/products/planet-imagery/
13. Tu, Y.H., Johansen, K., Aragon, B., El Hajj, M.M., McCabe, M.F.: The radiometric accuracy of the 8-band multi-spectral surface reflectance from the planet SuperDove constellation. Int. J. Appl. Earth Obs. Geoinf. **114**, 103035 (2022). https://doi.org/10.1016/j.jag.2022.103035
14. Irawadi, D., et al.: Smart detection of illicit cannabis plantations using remote sensing technology and machine learning. Geogr. Environ. Sustain. J. (2024). https://ges.rgo.ru/jour
15. Tacconi, L., Rodrigues, R.J., Maryudi, A.: Law enforcement and deforestation: lessons for Indonesia from Brazil. Forest Policy Econ. **108** (2019). https://doi.org/10.1016/j.forpol.2019.05.029

Assessing the Potential of LAPAN A3 Satellite Imagery for Land Cover Monitoring

Rizatus Shofiyati[1](✉), Mohammad Ardha[1], Patria Rachman Hakim[2], Agus Herawan[2], Satriya Utama[2], Muhammad Iqbal Habibie[3], Samijan Samijan[4], Sri Minarsih[4], and Wahyudi Hasbi[2]

[1] Research Center for Geoinformatics, National Research and Innovation Agency (BRIN), Bandung, Indonesia
rizatus.shofiyati@brin.go.id
[2] Research Center for Satellite Technology, BRIN, Bogor, Indonesia
[3] Research Center of Environmental and Clean Technology, BRIN, Jakarta, Indonesia
[4] Research Center for Food Crops, BRIN, Bogor, Indonesia

Abstract. This study evaluates the capabilities of LAPAN A3, a small Indonesian satellite with a spatial resolution of 15 m and four multispectral bands (R, G, B, NIR), in land cover identification. By comparing its imagery with Landsat 8 and Sentinel 2, the study assesses LAPAN A3's effectiveness using the Normalized Difference Vegetation Index (NDVI) to classify various land cover types, including water, green vegetation, yellowing vegetation, and bare land. The results show that LAPAN A3's imagery is comparable in resolution and quality to those of Landsat 8 and Sentinel 2, making it suitable for operational land monitoring. However, weak correlations were observed between the spectral bands of LAPAN A3 and the other satellites, with the highest R2 value of 0.3054 in the NIR band compared to Landsat 8, and 0.1711 in the red band compared to Sentinel 2. Additionally, the imagery from LAPAN A3 showed a significant 3 km georeferencing shift, necessitating manual rectification for accurate spatial analysis. While the satellite demonstrates the potential for filling data gaps caused by cloud cover in other satellites, further research is required to improve sensor calibration, georeferencing accuracy, and overall reliability for land cover monitoring in Indonesia.

Keywords: LAPAN A3 · Landsat 8 · Sentinel 2 · NDVI · land cover identification · multispectral satellite

1 Introduction

LAPAN A3, a small Indonesian-made satellite with a weight of 115 kg and dimensions of 667 × 574 × 930 mm, produces imagery with a spatial resolution of 15 m and four multispectral bands (R, G, B, NIR) [1, 2]. The satellite's mission includes both experimental and operational aspects, and its potential for land cover identification on the Earth's surface needs to be assessed. Testing was conducted by comparing LAPAN A3 imagery with Landsat 8 and Sentinel 2, operational satellites widely used in Indonesia [3,

E. Fitrianingsih et al. (Eds.): ISAST 2024, SPPHY 416, pp. 95–104, 2025.
https://doi.org/10.1007/978-981-96-1344-1_10

4]. LAPAN A3 can be utilized for various applications, including urban planning, agriculture monitoring, environmental assessment, and disaster management. Its versatile imaging capabilities make it a valuable tool for a wide range of many Earth observation needs [5].

The Normalized Difference Vegetation Index (NDVI) method was used to identify land covers such as buildings, bare land, rice fields, and others. NDVI is an index that can be utilized with LAPAN A3, which features both visible and near-infrared bands. Additionally, the characteristics of each band from LAPAN A3 were assessed by analyzing their patterns. This evaluation helps to determine the effectiveness of each band in different applications.

This study aims to assess the image capabilities of LAPAN A3 in identifying various types of land cover by comparing its results with those from Landsat 8 and Sentinel 2 satellites. LAPAN A3 achieves a similar overall effectiveness as the established satellites in terms of land cover identification.

2 Methodology

2.1 Data Used

The selected study location was Center Java Province (Fig. 1). Central Java province was chosen as the study location because it has a variety of land covers.

Fig. 1. Study location of Center of Java Island

Data used in this study consists of Landsat 8, Sentinel 2 and LAPAN A3. Their specifications are shown in Table 1 [3, 4]. Image dataset acquisition is selected based on the date of ground data collection.

The method used in this assessment is the Normalized Difference Vegetation Index (NDVI), along with the spectral patterns from the four bands (Red Green, Blue and Near Infra-Red) of Landsat 8, Sentinel 2 and LAPAN A3. NDVI analysis was processed using the Google Earth Engine platform. NDVI is a commonly employed technique for land cover analysis. NDVI and spectral patterns are used to identify various land cover types, including buildings, bare land, rice fields, and others. The multispectral image data generated by the LAPAN A3 satellite is compared with those from Landsat

Table 1. LAPAN A3, Landsat-8 and Sentinel-2 Specification

Characteristics	Landsat-8	Sentinel-2	LAPAN A3
Spatial Resolution	15-100 m (multispectral)	10-60 m (multispectral)	15 m (multispectral)
Swath-width	185 km	290 km	120 km
Altitude	705 km	786 km	510 km
Temporal Resolution	16 days	10 days	21 days
Radiometric Resolution	12 bits	12 bits	16 bits
Spectral Resolution	13 bands	13 bands	4 band

Characteristics	Landsat-8		Sentinel-2		LAPAN A3	
Band	Resolution (m)	Wavelength (μm)	Resolution (m)	Wavelength (μm)	Resolution (m)	Wavelength (μm)
1 (Coastal)	30	0.435-0.451	60	0.442	-	-
2 (blue)	30	0.452-0.512	10	0.492	15	0.41-0.49
3 (Green)	30	0.533-0.590	10	0.559	15	0.51-0.58
4 (Red)	30	0.636-0.673	10	0.664	15	0.63-0.70
5 (NIR)	30	0.851-0.879	10	0.833	15	0.77-0.90
6 (SWIR-1)	30	1.566-1.651	20	1.610	-	-
7 (SWIR-2)	30	2.107-2.294	20	2.185	-	-
9 (Cirrus)	30	1.363-1.384	60	1.377	-	-

8 and Sentinel 2. NDVI was selected as the primary index for comparison because all three images used in this study contain the necessary bands for NDVI calculation. Other indices, such as the Normalized Difference Built-up Index (NDBI), which could also be used to assess land cover conditions, but they were not utilized. This is because calculating NDBI requires the Shortwave Infrared (SWIR) band, which is unavailable in the LAPAN A3 satellite imagery. Therefore, NDVI was chosen as the most suitable index for this analysis.

3 Results and Discussion

LAPAN-A3 multispectral images have been systematically corrected for both radiometric and geometric errors; however, there are still image geolocation errors due to the imperfection of satellite nadir pointing stability and the error of satellite orbit calculation (Fig. 2). Therefore, further manual local georeferencing is needed to ensure that the area investigated has perfect image geolocation so that NDVI comparison and each band correlation to other imagery can be appropriately conducted.

The initial georeferencing of the LAPAN A3 imagery revealed a significant discrepancy, with the image shifted by approximately 3 km. This shift highlights a substantial error in the automated georeferencing system, which could affect the accuracy of subsequent spatial analyses. Such a misalignment underscores the need for manual rectification to ensure the imagery correctly corresponds to the actual geographic location, particularly when integrating it with other spatial datasets like Sentinel 2 and Landsat 8.

LAPAN-A3 multispectral images have been corrected for geometric and radiometric systematic errors (Table 2). For the geometric aspect, the co-registration inter-channel mistake due to each detector position difference was corrected using the image-matching approach. At the same time, both direct and manual georeferencing were applied to calculate image location on the Earth's surface.

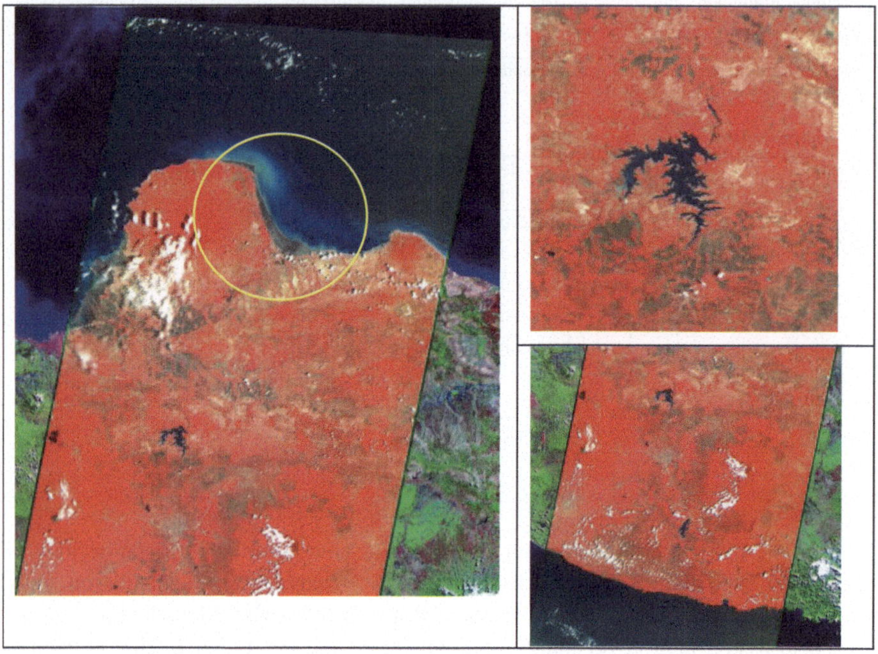

Fig. 2. Georeferencing Errors in LAPAN A3 Imagery (right) and Corrected Georeferencing of LAPAN A3 Imagery Manually Re-rectification (left).

Meanwhile, for the radiometric aspect, vignetting errors due to lens distortion and detector responsivity have been corrected by using pre-flight laboratory calibration data. However, the final product of the satellite images is still in digital number (DN) format, so to convert the images into radiance-value images, the DN-value of the images needs to be divided by some constant factor which depends on the imager electronic setting when the images are captured. For images used in this research, Table 2 shows the constant factor for each color channel of LAPAN-A3 multispectral images, which are used to convert DN-value images into radiance-value images [6]. Similar to the findings of Setiawan et al. [7], this study concludes that image pre-processing is necessary to improve the characterization of LAPAN A3 data. Additionally, developing appropriate algorithms or methods for geometric and atmospheric correction is crucial to enhance the accuracy of the data.

The analysis results show that reflectance patterns in the red, green, blue (RGB), and near-infrared (NIR) bands differ between Sentinel 2 and LAPAN A3. In Sentinel 2 imagery, the reflectance values align well with the expected characteristics of various

Table 2. Geometric and radiometric systematic error correction of LAPAN A3

Band	Correction
Blue	34,158
Green	56,438
Red	79,265
NIR	35,188

objects, such as water, green vegetation, yellowing vegetation, and bare soil, in response to the respective RGB and NIR wavelengths. It allows for clear differentiation between these objects based on their spectral signatures. However, in LAPAN A3 imagery, not all objects exhibit the expected patterns (Fig. 3). Certain features may not display distinct reflectance behavior, leading to difficulty distinguishing between different land cover types. A more in-depth study is needed to understand the reflectance patterns of LAPAN A3 better, as this could improve its applicability for identifying and differentiating land cover types.

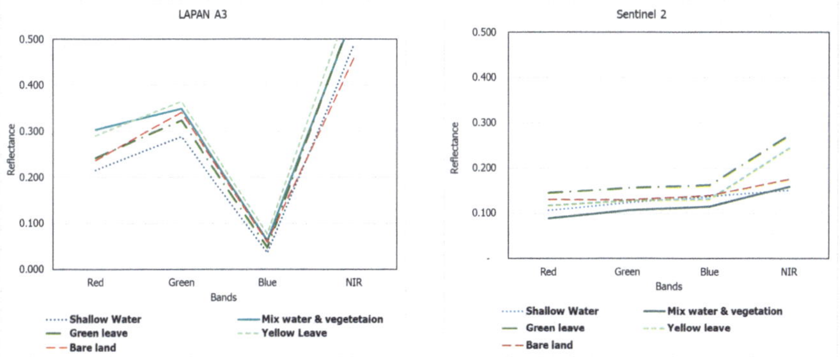

Fig. 3. Comparison of Spectral Pattern of RGB and NIR Bands between LAPAN A3 and Sentinel 2 Images.

In Fig. 4, the spectral values in the Red, Green, and NIR bands from the LAPAN A3 satellite are noticeably higher compared to those from both Landsat 8 and Sentinel 2. This indicates that LAPAN A3 captures stronger reflectance in these bands, possibly due to sensor characteristics or calibration differences. However, in the blue band, the spectral values of LAPAN A3 are lower than those of Landsat 8 and Sentinel 2. This could suggest that LAPAN A3's sensor is less sensitive in the shorter wavelengths, or it may be due to atmospheric effects like scattering, which are more pronounced in the blue spectrum and less effectively corrected in LAPAN A3 imagery.

The higher spectral values in the Red, Green, and NIR bands of LAPAN A3 compared to Landsat 8 and Sentinel 2 are likely due to differences in sensor calibration, radiometric sensitivity or less refined atmospheric corrections in LAPAN A3 imagery.

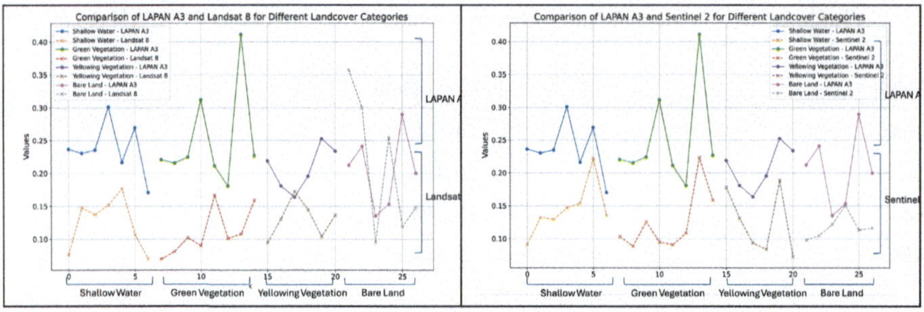

(a)

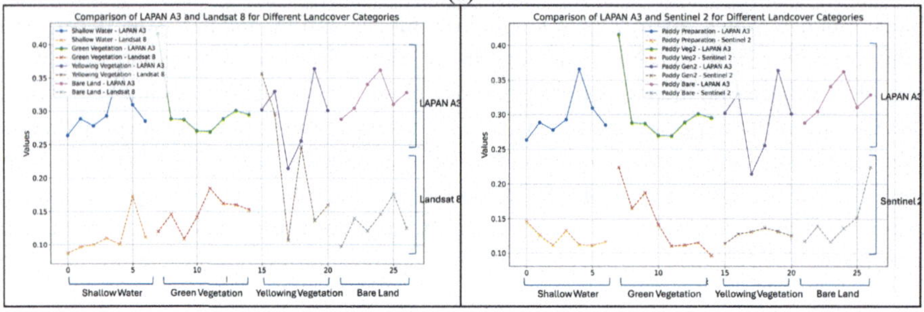

(b)

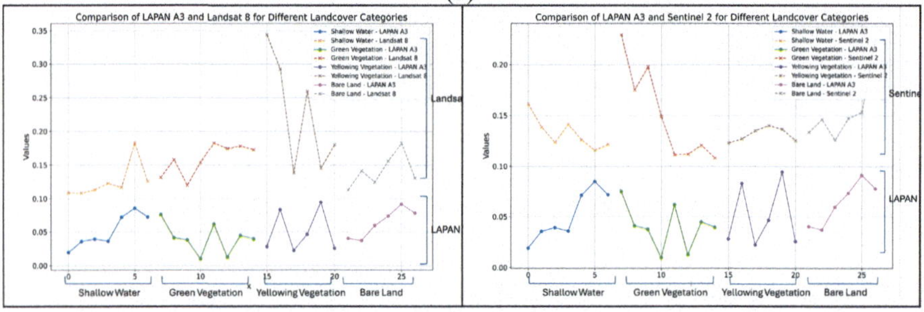

(c)

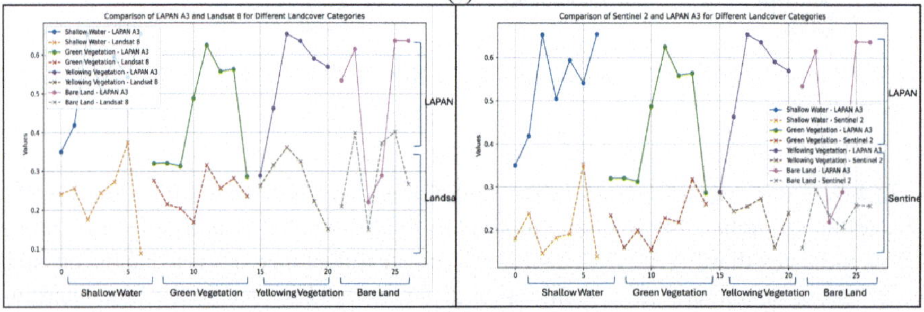

(d)

Fig. 4. Reflectance of Red (a), Green (b), Blue (c) and NIR (d) Bands Comparison between LAPAN A3 and Landsat 8 (left)/Sentinel 2 (right)

The lower spectral value in the blue band of LAPAN A3 likely results from reduced sensor sensitivity to shorter wavelengths and greater susceptibility to atmospheric scattering and aerosol interference. These effects, coupled with less advanced correction methods, explain the reduced reflectance compared to Landsat 8 and Sentinel 2.

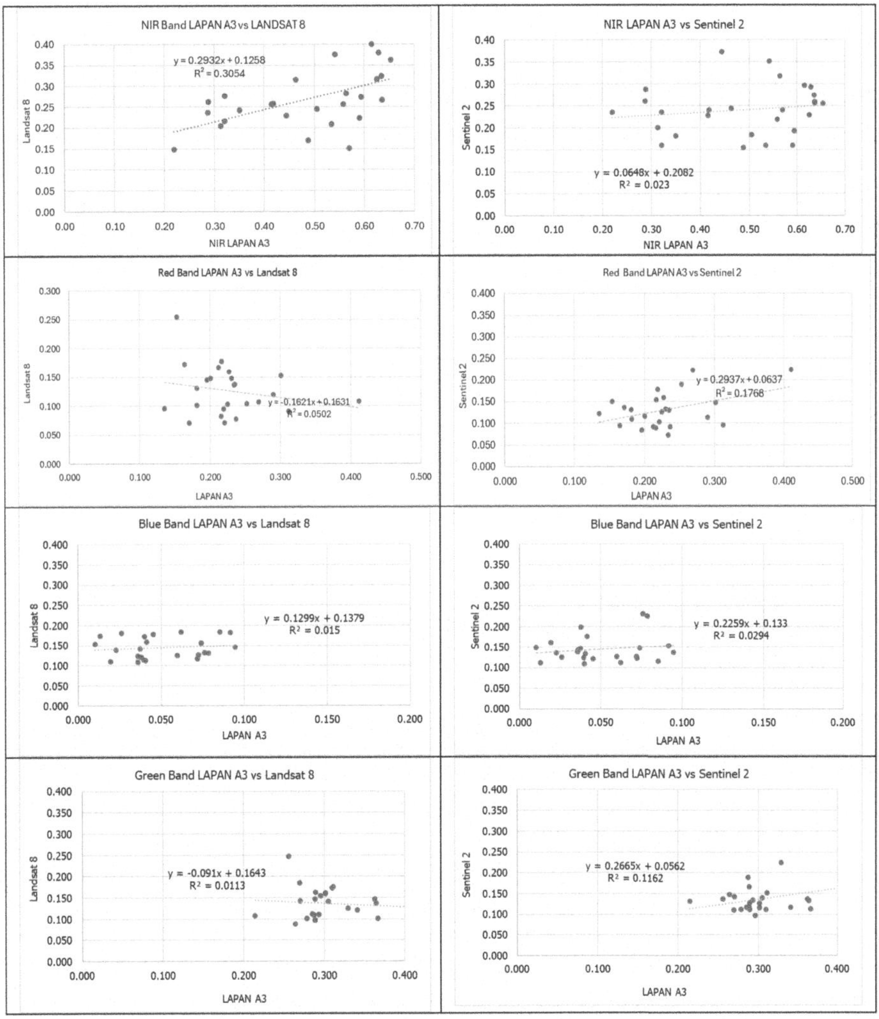

Fig. 5. Correlation Analysis of RGB and NIR Bands between LAPAN A3 and both Sentinel 2 and Landsat 8.

The comparison of spectral pattern across different bands between LAPAN A3, Sentinel 2, and Landsat 8 did not reveal a strong correlation without distinguishing between object types (water, green vegetation, yellowing vegetation, and bare soil), did not reveal a strong correlation. The highest R^2 value obtained was 0.3054 for the NIR band between LAPAN A3 and Landsat 8, followed by 0.1711 for the red band between LAPAN A3 and Sentinel 2. The other bands showed even lower correlations, with R^2 values ranging from 0.015 to 0.116. These results indicate a weak alignment of reflectance patterns between LAPAN A3, Sentinel 2 and Landsat 8 (Fig. 5).

Figure 6 shows the NDVI derived from LAPAN A3 demonstrated effective classification for various land cover types, such as water, green vegetation, yellowing vegetation, and bare land. The NDVI values from LAPAN A3 imagery in several sample locations were found to be nearly identical to the NDVI values from Sentinel 2 imagery across three types of landcover, except for shallow water. However, in some sample points, LAPAN A3 showed higher NDVI values. This variation suggests that while LAPAN A3 performs similarly to Sentinel 2 in landcover type, there may be factors such as sensor calibration or local conditions that cause higher NDVI readings in certain areas.

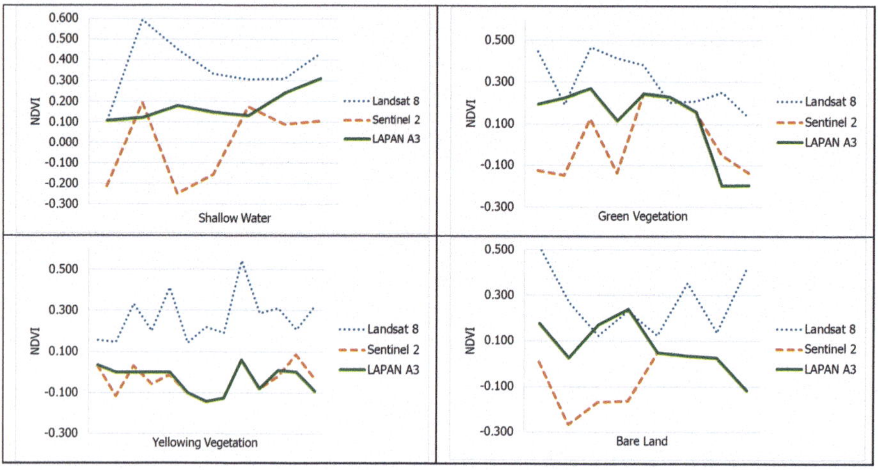

Fig. 6. NDVI Comparation for each land cover from Landsat 8, Sentinel 2 and LAPAN A3.

Compared to Landsat 8, the NDVI values from LAPAN A3 were generally lower for all landcover types and at almost every sample point. This indicates a consistent disparity between the two datasets, which could be attributed to differences in sensor calibration, resolution, or atmospheric correction techniques.

In Fig. 7, the NDVI patterns of four landcover types are compared. The results show that LAPAN A3 and Sentinel 2 have similar NDVI values for yellowing vegetation and bare land. However, LAPAN A3 shows higher NDVI values for shallow water and green vegetation, even though it follows the same overall pattern as Sentinel 2. Despite these findings, all four landcover types consistently exhibit lower NDVI values in LAPAN A3 than in Landsat 8.

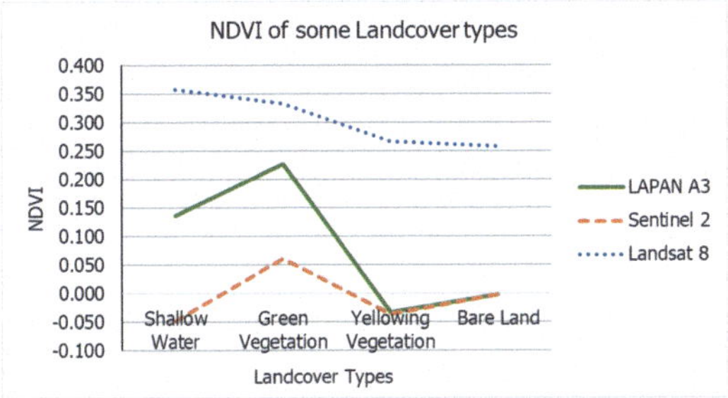

Fig. 7. NDVI patterns of Landsat 8, Sentinel 2 and LAPAN A3.

4 Conclusion

The NDVI values from LAPAN A3 are closely aligned with Sentinel 2 at several sample points for Green Vegetation, Yellowing Vegetation, and Bare Land, except for Shallow Water, while consistently lower than those from Landsat 8. However, the four bands of LAPAN A3 do not show a strong correlation compared to the same bands from Sentinel-2 and Landsat 8. High pixel values in LAPAN A3 imagery require calibration before use, with different calibration values needed for each scene. Further research is necessary to improve the sensors on future satellites.

A 3 km shift in LAPAN A3 imagery, especially in outer areas, highlights the need for georeferencing improvements and manual re-rectification.

The images produced have resolution and quality comparable to those from Landsat 8 and Sentinel 2, making them reliable for operational purposes, with the results of this study indicating their suitability for land cover monitoring. Further research and calibration of pixel values from the LAPAN A3 imagery, along with enhancements to the georeferencing system, are necessary.

Acknowledgement. Thanks to The Aeronautics and Space Research Organization of the National Research and Innovation Agency (BRIN) for providing satellite data of LAPAN A3.

References

1. Hasbi, W., Suhermanto, S.: Development of LAPAN-A3/IPB satellite an experimental remote sensing microsatellite. In: 34th Asian Conference on Remote Sensing 2013 (2013)
2. Anggaria, E.A., Hakim, P.R.: Study of potential applications of LAPAN-A3/IPB satellite's multispectral imager. In: Setiawan, Y., Prasetyo, L.B., Murayama, Y., Kanniah, K.D., Perez, G.J., Dat, P.T. (eds.) Sixth International Symposium on LAPAN-IPB Satellite, vol. 11372, pp. 1–13. SPIE (2019). https://doi.org/10.1117/12.2542769
3. Tang, L., Hupy, J.P., Wang, Y.: A commentary review on the use of normalized difference vegetation index (NDVI) in the era of popular remote sensing. J. For. Res. **32**(5) (2020)
4. Pettorelli, N., Ryan, S.J., Mueller, T., Bunnefeld, N.: The normalized difference vegetation index (NDVI): unforeseen successes in animal ecology. Climate Res. **46**, 15–27 (2011)
5. Kushardono, D., Wirawan, R., Mujtahid, Z., Mukayadi, M.: Teknologi LAPAN A3 LISA dan Pemanfaatan Citranya. Inderaja **3**(10), 23–29 (2017)

6. Hakim, P.R., Syafrudin, A.H., Salaswati, S., Utama, S., Hasbi, W.: Development of systematic image preprocessing of LAPAN-A3/IPB multispectral images. Int. J. Adv. Stud. Comput. Sci. Eng. **7**(10), 9–18 (2019)
7. Setiawana, Y., et al.: Pemanfaatan Fusi Data Satelit LAPAN-A3/IPB Dan Landsat 8 Untuk Monitoring Lahan Sawah. Jurnal Pengelolaan Sumberdaya Alam dan Lingkungan **8**(1), 67–76 (2018)

Object-Based Random Forest Classification for Rapid Mapping of Recent Volcanic Material Deposits

Imam Santoso[1(✉)], Rido Dwi Ismanto[2], Yenni Vetrita[1], Suwarsono[1], Arum Tjahyaningsih[1], Farikhotul Chusnayah[3], Rahmadi[1], Mamat Suhermat[4], and Donna Monica[1]

[1] Research Center for Geoinformatics, BRIN, Bogor, Indonesia
imam.santoso@brin.go.id
[2] Research Center for Computing, BRIN, Bogor, Indonesia
[3] Research Center for Hydrodynamics Technology, BRIN, Surabaya, Indonesia
[4] Research Center for Geological Disaster, BRIN, Bandung, Indonesia

Abstract. Rapid post-eruption mapping is needed to provide initial information about the impact of an eruption. Rapid mapping of eruption impacts using optical data is often tricky due to cloud cover, so combining optical data with SAR data is essential. This paper presents an object-based machine-learning procedure that utilizes Synthetic Aperture Radar (SAR) from Sentinel-1 and optical data from Sentinel-2 data to rapidly map recent volcanic deposits in the 2022 Semeru Volcano eruption. This study aims to test the algorithm's performance in identifying recent volcanic deposits during the eruption. NDVI (Normalized Difference Vegetation Index) and NHI (Normalized Hotspot Index) spectral index changes combined with Sentinel-1 SAR information in the form of interferometric coherence, VV and VH beta nought values, and polarimetric decomposition SAR features are used as model inputs. The result indicated that combined SAR and optical data use in an object-based Random forest algorithm provides good recent volcanic deposit detection. The algorithm's performance detecting the recent volcanic deposit in the Semeru Volcano case shows a good Dice Coefficient of 0.73 and an F1-score of 0.83. This straightforward algorithm will improve the rapid mapping process during and after volcanic eruptions.

Keywords: Volcanic Deposit Mapping · Multi-Sensor · Object Based-Random Forest

1 Introduction

Disasters caused by volcanic activity are one of the most dangerous disasters in the world. Volcanic hazards, namely lava flows, pyroclastic density current, Volcanic gas emissions, and volcanic material ejection [1], potentially affect society's livelihoods, damage the agricultural and infrastructure [2], and even cause human death [3]. With the tendency for the population around volcanoes to increase faster than the average global population growth [4, 5], the impact caused by volcanic eruptions is also increased even by activities with a small scale (VEI < 3) [6].

E. Fitrianingsih et al. (Eds.): ISAST 2024, SPPHY 416, pp. 105–112, 2025.
https://doi.org/10.1007/978-981-96-1344-1_11

Developments in volcanology have resulted in a better understanding of the character or behaviour of eruptions, recognition of precursors, and eruption forecasting [7], which have been able to reduce eruption event fatalities since the 18th century [8]. Besides that, a quick assessment of the impact of an eruption plays a vital role in a quick response during an emergency to provide information to the authorities in making policies [9, 10].

Spatial information on areas affected by eruptions is crucial when disasters happen. This information can be obtained through ground-based monitoring [11] or a more sophisticated way using UAVs to collect data on areas affected by eruptions [12, 13]. However, the lava field during the eruption is difficult to approach. Consequently, both approaches are complicated. Therefore, remote sensing data can be an alternative way of providing time-relevant, large-scale, regular, and safe data for humans to map the impact of an eruption [14] quickly.

The volcanic material in a lava flow, pyroclastic (sand, lapilli, bomb) current, and volcanic ash ejected by volcanoes usually lead to land surface changes. Land surface changes after the volcanic eruption are conventionally detected through optical remote sensing and SAR data simply through visual interpretation. A study on optical data fusion and SAR for detecting post-eruption land cover changes is essential, especially for volcanoes in Indonesia, where detection with optical images is often limited by cloud coverage and even volcanic ash [15]. SAR data is also important because of the limitations of optical imagery in detecting new material over older material due to the low contrast between the object and the background [16]. With its sunlight-independent operations and ability to penetrate clouds [17], SAR is a good alternative for obtaining rapid information about the impact of volcanic eruptions.

However, using SAR data also has challenges, such as salt-and-paper or speckle noise disadvantages. The speckle noise, inherently present in SAR data, often leads to the detection of false changes [18]. The object-based approach is an alternative that can be applied to address this problem. Through this research, we propose an object-based and pixel-based Random forest classification approach to detect post-eruption surface changes and try to compare the accuracy of the detection generated by those two approaches.

2 Materials and Method

2.1 Location

Semeru. Semeru volcano is located on the most populated island in Indonesia. Since 1967, it has been characterized by strombolian and vulcanian explosions, lava flow, pyroclastic flow, lahars, and dome collapse. Its explosive eruptions often result in fatalities among communities and losses of the infrastructure around the closest proximity region to the summit.

The event we are investigating occurred on 4 December 2022. CVHGM reported an eruption at 02.46 WIB, with the ash column height reaching 1500 m above the summit. The source of the avalanche comes from a pile of material at the tip of the lava tongue, which is about 800 m from the summit (Jonggring Seloko Crater). The avalanche flow continued until 06.00 WIB and reached 7 km from the summit towards Besuk Kobokan in the southeast direction. Pyroclastic flows reached 13 km from the summit due to heavy rains. These eruptions could have been triggered by heavy rains that eroded the lava dome, leading to its collapse and consequent major explosion [19].

Sinabung. Sinabung Volcano is located in North Sumatra, Indonesia. The volcano had its first confirmed Holocene eruption in 2010. Several activities have been recorded since 2010, which included lava flow, pyroclastic flow, block avalanches, and multiple explosions with ash plumes. The considered eruption was reported each month from March to May 2021. MAGMA Indonesia recorded several explosions accompanied by a column of ash 1000 m above the summit and drifted to the East and Southwest. Centre of Volcanology and Geological Hazard Mitigation (CVGHM, also known as PVMBG) reported a series of pyroclastic flows travelling to the east and southeast up to 3.7 km from the summit [20].

2.2 Standardization of Data

Sentinel-1 SAR data and Sentinel-2 A/B multispectral data were used in the analysis process. The Sentinel satellite is a mission launched by the European Space Agency (ESA) for earth observation, and currently, data from both satellite missions can be downloaded free of charge via the Copernicus web open access hub (https://dataspace.copernicus.eu).

The 90-day pre-event and last available post-event data was used to construct the Normalized Difference Vegetation Index (NDVI) composite. NDVI was calculated using bands 4 (red) and 8 (near infrared). Sentinel-2 optical data was also calculated to derive the normalized hotspot index (NHI) [21] to identify the presence of thermal anomalies.

Sentinel-1 data was downloaded in Interferometric Wide (IW) swath format, Single Look Complex (SLC), which covers the three post-event and three pre-event scenes. Sentinel-1 data was pre-processed using the sentinel-1 Toolbox (S1TBX) on the Sentinel Application Platform (SNAP). As input to machine learning, we derived several Sentinel-1 SAR information in interferometric coherence, VV and VH beta nought values, and polarimetric decomposition SAR features [18].

2.3 SLIC Segmentation and Random Forest Image Classification

Simple linear iterative clustering (SLIC) is an adaptation of k-means for superpixel generation. SLIC clusters pixels in the combined five-dimensional color and image plane space to efficiently generate compact, nearly uniform superpixels. SLIC works with two important distinctions of parameter setting: the number of distance calculations of N-pixels, and weighted distance provides control over superpixel size and compactness [22].

The classification process was done using the random forest classifier non-parametric method. This method used a decision tree set, where each tree depends on the pixel value from each vector that was randomly and inde-pendently taken [23]. The random forest does not tend to overfit and can run the process rapidly, so it is possible to process as many trees as desired [24]. The random forest can also handle thousands of inputs and evaluate each variable's importance value. Random forests can be used to rank the importance of variables in regression or classification problems using a significance metric known as

Mean Decrease Impurity [23]. MDI is based on the total decrease in node impurity from splitting on the variable, averaged over all trees. Therefore, this classification method can effectively classify the remote sensing data. We used spectral indices derived from Sentinel-2 images (Normalized Difference Vegetation Index and Normalized Hotspot Index) combined with Sentinel-1 Synthetic Aperture Radar features (interferometric coherence, VV and VH beta nought values, and polarimetric decomposition) as input model.

We visually interpreted Sentinel-2 images and selected the segments as a training sample. Approximately 0.31% (1,159 out of 362,599) were chosen as the samples. 70% of the samples were employed as training points, while the remaining 30% were used as validating points (Ntree = 100). We labeled training segments as change and non-change segments, which were randomly stratified.

2.4 Accuracy Assessment

To measure accuracy through an Area-based assessment, we compared it with references from visual interpretation of Sentinel-2, Sentinel-1, and Landsat 8 and Landsat 9 imageries before and after the incident. We used $F1 - Scored$ point-based accuracy in Eq. (1) assessment and area-based spatial assessment Dice Coefficient (DC), Eq. (2) to measure the area of overlap between products and references. The Omission Error (OE), Eq. (3), and Commission Error (CE) in Eq. (4) were also calculated to see the source of error in each case. As a reference, the dataset generated from the visual interpretation was performed by considering visual changes in optical and SAR data after the eruption event.

$$F1 - Scored = \frac{True\ Positive}{True\ Positive + \frac{1}{2}(False\ positive + False\ Negative)} \tag{1}$$

$$DC = 2 \times \frac{Area\ of\ Overlap}{Total\ Area} \tag{2}$$

$$OE = \frac{false\ negatif}{total\ area} \tag{3}$$

$$CE = \frac{false\ positif}{total\ area} \tag{4}$$

3 Result and Discussion

We used satellite data as soon as possible after the incident to test whether the available free, open-source data could be utilized for rapid post-eruption mapping purposes. The two eruption cases investigated demonstrate the algorithm's ability to detect volcanic material deposits under two conditions. Large areas of unvegetated older volcanic material background were found before the eruption in the case of Sinabung, and narrow channels with vegetation land cover were found in the case of Semeru.

Figure 1 shows that the algorithm can detect the recent material distribution in both cases. Major eruptions at Sinabung were reported to have occurred throughout March

2021. According to the [20], on 2 March 2021, pyroclastic flows traveled 2–5 km from the summit to the ESE flank. Semeru eruption shows a similar type of eruption material to Sinabung but moves in narrow valley topography. A series of pyroclastic flows were detected on 4 December as materials collapsed on the SE flank. These flows primarily proceeded 5–13 km SE and S and as far as 19 km in the exact directions [20].

The object-based random forest algorithm performs relatively well in rapidly identifying the recent volcanic deposit of the Sinabung case. An accuracy assessment shows an F1-scored value of 0.85 and a DC value of 0.75. The Semeru case classification shows lower F1-Scored and DC values of 0.83 and 0.73, respectively (Table 1). The RF model produced relatively better DC values than other methods at the exact location [18]. Unlike the point-based F1-scored value, the DC value was obtained from comparing overlapping areas. The DC value in this method shows a low value, especially since the method used still lacks detection in the boundary area between deposits and non-deposits.

Table 1. Area-based accuracy assessment of object-based approach.

Location	F1-Scored	DC	OE	CE
Semeru	0.83	0.73	0.3	0.23
Sinabung	0.85	0.75	0.2	0.05

In the case of Sinabung, the detection errors were mainly caused by the lack of detection around the boundary area between the recent deposit and the older volcanic materials (OE = 0.2 and CE = 0.05). Meanwhile, the values of OE and CE in the case of Semeru were respectively 0.3 and 0.23. The detection errors in the Semeru case were mainly found in the boundary area, and errors were also found in the narrow channel topography. The limited spatial resolution of the imagery used may cause this underestimation. Over-detection (CE) was found in the non-vegetated areas, always covered by clouds around the summit.

These results demonstrated the advantage of SAR data in detecting land surface changes due to overlaps on top of older material. In the case of Sinabung, the recent material that erupted mostly covered the older material, as stated by [26]. It is a challenge for optical sensors to detect volcanic deposits properly, mainly if the available images are obtained long after the eruption. Using SAR coherence parameters and SAR polarimetric decomposition provides a better way to detect recently erupted material covering older material. Using object-based classification, one superpixel object can be created from the properties of a group of related pixels from complex SAR data. This reduces the speckle effect, which improves the classification of volcanic deposits [27].

However, in general, the classification carried out on both eruption cases obtained underestimated results. This underestimation has already been reported by [17]. The involvement of the last pixel available of the NDVI mosaic as input has weaknesses for rapid mapping. The ever-present clouds around the summit affect the number of detection errors. We will consider including post-event cloud-free and cloud-shadow-free optical data for future development to improve change detection. The other development is to

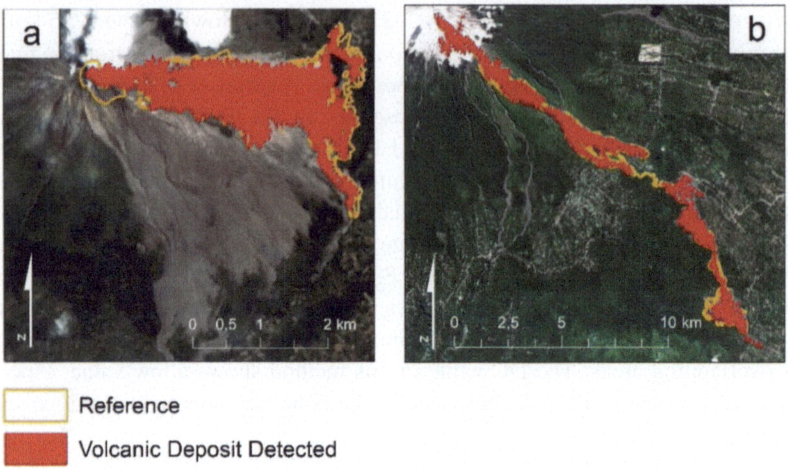

Fig. 1. Random forest results, a are the results for the object-based approach on (left) Sinabung volcano and (right) Semeru volcano.

study the performance of the developed RF training model transfer on new volcanic deposits in future eruptions. The variety of eruption types in the 125 active volcanoes in Indonesia provides excellent potential for carrying out this test in various types of eruptions and types of material ejected. The accuracy of the model transfer and the models built can help the Indonesian Government respond quickly after a large eruption.

Exploration of combined sentinel-1 SAR and Sentinel-2 MSI data for rapid mapping after volcanic eruptions is urgently needed in tropical countries like Indonesia. The existence of cloud cover, especially during the rainy season, has always been a significant obstacle to obtaining near real-time information in quick response to disaster management.

4 Conclusion

In this study, we explore an object-based random forest classification approach utilizing synthetic aperture radar (SAR) and optical data to rapidly map recent volcanic deposits in Semeru Volcano and Sinabung Volcano eruptions. We use SLIC segmentation to create superpixel objects so that random forest classification runs segment-based. The algorithm's performance detecting the recent volcanic deposit in both cases shows good accuracy. The use of SAR data helps in rapid mapping when optical data is not available. Future development is required to improve the accuracy in boundary areas, peak areas, and areas with narrow channel topographic dimensions. This straightforward algorithm will improve the rapid mapping process during and after volcanic eruptions.

References

1. Auker, M.R., Sparks, R.S.J., Siebert, L., Crosweller, H.S., Ewert, J.: A statistical analysis of the global historical volcanic fatalities record. J. Appl. Volcanol. **2**, 2 (2013). https://doi.org/10.1186/2191-5040-2-2

2. Brown, S.K., Jenkins, S.F., Sparks, R.S.J., Odbert, H., Auker, M.R.: Volcanic fatalities database: analysis of volcanic threat with distance and victim classification. J. Appl. Volcanol. **6** (2017). https://doi.org/10.1186/s13617-017-0067-4
3. Doocy, S., Daniels, A., Dooling, S., Gorokhovich, Y.: The human impact of volcanoes: a historical review of events 1900–2009 and systematic literature review. PLoS Curr. (2013). https://doi.org/10.1371/currents.dis.841859091a706efebf8a30f4ed7a1901
4. Barclay, J., et al.: Livelihoods, wellbeing and the risk to life during volcanic eruptions. Front. Earth Sci. (Lausanne) **7** (2019). https://doi.org/10.3389/feart.2019.00205
5. Freire, S., Florczyk, A.J., Pesaresi, M., Sliuzas, R.: An improved global analysis of population distribution in proximity to active volcanoes, 1975–2015. ISPRS Int. J. Geoinf. **8** (2019). https://doi.org/10.3390/ijgi8080341
6. Mani, L., Tzachor, A., Cole, P.: Global catastrophic risk from lower magnitude volcanic eruptions. Nat. Commun. **12** (2021)
7. Sparks, R.S.J.: Journal of volcanology and geothermal research. Phys. Earth Planet. Inter. **13**, 161 (1976). https://doi.org/10.1016/0031-9201(76)90082-0
8. Donovan, A., Oppenheimer, C., Bravo, M.: Bull. Volcanol. **74**, 1005–1022 (2012). https://doi.org/10.1007/s00445-012-0581-5
9. Carr, B.B., Clarke, A.B., Vanderkluysen, L., Arrowsmith, J.R.: Mechanisms of lava flow emplacement during an effusive eruption of Sinabung Volcano (Sumatra, Indonesia). J. Volcanol. Geoth. Res. **382**, 137–148 (2019). https://doi.org/10.1016/j.jvolgeores.2018.03.002
10. Andreastuti, S., Paripurno, E.T., Gunawan, H., Budianto, A., Syahbana, D., Pallister, J.: Character of community response to volcanic crises at Sinabung and Kelud volcanoes. J. Volcanol. Geoth. Res. **382**, 298–310 (2019). https://doi.org/10.1016/j.jvolgeores.2017.01.022
11. Turner, N.R., Perroy, R.L., Hon, K.: Lava flow hazard prediction and monitoring with UAS: a case study from the 2014–2015 Pāhoa lava flow crisis Hawai'i. J. Appl. Volcanol. **6** (2017). https://doi.org/10.1186/s13617-017-0068-3
12. Darmawan, H., Walter, T.R., Troll, V.R., Budi-Santoso, A.: Structural weakening of the Merapi dome identified by drone photogrammetry after the 2010 eruption. Nat. Hazards Earth Syst. Sci.Sci. **18**, 3267–3281 (2018). https://doi.org/10.5194/nhess-18-3267-2018
13. Gomez, C., et al.: LiDAR and UAV SfM-MVS of Merapi volcanic dome and crater rim change from 2012 to 2014. Remote Sens. (Basel) **14** (2022). https://doi.org/10.3390/rs14205193
14. Cigna, F., Tapete, D., Lu, Z.: Remote sensing of volcanic processes and risk. Remote Sens. (Basel) **12** (2020)
15. Aldeghi, A., Carn, S., Escobar-Wolf, R., Groppelli, G.: Volcano monitoring from space using high-cadence planet CubeSat images applied to Fuego volcano, Guatemala. Remote Sens. (Basel) **11** (2019). https://doi.org/10.3390/rs11182151
16. Corradino, C., Ganci, G., Cappello, A., Bilotta, G., Hérault, A., Del Negro, C.: Mapping recent lava flows at Mount Etna using multispectral Sentinel-2 images and machine learning techniques. Remote Sens. (Basel) **11** (2019). https://doi.org/10.3390/rs11161916
17. Bignami, C., Chini, M., Amici, S., Trasatti, E.: Synergic use of multi-sensor satellite data for volcanic hazards monitoring: the Fogo (Cape Verde) 2014–2015 effusive eruption. Front. Earth Sci. (Lausanne) **8** (2020). https://doi.org/10.3389/feart.2020.00022
18. Orynbaikyzy, A., Plank, S., Vetrita, Y., Martinis, S., Santoso, I.: Joint use of Sentinel-2 and Sentinel-1 data for rapid mapping of volcanic eruption deposits in Southeast Asia. Int. J. Appl. Earth Obs. Geoinf. **116**, 103166 (2023). https://doi.org/10.1016/j.jag.2022.103166
19. Handley, H.: Mount Semeru's deadly eruption was triggered by rain and storms, making it much harder to predict (2021)
20. Global Volcanism Program: Report on Sinabung (Indonesia). Bulletin of the Global Volcanism Network (2021)

21. Genzano, N., Pergola, N., Marchese, F.: A google earth engine tool to investigate, map and monitor volcanic thermal anomalies at global scale by means of mid-high spatial resolution satellite data. Remote Sens. (Basel) **12**, 1–22 (2020). https://doi.org/10.3390/rs12193232
22. Achanta, R., Shaji, A., Smith, K., Lucchi, A., Fua, P., Süsstrunk, S.: SLIC superpixels (2010)
23. Breiman, L.: Random forests. Mach. Learn. 5–32 (2001)
24. Cutler, D.R., et al.: Random forests for classification in ecology. Ecology **88**, 2783–2792 (2007). https://doi.org/10.1890/07-0539.1
25. Sennert, S. (ed.): Global Volcanism Program. Report on Semeru (Indonesia). Weekly Volcanic Activity Report (2022). 30 November–6 December 2022
26. Corradino, C., Amato, E., Torrisi, F., Del Negro, C.: Towards an automatic generalized machine learning approach to map lava flows. In: International Workshop on Cellular Nanoscale Networks and their Applications. IEEE Computer Society (2021)
27. Santoso, I., Ismanto, R.D., Chusnayah, F., Tjahjaningsih, A., Suwarsono, Vetrita, Y.: Lava flow mapping Karangetang Volcano during 2019 eruption using Sentinel-2 Images and Random Forest model. In: IOP Conference Series: Earth and Environmental Science. Institute of Physics (2022)

Effect of Atmosphere Correction for Unsupervised Classification in Multispectral Camera LAPAN A3

Ega Asti Anggari[1,2](✉), Sigit Heru Murti[2], and Wahyudi Hasbi[2]

[1] Department of Information Geographic Science, Faculty of Geography, Universitas Gadjah Mada, Yogyakarta, Indonesia
egaastianggari@mail.ugm.ac.id

[2] Research Center for Satellite Technology, BRIN, Bogor, Indonesia

Abstract. On their way to the Earth's surface, electromagnetic waves emitted by the Sun are subject to atmospheric distortion in the form of refraction, scattering, and absorption events. As a remote sensing satellite, LAPAN A3 imagery is not immune to atmospheric disturbance. To produce images free of atmospheric disturbance, corrections are needed to remove atmospheric effects. In this study, a comparison was made between three atmospheric corrections, namely histogram adjustment, FLAASH and 6S. The research was conducted in the Batu Jamus rubber plantation area owned by PT Perkebunan Nusantara I Regional 3 and the Karanganyar area. Atmospherically corrected images were used for unsupervised classification using the ISODATA method. Visual analysis showed that atmospheric correction of the LAPAN A3 image resulted in a brighter image in the FLAASH method. Spectral reflectance curve analysis shows that the histogram adjustment method has a similar spectral reflectance shape to the reference. The classification results show no change in the histogram adjustment method, which is indicated by the accuracy value reaching 100%. On the other hand, for the FLAASH and 6S methods, there is a change in the classification, indicated by a decrease in accuracy of 44.0514% and 48.8370%, respectively.

Keywords: LAPAN A3 · Atmospheric Correction · Atmospheric Disturbance · Unsupervised Classification

1 Introduction

LAPAN A3 is a microsatellite developed by the Research Center for Satellite Technology, BRIN. The LAPAN A3 has been designed to monitor the Earth. To support this mission, LAPAN A3 carries a multispectral camera payload commonly called the Line Imager Space Application (LISA). The LISA has a spatial resolution of 15, which can be applied to monitor agricultural activities, droughts, and land cover classification [1].

In order to employ LISA data for a broader range of applications, pixel information must be converted to surface reflectance. Nevertheless, the pixel value in LISA still contains information in the Digital Number (DN). To convert the DN value to surface

E. Fitrianingsih et al. (Eds.): ISAST 2024, SPPHY 416, pp. 113–122, 2025.
https://doi.org/10.1007/978-981-96-1344-1_12

reflectance, it is first necessary to convert the pixel value to radiance and then to the reflectance in the atmosphere. This process continues until the final conversion to surface reflectance is complete. The conversion of DN to radiance involves multiplying the DN values by a radiance constant [2]. The radiance value can be converted to atmospheric reflectance with the sun zenith angle and solar irradiance information in the satellite metadata. To obtain the surface reflectance value from the atmospheric reflectance value, atmospheric correction is required.

Atmospheric correction is a method used to eliminate disturbances in the atmosphere. Atmospheric disturbances can be classified into three main categories: refraction, scattering, and absorption phenomena. The presence of refraction causes the bending of light waves because they pass through different mediums. When light hits particles in the atmosphere, the light will scatter irregularly. This event is called scattering. Absorption is an event where radiation energy from light will be absorbed and converted into another form of energy. This process can result in the loss of information regarding the object's reflection, as observed by satellites.

Numerous researchers have conducted extensive research on atmospheric correction [3–6]. For example, Muchsin (2015) applied the FLAASH module to perform atmospheric correction on Landsat 7 data in Jakarta, comparing urban and maritime models. The study successfully mitigated atmospheric effects, aligning the spectral patterns with Landsat data. In another study [5], researchers explored atmospheric correction methods for Landsat 8, comparing FLAASH, 6S, DOS, TOA+BRDF, and TOA methods. Their findings indicated that FLAASH and 6S methods exhibited similar spectral responses to surface features, validated against surface object spectral responses and NDVI values.

Based on this theory, it is claimed that the results of the atmospheric correction are not needed when used for classification purposes [7]. However, this research aims to validate whether this theory applies to LAPAN A3 satellite data. The data were corrected using the histogram adjustment method, FLAASH, and 6S. The corrected results were then used for land cover classification using the unsupervised ISODATA method.

2 Material and Method

2.1 Data and Location

This study uses LAPAN A3 imagery from the Research Center for Satellite Technology, BRIN. The imagery utilizes data captured on 20 August 2022 at 00.22.02 UTC. Image specifications are outlined in Table 1, while additional parameters required for this study are listed in Table 2. These parameters are obtained from image metadata.

The research was conducted in the Batu Jamus rubber plantation area owned by PT Perkebunan Nusantara I Regional 3 and the Karanganyar area. The atmospheric data used is MODIS data at the same time as the image. The RBI 1:25,000 map was used for image georeferencing. The research location and the research flowchart are illustrated in Fig. 1.

Table 1. LAPAN A3 Band data.

Bands	Wavelength
Blue	410–490
Green	510–580
Red	630–700
NIR	770–990

Table 2. Parameter from satellite metadata.

Parameter	Angle
Solar Zenith Angle	66.5376
Solar Azimuth Angle	73.1532
Sensor Zenith Angle	0
Sensor Azimuth Angle	0

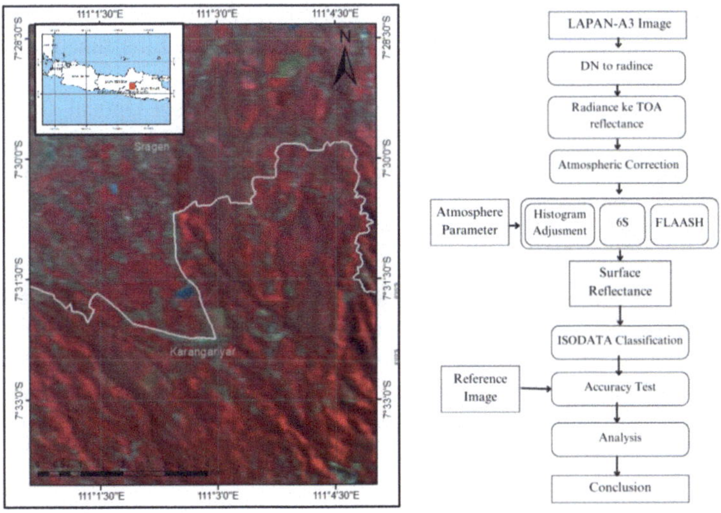

Fig. 1. Study area and flowchart.

2.2 Atmospheric Correction

The pixel information of the LAPAN A3 image is still in Digital Number, so it needs to be converted into radiance value for FLAASH and 6S method input. The algorithm for converting DN to radiance value $L(\lambda)$ is shown in Eq. (1). The radiance coefficient value KA is required, obtained from previous research [2].

Meanwhile, for the histogram adjustment method, the input is in the form of a reflectance value. The algorithm for obtaining the reflectance value $R(\lambda)$ is shown in Eq. (2). $E(\lambda)$ is the solar irradiance whose value is obtained from Thuillier's research [9], π has a value of 3.14159, and cos θ is the solar zenith angle.

$$L(\lambda) = K_A DN \tag{1}$$

$$R(\lambda) = \frac{\pi L(\lambda)}{E(\lambda)\cos\theta} \tag{2}$$

Histogram Adjustment Method. Histogram adjustment assumes the minimum value of each band is 0. If the minimum value is not 0, then the value is atmospheric interference and needs to be removed. The histogram adjustment algorithm is shown in Eq. (3). *Input BV* is the input pixel, the offset is the minimum value of each band, and the *output BV* is the result of the histogram adjustment.

$$output\ BV_{ijk} = input\ BV_{ijk} - offset \tag{3}$$

FLAASH Method. The FLAASH method is an atmospheric correction method using the radiative transfer model from MODTRAN4. FLAASH can be used on multispectral and hyperspectral satellites. The inputs that need to be filled in are the center position of the scene, sensor height, object height, date, and time of image acquisition. The sensor type is filled in with unknown-msi because the LAPAN A3 image is not in the software database, so the image spectral library is input manually in the filter function file menu. The FLAASH parameters are shown in Table 3.

Visibility parameters were obtained by converting Aerosol Optical Depth (AOT) values from Modis information. The Modis data used was the Modis L2 Aerosol 5Min L2 Swath 10km specifically on AOT data at 0.55 micron for both ocean (best) and land (corrected) with best quality data (Quality flag = 3)". The CO2 mixing ratio was taken from model data on the web https://data.giss.nasa.gov/modelforce/ghgases/Fig.1A.ext.txt, which is added by 20.

6S Method. The 6S is an atmospheric correction method that utilizes a radiative transfer model on electromagnetic signals passing through the atmosphere. In this study, the 6S method was processed using Python using the Py6S module. The 6S method considers various parameters to correct the model, such as atmospheric conditions, surface conditions, satellite conditions, and spectral conditions [6]. The 6S parameters selected in this study are shown in Table 3.

2.3 ISODATA Classification

In this study, the results of atmospheric correction were applied to unsupervised classification. In unsupervised classification, land cover is not previously known due to limited information not properly identified [4]. After the class was classified, the analyst renamed and grouped it into classes. The ISODATA classification method uses a technique for calculating the average of evenly distributed classes and then repeatedly grouping pixels with a minimum distance technique [7]. In this method, it is possible to merge clusters if they do not meet the threshold criteria and separate clusters if they exceed the threshold criteria.

Table 3. FLAASH and 6S parameters used.

Parameter FLAASH	Description	Parameter 6S	Description
Atmospheric model	Tropical	Atmospheric Profile	Tropical
Water retrieval	No	Aerosol Profile	Urban
Water column multiplier	1	Ground Reflectance	Homogeneous Lambertian
Aerosol Model	Urban	visibility	12 km
Aerosol Retrieval	2-Band K-T (upper KT NIR, lower KT Blue	Sensor altitude	Sensor satellite level
Initial Visibility	12 km	Target altitude	0.35 km
CO2 mixing Ratio	425	spektral	Depending on the wavelength band used (μm)
Modtran resolution	15 cm^{-1}	geometrical settings	According to image metadata

3 Result and Discussion

3.1 Visual Analysis Before and After Correction

The images were analyzed visually before and after correction (Fig. 2). The result shows that there are no significant visual differences. However, in the FLAASH method, the image appears brighter than relative to the pre-correction state. In contrast, no visual difference was evident in the images subjected to other correction methods. After correction, the atmospheric effect will be reduced, thereby enhancing the quality [8].

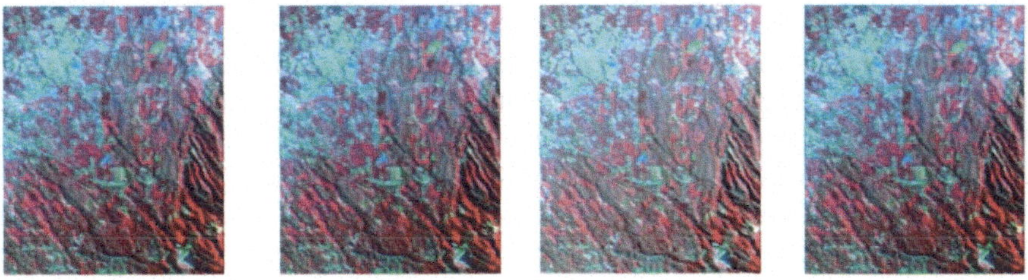

(a) Before correction (b) Histogram adjustment (c) FLAASH (d) 6S

Fig. 2. Image before correction (a) and after correction (b), (c), (d).

3.2 Spectral Reflectance Curve Analysis

The spectral curves of each land cover from each correction method were compared with the reference image curve. Afterward, the method that most closely approximates the reference image was identified. The reference uses data from the USGS [10] because there is no field reflectance data. The land cover reflectance curve can be seen in Table 4.

The water object is the first object to be discussed. In the histogram adjustment and 6S methods, the shape of the water curve is similar to that of the reference. The curve will increase drastically from the blue band to the green band. After reaching the peak, the value will decrease in the red band, and the curve will continue to decrease until the NIR band. This is different from the shape in 6S, which increases drastically from the blue band to the green band. From the green band, the curve rises slightly to the red band and continues to the NIR band.

Continued on the bareland object where the reference increases from 400–800 nm wavelength, the curve decreases at 800–1000 nm wavelength. Then, after that, the value tends to increase. Judging from the curve pattern, the histogram adjustment and 6S methods have the same pattern as the reference. The difference is that in the histogram adjustment, the curve increases sharply in the NIR band, while in 6S, the curve increases sharply in the blue band; after that, the curve continues to increase until NIR.

On the rubber object, the reference curve increases slightly in the green band and then decreases in the red band, and then the curve increases sharply in the NIR band. We found that all methods have similar patterns from the curves, but the histogram adjustment has the closest pattern to the reference. The reference spectral curve on other vegetation is similar to rubber, and the results of the three correction methods are similar to the correction results on rubber. The reflectance values are not much different from those of other vegetation and rubber.

There are discrepancies between the roof spectral curve in the three corrections and the reference. In the reference, the reflectance value increases continuously from a reflectance of 400–1000 nm linearly. However, the correction results show a curve on the roof similar to rubber and other vegetation. This is because the sample is not a pure pixel containing only roof information. The LAPAN A3 pixel has a value of 15 m, so the area of a pixel is 225 m^2. For that area in Indonesia, it is rare for a building to have an area that large. So, it is very likely that the roof was mixed with other objects in one pixel. The objects whose spectral curves are observed are water, bare land, rubber, other vegetation, and roof. The adjustment histogram generally has a spectral reflectance shape similar to the reference. In the FLAASH method, the shape of the curve is similar before correction, but the reflectance value is reduced. While in 6S, the reflectance value is generally also reduced compared to before correction, the value in the blue band is too low compared to the other bands. The results of the atmospheric correction using the FLAASH and 6S methods are highly dependent on the parameters selected. The parameters of the research location, satellite conditions, and atmospheric conditions, when the satellite records the image affect the correction results. So, choosing the suitable parameters will make a good model for atmospheric correction. Inappropriate parameters still cause the correction results to be different from the reference.

Table 4. The spectral curve of land cover.

Object	LAPAN A3				Reference
	Before	Histogram Adjustment	FLAASH	6S	
Water	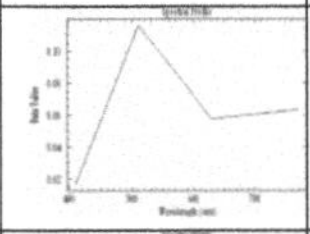	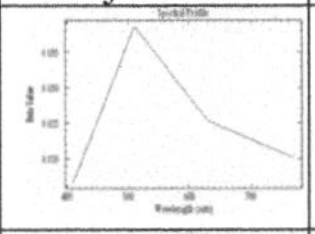	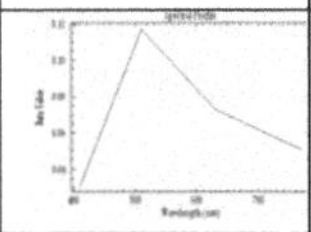	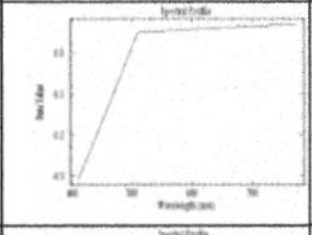	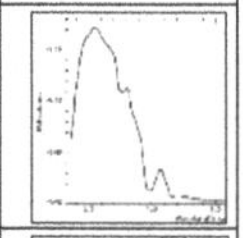
Bareland	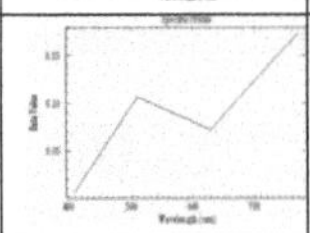	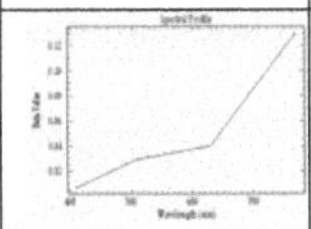	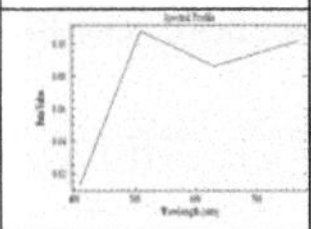	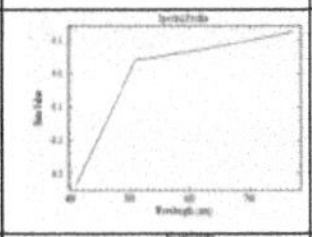	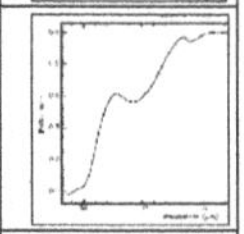
Rubber	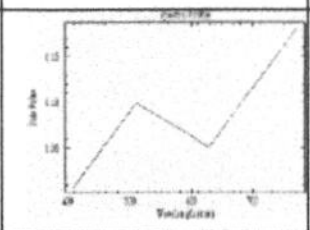	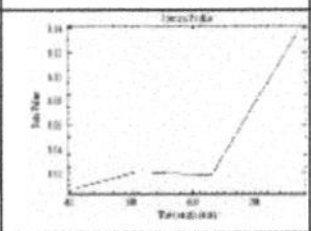	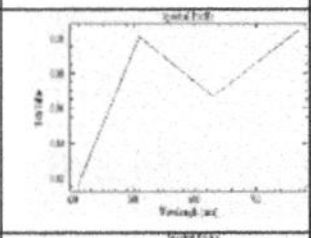	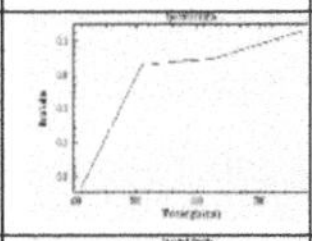	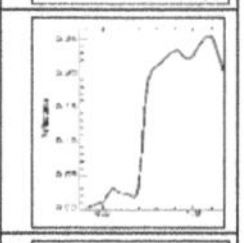
Vegeta-tion	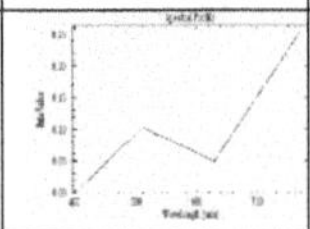	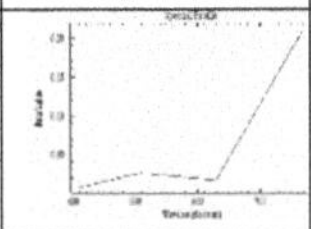	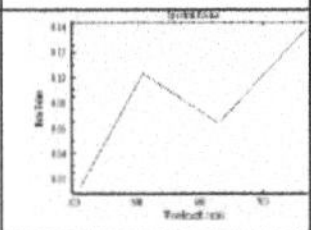	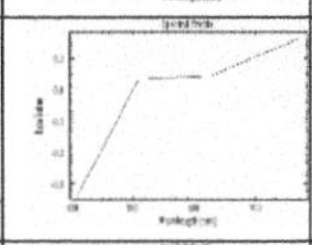	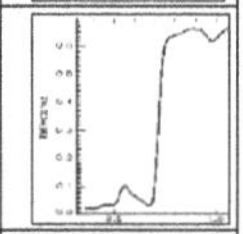
Roof	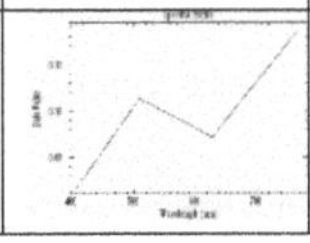	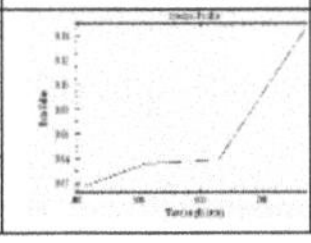	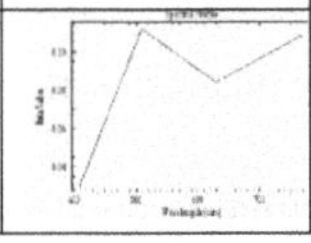	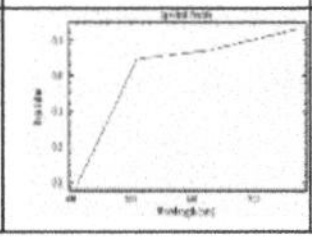	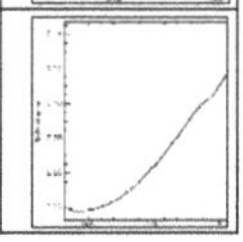

3.3 ISODATA Classification

The correction results were then used to classify the rubber plantation area and its surrounding environs using the unsupervised ISODATA method. This method was used because it can classify without training samples but only based on pixel values. The classification was applied to uncorrected images and corrected images. Each image was divided into several classes between five and ten, with ten iterations. The result is the formation of five classes in the image before and after correction. The identified classes include water, vegetation, roof, rubber, and bareland.

The classification results visually show no difference between the classification results before correction and the corrections in histogram adjustment. This occurs because the reflectance value scale remains the same before and in the histogram adjustment correction. However, the classification results on the FLAASH and 6S methods show different results from those before correction. The difference occurs due to changes in the reflectance value scale of the FLAASH and 6S correction results. The classified image can be seen in Fig. 3.

Quantitatively, the classification results before correction compared after correction can be seen in the class distribution summary table (Table 5) and accuracy (Table 6). In the class distribution, it can be seen that there is no difference in the distribution before correction and histogram adjustment. This is reinforced by the accuracy value reaching 100%. So that the histogram adjustment correction does not change the classification results even though it has been corrected. In the FLAASH method, the class distribution is different from before correction. The distribution values of the water, vegetation, and roof classes decreased compared to before the correction. In contrast, the class distribution of rubber and bare land increased compared to before the correction. The accuracy between before correction and FLAASH also decreased drastically to 44.0514%. The results of the class distribution on 6S are also in line with FLAASH. Its accuracy compared to before correction only reached 48.8370%.

Table 5. Class distribution summary.

	Before correction (%)	Histogram adjustment (%)	FLAASH (%)	6S (%)
water	7.261	7.261	6.946	8.810
vegetation	35.511	35.511	25.336	25.475
roof	34.290	34.290	25.767	33.030
rubber	17.343	17.343	30.942	24.549
bareland	5.594	5.594	11.011	8.136

Table 6. Accuracy image before correction against after correction.

	Before correction	Histogram adjustment	FLAASH	6S
Before correction	100.00	100.00	44.0514	48.8370
Histogram adjustment	100.00	100.00	44.0514	48.8370
FLAASH	44.0514	44.0514	100.00	82.2055
6S	48.8370	48.8370	82.2055	100.00

3.4 Discussion

The developing theory states that atmospheric correction is unnecessary for land cover classification and identification. This study aims to ascertain whether the theory applies to LAPAN A3 satellite data. The results show no difference in classification results using the histogram adjustment method. However, the FLAASH and 6S methods have different classification results. The difference is mainly due to variations in the reflectance value scale in the two methods. There are several points that we are concerned about in improving the classification results that have not been carried out in this study, such as the spectral reflectance sample of objects in the field, topographic correction, and accuracy testing with land cover references in the field. The spectral reflectance value of the object should use the value measured in the field at the same time as the image recording time. This is useful for increasing the spectral confidence of the object in the image. The image used has a difference in height on the southeast side. This affects the classification results, so topographic correction is needed in the image. To test the classification results, land cover data from the field is needed so that it is known whether the resulting classification is correct or needs improvement.

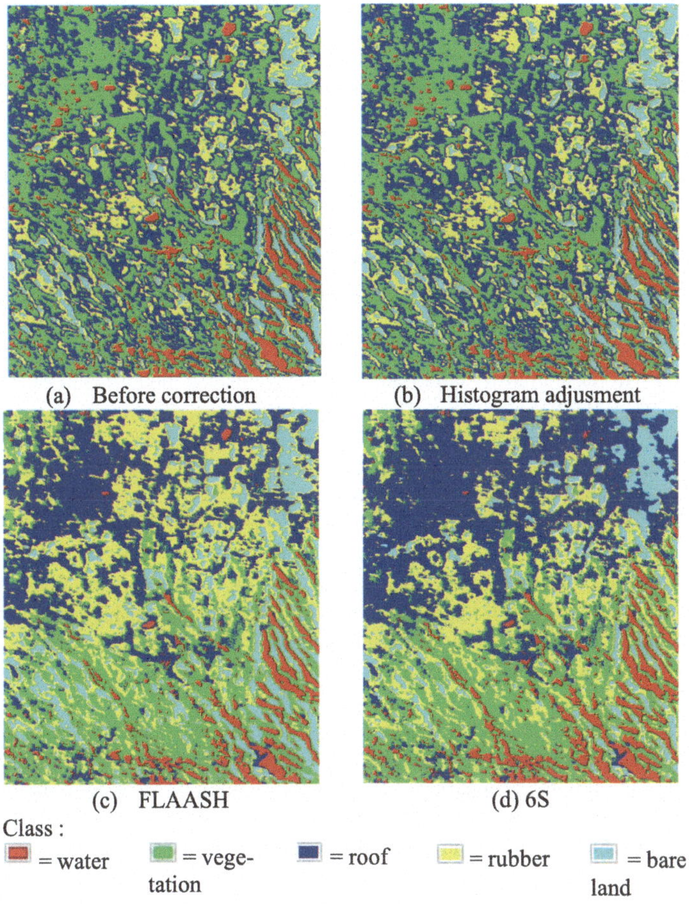

Fig. 3. Classification result before correction (a) and after correction (b), (c), (d).

4 Conclusion

LAPAN A3 image produces brighter images using the FLAASH method. The spectral reflectance curve analysis shows that the histogram adjustment method has a spectral reflectance shape similar to the reference. In the FLAASH method, the curve shape is similar before correction, but the reflectance value is reduced. While in 6S, the reflectance value is generally also reduced compared to before correction, the value in the blue band is too low compared to other bands. The classification results show no change in the histogram adjustment method, as indicated by the accuracy value of 100%. Meanwhile, in the FLAASH and 6S methods, there is a change in classification, which is indicated by a decrease in accuracy of 44.0514% and 48.8370%. To improve classification results, it is necessary to take spectral reflectance samples of objects in the field, topographic correction, and accuracy tests with land cover reference in the field.

References

1. Anggari, E.A., Hakim, P.R., Herawan, A., Salaswati, S., Hasbi, W.: Relationships of class number variation and image classification accuracy in the LAPAN-A3 multispectral imager. In: Proceedings of the 2022 IEEE International Conference on Aerospace Electronics and Remote Sensing Technology, ICARES 2022 (2022). https://doi.org/10.1109/ICARES56907.2022.9992294
2. Salaswati, S., Hakim, P.R., Syafrudin, A.H., Hartono, R., Utama, S.: Vicarious radiometric calibration of Lapan-A3/IPB satellite multispectral imager in Jaddih Hill Madura. J. Aerosp. Technol. **20**, 31–42 (2020)
3. Muchsin, F.: Koreksi atmosfer citra Landsat-7 menggunakan modul ENVI FLAASH. Pros. SNSA **24**, 93–100 (2015)
4. Kondraju, T.T., Mandla, V.R., Chokkavarapu, N., Peddinti, V.S.S.: A comparative study of atmospheric and water column correction using various algorithms on Landsat imagery to identify Coral Reefs. Reg. Stud. Mar. Sci. **49**, 102082 (2022)
5. Muchsin, F., et al.: Comparison of the radiometric correction of Landsat-8 images based on object spectral response. Int. J. Remote Sens. Earth Sci. **18**, 177–188 (2022)
6. Sartika, S., et al.: Determining the precision of spectral patterns arising from atmospheric correction utilizing MODTRAN-FLAASH and 6S approaches on high-resolution SPOT6 imagery. In: 2023 IEEE International Conference on Aerospace Electronics and Remote Sensing Technology, ICARES 2023, pp. 1–7 (2023). https://doi.org/10.1109/ICARES60489.2023.10329791
7. Jensen, J.R.: Introductory Digital Image Processing A Remote Sensing Perspective. Pearson Education Limited, Harlow, England (2015)
8. Dewi, E.K., Trisakti, B.: Comparing atmospheric correction methods for Landsat OLI data. Int. J. Remote Sens. Earth Sci. **13**, 105–120 (2016). Base Spectra (splib07a). https://crustal.usgs.gov/speclab/QueryAll07a.php. Accessed 25 July 2024
9. Gueymard, C.A.: The sun's total and spectral irradiance for solar energy applications and solar radiation models. Solar Energy **76**(4), 423–453 (2004). https://doi.org/10.1016/j.solener.2003.08.039
10. Base Spectra (splib07a). https://crustal.usgs.gov/speclab/QueryAll07a.php. Accessed 25 July 2024

Enhancing the Accuracy of Land Subsidence Monitoring: Correlation and Validation of PS-InSAR with GNSS Data Using the Nash-Sutcliffe Efficiency Matrix

Edy Trihatmoko[1(✉)], Joko Widodo[1], Nurlinda[2], and Shinichi Sobue[3]

[1] Research Center for Geoinformatics, National Research and Innovation Agency, Jakarta 10340, Indonesia
edytrihatmoko@gmail.com
[2] Department of Geography, Universitas Negeri Semarang, Semarang 50229, Indonesia
lindassi@students.unnes.ac.id
[3] Japan Aerospace Exploration Agency, Tokyo 181-0015, Japan
sobue.shinchi@jaxa.jp

Abstract. Persistent Scatterer Interferometric Synthetic Aperture Radar (PS-InSAR) is an effective technique for monitoring land subsidence due to its high spatial resolution and capability to measure ground displacement over large areas. However, validating PS-InSAR measurements with Global Navigation Satellite System (GNSS) data is essential to ensure accuracy. This study examines the correlation between PS-InSAR and GNSS measurements by analyzing the proximity of 255 GNSS points to 14,800 PS-InSAR records. To avoid high error rates in assessing land subsidence, PS-InSAR records were validated against 255 GNSS points using the following method: (1) employing a near tool with a 100 m distance criterion from the GNSS points and a vertical deviation threshold of −1 to 1 cm between the GNSS points and the nearest PS-InSAR records within 100 m; (2) applying the Nash-Sutcliffe Efficiency (NSE) matrix to compare the residual variance of the model (PS-InSAR) to the variance of the measured data (GNSS). The data treatment procedures resulted in 21 matching data points from both GNSS and PS-InSAR datasets, with a high NSE value of 0.746, indicating a strong correlation. This approach demonstrates that careful data selection based on proximity and refinement using statistical measures can significantly enhance the reliability of PS-InSAR data for more detailed analysis.

Keywords: PS-InSAR · GNSS · NSE

1 Introduction

Persistent Scatterer Interferometry Synthetic Aperture Radar (PS-InSAR) has become an extensively used method for measuring surface deformation [1–4]. This technique is valued for its ability to conduct planar measurements across extensive areas with low cost and high accuracy [5, 6]. The method examines SAR data over time to identify

E. Fitrianingsih et al. (Eds.): ISAST 2024, SPPHY 416, pp. 123–130, 2025.
https://doi.org/10.1007/978-981-96-1344-1_13

"Persistent Scatterers (PS)," which are points that exhibit consistent radar wave phase changes across various SAR images and do not experience temporal decorrelation [6]. By leveraging PS, it is possible to accurately assess vertical land displacement.

InSAR is a sophisticated high-resolution technique that is particularly sensitive to noise due to its ability to detect very small movements [7]. Consequently, terrestrial validation of interferometric products is essential when applying this method to new study areas. Over the past 15 years, validating PSI data has been a critical area of research [8].

PS-InSAR can generate tens of thousands of deformation data points, whereas GNSS data is considerably more limited, typically only numbering in the hundreds within a city. This disparity in data volume poses a challenge during the PS-InSAR validation process. A common issue is determining the most effective method for subsetting datasets so that PS-InSAR data can be appropriately matched with GNSS ground truth for validation.

Current PS-InSAR validation methods are varied, with insufficient documentation of technical accuracy in existing literature. This study introduces a PS-InSAR validation technique using GNSS, supported by statistical tests to ensure model accuracy. This approach addresses the current gap in validation techniques by providing a robust framework for assessing the accuracy of PS-InSAR data. In this context, this paper employs the near tool and vertical deviation threshold to choose 14,800 PS-InSAR data points, which will then be validated against 255 GNSS data points in the Jakarta area, Indonesia. Using the Nash-Sutcliffe Efficiency (NSE) matrix, a subsetting model is evaluated by comparing the predicted values (PS-InSAR) with the measured values (GNSS). This procedure results in the creation of a reliable technical reference for subsetting PS-InSAR datasets.

2 Material and Methods

2.1 Materials

PS-InSAR Data. The PS-InSAR data utilized in this study consist of 14,800 Persistent Scatterer (PS) points derived from applying the PS-InSAR technique to ALOS-2 PALSAR-2 spaceborne L-band SAR images from 2017 to 2022 in the Jakarta area, Indonesia (Fig. 1). ALOS-2 PALSAR-2 data were obtained from the Japan Aerospace Exploration Agency (JAXA) in single-look complex (SLC) format. SAR time-series data were analyzed to generate an interferogram, which captures changes in radar wave phases due to ground movement. This interferogram was then processed using Sarproz software to identify PS points, which represent surface locations exhibiting consistent radar wave changes throughout the acquisition period.

GNSS Data. The GNSS data used in this study consist of 255 X, Y, and elevation coordinate points distributed evenly across Jakarta. These data were obtained from a Global Positioning System Real-Time Kinematic (GPS RTK) survey conducted in Jakarta in 2021.

2.2 Methods

Near Tool. The Near tool is a spatial analysis tool that measures the distance between specified input features and nearby features. This process generates a new shapefile with

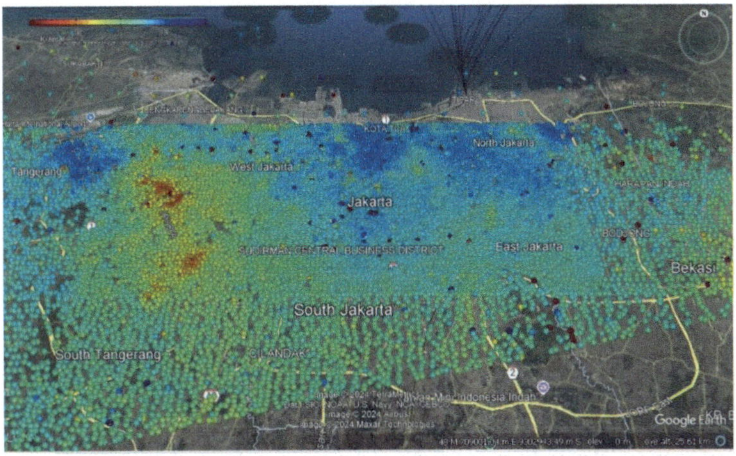

Fig. 1. PS-InSAR data of Jakarta and surroundings area.

selected near features that meet the predefined distance criteria from the input features. The tool is designed to select PS-InSAR data based on spatial proximity (horizontal X, Y filter). By using this tool, subsets of PS-InSAR data that are spatially close to GNSS data are created.

Vertical Deviation Threshold. Both PS-InSAR and GNSS data include elevation attributes. We perform Z-based subsetting using a vertical deviation threshold. This tool selects PS-InSAR data within 100 m of GNSS locations based on elevation differences. It compares the elevation of PS-InSAR data with that of GNSS data. By using this tool, subsets of PS-InSAR data will be created where the elevation deviation from GNSS data is minimal.

NSE. The data produced from the subsetting process is assessed using the NSE matrix. NSE compares the residual variance of the model (PS-InSAR) to the variance of the measured data (GNSS). An NSE value of 1 signifies a flawless alignment between the PS-InSAR model and GPS RTK data, while an NSE value of 0 suggests that the PS-InSAR model is as accurate as the mean of GPS RTK data. The model is defined as Eq. 1.

$$NSE = 1 - \frac{\sum_{t=1}^{T}\left(Q_o^t - Q_m^t\right)^2}{\sum_{t=1}^{T}\left(Q_o^t - \overline{Q}_o\right)^2} \tag{1}$$

$\overline{Q}_o$ represents the mean of the observed discharges, while Q_m signifies the modeled discharge. Q_o^t denotes the observed discharge at time t. Table 1 shows the detailed properties and values of the NSE.

Table 1. NSE properties and values [9].

Properties	Value
Very good	0.75 < NSE < 1.00
Good	0.65 < NSE < 0.75
Satisfactory	0.50 < NSE < 0.65
Unsatisfactory	NSE < 0.50

3 Results

3.1 Proposed Technical Procedure of PS-InSAR Pre-validation Subsetting

Spatial Subsetting Procedure. Our Near tool is configured to operate exclusively within a 100-m distance. GNSS points serve as the Input Features, while PS-InSAR points are the Near Features. This process effectively selected 3,479 points out of 14,800 PS-InSAR points.

There is no scientific foundation for the 100-m distance since similar research has not been conducted before. Initially, we used a 50-m distance, but the NSE results were excessively accurate, and only a few PS-InSAR data points were selected. When we extended the distance to 100 m, the NSE results were satisfactory, and more PS-InSAR data points were included. Increasing the distance beyond 100 m would select even more PS-InSAR data, but it would also increase the potential for errors due to the greater distance from the GNSS points. Therefore, we recommend that 100 m is a suitable distance.

Deformation Subsetting Procedure. The data selected based on spatial proximity is further refined using standard deviation. We apply a vertical deviation threshold ranging from −1 to 1 cm to identify PS-InSAR data (3,479 points) with elevation deviations from GNSS points within this range. This process involves subtracting the PS-InSAR elevations from the GNSS elevations in Microsoft Excel and then filtering the data according to the deviation threshold. This subsetting process resulted in 21 matching points (centimeter) (Table 2).

3.2 Models' Integrity

Twenty-one matching points from the final PS-InSAR subsets were validated using NSE, which yielded a strong modeling result of 0.746 (Fig. 2). This significant result is based on the integration of 21 data points, including both GNSS and PS-InSAR data. The high NSE value highlights the effectiveness of the subsetting techniques in accurately filtering PS-InSAR data, making it highly suitable for validation with the specific GPS RTK data.

The twenty-one matched points are not utilized for subsequent analysis but serve solely to validate the quality of the full set of 14,800 PS-InSAR data points. If the Nash-Sutcliffe Efficiency (NSE) analysis yields a favorable result, the entire dataset of 14,800

Table 2. Valid GNSS and PS-InSAR data.

No.	GNSS	PS-InSAR	No.	GNSS	PS-InSAR
1	−2.90	−1.90	12	−6.10	−5.62
2	−3.30	−2.33	13	−0.90	−0.45
3	−2.00	−1.08	14	−3.30	−2.86
4	−2.00	−1.09	15	−0.90	−0.68
5	−2.50	−1.65	16	−0.90	−0.74
6	−3.30	−2.50	17	−0.90	−0.83
7	−2.80	−2.03	18	−0.90	−0.95
8	−3.25	−2.55	19	−2.00	−2.20
9	−1.90	−1.22	20	−3.10	−3.32
10	−3.78	−3.10	21	−3.70	−4.50
11	−3.78	−3.20			

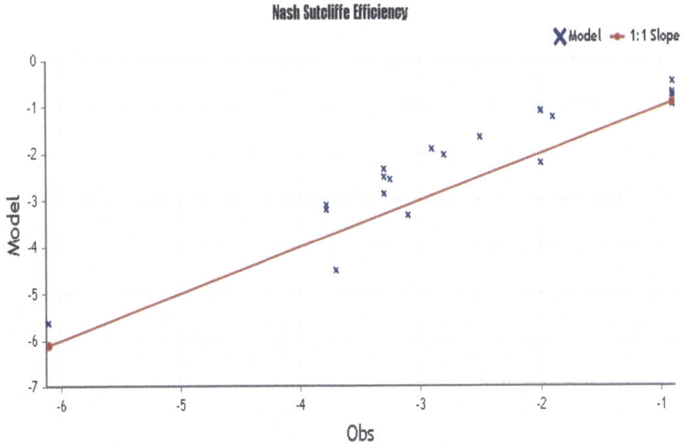

Fig. 2. Nash Sutcliffe Efficiency model result with a 0.746 (very good) value between Ps-InSAR as a model and GNSS data as an observation (Obs).

PS-InSAR points is deemed suitable for land subsidence analysis. Conversely, if the NSE result is suboptimal, the dataset may not provide reliable information for accurate land subsidence analysis.

4 Discussion

4.1 Current Techniques on Pre-validation Treatment

Pixel-Based. [10] only selected InSAR data within a 15 × 15 pixel area around each GNSS site for interferogram correction. Since their primary focus was on using GNSS for interferogram correction, they did not detail the quality of their interferogram selection. [10] did not specify the SAR data used to generate the interferogram. Given that remote sensing satellite images vary in spatial resolution, a 15 × 15 pixel area will differ between images. Therefore, a uniform 15 × 15 pixel selection cannot be applied to all images. These two limitations render the pixel-based approach used by [10] insufficient according to remote sensing principles.

Grid-Based. [11] grouped 122 continuous Global Positioning System (cGPS) stations, which have a spatial resolution ranging from 5 to 50 km, into equally sized grids of ~400 km^2. From each grid, they randomly selected one station for the correction and validation of InSAR displacement. This approach does not take into account how well the selected stations represent the grid. Ideally, the chosen stations should be those that best cover all areas within each grid (i.e., located centrally within the grid).

Distance-Based. [12] compared GNSS and PS points that were up to 200 m apart. This method is inadequate for selecting suitable PS points for GNSS validation. Many cGNSS stations in urban areas can easily find nearby PS points, whereas non-permanent GNSS sites in vegetated areas may struggle to locate PS points nearby. Consequently, a distance of 200 m may be too large and should be reconsidered due to variations in topography and PS density.

Journal articles on PS-InSAR validation with GNSS primarily focus on the validation process itself. However, there are no comprehensive papers that thoroughly address the technical workflow and scientific considerations for selecting PS-InSAR subsets to match with specific GNSS for validation. While some articles briefly outline subsetting techniques, they often overlook crucial factors such as varying spatial resolutions of satellite imagery, the need for GNSS to adequately represent PS-InSAR points within a grid (i.e., located centrally within the grid), differences in topography, and PS density. Additionally, the effectiveness of these techniques across different applications has not been well-documented. As a result, the current methodologies are insufficient as a standard reference for PS-InSAR subsetting in the international community.

4.2 Further Outlook for Our Proposed PS-InSAR Pre-validation Subsetting

The criteria we applied in spatial proximity analysis using the Near tool, which sets a distance of 100 m from each GNSS point, are the result of extensive trial and error. While these criteria yield satisfactory integrity values, additional experiments are needed to identify the optimal distance for selecting the nearest PS-InSAR points from GNSS points. Future research should focus on developing methodologies for determining distances that explain a strong spatial correlation between PS-InSAR and GNSS points. Such research will gain greater scientific and technical validation by incorporating continuous GPS data [13].

5 Conclusion

Previous studies have employed pixel-based, grid-based, and distance-based subsetting methods to select PS-InSAR data subsets for validation with GNSS data. Many of the discussed methods have potential limitations related to image spatial resolution, the degree to which GNSS data represents PS-InSAR data, and the density of PS-InSAR points. Therefore, the development of general standards should also be investigated.

We have developed a subsetting technique for the pre-validation treatment of PS-InSAR with GNSS data. This technique is based on spatial proximity and vertical deviation between PS-InSAR and GNSS data. We use the Near tool to select PS-InSAR points within a 100-m radius of the GNSS points and apply a vertical deviation threshold of −1 to 1 cm between the GNSS points and the nearest PS-InSAR records within 100 m. Applying this procedure to a large stack of 14,800 PS-InSAR points in Jakarta, we effectively selected 21 out of 14,800 PS-InSAR points for validation with GNSS points, achieving a high NSE value of 0.746, indicating a strong correlation. Our analysis shows that our proposed technique can enhance the reliability of PS-InSAR data for more detailed analysis. However, the 100-m criterion and the −1 to 1 cm threshold are based on our trial-and-error process. Continuous research that focuses on distance assessment and evolving subsetting techniques, both spatial and vertical deviation-based, is highly recommended to enhance understanding of the pre-validation treatment of PS-InSAR and GNSS data.

References

1. Yan, H., et al.: A method for correcting InSAR interferogram errors using GNSS data and the K-means algorithm. Earth Planets Space **76**(1) (2016)
2. Milillo, P., Giardina, G., DeJong, M.J., Perissin, D., Milillo, G.: Multi-temporal InSAR structural damage assessment: the London crossrail case study. Remote Sens. **10**(2) (2018)
3. Khorrami, M., Alizadeh, B., Ghasemi Tousi, E., Shakerian, M., Maghsoudi, Y., Rahgozar, P.: How groundwater level fluctuations and geotechnical properties lead to asymmetric subsidence: a PSInSAR analysis of land deformation over a transit corridor in the Los Angeles metropolitan area. Remote Sens. **11**(4) (2019)
4. Grzempowski, P., Badura, J., Milczarek, W., Blachowski, J., Głowacki, T., Zając, M.: Determination of the long-term ground surface displacements using a PSI technique—Case study on Wrocław (Poland). Appl. Sci. **10**(10) (2020)
5. Chen, B., et al.: Spatial–temporal evolution patterns of land subsidence with different situation of space utilization. Nat. Hazards **77** (2015)
6. Hussain, M.A., et al.: PS-InSAR based monitoring of land subsidence by groundwater extraction for Lahore metropolitan city, Pakistan. Remote Sens. **14**(16) (2022)
7. Armas, I., Gheorghe, M., Ana-Maria, G.-L., Dumitru, P.D., Bădescu, O., Alexandru, C.: InSAR validation based on GNSS measurements in Bucharest. Int. J. Remote Sens. **37**, 5565–5580 (2016)
8. Crosetto, M., Monserrat, O., Cuevas-González, M., Devanthéry, N., Crippa, B.: Persistent scatterer interferometry: a review. ISPRS J. Photogramm. Remote Sens. **115**, 78–89 (2016)
9. Neely, W.R., Borsa, A.A., Silverii, F.: GInSAR: a cGPS correction for enhanced InSAR time series. IEEE Trans. Geosci. Remote Sens. **58**(1), 136–146 (2019)
10. Yan, H., et al.: A method for correcting InSAR interferogram errors using GNSS data and the K-means algorithm. Earth Planets Space **76**(1) (2024)
11. Aye, P.P., Koontanakulvong, S., Tran, L.T.: Estimation of groundwater flow budget in the upper central plain, Thailand from regional groundwater model. Internet J. Soc. Soc. Manag. Syst. **11**(1) (2018)

12. Fabris, M., Battaglia, M., Chen, X., Menin, A., Monego, M., Floris, M.: An integrated InSAR and GNSS approach to monitor land subsidence in the Po river delta (Italy). Remote Sens. **14**(21) (2022)
13. Lu, C.H., Ni, C.F., Chang, C.P., Yen, J.Y., Hung, W.C.: Combination with precise leveling and PSInSAR observations to quantify pumping-induced land subsidence in central Taiwan. In: Proceedings of IAHS, vol. 372, pp. 77–82 (2015)

Multisource Satellite Data Merge and Ensemble Machine Learning for Improved LULC Mapping Accuracy

Vicca Karolinoerita[1(✉)], Fadhlullah Ramadhani[1], Destika Cahyana[2(✉)], Misnawati[3], Muhammad Ramdhan[1], Dino Gunawan Pryambodo[4], and Suria Darma Tarigan[5]

[1] Research Center for Geoinformatics, BRIN, Bandung, Indonesia
{vicca.karolinoerita,fadh004,muha307}@brin.go.id
[2] Research Center for Food Crops, BRIN, Cibinong, Bogor, Indonesia
dest009@brin.go.id
[3] Research Center for Climate and Atmosphere, BRIN, Bandung, Indonesia
misn003@brin.go.id
[4] Research Center for Limnology and Water Resources, BRIN, Cibinong, Bogor, Indonesia
dino001@brin.go.id
[5] Department of Soil Science and Land Resource Management, Faculty of Agriculture, IPB University, Bogor, Indonesia
sdtarigan@apps.ipb.ac.id

Abstract. This study introduces a method to improve the accuracy of Land Use and Land Cover (LULC) mapping by addressing the limitations of single-source satellite data. The approach combines data from multiple sources—Landsat-8, Landsat-9, and Sentinel-2 (30m) and applies an ensemble machine learning technique with majority voting. This integration results in a more comprehensive dataset for classification algorithms. The ensemble approach, which combines methods such as Random Forests, Support Vector Machines, and Neural Networks through majority voting, leverages the strengths of multiple predictive models to enhance the reliability and accuracy of LULC classifications. The combination of multisource data merging and the majority voting technique significantly improved LULC mapping accuracy compared to traditional single-source methods, with an overall accuracy increase of 11.28% and a standard deviation of ±6.82%. These findings suggest that integrating multisource satellite imagery with ensemble machine learning models using majority voting is an effective strategy for improving the quality and accuracy of LULC maps.

Keywords: LULC · Multisource Satellite Imagery · Machine Learning · Citarum Watershed

1 Introduction

Land Use and Land Cover (LULC) mapping is an important tool in environmental science that provides essential data for various applications, including environmental monitoring, urban planning, and natural resource management. The accuracy of LULC maps

E. Fitrianingsih et al. (Eds.): ISAST 2024, SPPHY 416, pp. 131–140, 2025.
https://doi.org/10.1007/978-981-96-1344-1_14

is pivotal, as these maps directly influence policymaking, land management decisions, and sustainability initiatives [1]. As a part of the advance of satellite technology, remote sensing has become an invaluable method for obtaining detailed and recurrent observations of the Earth's surface, enabling more dynamic and up-to-date LULC assessments. As such, enhancing the precision of these maps through advanced satellite imagery and analytical techniques is essential to both scientists and policymakers [2].

Research leveraging satellite imagery for LULC mapping has increasingly incorporated data from advanced Earth observation systems such as Sentinel-2, Landsat-9, and Landsat-8. These satellites are valued for their high-resolution optical imagery, which is essential for detailed environmental monitoring and mapping. For example, Sentinel-2 is known for its multispectral imaging capabilities and high temporal resolution, making it ideal for distinguishing vegetation types [3]. In contrast, Landsat-8 and Landsat-9 have played a significant role in historical land cover change analyses due to their extensive data archives, although their temporal resolution is lower than that of Sentinel-2.

Several studies have demonstrated the benefits of using data from these satellites individually. For example, Landsat data has been extensively used to track changes in forest cover and urban expansion over several decades, providing a historical context invaluable for long-term environmental studies [4]. Meanwhile, Sentinel-2, with its more frequent revisit times and higher resolution, has enabled more detailed analyses of seasonal crop types and conditions, making significant contributions to agricultural monitoring. However, individual datasets may have limitations in tropical areas, such as cloud cover, which can obstruct precise observations, and the spectral limitations of each sensor, which may not capture all necessary information for comprehensive LULC classification.

Recent research has shifted towards a multisource approach that synergistically combines Landsat-8, Landsat-9, and Sentinel-2 data to overcome the obstacle above. This approach maximizes the strengths of each satellite's data offerings—improving the temporal resolution, spectral diversity, and spatial coverage of the observational data, enhancing the accuracy and reliability of LULC maps. Studies utilizing multisource data fusion techniques have shown improved classification results, particularly in complex landscapes where the diversity of land cover types can make accurate classification challenging [5]. These advancements highlight the potential of integrated satellite data systems to provide more reliable datasets for environmental monitoring and land use planning.

This study aims to enhance the accuracy and reliability of LULC mapping significantly. This is achieved by integrating data from multiple satellite sources with advanced ensemble machine-learning techniques using majority voting. The research addresses the challenges of using single-source imagery by leveraging the combined capabilities of the Landsat-8, Landsat-9, and Sentinel-2 satellites. The proposed approach creates a more detailed dataset for LULC mapping, potentially reducing errors and increasing the precision of land cover classification methods.

2 Material and Methods

2.1 Material

The study was conducted on the Citarum upper watershed, a vital geographic area in West Java, Indonesia. This region was selected due to its significant environmental, economic, and social importance. For this study, high-resolution satellite images from Sentinel-2, Landsat-8, and Landsat-9 from Google Earth Engine [6–8] pre-processed from July 2023, are used to capture the detailed characteristics of the watershed. Especially for Sentinel-2, the final product was upscaled to 30m to match the Landsat images. These satellites were selected for their complementary capabilities, such as varying spectral bands and revisit times, which are crucial for obtaining a comprehensive view of the region. The images are processed and analyzed to create a multi-temporal dataset that reflects changes in the landscape over time, providing a dynamic view of land cover and usage [9, 10].

2.2 Methods

This study utilized an ensemble machine learning approach to improve the accuracy and robustness of LULC mapping using multisource satellite data. This methodology leverages the diverse strengths of various algorithms to produce more stable and reliable classification outcomes. For the analysis, we selected several well-established algorithms, including Random Forests (RF), Support Vector Machines (SVM), and Neural Networks (NN). The rationale behind using an ensemble instead of a single model is to exploit the complementary learning strategies of different algorithms; for example, while NN is adept at handling nonlinear relationships and high-dimensional data, RFs can efficiently manage missing data and maintain accuracy even with a large number of variables [11]. SVMs, on the other hand, are effective in high-dimensional spaces and versatile in their capability to implement both linear and nonlinear classification through kernel functions, making them ideal for the varied spectral and textural data derived from satellite images [12].

Model training and validation are essential processes for LULC mapping. The data was split into training and testing sets to ensure reliability and accuracy. Typically, 70% of the data was allocated for training, while the remaining 30% was reserved for testing, used exclusively to assess the model's performance and prevent bias from overfitting. This data-splitting strategy ensures that the ensemble models are trained on a comprehensive subset of the data while maintaining an independent evaluation dataset [13]. Although the model training and validation strategy is effective, incorporating in-situ data to validate LULC mapping outputs directly would further enhance the assessment. Due to the unavailability of such data, this study used high-resolution satellite imagery and widely accepted reference datasets, which have been well-validated in prior research and serve as reliable proxies for accuracy assessment.

The performance of the ensemble models was evaluated using several metrics to ensure a comprehensive assessment of their effectiveness in LULC classification. The primary metric used was overall accuracy, which provides a robust framework for assessing the precision and reliability of the ensemble learning models in classifying land use and land cover from satellite imagery. The workflow of the methodology is illustrated in Fig. 1.

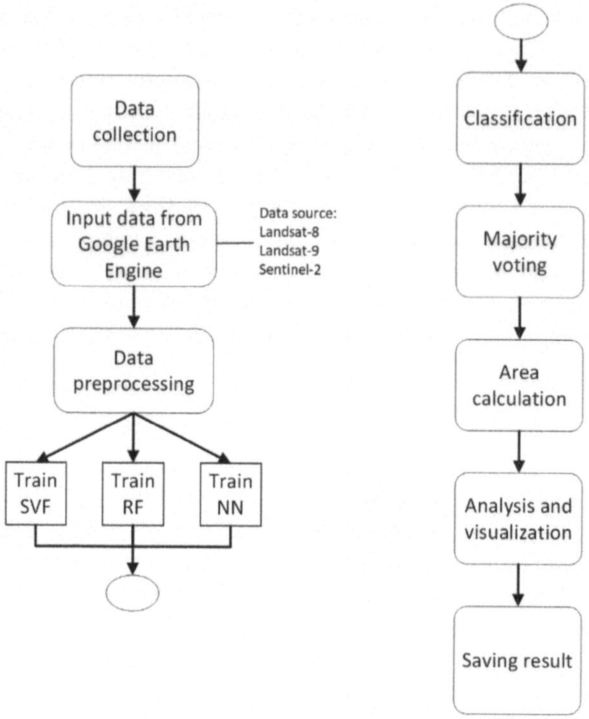

Fig. 1. Research methodology.

3 Result and Discussion

A thorough statistical analysis was conducted to quantify the improvement in model performance due to the multisource approach. Key metrics used in this analysis included overall accuracy. Overall accuracy with majority votes is 98.71% in multisource models.

3.1 LULC Areas Across Classifiers and Satellites

The data highlight distinct differences in classifier performance across various satellite platforms. For the urban class, SVM consistently produces higher area estimates than NN and RF, with the highest estimate reported by Landsat-9 using SVM (71.299,44 ha). It suggests that SVM has a higher sensitivity or preference for identifying urban structures within satellite imagery. In contrast, NN provides higher area estimates in more natural environments like forests, particularly with the Sentinel-2 (30m) platform (65.101,59 ha), indicating its effectiveness in capturing broader vegetative patterns, as shown in Fig. 2.

This variance highlights the importance of classifier selection based on specific LULC characteristics and the qualities of satellite data.

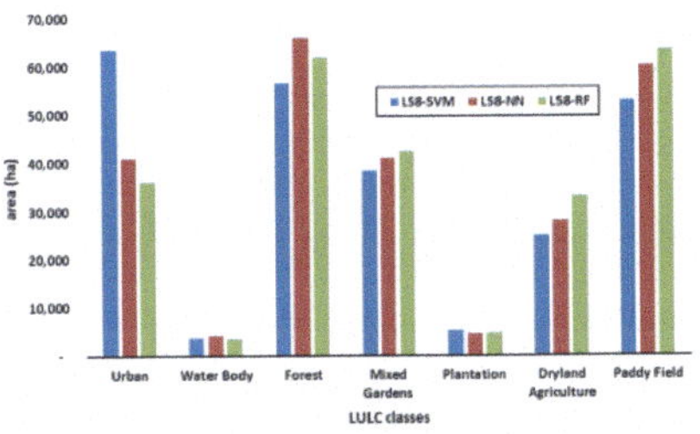

(a) Landsat-8

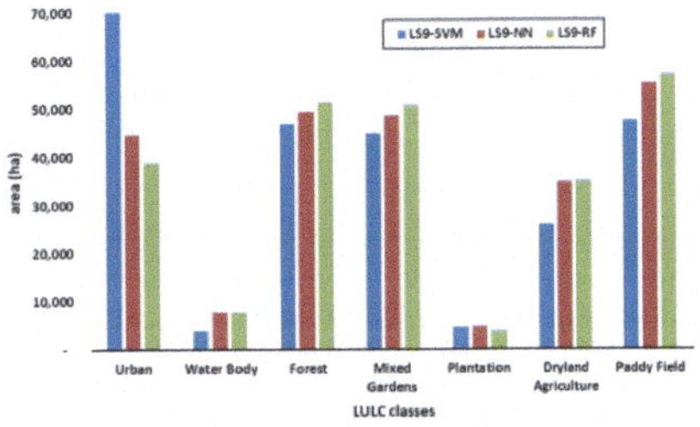

(b) Landsat-9

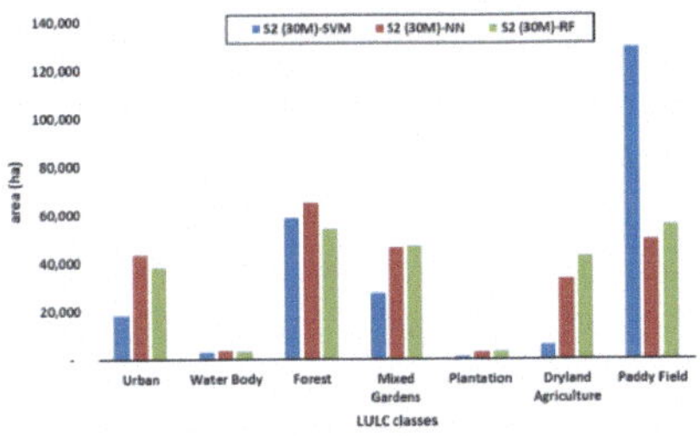

(c) Sentinel-2 (30m)

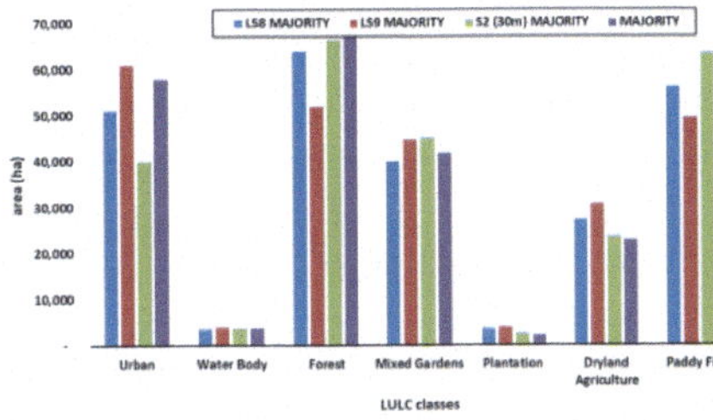

(d) Majority Votes

Fig. 2. Area by class for each classifier (a) Landsat-8's results, (b) Landsat-9's results, (a) Sentinel-2 (30m)'s results, (d) Majority votes results.

In analyzing specific LULC areas, such as plantation and dryland agriculture, significant discrepancies emerge among the classifiers. For example, NN reports significantly higher areas in Landsat-9 for plantation and dryland agriculture, potentially due to its pattern recognition capabilities that differ from those of SVM and RF. Such variations can significantly affect LULC analysis outcomes, particularly in agricultural planning and environmental monitoring. Furthermore, the high area reported for the paddy field by Sentinel-2 (30m) using SVM (129.448,26 ha) suggests potential overestimation issues or misclassifications, likely influenced by the specific spectral signatures recognized by the SVM model.

The majority votes chart, which synthesizes outputs from multiple classifiers, tends to show moderated values that provide a more balanced representation of each class across different landscapes. For example, majority votes often align closely with individual classifier outputs but adjust to a mean value that may better reflect the ground truth by mitigating individual classifier biases. For critical classes like forest and paddy fields, where environmental and agricultural management decisions are crucial, these majority values offer a reliable basis for analysis, suggesting a strategic integration of multiple classifiers in operational LULC mapping workflows. These insights highlight the complexity and importance of selecting appropriate classifiers and combining their strengths through majority voting to enhance the reliability and accuracy of LULC classifications derived from satellite imagery.

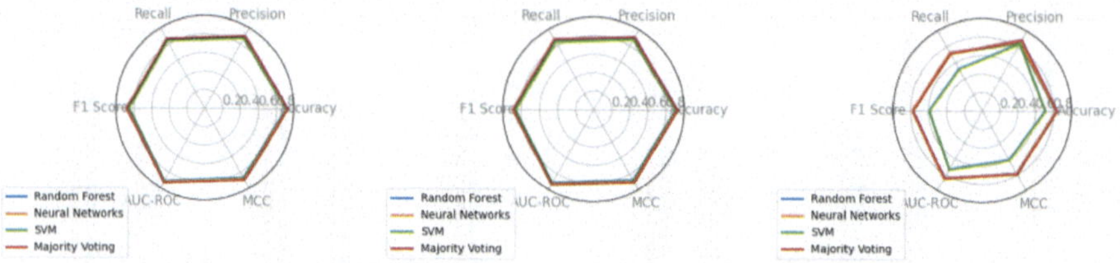

(a) Landsat-8 (b) Landsat-9 (c) Sentinel-2 (30m)

Fig. 3. Classifier performance accuracy on three sensors: (a) Landsat-8, (b) Landsat-9, and (c) Sentinel-2 (30m).

Figure 3 compares classifier performance among the SVM, RF, and NN classifiers using data from Landsat-8, Landsat-9, and Sentinel-2 (30m). The performance varies depending on the type of land cover analyzed and the specific satellite data used. The SVM demonstrates higher accuracy for urban areas, particularly with Landsat-9, except for Sentinel-2 (30m). Meanwhile, the NN classifier performs better for forested areas, especially with Sentinel-2 data. The RF classifier shows moderate performance across various LULC classes and satellite platforms. Integrating multisource satellite data enhances overall performance by providing a more comprehensive dataset, resulting in more accurate and reliable LULC classifications. The ensemble approach with majority voting improves classification results by reducing biases from individual classifiers.

Table 1 presents a detailed comparison of producer and user accuracy for various individual LULC classes to enhance the understanding of overall accuracy. Producer accuracy measures how well real-world courses are represented in the map, while user accuracy shows the probability that a classified pixel accurately represents the corresponding ground category. The results reveal significant variations among classifiers and data sources. For example, SVM consistently achieves high producer accuracy for urban areas, indicating effective identification of urban structures with minimal omission errors. Conversely, NN shows high producer accuracy for paddy field areas, demonstrating its ability to detect vegetative patterns accurately. User accuracy also varies, with high values for urban areas achieved by SVM with Landsat-9 and Landsat-8, reflecting reliable classification with minimal commission errors. However, lower user accuracy in some cases, such as dryland agriculture with Sentinel-2 (30m), suggests potential misclassifications. Integrating multisource satellite data from Landsat-8, Landsat-9, and Sentinel-2 (30m) generally improves producer and user accuracy, overcoming the limitations of single-source data and providing a richer dataset. Ensemble machine learning techniques, particularly majority voting, further enhance the robustness and reliability of LULC classifications by leveraging the strengths of multiple classifiers.

Table 1. Comparison of each sensor's producer and user accuracy for LULC classes.

Class	Random Forest		Neural Network		Support Vector Machine		Majority Vote	
	PA	UA	PA	UA	PA	UA	PA	UA
Landsat-8								
	OA: 0.919		OA: 0.915		OA: 0.908		OA: 0.924	
Urban	0.993	0.991	0.998	0.994	0.900	0.997	0.962	0.996
Water Body	0.974	0.912	0.972	0.915	0.972	0.919	0.984	0.922
Forest	0.899	0.933	0.875	0.948	0.902	0.924	0.890	0.946
Mixed Gardens	0.797	0.712	0.793	0.645	0.778	0.694	0.823	0.708
Plantation	0.857	0.826	0.866	0.796	0.860	0.774	0.898	0.814
Dryland Agriculture	0.753	0.719	0.806	0.677	0.753	0.636	0.815	0.703
Paddy Field	0.945	0.963	0.938	0.966	0.948	0.958	0.952	0.969
Landsat-9								
	OA: 0.923		OA: 0.921		OA: 0.910		OA: 0.928	
Urban	0.992	0.992	0.996	0.992	0.887	0.997	0.958	0.996
Water Body	0.976	0.922	0.963	0.941	0.973	0.933	0.979	0.931
Forest	0.906	0.944	0.914	0.944	0.910	0.938	0.904	0.952
Mixed Gardens	0.839	0.748	0.912	0.661	0.833	0.713	0.886	0.731
Plantation	0.854	0.809	0.827	0.853	0.876	0.784	0.891	0.833
Dryland Agriculture	0.759	0.731	0.705	0.775	0.742	0.613	0.793	0.745
Paddy Field	0.938	0.959	0.929	0.964	0.942	0.950	0.946	0.965
Sentinel-2 (30m)								
	OA: 0.831		OA: 0.833		OA: 0.708		OA: 0.837	
Urban	0.987	0.850	0.989	0.849	1.000	0.421	0.989	0.852
Water Body	0.981	0.751	0.983	0.748	0.980	0.668	0.990	0.748
Forest	0.702	0.920	0.700	0.926	0.703	0.838	0.700	0.945
Mixed Gardens	0.734	0.591	0.752	0.573	0.794	0.330	0.787	0.556
Plantation	0.878	0.667	0.913	0.615	0.967	0.316	0.942	0.601
Dryland Agriculture	0.743	0.432	0.763	0.496	0.839	0.163	0.893	0.383
Paddy Field	0.924	0.911	0.928	0.917	0.642	0.939	0.917	0.934

Note: OA: Overall Accuracy, PA: Producer's accuracy, and UA: User's Accuracy

3.2 Comparison of Satellite Data in Majority Voting

The analysis of LULC areas from different satellite sources reveals varying accuracy in classification across various environments and land uses. Landsat-8 and Landsat-9, notably, demonstrate higher resolution for detecting urban regions, with Landsat-9

(60.856,2 ha) reporting a larger area than Landsat-8 (51.048,9 ha), as shown in Table 1. It may indicate differences in sensor sensitivity or temporal variations in data acquisition. Sentinel-2's 30m resolution captures larger extents of natural and semi-natural landscapes, as reflected in the significantly higher area measurements for forest and paddy fields. It suggests a more effective performance in capturing broader vegetation and water body classifications due to its enhanced spectral resolution.

In the forest category, which represents forested areas, there is a significant disparity in the area covered, with Sentinel-2 (30m) showing the most extensive area (66.404,43 ha), likely due to its comprehensive coverage and spectral sensitivity to chlorophyll. In contrast, the plantation category has the lowest recorded areas across all datasets, with the smallest area measured by Sentinel-2 (30m) at 2.774,61 ha, indicating potential challenges in differentiating this specific crop type from surrounding vegetation. Both this category and dryland agriculture show varying figures, possibly reflecting differences in seasonal cover or agricultural practices captured during satellite passes.

This integrated approach can reduce individual sensor biases and provide a more stable basis for land management and urban planning decision-making. For example, the balanced figures in paddy fields and urban categories suggest a more accurate middle ground that could be useful for regulatory and development initiatives aimed at sustainable land use practices. These insights highlight the importance of selecting appropriate satellite data sources based on specific LULC classification goals and demonstrate the potential benefits of integrating multiple data streams to enhance environmental monitoring and planning accuracy and reliability.

The interpretation of results from this study indicates that integrating multisource satellite data with ensemble machine-learning techniques significantly enhances LULC classification performance. Combining data from Landsat-8, Landsat-9, and Sentinel-2 (30m) provided a richer dataset that captured a more comprehensive spectral and temporal landscape view. This integrated approach was essential for improving classification accuracy, as it allowed the ensemble models to effectively leverage each satellite system's unique strengths. For example, the frequent revisits of Sentinel-2 enabled more timely updates. At the same time, the long-term data continuity from Landsat-8 and Landsat-9 contributed to a better understanding of seasonal and annual changes in land cover [7].

The classification results using majority voting can be seen in Fig. 4. This study revealed several limitations that should be considered in future research. A key challenge was the inherent differences in the spatial and temporal resolutions of the satellite data used. Although integrating multisource data is intended to mitigate these variations, differences in resolution can still lead to inconsistencies, potentially affecting the precision of LULC classification.

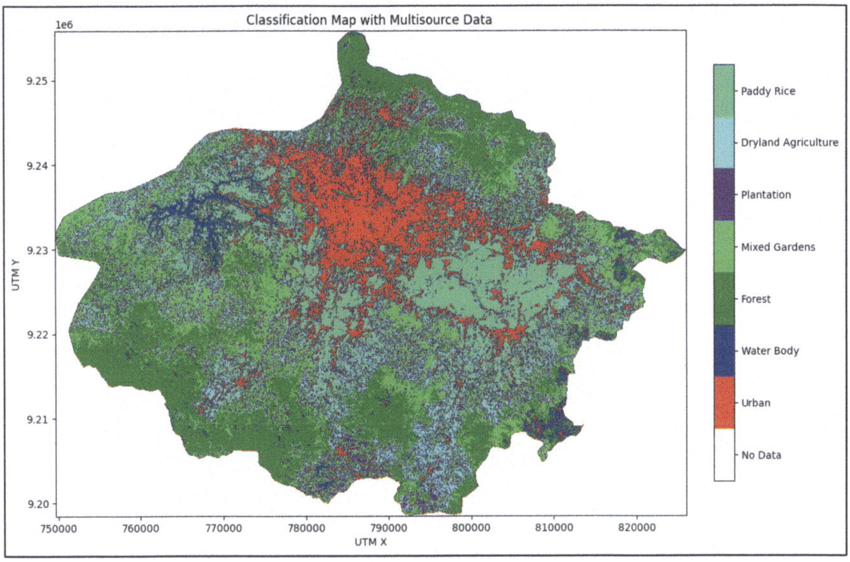

Fig. 4. Classification map with multisource data.

4 Conclusion

The combination of Random Forest (RF), Support Vector Machine (SVM), and Neural Networks (NN) with majority voting, along with multisource data merging, significantly enhances the Land Use and Land Cover (LULC) classification process. This integrated approach reduces the likelihood of overfitting. It increases robustness against variability in the data, resulting in an overall accuracy increase of 11.28% and a standard deviation of $\pm 6.82\%$ compared to traditional single-source methods. Such a technique could be particularly beneficial for studying changes in coastal areas, wetlands, and forest ecosystems, where complex dynamics require high-resolution, frequent observations. Future research could further explore integrating additional data sources, such as drone imagery or higher-resolution commercial satellite data, to improve the detail and accuracy of LULC classifications.

Acknowledgment. This research was supported by the National Research and Innovation Agency (BRIN) through the DSS Software Prototype for Watershed Management Research Program, 2024. We thank BRIN for providing the Mahameru High-Performance Computing Facility and the reviewers for their valuable feedback.

References

1. Stehman, S., Foody, G.: Key issues in rigorous accuracy assessment of land cover products. Remote Sens. Environ. (2019). https://doi.org/10.1016/j.rse.2019.05.018

2. Thapa, R.B., Murayama, Y.: Accuracy of land use and land cover mapping methods. In: Murayama, Y., Thapa, R. (eds.) Spatial Analysis and Modeling in Geographical Transformation Process. GEJL, vol. 100, pp. 135–151. Springer, Dordrecht (2011). https://doi.org/10.1007/978-94-007-0671-2_9
3. Forkuor, G., Dimobe, K., Serme, I., Tondoh, J.E.: Landsat-8 vs. Sentinel-2: examining the added value of sentinel-2's red-edge bands to land-use and land-cover mapping in Burkina Faso. GIScience Remote Sens. **55**, 331–354 (2018). https://doi.org/10.1080/15481603.2017.1370169
4. Ghayour, L., et al.: Performance evaluation of Sentinel-2 and Landsat 8 OLI data for land cover/use classification using a comparison between machine learning algorithms. Remote Sens. **13**, 1349 (2021). https://doi.org/10.3390/rs13071349
5. Wang, Q., et al.: Fusion of Landsat 8 OLI and Sentinel-2 MSI data. IEEE Trans. Geosci. Remote Sens. **55**(7), 3885–3899 (2017). https://doi.org/10.1109/TGRS.2017.2683444
6. Ganjirad, M., Bagheri, H.: Google Earth Engine-based mapping of land use and land cover for weather forecast models using Landsat 8 imagery. Ecol. Inform. **80**, 102498 (2024). https://doi.org/10.1016/j.ecoinf.2024.102498
7. Habibie, M.I., et al.: The development land utilization and cover of the Jambi district are examined and forecasted using Google Earth Engine and CNN1D. Remote Sens. Appl.: Soc. Environ. **34**, 101175 (2024). https://doi.org/10.1016/j.rsase.2024.101175
8. Zhao, Z., et al.: Comparison of three machine learning algorithms using google earth engine for land use land cover classification. Rangeland Ecol. Manag. **92**, 129–137 (2024). https://doi.org/10.1016/j.rama.2023.10.007
9. Yulianto, F., Maulana, T., Khomarudin, M.R.: Analysis of the dynamics of land use change and its prediction based on the integration of remotely sensed data and CA-Markov model, in the upstream Citarum watershed, West Java, Indonesia. Int. J. Digit. Earth **12**(10), 1151–1176 (2018). https://doi.org/10.1080/17538947.2018.1497098
10. Yulianto, F., Suwarsono, Nugroho, U., Nugroho, N., Sunarmodo, W., Khomarudin, M.: Spatial-temporal dynamics land use/land cover change and flood hazard mapping in the upstream Citarum watershed, West Java, Indonesia. Quaestiones Geographicae **39**(1), 125–146 (2020). https://doi.org/10.2478/quageo-2020-0010
11. Tassi, A., Vizzari, M.: Object-oriented LULC classification in Google Earth Engine combining SNIC, GLCM, and machine learning algorithms. Remote. Sens. **12**(3776) (2020). https://doi.org/10.3390/rs12223776
12. Jozdani, S.E., Johnson, B., Chen, D.: Comparing deep neural networks, ensemble classifiers, and support vector machine algorithms for object-based urban land use/land cover classification. Remote Sens. **11**, 1713 (2019). https://doi.org/10.3390/rs11141713
13. Barman, P., Mustak, S., Kuffer, M., Singh, S.K.: Transfer-ensemble learning: a novel approach for mapping urban land use/cover of the Indian metropolitans. Sustainability (2023). https://doi.org/10.3390/su152416593

The Relationship Between Oceanographic Parameters and Tuna Landings in Indonesian Waters

Aulia Try Atmojo[1,1(✉)], Maifha Linda Rusli[1], Emiyati[2], Agung Mahadi Putra Perdana[1], Muhammad Ario Eko Rahadianto[1], Tri Kies Welly[1], Redho Surya Perdana[1], and Ongky Anggara[1]

[1] Sumatra Institute of Technology, South Lampung, Indonesia
aulia.atmojo@gt.itera.ac.id

[2] National Research and Innovation Agency (BRIN), Cibinong, Indonesia

Abstract. Indonesia, the world's largest tuna producer, contributed 17% (1.12 million metric tons) to global tuna production in 2016. Mapping fisheries potential in Indonesia's vast waters presents challenges, but remote sensing technology enables effective analysis of oceanographic parameters. This study uses Multiple Linear Regression, Correlation Analysis, and T-Tests to examine the relationship between chlorophyll-a, sea surface temperature (SST), and depth with tuna catches. Results show a strong correlation for bigeye tuna, while frigate, skipjack, yellowfin, and mackerel-tunas-bonitos exhibit strong relationships. Longtail tuna shows a weak correlation. F-tests indicate that oceanographic parameters significantly affect the catches of bigeye, frigate, skipjack, yellowfin, and mackerel-tunas-bonitos, but not longtail tuna. T-tests reveal that chlorophyll-a and depth influence frigate and mackerel-tunas-bonitos, while depth alone impacts bigeye, skipjack, and yellowfin tuna. No oceanographic parameters significantly affect longtail tuna. The most productive fishing areas are in Eastern waters for bigeye, frigate, and longtail tuna and in Indian waters for skipjack and yellowfin tuna. Central waters show the highest potential for mackerel-tunas-bonitos.

Keywords: Tuna Landings · SST · Chlorophyll-a · Depth · Multiple Linear Regression

1 Introduction

It is known as an archipelagic country, making Indonesia the largest tuna-producing country in the world, contributing 17% or 1.12 million metric tons (MT) of total global tuna production in 2016 [1]. There remains a significant gap between Indonesia's fisheries' utilization and its full potential. Of the 12 million tons per year of fisheries potential, only 7.2 million tons, or 60.3%, are currently utilized [2]. The fisheries sector also contributes minimally to GDP compared to other sectors [3]. This disparity is partly due to the low quality and quantity of catches, linked to limited accessibility and insufficient information about fisheries potential [4]. With Indonesia's vast marine territory of

E. Fitrianingsih et al. (Eds.): ISAST 2024, SPPHY 416, pp. 141–150, 2025.
https://doi.org/10.1007/978-981-96-1344-1_15

3,257,357 km^2, mapping fisheries potential is challenging. Remote sensing technology could provide a solution by offering comprehensive oceanographic information over large areas.

The existence of fish is closely related to various oceanographic parameters, such as temperature, salinity, currents, water depth, and chlorophyll concentration [5–8]. Climate change is one of the factors of oceanographic dynamics [9]. La Nina and El Nino phenomena play a role in changing oceanographic parameters, and their changes can influence fish life and growth, such as eating speed, spawning, and metabolism. As a result, the number of fish in Indonesian waters is not definite every year [10–12]. Various technologies have been used to estimate oceanographic parameters either by direct measurements or in the case of vast areas such as remote sensing and global ocean models [13–15]. By analyzing oceanographic parameters like sea surface temperature (SST), chlorophyll-a, and depth from secondary fish catch data at the study site, correlations can be established between these factors and tuna catch rates. These findings can serve as a guide for optimizing fish aggregating devices or directing fishermen to improve their catch efficiency.

2 Material and Methods

2.1 Research and Data Collection

The research was conducted in Indonesian waters, which are divided into three regions: Central Indonesian waters, East Indonesian waters, and the Indian Ocean (Fig. 1). The data used in this study is described in Table 1 below.

Table 1. Research data types, time ranges, and sources.

Data	Time Range	Source
MODIS Terra Sea Surface Temperature and Chlorophyll Concentration	2008–2018 (Monthly)	NASA Ocean Color Data
MODIS Aqua Sea Surface Temperature and Chlorophyll Concentration		
SNPP - VIIRS Sea Surface Temperature and Chlorophyll Concentration		
Bathymetry Data	2022	GEBCO
Fish Catch Number	2008–2018	Sea Arround Us

Tuna catch data from 2008 to 2018 were obtained from the Sea Around Us website [16]. Chlorophyll-a concentration and SST data were derived from the Aqua-Terra Moderate Resolution Imaging Spectroradiometer (MODIS) and the Visible Infrared Imaging Radiometer Suite (SNPP-VIIRS) multi-sensor satellites in Net Common Data File format. Depth data for 2022 were sourced from GEBCO and can be accessed via the GEBCO website [17]. MODIS is an instrument aboard the Terra and Aqua satellites

(EOS AM-1 and EOS PM-1, respectively) and has been operational since 2000. MODIS crosses the equator at consistent local times (10:30 for Terra and 13:30 for Aqua). Utilizing its 36 spectral bands, MODIS generates various products, such as land use and land cover (LULC) maps, the Normalized Difference Vegetation Index (NDVI), aerosol optical depth, and more. The VIIRS (Visible Infrared Imaging Radiometer Suite) plays a critical role in operational environmental monitoring and numerical weather forecasting, utilizing 22 imaging and radiometric bands that cover wavelengths ranging from 0.41 to 12.5 μm. This sensor provides data for over twenty environmental parameters, including cloud cover, sea surface temperature, ocean color, polar wind, vegetation fraction, aerosol concentrations, fire detection, snow and ice cover, vegetation health, and other key applications.

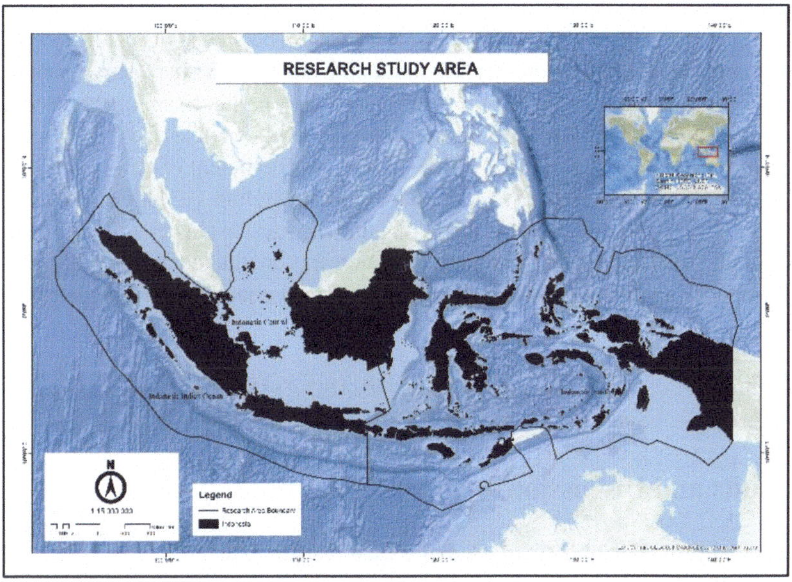

Fig. 1. Maps of Indonesian waters.

In this study, Level 3 (11 μ daytime) data (Chlorophyll-a and SST data) with a 4 km monthly resolution, available from the Ocean Color Website [18], were utilized. Level 3 (L3) data refers to spatially and temporally aggregated data products. In the case of MODIS Atmosphere L3 Monthly Products, the sensor collects raw observations that are processed and averaged over one month to generate global-scale grids. For SST, VIIRS uses infrared bands to detect thermal radiation from the ocean surface, while Chl-a uses visible and near-infrared bands to monitor ocean color changes. Both datasets are processed into Level 3 products, offering aggregated insights into ocean conditions [19].

2.2 Data Analysis

Python was used to derive and process monthly chlorophyll-a and sea surface temperature data from multi-sensor satellites. These data from the three sensors were integrated, contributing to the novelty of this research. Since the data were initially in monthly format, they were subsequently aggregated and converted into annual data. Previous studies, such as those by Khan et al., (2020) and Semedi et al., (2021), utilized data derived solely from the Terra satellite for SST and Chl-a assessments [20, 21], while on the other hand, Kunarso et al. (2021), obtained SST and Chl-a data exclusively from the Aqua satellite [22]. The processing and plotting of annual chlorophyll-a concentration and sea surface temperature data were performed using spreadsheet software. Tuna landings data were processed by filtering (separating the data for the three regions) and averaging the annual catch from 2008 to 2018. The species of tuna included in the study were bigeye tuna, frigate tuna, longtail tuna, yellowfin tuna, and skipjack tuna, as well as mackerels, tunas, and bonitos.

Depth data was processed by clipping the study area using GIS software and extracting average depth values through Python. A numerical analysis program was employed for multiple linear regression to analyze the relationship between tuna catch and oceanographic parameters (SST, chlorophyll-a, and depth) (Eq. 1). It was done by testing the independent variables (x_1 = SST, x_2 = chlorophyll-a, x_3 = depth) against the dependent variable (Y = tuna catch). The multiple correlation coefficient assessed the simultaneous relationship between x_1, x_2, and x_3 with Y, while the coefficient of determination measured the proportion of variance in Y explained by x_1, x_2, and x_3. An F-test determined whether x_1, x_2, and x_3 collectively influence Y, and the t-test evaluated the significance of each variable's effect. Finally, spatial interpolation was conducted using GIS software to model tuna species distribution based on the regression equation. This analysis process is depicted in Fig. 2.

$$Y = a + b_1x_1 + b_2x_2 + b_3x_3 \tag{1}$$

3 Results and Discussion

3.1 Interannual Variability of Sea Surface Temperature and Chlorophyll-A Concentration

Figure 3 illustrates the interannual variability of average SST in Indonesian waters from 2008 to 2018, with temperatures ranging from 27 °C to 32 °C. Similarly, Fig. 4 depicts the variability of chlorophyll-a concentrations during the same period, with values between 0.2 mg/m^3 and 0.5 mg/m^3. The data suggests a reduction in upwelling intensity in the Indonesian Indian Ocean and Eastern Indonesia, leading to warmer waters and lower fertility, as indicated by declining chlorophyll-a levels. This aligns with Ramadhan et al. (2021), who found that La Niña events combined with negative IOD decrease cold water and nutrient supply, resulting in warmer SST and reduced chlorophyll-a [23].

3.2 Analysis of the Relationship Between Oceanographic Parameters and Tuna Fisheries in Indonesian Waters (2008–2018)

The regression statistics, ANOVA, and regression coefficients (Table 2) highlight the relationship between tuna catch and the explanatory variables (x_1, x_2, and x_3). With 95%

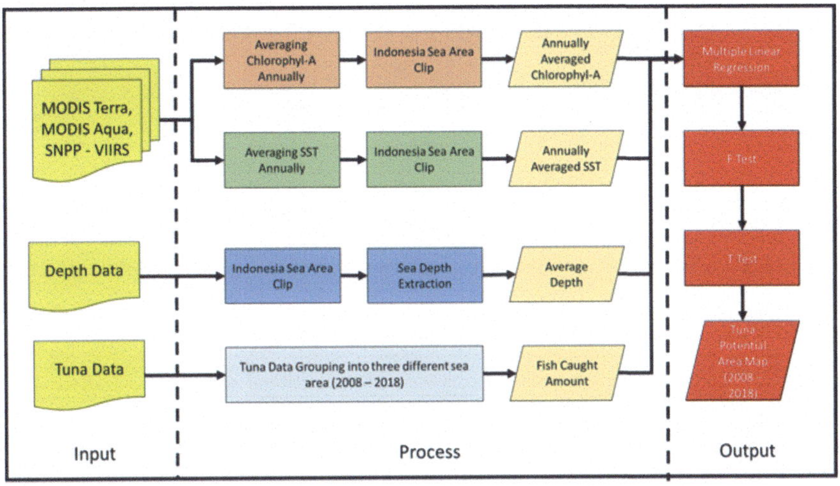

Fig. 2. The tuna potential area analysis process combines remote sensing satellite data, ocean depth, and tuna data.

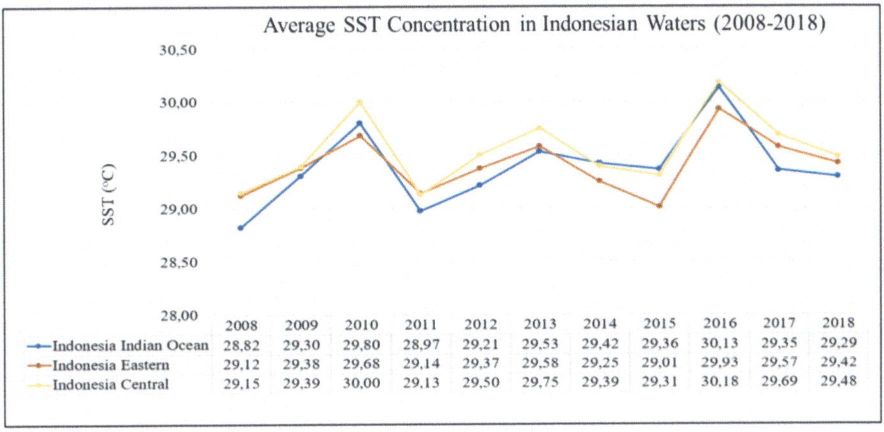

Fig. 3. Average SST Concentration in Indonesian Waters (2008–2018).

confidence and a 0.05 margin of error, the multiple linear regression analysis results indicate that the model's independent variables can explain only 8% (R2 = 0.08) of longtail tuna catch variations. In comparison, the remaining 92% are due to other oceanographic factors and random errors. The F-significance value of 0.68 (greater than 0.05) confirms a weak correlation. These findings suggest that additional factors, such as ocean depth, oceanic fronts, and prey availability (e.g., anchovies and sardines), contribute to CPUE fluctuations, reflecting the complexity of environmental influences beyond just SST and chlorophyll-a [24].

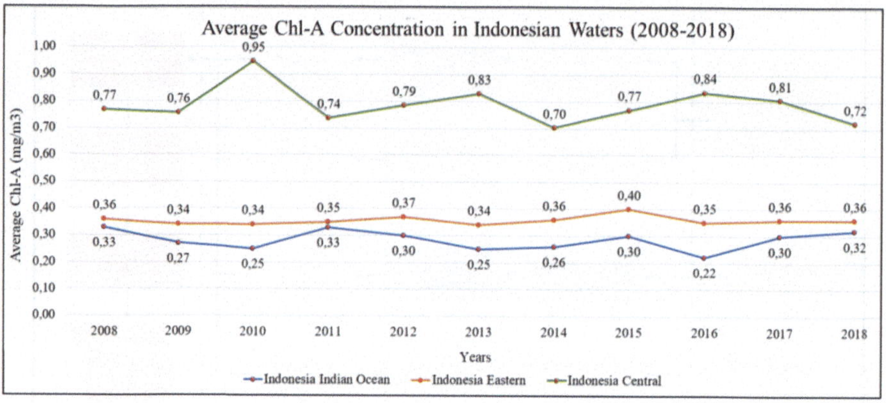

Fig. 4. Average Chl-A Concentration in Indonesian Waters (2008–2018).

Conversely, the oceanographic parameters show a significant simultaneous effect on the catch of bigeye tuna, frigate tuna, skipjack tuna, yellowfin tuna, and mackerels-tunas-bonitos. Multiple correlation coefficients indicate a strong to very strong relationship between tuna catch rates and these parameters, with R values ranging from 0.78 to 0.97. Previous studies have confirmed the correlation between CPUE and key environmental factors such as SST and chlorophyll-a, highlighting their importance in predicting tuna catch rates and optimizing fishing efforts [24]. Skipjack tuna exhibits the highest coefficient of determination (R2 = 0.95), while frigate tuna has a lower R2 value of 0.62, just above longtail tuna.

Table 2. Regression Statistic, ANOVA (Analysis of Variance) and Regression Coefficients Results

Tuna Type	R	R^2	F Significance	SPL (x_1) P-value	Chlorophyll-a (x_2) P-value	Depth (x_3) P-value
Bigeye Tuna	0.87	0.77	0.00	0.77	0.61	0.00
Frigate Tuna	0.78	0.62	0.00	0.21	0.00	0.00
Skipjack Tuna	0.97	0.95	0.00	0.56	0.42	0.00
Yellowfin Tuna	0.90	0.81	0.00	0.91	0.42	0.00
Mackerel-Tunas-Bonitos	0.83	0.70	0.00	0.12	0.00	0.00
Longtail Tuna	0.28	0.08	0.68	0.76	0.89	0.50

3.3 Estimating Potential Tuna Fishing Locations in Indonesian Waters

Table 3 presents the multiple linear regression equations, which will be recalculated by re-entering each value of SST, chlorophyll-a, and depth into these equations for each tuna species to estimate potential tuna fishing locations. The results of these estimations will be visualized in Figs. 5, 6, and 7. The subsequent section will discuss the results of the estimated potential tuna fishing locations for each species within Indonesian waters (Central Indonesian Waters, East Indonesian Waters, and the Indian Ocean).

Table 3. Multiple linear regression equation for each tuna species.

Tuna Type	Multiple Linear Regression Equation
Bigeye Tuna	$Y = 6.887{,}49 - 324{,}83\, x_1 + 7.544{,}99\, x_2 - 1{,}86\, x_3$
Frigate Tuna	$Y = - 466.334{,}55 + 11.790{,}64\, x_1 + 437.869{,}74\, x_2 - 25{,}27\, x_3$
Skipjack Tuna	$Y = - 101.301{,}73 + 5.037{,}08\, x_1 - 90.481{,}05\, x_2 - 27{,}06\, x_3$
Yellowfin Tuna	$Y = - 3.815{,}40 - 1.283{,}67\, x_1 + 113.975{,}41\, x_2 - 21{,}00\, x_3$
Mackerel-Tunas-Bonitos	$Y = - 37.514{,}16 + 926{,}56\, x_1 + 35.233{,}44\, x_2 - 1{,}48\, x_3$
Longtail Tuna	$Y = 5.499{,}07 - 188{,}19\, x_1 + 1.068{,}74\, x_2 - 0{,}17\, x_3$

The regression function estimation identifies critical areas for tuna fishing in Indonesian waters:

- Bigeye Tuna: The most promising areas are near the coastlines, notably around Kalimantan, Sumatra, and Enggano Island. Eastern Indonesian waters, particularly near Papua, show the highest potential, while the Indian Ocean offers lower potential.
- Frigate Tuna: Potential fishing areas are concentrated near coastal regions like Kalimantan, Bangka Island, and between Sumatra and Java. The highest potential is in Eastern Indonesian waters, with the Indian Ocean also showing promising coastal areas.
- Skipjack Tuna: Central Indonesian waters offer significant potential, especially near the Natuna Sea. In Eastern waters, areas near Papua and Aru Islands show high potential, while the Indian Ocean presents the most promising overall area.
- Yellowfin Tuna: Coastal areas near Kalimantan, Sumatra, and the Indian Ocean show high potential, with red indices extending from nearshore to oceanic boundaries. Eastern waters like the Timor and Arafura Seas show lower potential.
- Mackerel-Tunas-Bonitos: Coastal regions of Kalimantan, Sumatra, and Eastern Indonesian waters, especially near Irian Jaya and Aru Islands, show the highest potential. In the Indian Ocean, coastal areas display moderate potential.
- Longtail Tuna: Coastal regions across Central, Eastern, and Indian Ocean waters, indicated by blue indices, show moderate potential for fishing.
- Eastern and Central Indonesian waters, particularly along coastlines, consistently exhibit the highest potential for tuna fishing across species.

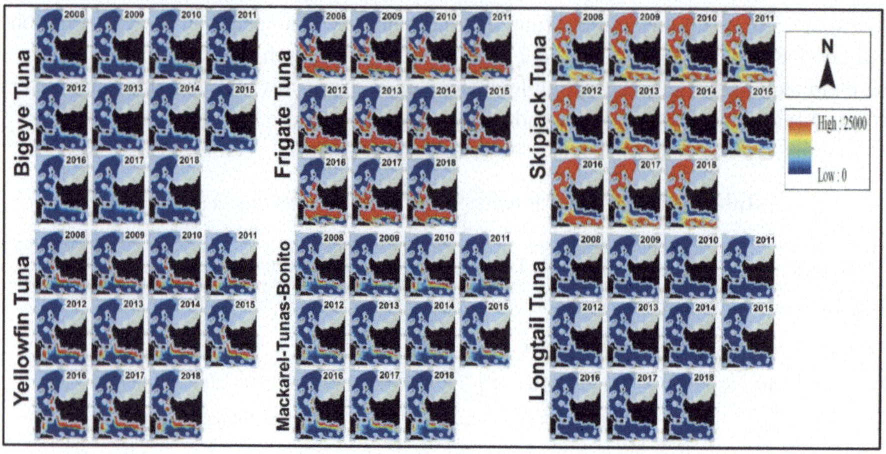

Fig. 5. Estimating potential tuna fishing locations in central Indonesian waters (2008–2018).

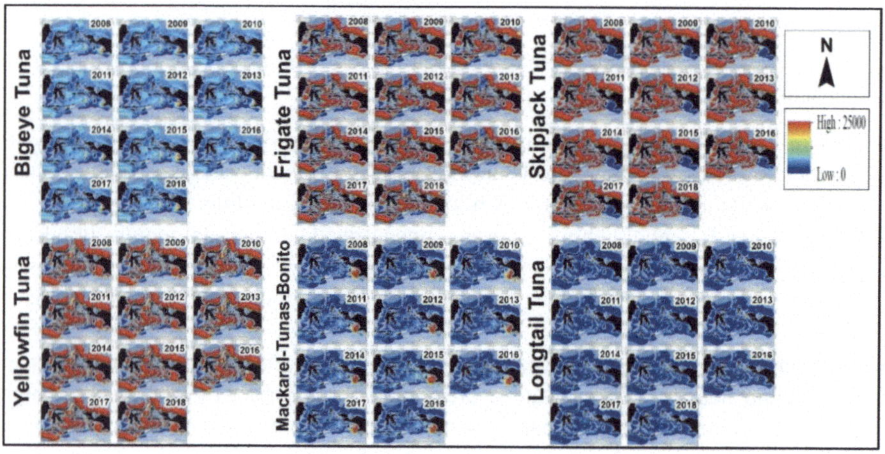

Fig. 6. Estimating potential tuna fishing locations in east Indonesian waters (2008–2018).

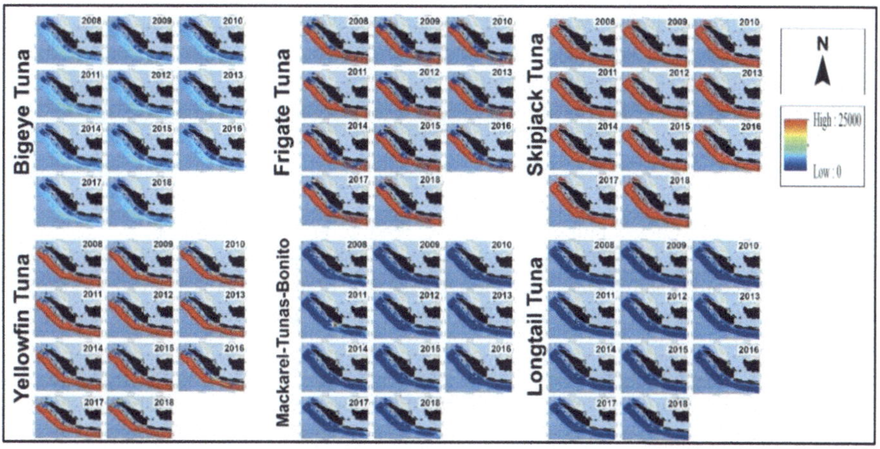

Fig. 7. Estimating potential tuna fishing locations in the Indian ocean (2008–2018).

4 Conclusion

Statistical analysis of oceanographic parameters showed strong correlations with tuna catch for frigate tuna (0.78), bigeye tuna (0.87), skipjack tuna (0.97), yellowfin tuna (0.90), and mackerels, tunas, and bonitos (0.83). In contrast, the correlation for longtail tuna was weak (0.28). The F-test indicated a significant effect of oceanographic parameters (SST, chlorophyll-a, depth) on the catch for most tuna species, except longtail tuna. The T-test revealed that chlorophyll-a and depth significantly influenced the catch of frigate tuna and mackerels, while depth alone affected bigeye, skipjack, and yellowfin tuna. No oceanographic parameters significantly influenced longtail tuna catch.

References

1. FAO: The State of World Fisheries and Aquaculture 2018 - Meeting the sustainable development goals. Food and Agriculture Organization of the United Nations (2018)
2. Marine and Fisheries Ministry: Fisheries Production. https://statistik.kkp.go.id/home.php. Accessed 12 Feb 2024
3. BPS-Statistics Indonesia: Statistical Yearbook of Indonesia (2024)
4. Umar, M.T., Safruddin, S., Zainuddin, M.: Potencial utilization of skipjack tuna (Katsuwonus pelamis) in Bone Bay **2**, 58–68 (2019)
5. Ashida, H.: Spatial and temporal differences in the reproductive traits of skipjack tuna Katsuwonus pelamis between the subtropical and temperate western Pacific Ocean. Fish. Res. **221**, 105352 (2020). https://doi.org/10.1016/j.fishres.2019.105352
6. Zhou, S., Zhang, N., Fu, Z., Yu, G., Ma, Z., Zhao, L.: Impact of Salinity Changes on the Antioxidation of Juvenile Yellowfin Tuna (Thunnus albacares). J. Mar. Sci. Eng. **11**, 132 (2023). https://doi.org/10.3390/jmse11010132
7. Djunarsjah, E., et al.: Utilization of marine geospatial data for determining optimal FAD locations in improving the living standards of the north Gorontalo coastal community. Sustainability **15**, 2242 (2023). https://doi.org/10.3390/su15032242

8. Hidayat, R., Safruddin, S.: Spatial and temporal distribution of skipjack and little tuna in fisheries management area 713. Jurnal IPTEKS PSP **10** (2023). https://doi.org/10.20956/jipsp.v10i1.23774
9. Cheng, L., et al.: Another year of record heat for the oceans. Adv. Atmos. Sci. **40**, 963–974 (2023). https://doi.org/10.1007/s00376-023-2385-2
10. Kasim, K., Widodo, A.A., Prasetyo, A.P.: Pengaruh episode la nina dan el nino terhadap produksi beberapa pelagis kecil yang didaratkan di Pantai Utara Jawa. Jurnal Penelitian Perikanan Indonesia **17**, 257 (2017). https://doi.org/10.15578/jppi.17.4.2011.257-264
11. Hartoko, A., Suradi, W.S., Ghofar, A.: Impact of climate variability on skipjack tuna (Katsuwonus pelamis) catches in the Indonesian Fisheries Management Area (FMA) 715. IOP Conf. Ser. Earth Environ. Sci. **800**, 012003 (2021). https://doi.org/10.1088/1755-1315/800/1/012003
12. Hsu, T.Y., Chang, Y., Lee, M.A., Wu, R.F., Hsiao, S.C.: Predicting skipjack tuna fishing grounds in the western and central pacific ocean based on high-spatial – temporal-resolution satellite data. Remote Sens. (Basel) **13**, 1–17 (2021). https://doi.org/10.3390/rs13050861
13. Semedi, B., Hardoko, H., Dewi, C.S.U., Syam's, N.D.S., Diza, N.F., Bayuaji, G.D.A.P.: Seasonal migration zone of Skipjack Tuna (Katsuwonus pelamis) in the South Java sea using multisensor satellite remote sensing. J. Mar. Sci. **2023**, 1–14 (2023). https://doi.org/10.1155/2023/1073633
14. Vaihola, S., Kininmonth, S.: Environmental factors determine tuna fishing vessels' behavior in Tonga. Fishes **8**, 602 (2023). https://doi.org/10.3390/fishes8120602
15. Alkoiri, M.F., Ramdhan, M., Setyamarta, G., Hayati, N.: Estimation of potential skipjack tuna fishing ground using MODIS data in 2020–2021 (case study: Bali Strait).(2024). https://doi.org/10.1007/978-981-97-0740-9_31
16. Pauly, D., Zeller, D., Palomares, M.L.D.: Sea Around Us. Tools & Data, https://www.seaaroundus.org/. Accessed 20 Dec 2020
17. GEBCO: Gridded Bathymetry Data Download. https://download.gebco.net/. Accessed 05 Dec 2020
18. NASA: Ocean Color. https://oceancolor.gsfc.nasa.gov/. Accessed 20 Dec 2020
19. Platnick, S., et al.: MODIS Atmosphere L3 Monthly Product. NASA MODIS Adaptive Processing System. https://doi.org/10.5067/MODIS/MOD08_M3.006. Accessed 02 May 2024
20. Khan, A.M.A., et al.: Oceanographic characteristics at fish aggregating device sites for Tuna
21. Semedi, B., Bakar Sambah, A., Septiani Agatha, L.: Oceanographic parameter analysis using GIS and GAM for potential fishing zone mapping of Bigeye Tuna (Thunnus obesus). J. Fish. Mar. Res. **5**, 164–171 (2021)
22. Kunarso, K., Arfa, M.Y., Setiyono, H., Rifai, A., Subardjo, P.: Respon Kesuburan dan Hasil Tangkapan Ikan Terhadap Variablitias ENSO dan IOD di Perairan Teluk Lampung Indonesia (2021)
23. Ramadhan, F., Wirasatriya, A., Maslukah, L., Handoyo, G.: Differences in depth and thickness of thermocline layers on ENSO, IOD and monsoon variability in the southern waters of Java (2021)
24. Klemas, V.: Fisheries applications of remote sensing: an overview. Fish. Res. **148**, 124–136 (2013). https://doi.org/10.1016/j.fishres.2012.02.027

Space Environment and Climate Change Interactions

D-region Ionosphere and Space Weather Phenomena: Latest Research and Impacts in AVON Project

Hiroyo Ohya[1](✉), Fuminori Tsuchiya[2], Masaharu Nakayama[1], Sora Maruyama[1], Akane Kubota[1], Yukihiro Takahashi[3], Emanuel Sungging Mumpuni[4], La Ode Muhammad Musafar Kilowasid[5], and Timbul Manik[4]

[1] Chiba University, Chiba 263-8522, Japan
ohya@faculty.chiba-u.jp
[2] Tohoku University, Sendai 980-8578, Japan
[3] Hokkaido University, Sapporo 060-0810, Japan
[4] National Research and Innovation Agency (BRIN), Bandung 40135, Indonesia
[5] National Research and Innovation Agency (BRIN), Pontianak 78241, Indonesia

Abstract. Although the D-region ionosphere is an important region for elucidating the coupling between the atmosphere, ionosphere, and magnetosphere, there are few observational methods and its characteristics are not fully understood. Due to the limited observational methods for the D-region, observational data cannot be obtained, and the International Reference Ionosphere (IRI) model is not realistic. However, because very low frequency (VLF, 3–30 kHz)/low frequency (LF, 30–300 kHz) waves are reflected at altitudes of low electron density in the D-region ionosphere, we can know up-down movement at equal electron density by VLF/LF observations. Since 2008, we have established the Asia VLF observation network (AVON) in five Southeast Asian countries (Indonesia, Taiwan, Thailand, Philippines, and Vietnam) for the purpose of observing lightning discharges and the D-region ionosphere. In this study, we will present the recent results of D-region ionospheric variations associated with solar flares, solar eclipses, ULF-modulated energetic particle precipitation, earthquakes, and the Tonga volcanic eruption. These VLF/LF wave observations can contribute to space weather.

Keywords: D-region Ionosphere · AVON · Solar Flares

1 Introduction

The D-region ionosphere, defined as the region of 60–90 km height, is a link between the solid Earth and E-/F-region ionosphere, magnetosphere, and solar activities. The D-region is an important region for various phenomena from above and below. However, for the D-region ionosphere observations, we cannot use usual observation tools such like ionosondes, GNSS-total electron content (TEC) data, and radars, because of high collision frequency for electrons with neutrals (10^5–10^8 s^{-1} at 60–90 km height, and 10^1–10^2 s^{-1} at 300 km height). Rocket measurements have high accuracy for

E. Fitrianingsih et al. (Eds.): ISAST 2024, SPPHY 416, pp. 153–160, 2025.
https://doi.org/10.1007/978-981-96-1344-1_16

measuring electron density in the D-region, although the location and time of rocket launches are limited. The D-region height is too high for balloon observations, and too low for usual tools of ionospheric observations. Traditionally, very low frequency (VLF, 3–30 kHz)/low frequency (LF, 30–300 kHz) waves that are radiated from lightning discharges and manmade transmitters have been used for studies of the D-region ionosphere. Routine observations of VLF/LF waves can be performed, which amplitude and phase respond sensitively to variations in electron density. Manmade narrowband VLF/LF transmitter signals also have been used for the D-region studies, for example, solar flares, energetic electron precipitations, earthquakes, volcanic eruptions and so on [1]. Currently, we have installed AVON (Asia VLF Observation Network) and OCTAVE (Observation of CondiTion of ionized Atmosphere by VLF Experiment) network. In this study, we will show some D-region phenomena observed by VLF/LF waves and discuss future directions of the D-region studies.

2 AVON Project

Observation sites of AVON are Tainan (TNN, Taiwan), Saraburi (SRB, Thailand), Pontianak (PTK, Indonesia), Los banos (LBS, Philippines) and Sontay (HNI, Vietnam) (Fig. 1). The top three lightning regions on the Earth are North/south America, Africa, and Southeast Asia. However, there were no lightning observation network in Southeast Asia, although NLDN (National Lightning Detection Network) and ACLENet (African Centres for Lightning and Electromagnetics) have been installed in the North America and Africa, respectively. Therefore, we constructed a new observation system for monitoring lightning in the Southeast Asia.

OCTAVE has been established in Japan, North America, and Europe by our research group with the following objectives: energetic particle precipitation from radiation belts [3–7], lightning effects on upper and middle atmospheres, solar X-ray flare and proton event effect on upper and middle atmospheres [8], effects of solar eclipse [9], atmospheric gravity waves [10], and sound wave on the lower ionosphere (Fig. 2) [11, 12], and detection of gamma ray burst. AVON in Southeast Asia is a part of the OCTAVE.

The D-region ionosphere is the lowest layer of the Earth's ionosphere located between altitudes of approximately 60 to 90 km (Fig. 2). This layer is responsible for the absorption of HF (High Frequency, 3–30 MHz) radio waves, which makes it an important factor in communication and navigation systems.

In Southeast Asia, the D-region ionosphere is affected by a number of different phenomena. One of the most significant is the occurrence of ionospheric irregularities, which are caused by fluctuations in the electron density of the ionosphere. These irregularities can disrupt radio communication and navigation systems, particularly in the equatorial region where they are most prevalent.

In addition to these phenomena, the D-region ionosphere in Southeast Asia is also affected by solar activity, particularly during periods of high solar flux. This can cause increased ionization in the ionosphere, which can lead to enhanced absorption of radio waves and disruptions to communication and navigation systems.

Overall, the D-region ionosphere in Southeast Asia is a complex and dynamic system that is influenced by a wide range of factors. Understanding these phenomena is important

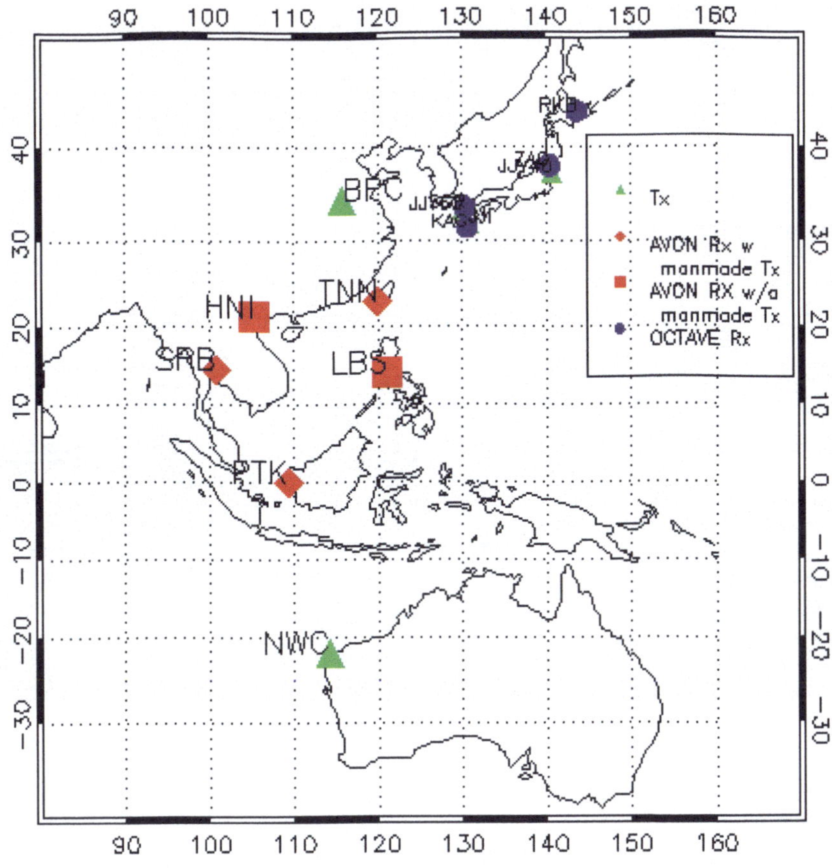

Fig. 1. Location of transmitters and OCTAVE/AVON receivers used in this study.

for ensuring the reliable operation of communication and navigation systems in the region.

3 Variations in the D-region Ionosphere

3.1 X-class Solar Flares

We investigated horizontal inhomogeneity for the D-region height during four X-class solar flares occurred from 2014 to 2022. Variations in VLF/LF amplitudes and phases were observed in 27 out of the 40 paths (Fig. 1). Note that observations at SGR and ZAO in Fig. 1 had already finished. The observational periods at SGR and ZAO were 27 November, 2014 – 26 December, 2017 and 10 October, 2007 – 21 February, 2010, respectively. Figure 3 shows an example of VLF/LF variations during a X-class flare. The amplitude of the variations in intensity (ΔA) was 2.65–14.73 dB, while the amplitude of the variations in phase (ΔP) was 31.0°–150.3°. Based on wave-hop method, we calculated reduction in the reflection height (Δh) from observed ΔA and ΔP. For each

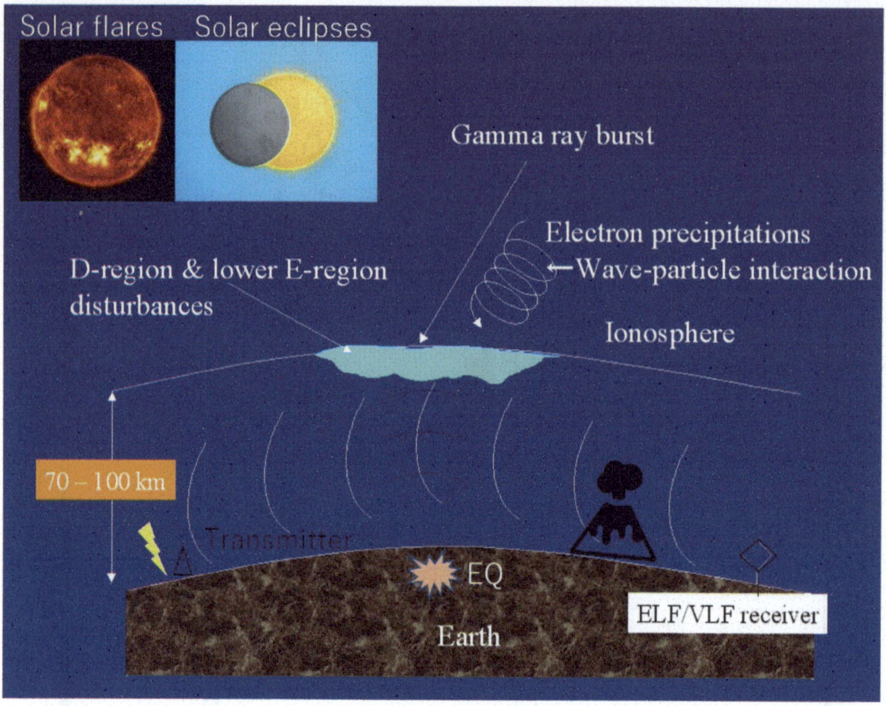

Fig. 2. Factors affecting the D-region ionosphere.

class of solar flare, the trend of Δh with solar zenith angle at the mid-point of each path was different. We showed that maximum rate of change in electron density ($\Delta N_{\max}$) for east-west propagation path was estimated larger than that of north-south one (Fig. 4). We also showed that reflection coefficient in the lower ionosphere for west to east propagation was larger than that for east to west one, which suggests that the attenuation of VLF/LF waves change due to the angle between the propagate direction and magnetic field lines. Therefore, we found that the amount of increase in electron density was underestimate for north-south and east to west propagations.

During the X5.89-class solar flare at 01:23 UT on May 11, 2024, the phase of the JJY60kHz-PTK path oscillated [13]. The oscillation period was about 550 s and the amplitude of the oscillation was about 20 dB (Fig. 5). Such oscillations during solar flares are a rare phenomenon, and usual VLF/LF variations follow a similar one to the X-ray flux. Solar protons observed by GOES-16 satellite fluctuated with the similar period. However, the cause of this LF phase variation is still unknown.

3.2 Solar Eclipse

During solar eclipses, electron density in the D-region ionosphere (60–100 km) decreases due to a decrease in solar radiation. This can cause communication satellite interference and GPS positioning inaccuracies, so it is important to understand the disturbances in the D-region ionosphere, such as variations in electron density. We analyzed for a hybrid

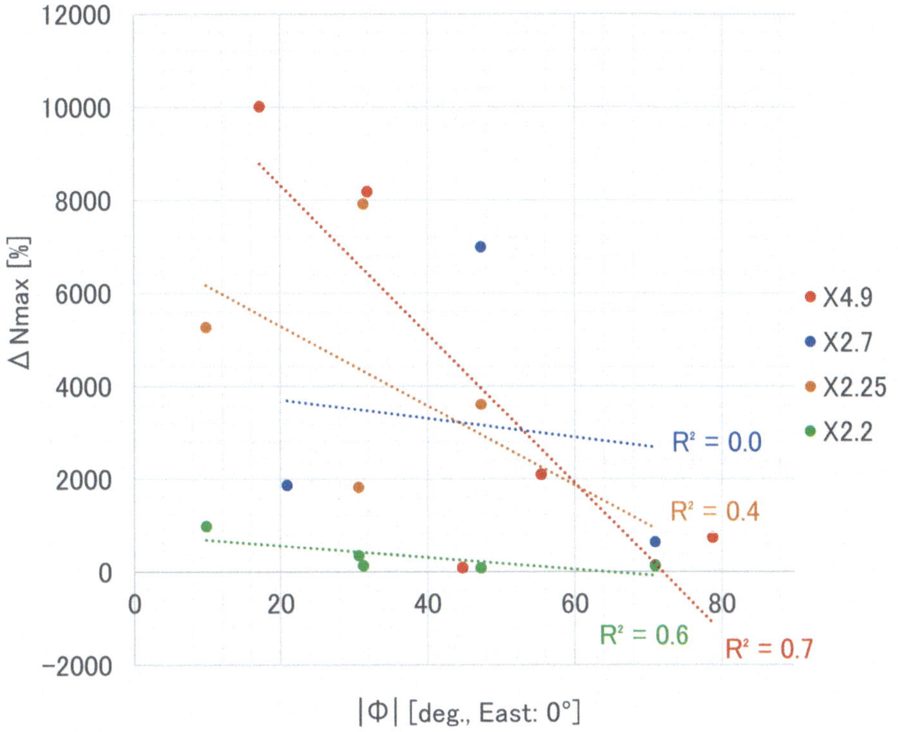

Fig. 3. From the top panel, X-ray flux derived with GOES-15 satellite, amplitude for JJY40-RKB, JJY60-RKB, and BPC-RKB during a X2.25 class solar flare on 20 April, 2022.

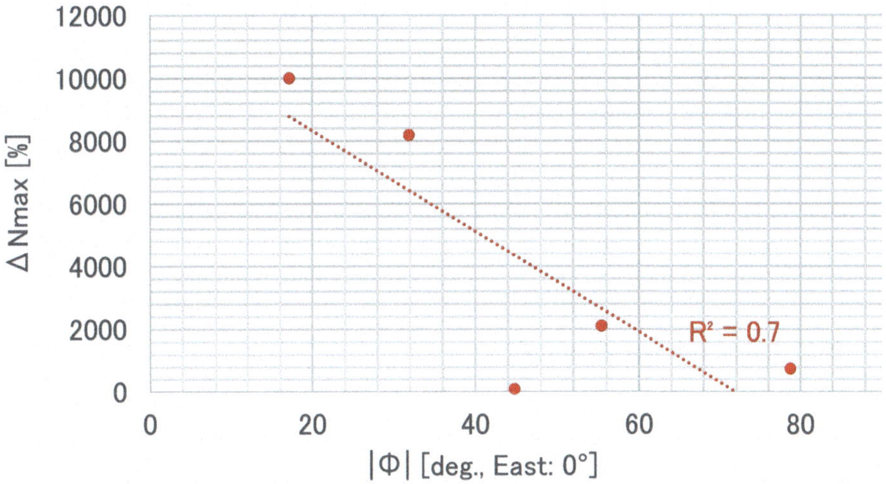

Fig. 4. $\Delta N_{\max}$ vs. azimuth at mid-point of paths.

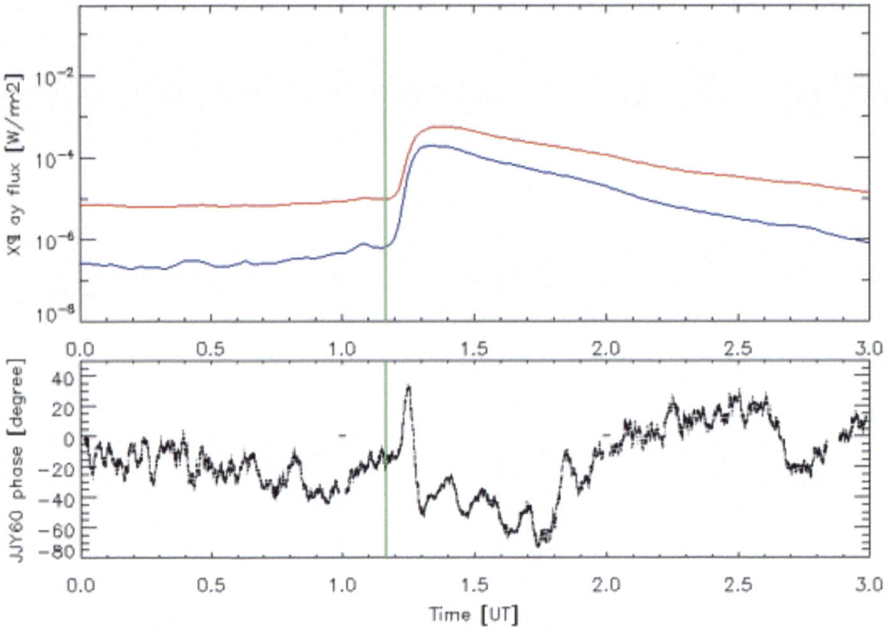

Fig. 5. (Upper) X-ray flux obtained from GOES-16 satellite and (lower) phase of JJY60kHz-PTK path on May 11, 2024. The green lines indicate onset of X-5.89 class solar flare.

solar eclipse occurred during 01:34:15–06:59:13 UT on 20 April, 2023. Variation in VLF amplitude of about 2 dB was observed for the NWC-KAG (North West Cape, Australia – Kagoshima, Japan) path (Fig. 6), which was less than that at sunset/sunrise time. Reflection height at the mid-point of the NWC-KAG path increased by 8 km compared with usual daytime. In addition, occurrence number of tweeks increased during the solar eclipse, which shows that electron density decreased and reflection coefficient in the lower ionosphere increased. There were no variations for the NPM-KAG (Lualualei Hawaii, USA – KAG) and JJY40-KAG (Mount Otakadoya, Japan – KAG) paths, because the eclipse magnitude was small to be less than 0.5 over the two paths. The peak time of the variation in VLF amplitude was close to the time when distance between the NWC and center of the Moon's umbra was minimum, suggesting that the eclipse magnitude at the NWC is one of the main causes of the VLF amplitude variation.

3.3 2022 Tonga Volcanic Eruption

The submarine volcano Hunga Tonga–Hunga Ha'apai erupted explosively on January 15, 2022, providing a chance to study interactions between the atmosphere and ionosphere caused by Lamb and Pekeris waves. Although Pekeris wave resonance had not been previously detected in the D-region ionosphere, this study used multi-point monitoring of the lower ionosphere (the AVON) and atmospheric electric field. Oscillations of 100–200 s were observed in LF transmitter signals and magnetic and atmospheric electric fields due to Pekeris waves, although no corresponding atmospheric pressure changes

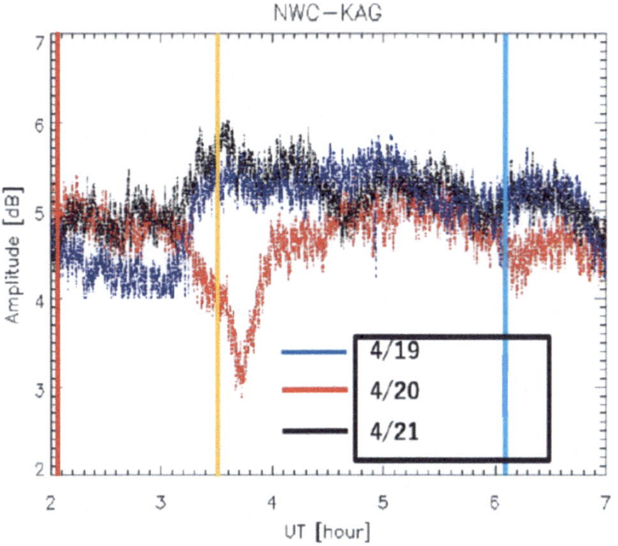

Fig. 6. Amplitude for NWC-KAG path during a hybrid solar eclipse on 23 April, 2023. The red line indicates solar eclipse day.

were detected on the ground. Simulations showed Pekeris waves oscillating near the mesopause, suggesting resonance, which projected onto the Earth's surface through a global electric circuit [2, not shown].

4 Conclusions

The D-region ionosphere varies due to various factors: solar flares, solar eclipses, energetic particle precipitations, earthquakes, and volcanic eruptions. The VLF/LF waves are useful tool for monitoring the D-region ionosphere. The AVON network is to monitor lightning discharges and the D-region ionosphere in the Southeast Asia. The AVON data can contribute on space weather research.

It is essential to develop the capabilities of motivated interdisciplinary scientists to enhance space weather monitoring and measurement in Southeast Asia. The AVON working group plans to organize workshops and mentorship programs to support and encourage young researchers and students in Southeast Asia in the future.

References

1. Nakayama, M., Ohya, H., Tsuchiya, F., Nozaki, K., Shiokawa, K., Nakata, H.: Horizontal inhomogeneity in the D-region ionosphere during an X-class solar flare determined by OCTAVE VLF observations. URSI Radio Sci. Lett. **4** (2022)
2. Ohya, H., Tsuchiya, F., Takamura, T., Shinagawa, H., Takahashi, Y., Chen, A.B.: Lower ionospheric resonance caused by Pekeris wave induced by 2022 Tonga volcanic eruption. Sci. Rep. **14**, 15659 (2024). https://doi.org/10.1038/s41598-024-65929-x

3. Hirai, A., et al.: Spatio-temporal characteristics of IPDP-type EMIC waves on April 19, 2017: implications for loss of relativistic electrons in the outer belt. J. Geophys. Res.: Space Phys. **128**, e2023JA031479 (2023). https://doi.org/10.1029/2023JA031479
4. Tanaka, K., et al: Ultra low frequency modulation of energetic electron precipitation in the D-region ionosphere in a magnetically quiet time using OCTAVE very low frequency and low frequency (VLF/LF) observations. special section on URSI-JRSM2022. Radio Sci. Lett. **4** (2022). https://doi.org/10.46620/22-0049
5. Miyashita, T., et al.: ULF modulation of energetic electron precipitations observed by VLF/LF radio propagation. Radio Sci. Bull. 372 (2020).http://www.ursi.org/content/RSB/RSB_372_2020_03.pdf
6. Hirai, A., et al.: Temporal and spatial correspondence of Pc1/EMIC waves and relativistic electron precipitations observed with ground-based multi-instruments on 27 march 2017 Geophys. Res. Lett. (2018). https://doi.org/10.1029/2018GL080126
7. Tsuchiya, F., et al.: Energetic electron precipitation associated with pulsating aurora observed by VLF radio propagation during recovery phase of substorm on 27 March 2017 Geophys. Res. Lett. (2018). https://doi.org/10.1029/2018GL080222
8. Nakayama, M., Ohya, H., Tsuchiya, F., Nozaki, K., Shiokawa, K., Nakata, H.: Horizontal inhomogeneity in the d-region ionosphere during an X-classsolar flare determined by OCTAVE VLF observations. special section on URSI-JRSM2022. Radio Sci. Lett. **4** (2022). https://doi.org/10.46620/22-0050
9. Ohya, H., et al.: Reflection height of daytime tweek atmospherics during the solar eclipse of 22 July 2009. J. Geophys. Res. **117**, A11310 (2012). https://doi.org/10.1029/2012JA018151
10. Nozaki, Y., Ohya, H., Tsuchiya, F., Nozaki, K., Nakata, H., Shiokawa K.: Local time and seasonal variations in the D-region ionosphere: does itreflect sudden stratospheric warming effects?. Special section on URSI-JRSM2022. Radio Sci. Lett. 4 (2022). https://doi.org/10.46620/22-0045
11. Akashi, T., Ohya, H., Tsuchiya, F., Nozaki, K., Nakata, H.: Variation in the D-region ionosphere after the 2015 Nepal earthquake using LF transmitter signals. J. Atmos. Electr. **40**(1), 1–9 (2021). https://doi.org/10.1541/jae.40.1
12. Ohya, H., Tsuchiya, F., Takishita, Y., Shinagawa, H., Nozaki, K., Shiokawa, K.: Periodic oscillations in the D region ionosphere after the 2011 Tohoku earthquake using LF standard radio waves. J. Geophys. Res.-Space Phys. **123**(6), 5261–5270 (2018). https://doi.org/10.1029/2018JA025289
13. AVON and list of VLF-transmitter. http://radio.gp.tohoku.ac.jp/db/LF/doc/avon/readme_LF_IDL.pdf. Accessed 31 July 2024

Ethnoastronomical Knowledge Surrounding Timau National Observatory

Sastri Sunarti[1], Helmina Kastanya[1], Thomas Djamaludin[2](✉), Abdul Rachman[3], Erwin Kembaren[1], and Primus Lake[4]

[1] Research Center for Manuscripts, Literature, and Oral Traditions, National Research and Innovation Agency (BRIN), Jakarta, Indonesia
sast003@brin.go.id

[2] Research Center for Space, National Research and Innovation Agency (BRIN), Bandung, Indonesia
thom001@brin.go.id

[3] Timau National Observatory, National Research and Innovation Agency (BRIN), Kupang, Indonesia

[4] Faculty of Social and Political Sciences, Universitas Nusa Cendana, Kupang, Indonesia

Abstract. Currently, the Indonesian government is constructing a national observatory on Mount Timau, Amfoang, Kupang. The selection of the Mount Timau area was based on the dry climate and sky conditions in the area which are still dark, far from light pollution, and it is easy to observe both the northern and southern skies. Behind the existence and construction of the national observatory, it turns out that the Amfoang people already have ethnoastronomical knowledge. Exploring this ethnoastronomical knowledge is the aim of this research, i.e., to minimize the potential conflict that could occur due to the impact of the construction of the Timau National Observatory of Indonesia (TNOI). The ethnoastronomical knowledge is still stored in people's memories and is still conveyed orally. We hope that this ethnoastronomical knowledge can support Astro-tourism that is free of light pollution and also support the existence of the national observatory. Primary data consists of ethnoastronomical knowledge obtained using in-depth interview techniques and field observations. To support data analysis, this research uses collective memory, ethnographic and ethnoastronomical approaches. Through this research, it is proven that the Amfoang people already have ethnoastronomical knowledge which is useful for determining farming seasons, rainy seasons, dry seasons, disaster prediction, and time period for harvesting honey. The ethnoastronomical knowledge of the Amfoang people was then compared with a sky map to confirm the names and positions of the stars in the sky. Identification of stars or constellations is also used to find links between people's beliefs and rational astronomical reasons.

Keywords: Ethnoastronomy · Timau National Observatory · Astro-tourism

E. Fitrianingsih et al. (Eds.): ISAST 2024, SPPHY 416, pp. 161–169, 2025.
https://doi.org/10.1007/978-981-96-1344-1_17

1 Introduction

Currently on Mount Timau, Amfoang districts area, Kupang Regency, East Nusa Tenggara (NTT), the Indonesian national observatory called Timau National Observatory of Indonesia (hereinafter abbreviated as TNOI, pronounced as /tinoi/) is being built. Mount Timau was chosen as the location for TNOI for several reasons. Firstly, Timau's sky is among the longest and darkest nights of the year nationally [1, 2]. Secondly, observations of the northern sky and southern sky can be done at the same time from the site due to its location near the equator. Apart from that, the construction of TNOI is a complement to the Bosscha Observatory in Lembang which is 100 years old now and has begun to experience severe light pollution from Bandung and the surrounding area [3]. The International Commission on Illumination (CIE) defines light pollution as the "sum total of all adverse effects of artificial light" [4]. Mander, et al. [5] revealed that light pollution has affected many aspects of life. It not only disrupts astronomical observations but also human health and the life of wild animals and plants. The direct impact felt is the loss of the beauty of the night sky in big cities because starlight is lost to light pollution from city lights.

To support the existence of TNOI which is free from light pollution, a cultural approach is needed, such as exploring ethnoastronomical knowledge in the Amfoang people. Ethnoastronomy is the knowledge of reading the stars in traditional communities which is useful as a marker of seasons, direction markers in fishing communities, also a marker of planting time, and a marker of lunar and solar eclipses [6]. This ethnoastronomical knowledge is generally possessed by rural and coastal communities, including the Amfoang people in the area around Mount Timau, especially in district of Central Amfoang.

The cultural approach referred to is an approach based on ethnoastronomical knowledge which is related to belief in the universe. As is the belief in *Uis Neno* (Lord of the Sky) and *Uis Pah* (Lord of the Earth) in the agrarian community in Amfoang which depends on the regularity of the planting season and climate. Uis Neno in this belief is described as a fertility god who regulated the seasons, gave rice and corn, and regulated nature. However, Uis Neno is also believed to bring long droughts which cause plants to die and bring pests and diseases to crops and livestock, as well as to humans [7–9].

The world view of the Amfoang indigenous people towards the universe, which originates from local beliefs, is still passed on to this day as knowledge that can be used to read the stars and the universe. Departing from the Amfoang indigenous people's belief in the universe, this research aims to synergize the development of TNOI with the Timau people's ethnoastronomical knowledge and belief in the universe. It is hoped that ethnoastronomical knowledge can strengthen the existence of TNOI, reduce potential conflicts that can occur due to land struggles, and advance the ethnoastronomical wealth of local communities; as well as the development of night sky tourism or Astro-tourism. Astro-tourism is a tourism activity with astronomical phenomena or facilities as the main attraction [10]. It will adopt a unique tourism concept that combines natural beauty and the strength of local culture with the beauty of the night sky that is free of light pollution. The condition of the night sky that is free of light pollution is known as "Dark Sky".

Several ethnoastronomical studies review cultural aspects from an astronomical perspective while others review astronomical aspects from a cultural perspective. Some examples are Sundanese culture viewed from astronomical aspects by astronomer [11] and Madurese culture related to astronomical phenomena [12]. Reflecting on previous research, it is hoped that exploring the ethnoastronomical knowledge of the Amfoang community in Timau can support the existence of TNOI.

2 Methods

On 6 until 10 June 2023, we conducted interviews and a focus group discussion with several Amfoang traditional leaders. The main research locations are in four villages, namely Amfoang, Bioba, Kauniki, Bitobe, and Fatu Monas, Central Amfoang District, Kupang Regency, NTT. The research team conducted in-depth interviews with key informants and supporting informants [13]. In-depth interviews were carried out with assistance from field assistants at each location to obtain primary data related to ethnoastronomical knowledge that was still stored in the community's memory. Interviews were conducted using audio recording which were then transcribed and transliterated as part of the work steps in oral tradition research. Meanwhile, ethnoastronomical information and terms in the Dawan language (the name of the local language) will be described and compared with modern astronomical terms (Indonesian and English).

To assist this research, the collective memory approach from [14] was used. Ong [15] and Sweeney [13] are several experts who studied oral traditions in Europe and Southeast Asia. The cultural aspect is related to the various messages contained in the oral tradition and the rules of implementation and symbols used to express that oral tradition. In this sense, oral traditions are an important source of knowledge for the formulation of cultural identity because the values contained in them play a role in the life of the people who own the oral traditions in question.

As a traditional society that depends on nature, the Amfoang people at the foot of Mount Timau are very familiar with the sky. Some constellations or celestial phenomena were associated with determining seasons for farming, honey harvesting or hunting, and certain other omens. Constellation analysis uses the night sky map published by the Planetarium & Observatory of Jakarta.

3 Results

3.1 Ethnoastronomical Knowledge: Stars as Markers of the Planting Season and Honey Harvest Time

As an agricultural society that lives in dry mountainous areas, agriculture is rain-fed. Therefore, they are very dependent on the accurate prediction of rainy season and dry season. From empirical experience of knowing nature, the Amfoang people are equipped with local astronomical knowledge to read signs from the stars (called in local language as *maklafu*) to determine the right time to start farming. The appearance of *Ha ma Gua* (the "four and two" stars) and *Maklafu Kotog* (a cluster of seven stars) is used as a sign that the rainy season is approaching. *Maklafu Ha ma Gua* indicates that planting can begin immediately. If *Maklafu Kotog* disappears accompanied by light rain, it is a sign that the start of the rainy season is about to happen. If the four stars disappear accompanied by several times of rain, it is a sign that the very heavy rainy season will arrive. If the two stars also disappear, it means the rainy season has started.

Amfoang is known as the region that has the best honey. The trees in the Amfoang forest are place for honeycombs that can produce 50–80 honeycombs in one tree. The tree where the bee's nest is called *Fanik*. The types of Fanik most often chosen by bees are the tree of *neke fiug* (forest cottonwood tree), *besak, bonak, anisa, taupi, nunuh noh anaek* (big leaf banyan) (interview with Mr. Yeremias Sonbai, 67 years old, in Kauniki, South Amfoang village, June 2023). Mr. Yeremias Sonbai further added that the most appropriate time to harvest honey is when the light of the *Sua Oni* (Evening Star, Venus) begins to disappear. When the light dims, i.e., only one yellow light appears, the ladder for climbing the honeycomb can be prepared. When the light has completely disappeared, honey harvesting can be done without having to worry about being stung by bees. Bees will be aggressive when the *Sua Oni* light is still visible. Honey harvest is carried out in the dark months of June and September. Before the honey harvest, traditional elders such as Mr. Yeremias are asked to carry out a honey harvest ritual called *Toit Oni*. The ritual is usually carried out under a *Fanik* tree which is full of beehives. The person in charge of harvesting honey is called *Ahel Oni*. An *Ahel Oni* must fulfil the main requirements to be successful in carrying out his duties. The conditions are clean intentions and behavior.

3.2 Stars as Markers of the Dry Season and the Rainy Season

Apart from marking the farming season, observing the stars in the sky for the Amfoang people is also used to mark the arrival of the dry season. For example, if the presence of *Maklafu Kotog* fades and disappears without rain, it is a sign that there will be a long dry season. This long dry season will also have an impact on famine and the death of livestock such as cows, horses and goats because the grasslands are dry and barren.

Another sign of famine can also be seen from the presence of clouds in the sky which are called in the Dawan language as *Bi afu neog*, a phenomenon of large clouds associated with a mother and small clouds associated with a child. If the mother cloud appears without a small cloud (child), it is a sign that starvation will occur. On the other hand, if it appears together (the mother holding the child), the opposite sign is that the harvest will be abundant (interview with Ina Meriana, June 2023, in Fatu Monas village, Central Amfoang District, Kupang).

4 Analysis

Various stories about the sky as gathered from the interview are then confirmed by certain constellations in the sky or by showing pictures of certain constellations according to the characteristics described. The narrative is then compared with astronomical phenomena to describe the practice of ethnoastronomy in the Amfoang community. Some can be linked logically to the position of celestial bodies in certain seasons, but there are also those that are simply public beliefs that are beyond scientific logic.

A common phenomenon that is usually clearly observed (from a pristine dark sky) is the appearance of the Milky Way Galaxy as seen from Earth, which looks like a hazy band of light. The Japanese calls it as *Amano Gawa* (sky river) or *Ginga* (silver river). It turns out that the Amfoang people also call it *Noel Neno* means river in the sky.

Scattered stars in the sky are generally called *maklafu*. Sometimes there are meteors that pass by which are called *Maklafu Naen* (running stars). The brightest star, the Morning Star or the East Star have special meaning in Amfoang culture. This bright star is actually the planet Venus. When it looks like a bright star on the western horizon after sunset, the Evening star, in Amfoang it is called *Sua Oni*. When it looks like a bright star on the eastern horizon before sunrise, it is called the Morning Star or *Fafnome timog*, which means the star that measures the night.

The setting of *Sua Oni* is used as a sign of the beginning of the dark night. In the honey harvest season during the dark moon in June and September, the honey harvest will not be started if *Sua Oni* is still visible, since it means the night is not dark enough yet. The honeycomb collectors (called *Ahel Oni*) are just getting ready by installing ladders. If the honey harvest is done before dark at night, it is feared that swarms of bees will attack. When the altitude of *Sua Oni* is low, this will mean that the time to harvest honey is longer, because *Sua Oni* sets more quickly. The Morning Star is usually called the star that measures the night [*Fafnome timog*], because its appearance is a sign that morning is approaching. That is the time to stop the honey harvest.

Among the *maklafu* there are several constellations or series of stars which are used as indicators of the seasons. An example for this case is what the Amfoang people calls *Ha ma Gua* (means the four and two) stars. From the description given by the source and confirmed by the shape of the constellation, the *Ha ma Gua* stars turn out to be the four stars of the Southern Cross constellation (Crux) and the two bright stars from the Centaurus constellation (the Alpha and Beta Centaury) (see Fig. 1). When *Ha ma Gua* has set in the south-southwest (SSW) at the beginning of the night, this is a sign of the start of the rainy season. Specifically, if *Fafnome haen* (the four stars of the Southern Cross constellation) is leaning to the west and approaching the skyline at the beginning of the night then it is a sign that rain is coming. Therefore, it is time to plant. Amfoang people calls the two bright stars as *Fafnome angua*. From astronomical analysis, in October the constellation of the Southern Cross and the two bright stars nearby set on the western horizon early in the evening in October, which is a sign of the rainy season (Fig. 2). When *Fafnome haen* (the constellation of the Southern Cross) begins to tilt westward in the morning, it is the beginning of April, which is the start of the dry season. In addition, if *Ha ma Gua* rises in the east at the beginning of the night then it is the early of March which means that dry season is coming.

Apart from constellations, there are also star clusters that are used as references for the seasons. The Seven Stars (Pleiades) which are used by several cultures [16] in relation to the seasons are also used by the Amfoang people. The Pleiades stars called *Maklafu Kotog* are a sign of the end of the dry season, the beginning of the rainy season. When it rises after sunset (at the beginning of evening), the planting season begins. If it rises around 9 PM, it is a sign that it is still the dry season. If it rises in the east early in the morning (obviously before sunrise), it is a sign that the air temperature is very cold. Astronomically, if it rises in the east at the beginning of the night, it occurs around November, a sign of the end of the dry season towards the start of the rainy season. If it rises around 9 PM then it is still in early October which is still the dry season. If it rises in the east early in the morning, it corresponds to June, in which the wind starts blowing from Australia which at the beginning of their early winter season.

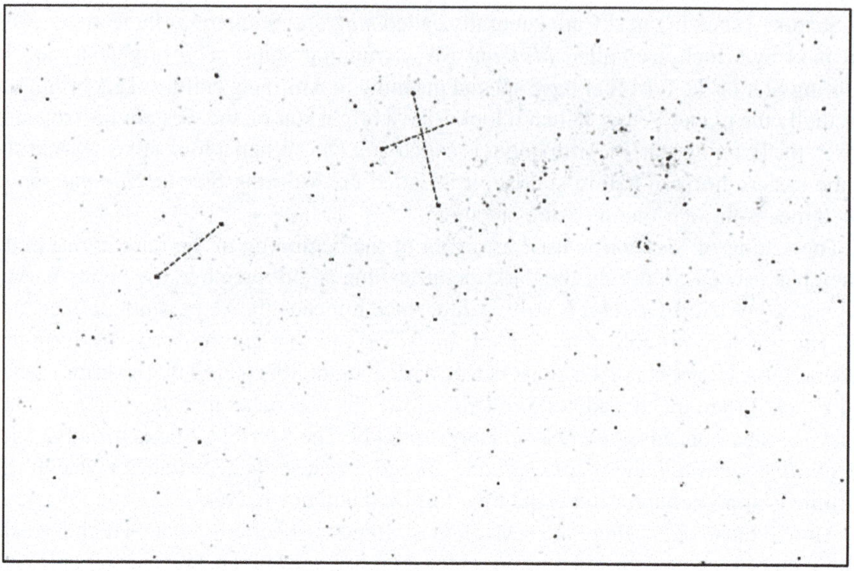

Fig. 1. *Ha ma Gua* (four and two) stars which are the four stars of the Southern Cross constellation on the top center of the image and the two stars (Alpha and Beta Centaury) on the left of the image. Connecting lines are added for easy identification. Photo by the research team (converted to negative photo for clarity).

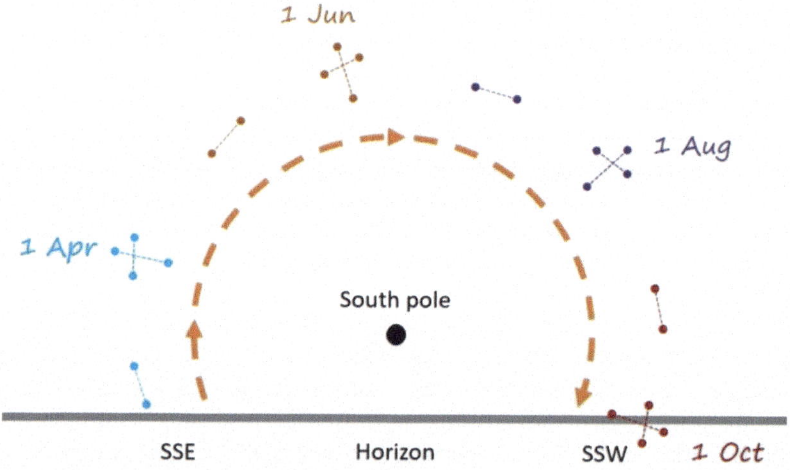

Fig. 2. The appearance of *Ha ma Gua* (four and two stars) at around 7 PM local time in the beginning of the month since April until October. Arrows indicate the apparent motion of the stars in the sky.

Table 1 shows a summary of the ethnoastronomical knowledge of the Amfoang people, analysis of modern astronomy, and its meaning for the Amfoang people.

Table 1. Ethnoastronomical knowledge of the Amfoang community, its analysis based on modern astronomy and its meaning according to the local people.

No	Ethnoastronomical Knowledge Of the Amfoang Community	Modern Astronomy	Meaning For the Amfoang Community
1	*Noel Neno* (river in sky)	Milky Way	Observable phenomena
2	*Maklafu*	General stars in sky	Observable phenomena
3	*Maklafu Naen* (running stars)	Meteor	Observable phenomena
4	*Fafnome timog*	"Morning Star" (Venus, above East horizon, before sunrise)	Night measuring star When it rises in the east, it is a sign that morning is approaching, a sign to stop honey harvest
5	*Sua Oni*	"Evening Star" (Venus, at beginning of the night). The setting of the Evening Star can be a sign of the onset of darkness. If the position is low, it means that the Evening Star is setting quickly, or that nighttime is considered early	If it doesn't disappear, it's a sign that it's not time to take the bees, just prepare by installing a ladder. If it is harvested the bees will attack. If the Evening Star is low, the honey harvest will take longer
6	*Ha ma Gua* stars (four and two stars): - *Fafnome Haen* (the four stars) - *Fafnome Angua* (the two stars)	Four stars of Southern Cross (Crux) and two bright stars (Alpha dan Beta Centaury) - In October the constellation of the Southern Cross and two bright stars nearby set on the western horizon early in the evening, as a sign of the rainy season - If it rises in the east at the beginning of the evening, it is at the beginning of March, towards the start of the dry season	When it sets in the west, it is a sign of the beginning of the rainy season When it rises in the east, the beginning of the dry season
7	*Maklafu Kotog*	Seven stars (Pleiades) - When it rises in the east at the beginning of the night, it occurs around November, a sign of the end of the dry season towards the start of the rainy season - If it rises around 9 PM, it's still early October, still the dry season -If it rises in the east in the morning, it occurs in June, when the wind starts blowing from the direction of Australia which enters the cold season	A sign of the end of the dry season, the beginning of the rainy season, when it rises after sunset, the planting season begins. If it rises around 9 PM, it is a sign that it is still the dry season If it rises in the east in the morning, it is a sign that it is very cold

5 Discussions

Jaafar and Khairuddin [6] said that behind each name of an astronomical object it symbolizes the relationship between humans, animals, plants and the weather in a certain seasonal window. This phenomenon can also be found in the Timau, Amfoang people who observe astronomical objects to mark certain times and seasons such as the dry season marked by the appearance of *maklafu Ha ma Gua* (the four stars of the constellation Southern Cross -- Crux and two bright stars which are Alfa and Beta Centaury) when rises in the east. *Maklafu Kotog* (Pleiades) marks the rainy season which is not heavy and the start of good farming. *Manas ntes* (the sunset) in the direction of the Termanu estuary, which is to the southwest of Amfoang, means that the sun is to the south of the west point, around February. This is a sign of the peak of the rainy season which will last seven-twelve days will be heavy rain. At this time the plants must be planted to get a good source of water for the dry land rice field.

Likewise, astronomical observations for the Amfoang people are useful in determining the best time to harvest honey. The result of observing the sky for the purpose of harvesting honey is when *Sua Oni* (planet Venus, evening bright star) has begun to disappear at night so the sky becomes dark. The darkness of the sky calms the bees whose honey is being taken so that they are not vicious and sting the honey harvesters who are near their nests (interview with Agustinus, Yeremias Sonbai, and Mama Yohanna Maumeley).

The same as the statement of Jaafar and Khairuddin [6], the Amfoang people in Timor, who have a tradition of farming rain-fed rice fields, rely on observing celestial objects such as the position of the stars, the Sun and the Moon to determine the right time when planting crops. For example, the position of the *Ha ma Gua* stars (four stars of Crux and two nearby Centaury stars) as sign of the beginning of dry or rainy seasons. Likewise, the position of the Pleiades star cluster, which is called *Maklafu Kotog* in the Dawan language, marks the end of the dry season and the weather starts to cool in the Timau mountains.

Often ethnoastronomical knowledge is also linked to the beliefs and religion of local communities. Likewise, among the Amfoang people, the ethnoastronomical knowledge they have is also closely related to the heritage of belief and religion that they inherited, namely the belief in Uis Neno [Lord of Heaven] and Uis Pah [Ruler of the earth]. Uis Neno is the supreme ruler and Uis Pah refers to rulers on Earth who are often associated with kings who have power over land, forests and their contents [7, 9, 17].

6 Conclusions

Ethnoastronomy studies of the Amfoang people around Mount Timau show that cultural aspects related with celestial bodies have influenced their lives. *Maklafu Kotog* or the Seven Stars (Pleiades) as well as the *Ha ma Gua* stars (the "four and two" stars) or the Southern Cross Constellation (four stars) and the two brightest stars of the Centaury constellation are used to mark the rainy season to start planting and mark the beginning of the dry season. The bright *Sua Oni* (evening Star) and *Fafnome Timog* (Morning Star), which are actually the planet Venus, are used as markers for the darkest night which is the time to harvest honey. The activity is carried out during the dark months (near new moon) in June and September. Astronomical analysis shows that some of the ethnoastronomical beliefs have scientific basis.

Local wisdom about the night sky, which still guides farming and honey harvesting and other activities, needs to be preserved. The appearance of *Noel Neno* (river in the sky) is also similar to other cultures which call the huge band of stars (Milky Way) in the dark night sky as *Amano Gawa* (the sky river according to Japanese culture). *Noel Neno*'s appearance needs to be protected, because it is becoming increasingly rare in cities that are already affected by light pollution.

Acknowledgments. This research was supported by research grant 2023 from National Research and Innovation Agency (BRIN).

References

1. Hidayat, T., Mahasena, P., Dermawan, B., Hadi, T.W., Premadi, P.W., Herdiwijaya, D.: Clear sky fraction above Indonesia: an analysis for astronomical site selection. Mon. Not. R. Astron. Soc. **427**(3), 1903–1917 (2012)
2. Priyatikanto, R., et al.: Characterization of Timau national observatory using limited in situ measurements. Mon. Not. R. Astron. Soc. **518**(3), 4073–4083 (2023)
3. Herdiwijaya, D., et al.: Measurements of sky brightness at Bosscha observatory. Indonesia Heliyon **6**(8), e04635 (2020)
4. CIE. e-ILV entry 17–29–177: Light Pollution (2022). https://cie.co.at/eilvterm/17-29-177. Accessed 27 June 2023
5. Mander, S., Alam, F., Lovreglio, R., Ooi, M.: How to measure light pollution—a systematic review of methods and applications. Sustain. Cities Soc. **92**, 104465 (2023). https://doi.org/10.1016/j.scs.2023.104465
6. Jaafar, N.F., Khairuddin, A.H.: Astronomical knowledge and practices of the Orang Asli of Malaysia. In: historical and cultural astronomy: exploring the history Southeast Asian astronomy, a review of current project and future prospects and possibilities, pp. 526–539. Springer Nature Switzerland AG 2021 (2021)
7. Taum, Y.Y.: Tradisi Fuah Pah: Ritus dan Mitos Agraris Masyarakat Dawan di Timor, in Bunga Rampai Bahasa Merajut Sastra Merunut Budaya. Jogyakarta. Downloaded from https://www.academia.edu/3231297/T. Accessed 31 Jan 2023
8. Benyamin, R.F.: Tradisi Fua Pah: Ritus dan Mitos Agraris Masyarakat Dawan di Timor (2007)
9. Manafe, Y.D.: Cara Pandang (world view) Orang Atoni Pah Meto dalam Perspektif Komunikasi ritual. Jurnal Scriptura, vol. 6, no. 2, pp. 48–56 (2016)
10. Kunjaya, C., Melany, A., Kumbaya, S., Arsono, T.: Possibility of astronomical phenomena to be used to support tourism industry. J. Phys.: Conf. Ser. **1231**, 012025 (2019)
11. Wiramihardja, S.D.: Ethnoastronomy: the sundanese of west java and their relation to Ethnoastronomy. CSEAS News Lett. (2014)
12. Fatima, S., Orchiston, W., dan Hidayat, T.: Ethnoastronomy in Madura, Indonesia: observations of the night sky and eclipses. Part Hist. Cult. Astron. Book Ser. (HCA) (2021)
13. Amin, S.: A Full Hearing: Orality and Literacy in The Malay World. Berkeley (1987)
14. Rubin, D.C.: Memory in Oral Traditions: The Cognitive Psychology of Epic, Ballads, and Counting -- Out Rhymes. University Press, Oxford (1995)
15. Ong, W.J.: Orality and Literacy: The Technologizing of the Word. Methuen, London and New York (1982)
16. McClure, B.: Pleiades – or 7 Sisters – Known Around the Worlds. https://earthsky.org/favorite-star-patterns/pleiades-star-cluster-enjoys-worldwide-renown/. Accessed 15 June 2023
17. Liubana, M.M.J., dan Neonhai, I.: Unsur kearifan Lokal Masyarakat atoni Pah Meto. Jurnal Onima: Pendidikan Bahasa dan Sastra, vol. 7, no.2. Downloaded from https://www.academia.edu/95014391/Unsur_Kearifan_Lokal_MasyarakatMeto_dalam_Legenda_Oepunu. Accessed 02 Feb 2023

Modeling of BRIsat Orbital Parameters Due to Solar Radiation Pressure

Nindhita Pratiwi[1,2](✉), Dhani Herdiwijaya[1], Nizam Ahmad[3], Taufiq Hidayat[1], and Muhammad Isnaenda Ikhsan[1,2]

[1] Department of Astronomy, Institut Teknologi Bandung (ITB), Bandung, Indonesia
nindhita.pratiwi@sap.itera.ac.id
[2] Atmospheric and Planetary Science Study Program, Institut Teknologi Sumatera (ITERA), Lampung, Indonesia
[3] Space Research Center, National Research and Innovation Agency (BRIN), Jakarta, Indonesia

Abstract. This study presents a comprehensive modeling approach to investigate the impact of solar radiation pressure (SRP) on the BRIsat's orbital parameters, Indonesia's geosynchronous satellite. To calculate the effect of SRP on BRIsat, we utilized data on satellite dimensions and geocentric vectors of both the satellite and the Sun, derived from Two-Line Element sets. Due to the lack of precise data, we relied on assumptions to estimate the satellite dimensions. Data processing was performed using FreeFlyer and Python. We compared the orbital parameters of BRIsat with and without the influence of SRP, demonstrating the effect of SRP. The results show that SRP has a significant impact on BRIsat's orbital parameters, particularly in the eccentricity and argument of perigee. The increasing distance between the two scenarios indicates the impact of SRP Over a period of almost one year, the distance between the satellite's scenarios with SRP and without SRP increases progressively, reaching up to 250 km. Over a span of more than two and a half years, this distance further expands, with a maximum observed value of 364 km. This study concludes the importance of considering SRP in the modeling of satellite orbits, particularly for long-term predictions and control of satellite trajectories.

Keywords: Solar Radiation Pressure · BRIsat · Orbital Parameter · Modeling

1 Introduction

Orbital perturbations are caused by the interaction between a satellite's surface and photons, plasma, or molecular gasses. The magnitude of these perturbations depends on solar activity and space weather, resulting in changes to the orbital parameters. These perturbations can be categorized into two types: gravitational and non-gravitational. Among non-gravitational orbital perturbations, solar radiation pressure (SRP) is the most significant factor [1]. This research focuses on the effects of SRP on orbital satellites. The influence of SRP on satellite orbital parameters manifests as long-period changes in eccentricity, inclination, right ascension of the ascending node, and argument of perigee [2]. The interaction between solar radiation and the satellite surface generates SRP. This

E. Fitrianingsih et al. (Eds.): ISAST 2024, SPPHY 416, pp. 170–178, 2025.
https://doi.org/10.1007/978-981-96-1344-1_18

interaction can be direct, where solar radiation directly impacts the satellite surfaces, or indirect, where solar radiation reflected by Earth's atmosphere affects the satellite surfaces [3]. Additionally, the satellite experiences eclipses during its orbit around the Earth. The calculation of the eclipse factor considers shadows from the satellite's current central body, except when the central body is the Earth or Moon, in which case shadows from both are considered.

This study models the influence of SRP on the orbital parameters of BRIsat, a geostationary communications satellite operated by Bank Rakyat Indonesia (BRI). Constructed by Space Systems/Loral (SS/L) for BRI, BRIsat was launched in June 2016 aboard an Ariane-5ECA rocket from Kourou and is situated at 150.5° east longitude. Weighing 3,540 kg, BRIsat is equipped with 9 Ku-band and 36 C-band transponders [4]. For our calculations, we considered the area of the satellite perpendicular to the disturbing force and assumed that the satellite's orbital orientation remains constant relative to the Sun.

2 Data and Methods

In this study, we used Two-Line Element (TLE) data supplied by Dr. T.S. Kelso as the main input [5, 6]. TLE data consists of two lines of Keplerian information about a satellite and its orbital parameters, allowing for the prediction of future or past satellite passes [7]. This data was essential for generating the geocentric vectors of both the satellite and the Sun, which are critical for calculating SRP [8]. To determine the acceleration and the resulting changes in orbital parameters due to SRP, we employed the software FreeFlyer [9]. Additionally, data processing was conducted using the Freeflyer and Python. The equation used by FreeFlyer to calculate the acceleration due to SRP is as follows [10]:

$$\vec{a}_{SRP} = \frac{S * AU^2 * C_R * A_{SRP}}{m * R^2 * c} \vec{U} v \tag{1}$$

C_R = reflectivity coefficient
AU = astronomical unit (distance between Earth and Sun)
R = distance from satellite to Sun
m = satellite mass
S = average solar flux at 1 AU
c = speed of light
A_{SRP} = cross-sectional area of satellite
U = unit vector from satellite to Sun
v = eclipse factor ($v = 0$ if satellite in umbra, $v = 1$ if satellite fully sunlight, $0 < v < 1$ if satellite in penumbra)

In this study, we employed SRP model to determine the SRP effects on BRIsat using FreeFlyer. The specific parameters used as initial inputs to generate the orbital parameters in FreeFlyer are detailed in Table 1.

The spherical model for SRP in Freeflyer is simpler and requires minimal setup. It's generally applied to predict SRP effects on objects with uncomplicated shapes. This model presumes the satellite is a perfectly round object with consistent reflectivity and absorptivity across its whole surface, leading to no change in the surface's orientation relative to sunlight [11].

Table 1. Specific inputs to FreeFlyer.

Parameters	Input
Gravitational source	Sun, Earth, Moon
Atmospheric density model	Analytic
Perturbator	SRP
SRP model	Spherical
Integrator	Runge Kutta 8(9)
Time step	300 s
Relative error tolerance	1e-9
Semimajor axis	42164 km
Eccentricity	0.0002268
Inclination	0.0190°
Right ascension of the ascending node	307.5717°
Argument of perigee	113.4813°
Mean anomaly	144.9618°
Satellite dry mass	1887 kg
SRP area	60 m2
Coefficient of reflectivity	1.4

3 Results and Discussion

In this section, we compare the orbital parameters of BRIsat with and without the influence of SRP.

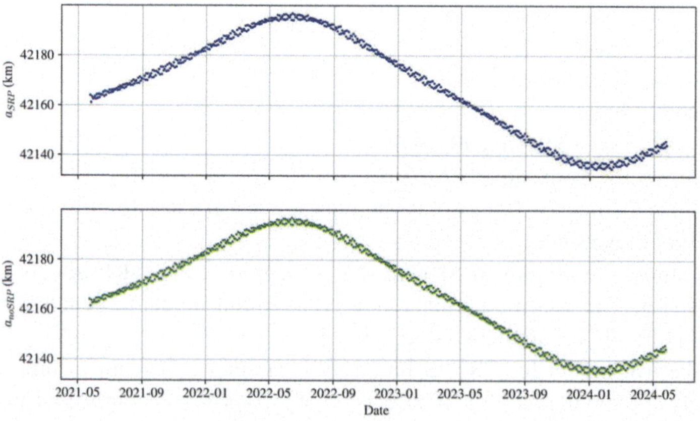

Fig. 1. Semi-major axis variation with SRP (blue) and without SRP (green) for BRIsat.

Figure 1 illustrates the effect of SRP on the semi-major axis (a) of BRIsat over a three-year period. The top plot shows the variation in the semi-major axis when SRP is accounted for (a_{SRP}), while the bottom plot shows the semi-major axis without considering SRP (a_{noSRP}). The x-axis represents the date range from May 2021 to May 2024, and the y-axis represents the semi-major axis in kilometers. Both plots exhibit a similar cyclical pattern, indicating that although SRP has a significant influence, the overall trend in the semi-major axis is consistent with or without SRP. These plots underscore the importance of accounting for SRP in satellite orbit determination, demonstrating how it can cause notable variations in the semi-major axis over time.

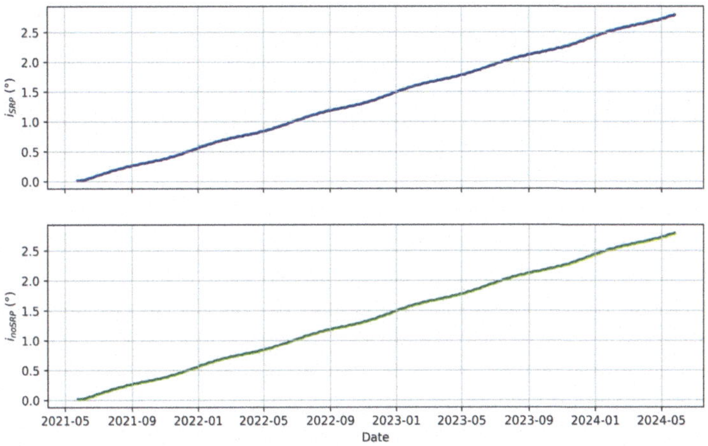

Fig. 2. Inclination variation with SRP (blue) and without SRP (green) for BRIsat.

Figure 2 consists of two subplots showing the variation in the inclination (i) of BRIsat over a three-year period from May 2021 to May 2024. The top plot (blue) shows the variation in inclination when accounting for SRP, where the inclination increases steadily from 0° to approximately 2.5°, indicating a gradual influence of SRP on the satellite's orbital inclination. The bottom plot (green) shows the variation in inclination without considering SRP. Similar to the top plot, the inclination increases steadily from 0° to approximately 2.5° over the same period, indicating a consistent trend in inclination increase. Both plots display a consistent increase in inclination over the three-year period. The rate of increase in inclination is similar in both plots, suggesting that while SRP does influence the inclination, the overall trend of inclination increase is comparable with and without SRP. The plots emphasize the importance of including SRP in satellite orbit determination, as the slight differences may accumulate over time and affect long-term orbital predictions. These plots highlight the impact of SRP on the satellite's inclination, demonstrating a gradual and continuous change that is crucial for precise orbit maintenance and prediction.

Figure 3 illustrates the effect of SRP on the eccentricity (e) of BRIsat over a three-year period. The eccentricity with SRP (blue) exhibits a clear cyclical pattern, indicating periodic changes. This cyclical behavior demonstrates that SRP significantly influences

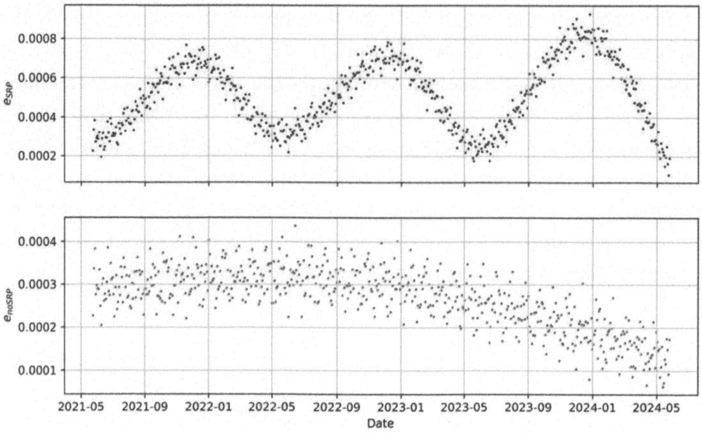

Fig. 3. Eccentricity variation with SRP (blue) and without SRP (green) for BRIsat.

the satellite's orbit, leading to regular variations in eccentricity. The consistent amplitude of these changes suggests a regular influence of SRP on the satellite's orbital dynamics. These periodic changes in eccentricity likely correspond to the orbital period of BRIsat or other recurring forces acting on the satellite. In contrast, the eccentricity without SRP (green) appears more scattered and lacks a clear cyclical pattern. There is a slight decreasing trend in eccentricity over time, indicating that, in the absence of SRP, other forces may be causing a gradual change in the orbit. The less pronounced variations in eccentricity in the absence of SRP underscore the significant impact of SRP on the satellite's orbital eccentricity compared to other factors.

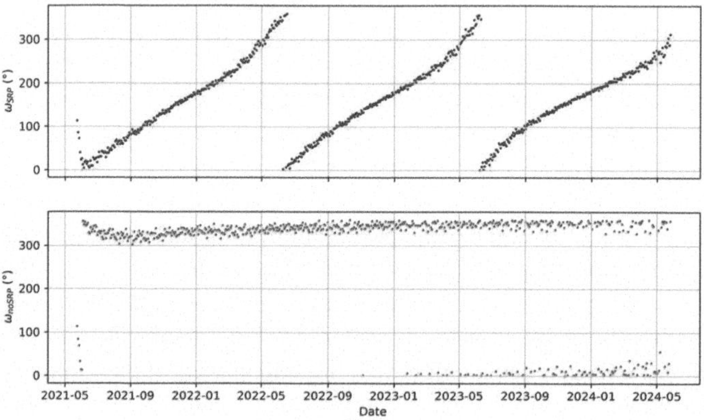

Fig. 4. Argument of perigee variation with SRP (blue) and without SRP (green) for BRIsat.

Figure 4 consists of two subplots showing the variation of the argument of perigee (ω) for BRIsat over time, with SRP (top plot) and without SRP (bottom plot). ω is a critical orbital element that indicates the orientation of the elliptical orbit within its orbital plane. In the presence of SRP (top plot), ω demonstrates a consistent, linear increase over time. The data is divided into three distinct segments with clear gaps, likely corresponding to periods of unrecorded data or operational anomalies. Each segment exhibits a similar slope, indicating a steady rate of change influenced by SRP. The periodic gaps might suggest a recurring operational or observational cycle. The steady, linear increase suggests that SRP exerts a continuous and significant influence on ω, contributing to its gradual change over time. In contrast, ω without SRP (bottom plot) shows much less variation. The data points remain relatively flat, with minor fluctuations around a constant value. The absence of a clear trend or significant changes indicates that, without SRP, ω remains relatively stable over time. The minor fluctuations do not follow a specific pattern, suggesting the influence of other minor forces or measurement noise. The stability of ω in the absence of SRP implies that other forces acting on the satellite do not significantly alter this orbital parameter. The comparison between the two plots highlights the dominant role of SRP in altering ω. With SRP, ω shows a predictable, linear increase, whereas without SRP, it remains almost unchanged, underscoring the significant impact of SRP on BRIsat's orbital dynamics.

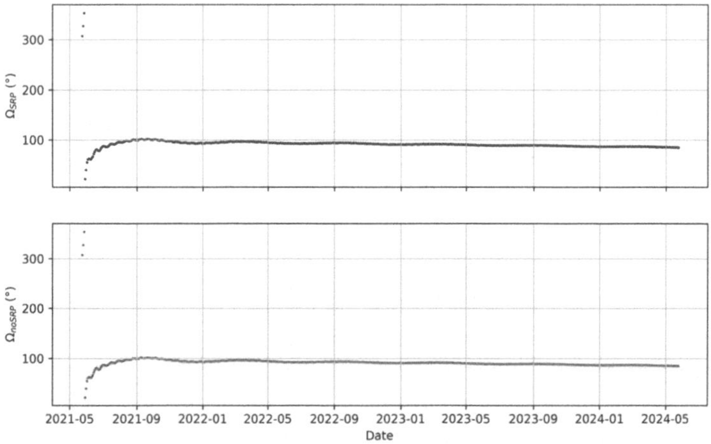

Fig. 5. Right ascension of the ascending node variation with SRP (blue) and without SRP (green) for BRIsat.

Figure 5 illustrates the effect of SRP on the right ascension of the ascending node (Ω) of BRIsat over a three-year period. In the top plot, there is initial variation in Ω at the start of the observation period. After this initial phase, Ω stabilizes and remains relatively constant. The densely packed and consistent data points indicate minimal variation in Ω over time. This initial variation may be due to the satellite adjusting to its orbit or the influence of SRP being more pronounced during the initial phases. The subsequent stabilization suggests that SRP does not significantly affect Ω over long periods, or that any influence is quickly counteracted by other forces or adjustments. Similarly, in the bottom plot, there is an initial increase in Ω, followed by stabilization at a constant value. After the initial phase, Ω remains almost unchanged, with data points showing very little fluctuation. The initial increase suggests that the satellite undergoes dynamic changes at the start of the observation period, which, without SRP, are quickly stabilized. The lack of significant variation in Ω without SRP implies that other forces acting on the satellite contribute to a stable orbital configuration. These plots illustrate that BRIsat's right ascension of the ascending node experiences initial variations followed by stabilization, both with and without SRP. This suggests that while SRP may influence the satellite's initial orbital behavior, its long-term impact on Ω is minimal, leading to a stable orbital configuration over time.

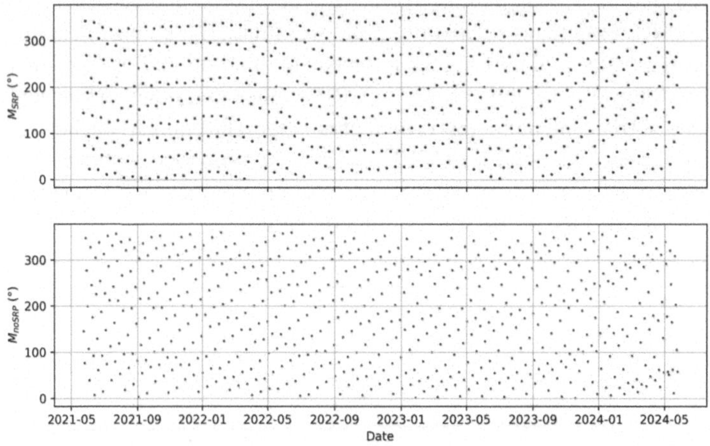

Fig. 6. Mean anomaly variation with SRP (blue) and without SRP (green) for BRIsat.

Figure 6 consists of two subplots showing the variation of the mean anomaly (M) of BRIsat over time, with and without SRP. The top plot (blue) indicates that the mean anomaly with SRP exhibits a periodic wave-like pattern over time, characterized by a consistent amplitude and regular frequency of oscillations. This pattern suggests that SRP has a significant and cyclical influence on the mean anomaly of BRIsat, making these changes predictable and useful for orbit maintenance. In contrast, the bottom plot (green) reveals that the mean anomaly without SRP also displays periodic variations, though they are less pronounced, with a smaller amplitude. This implies that while other forces contribute to the periodic changes in the mean anomaly, SRP exerts a more substantial impact. The smaller amplitude of oscillations without SRP indicates a more stable mean anomaly when SRP is not considered. Overall, the plots illustrate that BRIsat's mean anomaly undergoes periodic variations both with and without SRP. However, the influence of SRP results in more significant oscillations, highlighting its dominant role in affecting the satellite's orbital mean anomaly.

Figure 7 shows the variation of BRIsat's range over a three-year period, from mid-2021 to mid-2024. The range, defined as the distance between BRIsat with and without the influence of SRP, fluctuates significantly, demonstrating a clear periodic pattern. These cyclical variations exhibit distinct peaks and troughs, with each cycle lasting approximately one year. The amplitude of these oscillations changes over time: the first peak (around January 2022) reaches approximately 250 km, the second peak (around

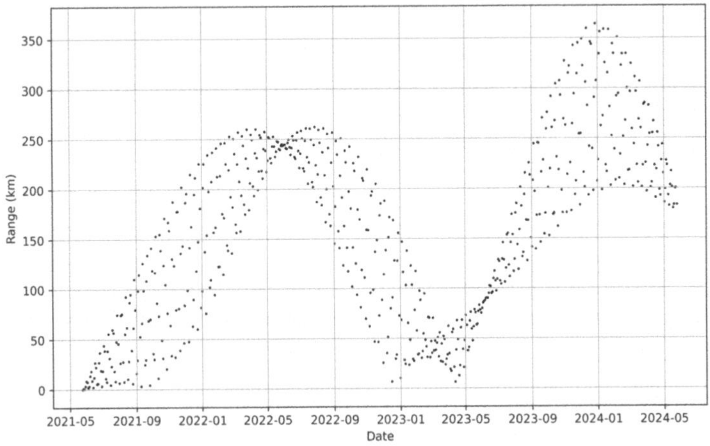

Fig. 7. Variation of BRIsat's range over time.

January 2023) is slightly higher, and the third peak (around January 2024) reaches around 350 km. The periodic changes in the range indicate the satellite's response to SRP, highlighting its significant influence on BRIsat's orbital dynamics.

4 Conclusion

The comparative analysis of BRIsat's orbital parameters with and without the influence of SRP reveals significant impacts. The SRP notably affects the eccentricity and argument of perigee leading to observable variations in the satellite's trajectory. Furthermore, the increasing distance between the satellite scenarios with and without SRP underscores the influence of SRP on BRIsat. Over a period of almost one year, the distance between the satellite's scenarios with SRP and without SRP increases progressively, reaching up to 250 km. Over a span of more than two and a half years, this distance further expands, with a maximum observed value of 364 km. These findings emphasize the importance of accounting for SRP in precise satellite orbit determination and prediction to ensure the accuracy and reliability of BRIsat's orbital parameters.

Acknowledgment. I would like to express my gratitude to Dr. T.S. Kelso for providing the TLE data, which was instrumental in completing this study. Additionally, I would like to thank Fakultas Sains Itera for the funding and all the support.

References

1. Baron, A.S.: Study the effect of solar radiation pressure at several satellite orbits. Baghdad Sci. J. **10**(4), 1253–1261 (2013)
2. Alessi, E.M., Schettino, G., Rossi, A., Valsecchi, G.B.: Solar radiation pressure resonances in low earth orbits. Mon. Not. R. Astron. Soc. **473**(2), 2407–2414 (2018)

3. Lupu, S., Zaharescu, E.: Effects of direct and indirect solar radiation pressure in orbital parameters of GPS Satelittes. Versita **22**, 141–150 (2014)
4. Krebs, Gunter D. "BRIsat". Gunter's Space Page. https://space.skyrocket.de/doc_sdat/brisat-1.htm. Accessed 5 May 2024
5. CelesTrak Homepage. https://celestrak.org. Accessed 20 Apr 2024
6. Vallado, D., Cefola, P.: Two-line element sets—practice and use. In: Proceedings of the 63rd International Astronautical Congress, pp. 5812–5825. Curran Associates Inc., NY (2012)
7. Hoots, F., Roehrich, R.: SpaceTrack Report No. 3—models for propagation of NORAD element sets. U.S. air force aerospace Defense command. Colorado Springs, CO, USA (1980)
8. Paris, J.A.: The effects of using solar radiation pressure to alleviate fuel requirements for orbit changing and maintenance of the DSCS II F-13 satellite. Master's thesis. Air Force Institute of Technology, Wright-Patterson Air Force Base, Ohio: Department of the Air Force, Air University (2006)
9. Binder, J.: Planning space missions with Freeflyer. Aerosp. Am. **43**(7), 24–25 (2005)
10. Vallado, D.A.: Fundamentals of Astrodynamics and Applications, 5th edn. Microcosm Press, Portland (2022)
11. Newman, C.P., Hollister, J.R., Davis, D.C., Zimovan-Spreen, E.M.: Investigating solar radiation pressure modeling for operations in near rectilinear halo orbit. In: Proceedings of the Astrodynamics Specialists Conference, p. 10382 (2022)

The Impact of Satellite Position and Derived Parameters on Electron Flux Prediction with Long Short-Term Memory Network

Rian P. Salasa(✉), Siska Filawati, Silmie V. Fani, and Dian Y. Risdianto

Research Center of Space, BRIN, Bandung, Indonesia
rian010@brin.go.id

Abstract. Satellites in the geostationary orbit are located in the outermost radiation belt area and are vulnerable to space weather conditions, such as high energetic electron interactions, which can cause damage to satellite components. It is important to predict electron flux in this orbit to reduce the negative impact of these phenomena. Prediction models have been developed using different approaches, such as empirical, probabilistic, physics-based, and machine learning models. One of the methods used is the Long Short-Term Memory (LSTM). LSTM process inputs in the form of multidimensional data points where one data point may have multiple parameters. This allows a large number of parameter combinations to be served as input. Previous studies have employed the integral value of electron flux observed by Geostationary Operational Environmental Satellite (GOES) satellites as the main LSTM input due to its autocorrelation characteristic, along with other parameters from various space weather phenomena such as solar wind and geomagnetic indices. However, more parameters can be derived from the GOES data and are yet to be employed. In this study, we use satellite position data and parameters derived from electron flux data (average, minimum, maximum, data sub-point, Day-of-Year, local time) as input parameters to determine the impact of each parameter on the performance of LSTM in predicting $\geq$ 2 MeV electron flux. We used data with 12-h and 6-h resolutions to predict the electron flux value 12 and 6 h ahead, respectively. The results show that the combination of average, minimum, maximum, and data sub-points as input parameters has a significant impact on increasing the prediction result. The results reached 0.845 prediction efficiency (PE) for 6-h resolution and 0.803 for 12-h resolution. We also used the logarithmic electron flux value instead of the actual value and got even higher scores with the average PE of 0.925 for 6-h resolution and 0.893 for 12-h resolution.

Keywords: Electron Flux · Time Series Prediction · Long Short-Term Memory

1 Introduction

The Earth's radiation belt is the high-energetic protons and electrons trapped on stable orbits within Earth's magnetosphere. Earth's electron radiation belt is divided into inner belt, spanning from 1.2 RE to 2.5 RE (RE is Earth's radii, 6378 km) and outer belts

E. Fitrianingsih et al. (Eds.): ISAST 2024, SPPHY 416, pp. 179–188, 2025.
https://doi.org/10.1007/978-981-96-1344-1_19

expand from 4 to 6 RE with the slot region that separates those belts [1]. Outer radiation belt spans across geostationary orbit which is ideal for certain kinds of communication satellites and meteorological satellites. A spacecraft in geosynchronous orbit appears to remain above Earth at a constant longitude and returns to the same point in the sky at the same time each day. Previous researches showed that electron flux with energy ≥ 2 MeV is associated with satellite deep-dielectric charging. This discharge causes anomalous behavior in satellite systems and results in temporary or permanent loss of functionality [2, 3].

There are many different processes regarding to the energization of magnetospheric electrons to relativistic energies in radiation belt. Each process does not operate by itself but often occurs simultaneously with other acceleration or transport processes as part of the solar wind-magnetosphere coupling to inner magnetosphere's as collective effects and leading to the observed various modes of responses. In the complex acceleration processes, the loss processes are also numerous and potentially complex and include deceleration due to waves, loss to the magnetopause, and interaction with ionospheric electrons and ions [4]. Often these complexities are analyzed in approximation and the time dependence of the drivers (solar activities, waves, compressions) is slow compared to the diffusion and loss time scales [4, 5].

Predicting electron flux is an important task to mitigate the potential risks from its impacts. Ideally, understanding the physical processes responsible for electron acceleration would unravel the task. However, our present understanding does not fully accommodate it. Meanwhile from the operational point of view, the needs of the prediction task cannot be ignored. While some studies attempt to make prediction models using physics-based approaches [6–11], others use statistical and probabilistic approaches [12–15] as well as neural networks [16–20] because of the less dependence on its physical understanding.

Deep learning is an advanced neural network built with many layers, operates in high dimensionality of data, and has been widely used to solve complex non-linear problems. Deep learning has the strength in learning patterns and relationships from data without needing to extract the features explicitly. However, its mechanism can be highly complex and difficult to interpret by humans, thus often referred to as a black box [21, 22]. Long Short-Term Memory (LSTM) is a type of deep learning designed for predicting sequential data. LSTM can process input in the form of multidimensional time series data where one data point may have multiple parameters [23–25]. This allows a large number of parameter combinations to be served as input. There have been studies that utilize LSTM to predict electron flux. Reference [26] developed an LSTM model to predict $\geq$ 2 MeV electron integral flux with different input combinations including daily integral of electron flux, geomagnetic indices, and solar wind parameters. In [27], the combination of LSTM and Empirical Mode Decomposition (EMD) algorithm was applied to predict the variation of the electron fluxes 1 day ahead. The input parameters include the daily integral of electron flux and geomagnetic indices. Reference [28] attempted to predict daily fluences of ≥ 2 MeV electron flux for the following three days using electron flux, geomagnetic indices, and solar wind parameters as input.

Because of the black box nature of deep learning, choosing input parameters for LSTM can be a challenging task. Previous studies have shown that historical data of electron flux has a great contribution to quantitative prediction because of its strong autocorrelations [11, 16, 19, 26], and it encourages us to explore deeper into the historical data of electron flux itself without any addition of external parameters. We use several combinations of satellite coordinates and parameters derived from $\geq$ 2 MeV electron flux data (average, minimum, maximum, data sub-point, Day-of-Year) as input for the LSTM model to determine the influence of each parameter. We highlight that the position of this study is in the early stage of the model development with the focus on finding out impacts of satellite position and parameters derived from the electron flux data only. We will develop the final model in the next researches with the addition of external parameters, such as solar wind and geomagnetic indices.

2 Data Preparation

The data used in this study include 5-min averaged $\geq$ 2 MeV electron flux observed by Geostationary Operational Environmental Satellite (GOES) satellites and the location data of the corresponding satellite. Both data are available on the National Centers for Environmental Information (NCEI) website. The data had been corrected from proton flux contamination by NCEI. We used electron flux data with good quality flag only. A pair of GOES satellites continuously orbit in two different longitudes: at around 75° W and 135° W longitude. However, the exact longitudes are varied in time. A GOES satellite may operate in standby mode, orbits outside 75° W or 135° W, and still record observation data. To limit the scope of this study, we only use data when the satellites are located in the range of 74° W to 77° W, which are GOES-8 (in period of May 1995-April 2003), GOES-12 (April 2003-April 2010), and GOES-13 (April 2010-November 2017). The satellite location data is recorded in two coordinate systems: Geosynchronous (GSE) and Geographic (GEO) coordinate system.

To derive parameters from electron flux data, we did some processing steps. Firstly, we concatenated the electron flux data from the three satellites in time order to create 5-min averaged electron flux data ranging from May 1995 to November 2017. Missing values were filled using linear interpolation method. We then resampled the data to 12-h and 6-h resolutions using 12 h and 6 h windows with base 0. We calculated the average, maximum, and minimum values from each window and used them as input parameters. We also divided each window into subsets of 1 h and calculated the average value for each subset from both 12 h and 6 h window to create 12 and 6 sub-points respectively. For the 12-h resolution data, we only included the second, fourth, 6th, 8th, 10th, and 12th sub-points as input parameters, and for the 6-h resolution data we used all of the six sub-points. The satellite coordinates were then added as parameters to each data point using left join technique. Finally, the Day-of-Year value for each data point was calculated from the timestamp data and used as a parameter to represent Earth's location relative to the Sun. At the end of data preparation, we had two datasets, 12-h resolution and 6-h resolution, with total data points of 16.500 and 33.000 respectively. Each data point on the two datasets is multidimensional and includes the following parameters as the dimensions: the average (*avg*), minimum (*min*), and maximum (*max*) values of $\geq$ 2 MeV electron flux, GSE coordinates (*gse_x, gse_y, gse_z*), GEO longitude (*geo_lon*), Day-of-Year (*DoY*), and the six sub-points (*sub*). We used two different resolutions to see if the method we use has a similar impact on datasets with different resolutions.

3 Method

3.1 Long Short-Term Memory (LSTM)

Neural networks are a composition of functions formed by learning patterns from data. Computation processes on neural networks are commonly visualized as interconnected nodes, arranged in ordered layers. Each node can receive input, perform computations, produce output, and pass it to nodes in the next layer. Recurrent Neural Networks (RNN) are neural networks designed to recognize patterns in sequential data. Apart from passing information to the next layer, RNN also transmits information back into itself. This mechanism allows them to learn from both current and previous data, thus allowing the network to maintain a kind of memory, enabling it to process sequences of data [25, 29].

LSTM is a variant of RNN with some improvements in handling long sequences of data over standard RNN. LSTM has structures called cell state, input gate, forget gate, and output gate. Cell state passes information unchanged to remember information over sequences, input gate decides which information is added into cell state, forget gate decides which information is discarded, and output gate decides the output of the LSTM. These gates are implemented using sigmoid functions, and its open or close state is trained using training data. The architecture of an LSTM unit is shown in Fig. 1. Mathematical expressions and further details can be found in [24].

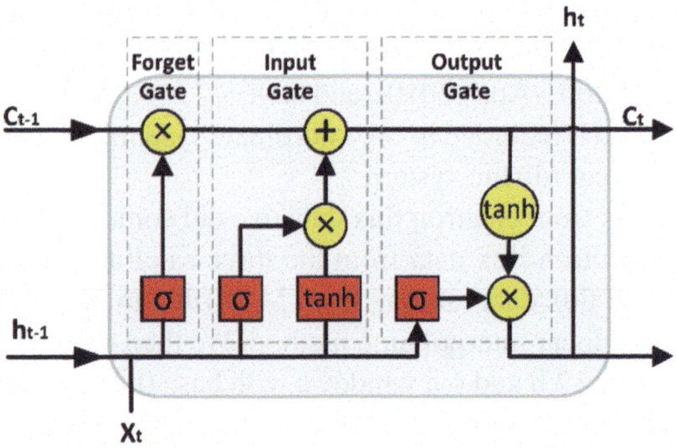

Fig. 1. LSTM architecture. Red boxes represent neural nodes, yellow circles represent pointwise operations, X is input, C is cell state and h is output.

3.2 Training Parameters and Model Evaluation

We built the model using a three-layer network containing an LSTM layer with 64 neurons, a feedforward layer with 16 neurons, and an output layer with 1 neuron. The training parameters were selected using several try-error experiments and iterations. The batch size of the training dataset was set to 64, and the learning rate was set to 0.001. The model was trained in 300 epochs with early stopping technique. Adam optimizer [30] was set to optimize the training process. We also tried several combinations of training parameters and network layers with no noticeable improvement. Implementation of the model was done using Keras framework. The output of the model is a value of 6 h averaged electron flux for the 6-h resolution data and a value of 12 h averaged electron flux for the 12-h resolution data.

The performance of the model is evaluated using two metrics, namely prediction efficiency (PE) and root mean square error (RMSE). PE is commonly used as a metric to evaluate prediction performance of the model, defined as:

$$PE = 1 - \frac{\sum_i^n (m_i - p_i)^2}{\sum_i^n (m_i - \overline{m})^2} \quad (1)$$

where n is the total number of samples, m_i is the i th observation, p_i is the i th prediction, and $\overline{m}$ is the mean value of all observation samples. The value of PE less than zero indicates that the model is worse than the average value of the observation, and one indicates the prediction is equal to the corresponding observation. RMSE is the common metric for accuracy fit. The smaller the RMSE indicates the better performance of the model. RMSE is defined as:

$$RMSE = \sqrt{\frac{1}{n}\sum_i^n (p_i - m_i)^2} \quad (2)$$

4 Results and Discussions

LSTM requires inputs from a certain offset time to predict the current data point. Offset time is the number of past data points before the current data point. To determine the optimum offset time used in model training, we conducted some initial tests by training the model using different offset time values, ranging from 2 to 12. The tests were carried out using only *avg* as input parameter. The tests were done using *k*-fold cross-validation as evaluation method with $k = 10$. The *k*-fold cross-validation is a model evaluation method where the dataset is split into *k* subsets, then the model is trained *k* times with each subset taken out from training to be reserved as validation/sample set once and the rest of the subsets served as training set. Each training result is then tested on the corresponding taken out subset. The mean PE value is then calculated and used as the test result. The initial tests results showed the optimum offset time for 12-h resolution and 6-h resolution data were different. The mean PE reached a maximum at an offset time of 10 and 8 for 12-h and 6-h resolution data, respectively (Fig. 2). These maximum offset time values were then used as the offset time of the model.

To determine the impact of satellite position and parameters derived from electron flux data, we trained the model using different input combinations. We used following input combinations: (*avg*, *gse_x*, *gse_y*, *gse_z*), (*avg*, *geo_lon*), (*avg*, *DoY*), (*avg*, *max*, *min*), (*avg*, *sub*), and (*avg*, *max*, *min*, *sub*). We also trained the model using the logarithmic values of electron flux instead of the actual values, with following input combinations: (log *avg*), (log *avg*, log *max*, log *min*), (log *avg*, log *sub*), and (log *avg*, log *max*, log *min*, log *sub*). The training was evaluated using *k*-fold cross-validation with $k = 10$. We used the best model trained with only *avg* parameter from the initial tests as baseline model. We then compared each model with the baseline model to determine the impact

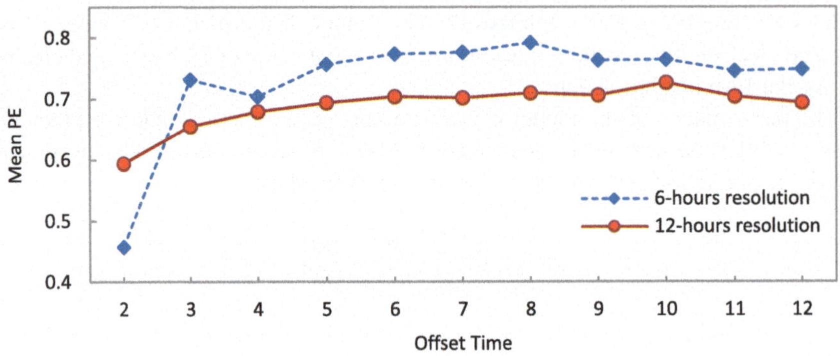

Fig. 2. Mean PE for each offset time value in initial tests.

of its input parameters on the prediction result. The 12-h resolution and 6-h resolution datasets were trained separately using the same method.

The evaluation score of the models is presented in Table 1. The result shows the (*avg, gse_x, gse_y, gse_z*) model, the (*avg, geo_lon*) model, and the (*avg, DoY*) model achieved relatively similar PE, which is all lower than the baseline model. This occurred in both 12-h and 6-h resolution. We did further examination by looking at results from all iterations in the *k*-fold cross-validation (Fig. 3). We found that the PE of the (*avg, gse_x, gse_y, gse_z*) model and (*avg, geo_lon*) model is dropped significantly on 3rd, 4th and 5th iteration for 6-h resolution and slightly lower than baseline model in almost all iterations for 12-h resolution. From the results, we can state that the addition of exact satellite coordinates does not improve the prediction accuracy and efficiency. Thus, it is safe to assume that the longitude of the satellites is around 75° W, and a slight difference will not make any noticeable effect on the prediction using LSTM.

On the contrary, models with electron flux derived parameters achieved better mean values of PE and RMSE than baseline model. The (*avg, max, min*) model has a slight improvement, while the (*avg, sub*) and (*avg, max, min, sub*) models have a noticeable increase of PE compared to baseline model. The conditions occurred in both 12-h and 6-h resolutions.

The *sub* parameter, which is consists of 6 subsets in higher data resolution, has a good contribution in increasing the prediction performance of the model. In our estimation, this is possibly because the model was able to learn hidden features in the higher resolution data which is not found in 12-h and 6-h resolutions. This is corroborated by results from the models that use log values. Models with the combination of log values of *avg, max, min,* and *sub* have similar patterns in the results, where (log *avg*, log *max*, log *min*) have slightly higher PE and both (log *avg*, log *sub*) and (log *avg*, log *max*, log *min,* log *sub*) models score significant PE improvement compared to (log *avg*) model.

We can also see from the result that the use of log value instead of actual value increases the PE score for all corresponding models, as shown in Fig. 4. The black circles, red boxes, and blue triangles denote the mean, minimum, and maximum PE values from the 10 iterations, respectively. Dashed lines represent the PE standard deviation of the models. Models with log values (B, C, D, E) have lower standard deviations compared

Table 1. Prediction efficiency and RMSE of the model

Input parameter	6-h resolution		12-h resolution	
	Mean PE	Mean RMSE (e/(cm^2 s sr))	Mean PE	Mean RMSE (e/(cm^2 s sr))
avg (baseline model)	0.793	1329	0.727	1394
avg, gse_x, gse_y, gse_z	0.784	1386	0.645	1630
avg, geo_lon	0.782	1392	0.671	1577
avg, DoY	0.784	1385	0.683	1599
avg, max, min	0.816	1183	0.760	1273
avg, sub	0.848	1170	0.797	1236
avg, max, min, sub	0.846	1080	0.803	1116
Input parameter	Mean PE	Mean RMSE	Mean PE	Mean RMSE
log(*avg*)	0.872	0.338	0.820	0.390
log(*avg*), log(*max*), log(*min*)	0.904	0.292	0.871	0.329
log(*avg*), log(*sub*)	0.926	0.258	0.898	0.294
log(*avg*), log(*max*), log(*min*), log(*sub*)	0.926	0.256	0.899	0.291

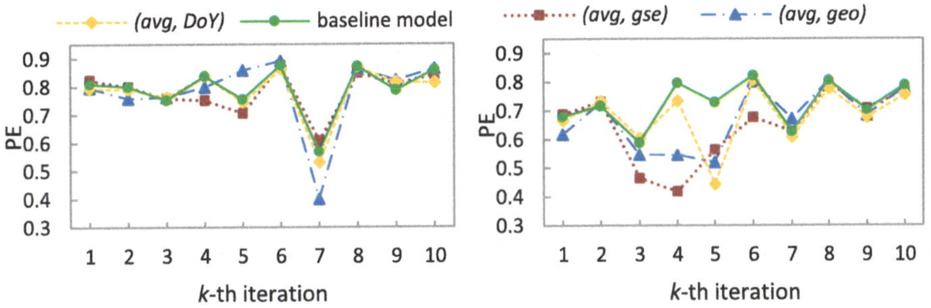

Fig. 3. The PE of each model in *k*-fold cross-validation iterations for models with satellite position parameters. Left: 6-h resolution. Right: 12-h resolution.

to baseline model (A), indicating better consistency in the prediction performance. The maximum and minimum PE of models with log values are also higher than those of baseline model. The (log *avg*, log *max*, log *min*, log *sub*) model achieved the highest PE value among all models at the iteration period of August 2006-November 2008 for both 12-h and 6-h resolution with the scores of 0.928 and 0.954, respectively. Lastly, we presented an example of a comparison between observation and prediction results from (log *avg*, log *max*, log *min*, log *sub*) model on the observation period of September 2006 in Fig. 5.

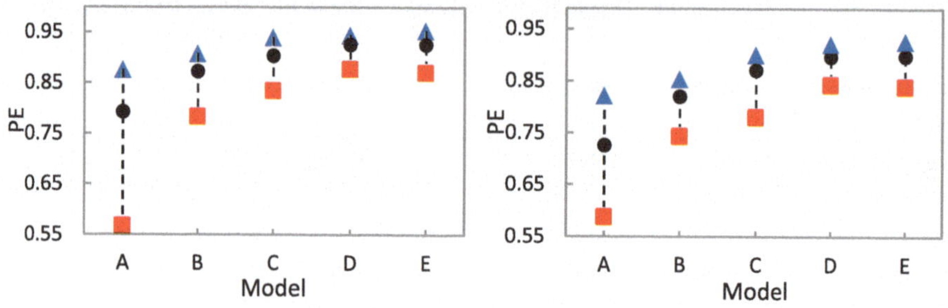

Fig. 4. Comparison of *k*-fold cross-validation performance. A is (*avg*), B is (log *avg*), C is (log *avg*, log *max*, log *min*), D is (log *avg*, log *sub*), and E is (log *avg*, log *max*, log *min*, log *sub*) model. Left: 6-h resolution, right: 12-h resolution.

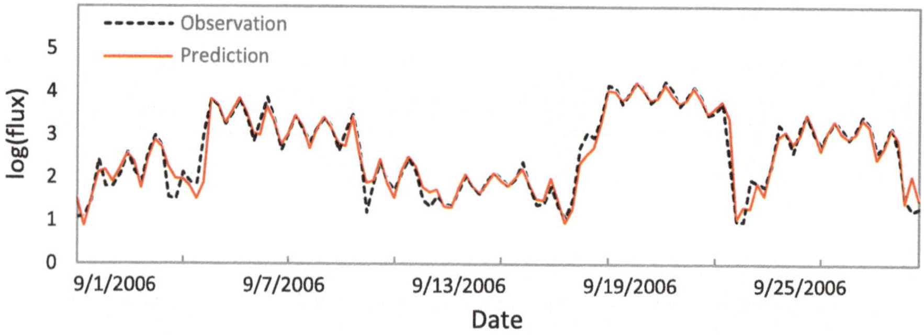

Fig. 5. Comparison between observation and prediction of (log *avg*, log *max*, log *min*, log *sub*) model on 6-h resolution in periods of September 2006.

5 Conclusion and Future Works

Combinations of parameters derived from electron flux and location of corresponding observation satellite data were used as input for LSTM prediction models. The prediction efficiency (PE) and root mean squared error (RMSE) score of each model were then compared to the baseline model to determine the impact of each parameter on the performance of the prediction. The results show that including the exact satellite coordinates in the input parameters does not improve the prediction performance, while including average, minimum, maximum values of electron flux and subsets with higher resolution improves the PE score. Moreover, the use of logarithmic values on electron flux derived parameters improves the score significantly. The results are confirmed in 6-h & 12-h resolution datasets.

From this study we have found that the impact of the electron flux data on prediction with LSTM can be extended with the addition of the derived parameters. In the next research we will improve the model by adding external parameters such as solar wind parameters and geomagnetic indices. However, in the situation where only the ≥ 2 MeV electron flux data is available, we consider the model is sufficient to perform the prediction with reasonable results. We also plan to develop the model further with multiple time-step prediction. The NOAA (National Oceanic and Atmospheric Administration) website provides near real-time GOES data services with the smallest time range of 6 h (https://services.swpc.noaa.gov/json/goes/primary/), thus our model can utilize the service as the data source to make a prediction as near real-time as possible.

References

1. Li, X., Temerin, M.A.: The electron radiation belt. Space Sci. Rev. **95**, 569–580 (2001)
2. Kataoka, R., Miyoshi, Y.: Flux Enhancement of radiation belt electrons during geomagnetic storms driven by coronal mass ejections and corotating interaction regions. Space Weather **4**(9), 1542–7390 (2006)
3. Allen, J.H.: Historical and recent solar activity and geomagnetic storms affecting spacecraft operations. In: Government Microcircuit and Applications Conference, Monterey, California (2002)
4. Vassiliadis, D.: Response of the radiation belt electron flux to the solar wind velocity: parameterization by radial distance and energy. J. Atmos. Solar Terr. Phys. **70**, 1810–1828 (2008)
5. Schulz, M., Lanzerotti, L.J.: Particle Diffusion in the Radiation Belts. Springer, New York (1974)
6. Nagai, T.: Space weather forecast: prediction of relativistic electron intensity at synchronous orbit. Geophys. Res. Lett. **15**(5), 425–428 (1988)
7. Baker, D.N., McPherron, R.L., Cayton, T.E. Klebesadel, R.W.: Linear prediction filter analysis of relativistic electron properties at 6.6 R. J. Geophys. Res.: Space Phys. **95**, 15133–15140 (1990)
8. Li, X., Temerin, M., Baker, D.N., Reeves, G.D., Larson, D.: Quantitative prediction of radiation belt electrons at geostationary orbit based on solar wind measurements. Geophys. Res. Lett. **28**, 1887–1890 (2001)
9. Li, X.: Variations of 0.7–6.0 MeV electrons at geosynchronous orbit as a function of solar wind. Space Weather **2**, 2003SW000017 (2004)
10. Lyatsky, W., Khazanov, G.V.: A predictive model for relativistic electrons at geostationary orbit. Geophys. Res. Lett. **35**, 2008GL034688 (2008)
11. Turner, D.L. Li, X.: Quantitative forecast of relativistic electron flux at geosynchronous orbit based on low-energy electron flux. Space Weather **6**, 2007SW000354 (2008)
12. Boynton, R.J., et al.: Electron flux models for different energies at geostationary orbit: goes electron flux models. Space Weather **14**, 846–860 (2016)
13. Wei, H.-L., et al.: Forecasting relativistic electron flux using dynamic multiple regression models. Ann. Geophys. **29**, 415–420 (2011)
14. Miyoshi, Y., Kataoka, R.: Probabilistic space weather forecast of the relativistic electron flux enhancement at geosynchronous orbit. J. Atmos. Solar Terr. Phys. **70**, 475–481 (2008)
15. Ukhorskiy, A.Y. et al. Data-derived forecasting model for relativistic electron intensity at geosynchronous orbit. Geophys. Res. Lett. **31**, 2004GL019616 (2004)
16. Fukata, M., Taguchi, S., Okuzawa, T., Obara, T.: Neural network prediction of relativistic electrons at geosynchronous orbit during the storm recovery phase: effects of recurring substorms. Ann. Geophys. **20**, 947–951 (2002)

17. Koons, H.C., Gorney, D.J.: A neural network model of the relativistic electron flux at geosynchronous orbit. J. Geophys. Res. Space Phys. **96**, 5549–5556 (1991)
18. Ling, A.G., Ginet, G.P., Hilmer, R.V., Perry, K.L.: A neural network-based geosynchronous relativistic electron flux forecasting model. Space Weather **8**(9), S09003 (2010)
19. Shin, D.-K., Lee, D.-Y., Kim, K.-C., Hwang, J., Kim, J.: Artificial neural network prediction model for geosynchronous electron fluxes: dependence on satellite position and particle energy. Space Weather **14**, 313–321 (2016)
20. Wang, J., et al.: Prediction of geosynchronous electron fluxes using an artificial neural network driven by solar wind parameters. Adv. Space Res. **71**, 275–285 (2023)
21. Montavon, G., Samek, W., Müller, K.-R.: Methods for interpreting and understanding deep neural networks. Digit. Signal Process. **73**, 1–15 (2018)
22. Guidotti, R., et al.: A survey of methods for explaining black box models. ACM Comput. Surv. **51**, 1–42 (2019)
23. Hochreiter, S., Schmidhuber, J.: Long short-term memory. Neural Comput. **9**, 1735–1780 (1997)
24. Yu, Y., Si, X., Hu, C., Zhang, J.: A review of recurrent neural networks: LSTM cells and network architectures. Neural Comput. **31**, 1235–1270 (2019)
25. Sherstinsky, A.: Fundamentals of Recurrent Neural Network (RNN) and Long Short-Term Memory (LSTM) Network. Physica D **404**, 132306 (2020)
26. Wei, L., et al.: Quantitative prediction of high-energy electron integral flux at geostationary orbit based on deep learning. Space Weather **16**, 903–916 (2018)
27. Zhang, H., et al.: A prediction model of relativistic electrons at geostationary orbit using the EMD-LSTM network and geomagnetic indices. Space Weather **20**, e2022SW003126 (2022)
28. Sun, X., et al.: Prediction models of $\geq$ 2 MeV electron daily Fluences for 3 Days at GEO orbit using a long short-term memory network. Remote Sens. **15**, 2538 (2023)
29. Benidis, K., et al.: Deep learning for time series forecasting: tutorial and literature survey. ACM Comput. Surv. **55**, 1–36 (2023)
30. Kingma, D.P. Ba, J.: Adam: a method for stochastic optimization. arXiv e-prints: arXiv:1412.6980 (2014)

Study of Sunspot's Proper Motion During X1.0 Solar Flare on 28th October 2021

Bryant Randolph(✉), Naufal Farras Nurindra, Firshanda Alvyanita, Fathia Rahmi Izzati, Dhani Herdiwijaya, and Fargiza Abdan Malikul Mulki

Department of Astronomy, Faculty of Mathematics and Natural Science, Institut Teknologi, Bandung, Indonesia
10321038@mahasiswa.itb.ac.id

Abstract. Many studies have been conducted on active region 12887 (AR12887), which caused the X1.0-class solar flare event on October 28, 2021, especially on the correlation of the solar flare with other solar activities such as magnetic evolution of the aforementioned region. In this work, we aim to further investigate the relationship between the motion of the sunspots and the solar flare. For that, we carried out an analysis regarding the proper motion of sunspots in AR12887 for all of its observable duration, in which the measurement was conducted using the Carrington heliographic coordinate system using JHelioviewer, based on Solar Dynamic Observatory (SDO) observation images. We measured a total of 7 days, starting from October 25th to October 31st, 2021, with a 1-h interval, totaling 166 images and 37 sunspots. The number of sunspots increased before the solar flare due to activities such as the emergence and fragmentation of sunspots, then decreased after the flare as there were no more new sunspots emerging. Sunspots' movements were far more erratic before the flare because of the high magnetic activity, especially shearing, and gradually slowed down after the flare. The flare itself did not significantly affect the movement direction of the spots inside its region, while sunspots located at the periphery of the flare have the tendency to move away from it.

Keywords: Sun Spot · Solar Flare · AR12887

1 Introduction

An active region (AR) is a feature in the sun's photosphere characterized by a strong and complex magnetic field, often associated by the presence of multiple sunspots with opposite polarity. A solar flare is a sudden and intense radiation burst coming out of the surface of the sun, occurring when a sufficient amount of free magnetic energy is accumulated. The energy is effectively accumulated during photospheric surface motion, such as the magnetic flux emergence and dissipation, magnetic shear, and other magnetic field evolution processes [1].

According to the National Oceanic and Atmospheric Administration (NOAA), solar flares can be classified based on their X-ray intensity: classes A, B, C, M, and X, ranked from lowest to highest, with X being the class with the highest intensity of X-rays [2].

E. Fitrianingsih et al. (Eds.): ISAST 2024, SPPHY 416, pp. 189–195, 2025.
https://doi.org/10.1007/978-981-96-1344-1_20

High-intensity solar flare radiation may cause disturbances in the earth's ionosphere region, disturbing the earth's communication and navigation systems, not to mention its hazards to astronauts in space [3].

AR12887 could be observed from October 23rd to November 1st, 2021. This active region caused an X1.0 solar flare on October 28th, 2021, 15:35 UT (see Fig. 1) [4]. Our aim is to study the correlation between the sunspot's presence, its movement in the photosphere and its flaring activity. Some studies regarding the same active region have also been conducted by Fang [5] and Song [6] on its magnetic structure and magnetic build-up leading to an X-class solar flare. Meanwhile, similar kinds of studies of sunspot movement have also been conducted by Ambastha & Bhatnagar (1988) on NOAA 2372 in 1980 [7] and Herdiwijaya & Permata (2019) on AR11429 in 2012 [8], showing a strong correlation between emerging solar flares at the corona layer and its foot points (sunspots) at the photosphere.

2 Data and Method

Data was taken from SDO Helioseismic and Magnetic Layer (HMI) continuum observation images, visualized using JHelioviewer. We divided the sunspots into four regions and measured each individual sunspot's position using JHelioViewer based on the Carrington heliographic coordinate system [9]. The measurement was done manually at 1-h intervals for each day from October 25th to October 31st, 2021 and is limited to -45 to 45 degrees in viewpoint to avoid inaccuracy due to spherical foreshortening. We separate AR12887 into 4 regions: A, B, C, and D (see Fig. 1).

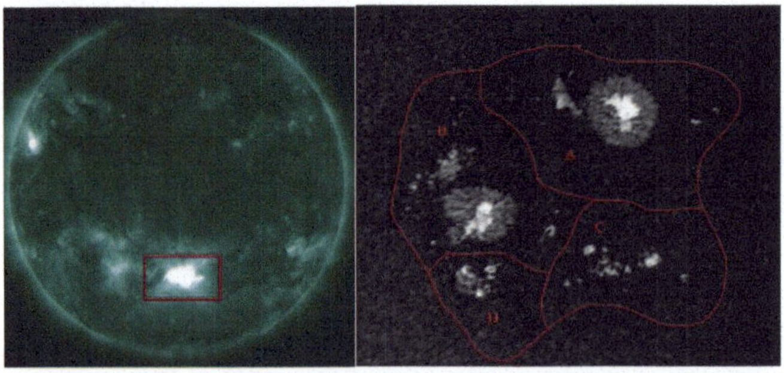

Fig. 1. AIA 94 solar image of the October 28, 2021, X1.0 flare at 15:35 UT (left). AR12887 sunspot region division (right). The naming of a sunspot is preceded by a region name, followed by a number identification. If the sunspot experiences fragmentation, the name is then followed by another number identification, for instance: A1.1.

3 Result and Discussion

We measured a total of 166 images, with 37 sunspots' movements, from October 25th until October 31st, 2021. We found out prior to the X1.0 flare, there were many sunspots experiencing fragmentation (see Fig. 2), especially in the southeast region (B and D),

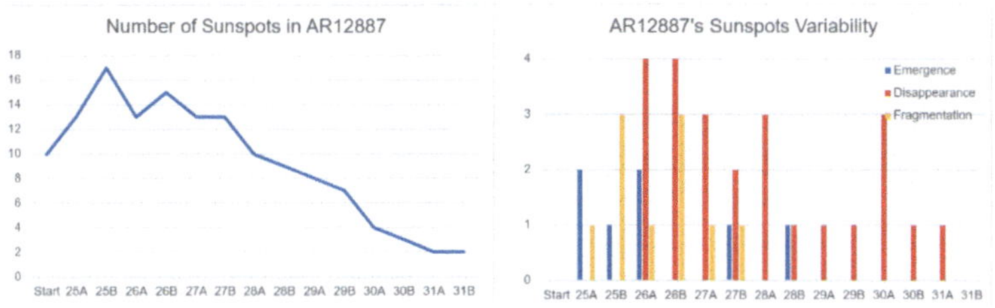

Fig. 2. Graph on the number of sunspots in AR12887 (left) and their cause for their variability (right) from October 25th until October 31st, 2021, with each day separated into 2 phases: A for 00:00–12:00 and B for 12:00–24:00. Flare is between 28A and 28B.

while sunspots in Region C continuously became smaller and eventually disappeared before the flare happened (see Fig. 3). After the flare, the AR's magnetic activities seem to be more inactive as there is no more sunspot emergence and fragmentation, leading to a decreasing number of sunspots in AR12887.

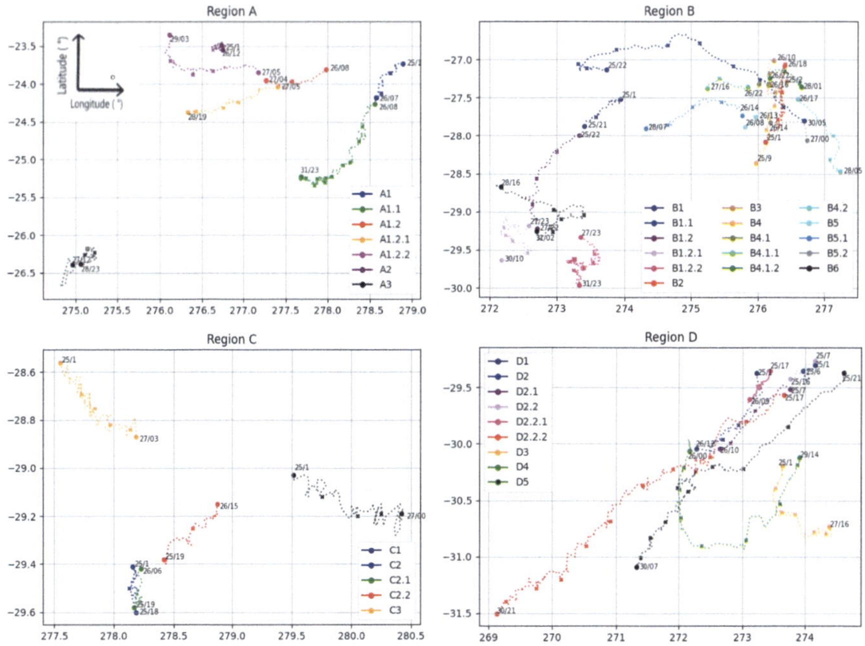

Fig. 3. Position map of sunspots' movement in AR12887, divided per region as in Fig. 1. The date is labeled for the emergence and disappearance of each spot, while the position of the spots is x-marked every 12 h after their emergence. Only B4.1 is not labeled to avoid overlapping text (emergence time is the same as B4.2, while disappearance time is right before the emergence of B4.1.2).

According to NOAA report, the area of AR12887 was 320 millionth of solar hemisphere (msh) [10], which is much smaller than average X-class-flare-producing ARs in the past 25 years (782 msh), not even reaching the average M-class-flare-producing ARs (504 msh) [6]. To compensate for its small area, AR12887 must have experienced a huge amount of magnetic activity to accumulate enough energy to release an X-class flare [11, 12]. The amount of magnetic activity measured prior to the flare (see Fig. 2) supports this fact. These magnetic activities may have also led to smaller flare occurrences, totalling 13 C-class and 2 M-class, in AR12887 prior to the X1.0 flare [13].

From the position measured, we then choose several spots that last from before the X1.0 flare until after the occurrence and make its velocity map (see Fig. 4) to observe the change in velocity throughout its movement. From all the measurements done, we found out that sunspots experiencing fragmentation will experience velocity transformation, both in speed and direction, as can be seen in the high initial speed on sunspots A1.1, A1.2.2, B1.2.1, and B1.2.2. Some of the sunspots, such as A1.1, B1.2.1, B1.2.2, D4, and D5, gradually slowed down after the flare, which is highly probable due to the lessened magnetic activity.

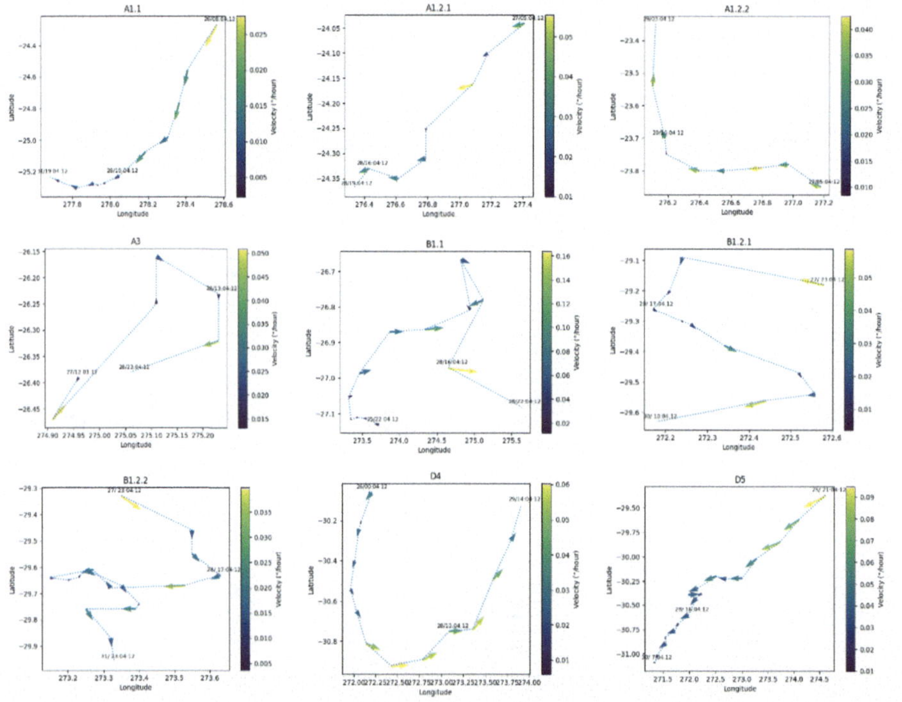

Fig. 4. Velocity map of the selected sunspots' movement in AR12887.

While the velocity map gives general insight on the sunspot change in velocity, to truly know what may cause those fluctuations in velocity, we performed a more detailed analysis by plotting their change in both latitude and longitude (see Fig. 5). We observed that sunspots which crossed paths with each other also experienced changes in their movement direction, shifting from northward to southward and vice versa This occurs due to the phenomenon called shearing, which is the twisting and tangling of magnetic lines in the sunspots, causing the buildup of additional magnetic energy and thus affecting the sunspot's proper motion. From the extreme motion change, we hypothesized that the crossing of spots B1.1 and A3 may play quite a significant role in the X1.0 flare event.

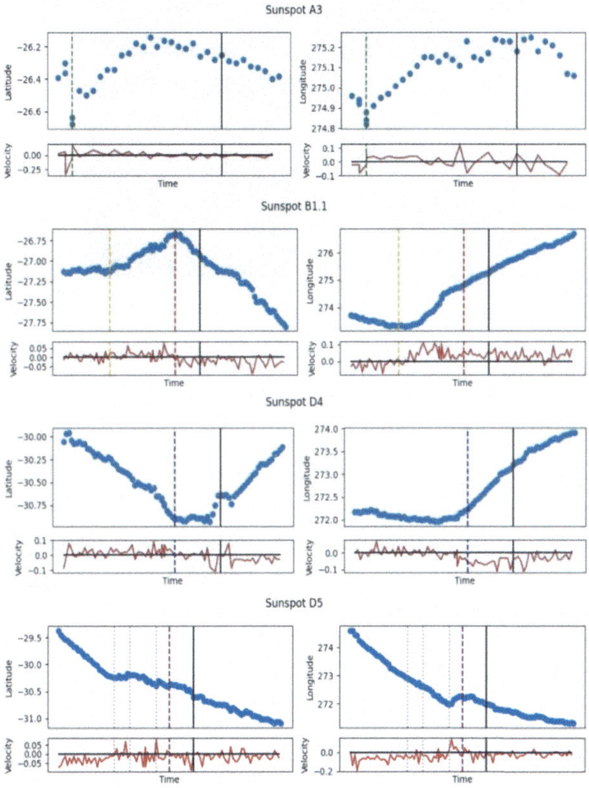

Fig. 5. Graphs on the position and velocity of sunspot A3, B1.1, D4, and D5. The black solid line shows the moment of the X1,0 flare, while the dashed line shows the moments of significant change in direction and speed. The yellow dashed line is the moment when B1.1 finished adjusting itself after B1's fragmentation. The green and red dashed lines are due to shearing between A3 and B1.1. Similarly, the blue and purple dashed lines are due to the shearing between D4 and D5. Dotted lines in D5 represent the moment where D5 interacts closely with other sunspots in the D region, causing fluctuations in its proper motion.

Comparing our result of the sunspot motion analysis to the reference, as expected, the location of B1.1 and A3 (see Fig. 6) is mentioned as a collision of polarities by Song [6], which will cause disruption in the magnetic filaments, triggering an X-class flare. A similar result is also reported by Fang [5]: the area where the collision occurred is the trigger for an X1.0 flare.

While spots A3, B1.1, D4, and D5, which are located right in the middle of the flare, show that the solar flare did not affect their proper motion that much compared to

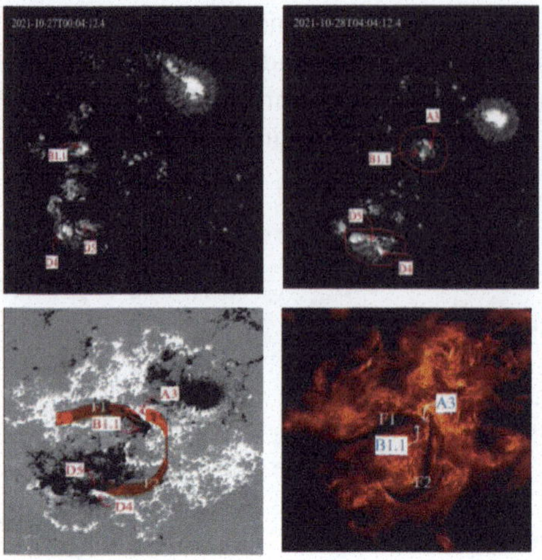

Fig. 6. The condition of sunspots in AR12887 on October 27, 2021 at 00:04:12.4 (top left) and sunspots on October 28, 2021 at 04:04:12.4 (top right), focused especially on sunspots experiencing shearing. The magnetogram image of AR12887 (bottom left) with positive polarity (white) and negative polarity (black), alongside the magnetic filament F1 and F2, which can be seen more clearly through the AIA304 filter (bottom right).

fragmentation and shearing, the same cannot be said for spots located at the periphery of the flare (see Fig. 7), as the flare seemed to push the sunspots away from it. Since the flare is located in the southeast of region A, spot A1.1, which originally was moving southeast, has its movement deflected northward, canceling its southward movement. Spot A1.2.2, which originally is only moving eastward, has its movement trajectory changed into moving northward.

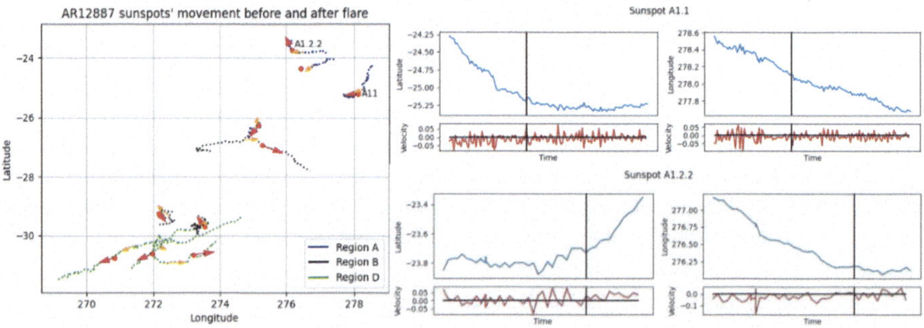

Fig. 7. Movement of spots in AR12887 throughout October 28, 2021, with the red dot denoting the event of solar flare while orange and red arrow representing its movement direction before and after the flare (left), focusing on A1.1 (top right) and A1.2.2 (bottom right) movements, with the solid black line denoting the flare event.

4 Conclusion

Prior to X1.0 flare, movements of the sunspots were far more erratic, particularly due to the presence of high magnetic activity such as the emergence, fragmentation, disappearance and shearing of sunspots (Fig. 2), which caused significant fluctuation in proper motion (Fig. 5). After the flare, the magnetic activity diminished, leading to a gradual slowdown noin the movement of some sunspots (Fig. 3).

Sunspots located within the flare region did not experience noteworthy change in their motion direction due to the flare, while sunspots located at the periphery of the flare have the tendency to move away from it (Fig. 7).

References

1. Jiang, C., et al.: Developments of a fundamental mechanism for initiation of solar eruptions. Rev. Mod. Plasma Phys. **8**, 18 (2024)
2. Solar Flares (Radio Blackouts) | NOAA / NWS Space Weather Prediction Center. https://www.swpc.noaa.gov/phenomena/solar-flares-radio-blackouts. Accessed 30 July 2024
3. Tsurutani, B.T., Verkhoglyadova, O.P., Mannucci, A. J., Lakhina, G.S., Li, G., Zank, G.P.: A brief review of "solar flare effects" on the ionosphere. Radio Sci. **44**, RS0A17 (2009)
4. SOHO/LASCO CME Catalog. https://cdaw.gsfc.nasa.gov/CME_list/. Accessed 30 July 2024
5. Fang, Y., Zhang, J., Bi, Y., Song, Z.: Dynamics of threads wrapping a filament's leg prior to the eruption on 2021 october 28. Astrophys. J. **955**(2), 87 (2023)
6. Song, Q., et al.: The triggering process of an X-class solar flare on a small quadrupolar active region. Astrophys. J. **959**(1), 53 (2023)
7. Ambastha, A., Bhatnagar, A.: Sunspot proper motions in active region NOAA 2372 and its flare activity during SMY period of 1980 April 4–13. J. Astrophys. Astron. **9**, 137–154 (1988)
8. Herdiwijaya, D., Permata, K.: Sunspot proper motion of individual sunspot in active region 11429 during X5.4 solar flare on March 7, 2012. J. Phys.: Conf. Ser. **1153**, 012134 (2019)
9. Müller, D., et al.: JHelioviewer - Time-dependent 3D visualisation of solar and heliospheric data. A&A **606**, A10 (2017)
10. Sunspot region AR12887 | Solar activity | SpaceWeatherLive.com. https://www.spaceweatherlive.com/en/solar-activity/region/12887.html. Accessed 30 July 2024
11. Lin, J., Wang, F., Deng, L., Deng, H., Mei, Y., Zhang, X.: Evolutionary relationship between sunspot groups and soft X-ray flares over solar cycles 21–25. Astrophys. J. **958**(1) (2023)
12. Atilgan, S.E., Kilcik, A.: Investigation of the relationship between solar flares and sunspot groups. Sun Geosphere **12**(1), 7–10 (2017)
13. Sunspot region AR12887 | Solar activity | solarmonitor.org. https://www.solarmonitor.org/index.php?date=20211028&region=&indexnum=1. Accessed 14 Aug 2024

Magnetic Field Analysis of the May 2024 Geomagnetic Storm Using Swarm Data and CHAOS Model

Dzikrina Silmi Syukria, Muhammad Isnaenda Ikhsan(✉), and Alka Budi Wahidin

Planetary and Atmospheric Science Study Program, Faculty of Science, Institut Teknologi Sumatera, South Lampung, Lampung, Indonesia
isnaenda.ikhsan@sap.itera.ac.id

Abstract. This study examines changes in the intensity and direction of the Earth's magnetic field using observational data from the Swarm-A satellite magnetometer during the May 2024 geomagnetic storm. This geomagnetic storm is one of the extremes occurring after several decades and is classified as G5, with a maximum Kp index of 9 and a Dst index reaching - 400 nT. In addition, the CHAOS geomagnetic field model was also employed in this study to evaluate the magnetometer measurement data, thereby facilitating the identification of anomalous data. The findings of this study show that the Earth's magnetic field undergoes significant alterations when affected by geomagnetic activity. The CHAOS model also demonstrates considerable accuracy in evaluating the Earth's magnetic field values as measured by the magnetometer. The error value, obtained through the calculation of Root Mean Square Error, only ranges less than 3% on the peak day of the storm, indicating that the accuracy is quite good.

Keywords: Earth's Magnetic Field · Geomagnetic Storm · Swarm-A · CHAOS Model

1 Introduction

The Earth's magnetic field plays a crucial role in protecting the Earth from radiation originating from external sources. The Earth's magnetic field constitutes a layer at the outermost part of the Earth's atmosphere, known as the magnetosphere [1]. The magnetosphere serves as a protective barrier, deflecting energetic particles from external sources, including those carried by the solar wind [2]. The Earth's magnetic field is formed due to the convection process, whereby charged fluid circulates in the Earth's core. This process is enabled by a dominant concentration of iron and nickel in the Earth's core. The magnetic field generated from those processes is known as the main magnetic field.

The non-stationary nature of Earth's magnetic field is attributed to several factors, such as the influence of interactions with other geomagnetic field shapers and the effect of Earth's rotation. Although the Earth's magnetic field is mainly generated within the Earth's core, which accounts for 97% of the total magnetic field strength [3], other factors

E. Fitrianingsih et al. (Eds.): ISAST 2024, SPPHY 416, pp. 196–204, 2025.
https://doi.org/10.1007/978-981-96-1344-1_21

also influence the formation of the main magnetic field, which are the external magnetic field and crustal field. The external field is formed by energetic particles outside the Earth interacting with the ionosphere, causing electric currents and generating magnetic flow. Then, the crustal field, also known as the magnetic anomaly field, is formed by variations in the magnetic mineral content of rocks in the Earth's crust [4].

The influence of both factors is relatively small due to the localized nature of the crustal field, and the external magnetic field is typically periodic, deriving from a phenomenon. The strength of the Earth's magnetic field generally ranges from 25,000 to 65,000 nT [5]. Among external magnetic field phenomena, geomagnetic storms significantly influence the main magnetic fields. This phenomenon will be the focus of this research project, which aims to elucidate the Earth's magnetic field changes due to such storms.

A geomagnetic storm is defined as a disturbance formed due to the interaction between energetic particles from solar activity, such as Coronal Mass Ejections (CMEs), carried by the solar wind towards Earth [6]. The interaction of those particles with the magnetosphere results in an asymmetry in the Earth's magnetic field lines of force. This asymmetry causes compressions along the day side of the magnetosphere and expansion along the night side. The impetus of the particle interactions influences the directionality of this compression and expansion [7].

The geomagnetic storm that impacted Earth on May 11, 2024, was the most extreme in recent decades. As demonstrated by the research of Y. Yamazaki et al., the geomagnetic storm that occurred in May 2024 was one of the most intense storm events in the modern era since 1985 [8]. The storm's strength reached a Kp index value of 9, representing the maximum value for measuring storm strength, while the Dst index reached a value of - 400 nT. The values in the geomagnetic indices were identified through SpaceWeatherLive[1] for the Kp index and WDC for Geomagnetism[2] for the Dst index. One of the effects of this geomagnetic storm was the appearance of the aurora up to low latitudes. In addition, these storms will likely disrupt satellites in low orbit. Given the considerable strength of this extreme storm, we undertook this study to ascertain the extent of the changes to the Earth's magnetic field during such events.

2 Data and Method

We used data from the swarm satellite magnetometers to analyze earth's magnetic field changes during the geomagnetic storm event in May 2024. Swarm is the satellite constellation developed by the European Space Agency (ESA), designed to study the dynamics of the Earth's magnetic field and its interaction with the Earth system [9]. The constellation comprises three identical Swarm satellites (Alpha, Bravo, and Charlie). The three satellites are positioned in near-polar orbit with different altitudes. Two satellites, swarm-alpha and Charlie, are placed in a side-by-side configuration at an altitude of 450 km with an inclination of 87.3°, while swarm-bravo is placed at a higher altitude of 530 km with an inclination of 87.7° [10].

[1] https://www.spaceweatherlive.com/.

[2] https://wdc.kugi.kyoto-u.ac.jp/.

The data utilized in this study are magnetic data obtained from the Swarm-Alpha (Swarm-A) satellite, specifically the observation data on the intensity and direction parameters of the Earth's magnetic field in local cartesian coordinates (North, East, Center). Magnetic data from Swarm A is preferable for obtaining a more accurate value of the Earth's magnetic field with a higher resolution due to its lower altitude than Swarm B. Furthermore, a geomagnetic field model is used to evaluate the values derived from the observations to provide precise anomaly changes during the geomagnetic storm period.

The geomagnetic field model used is the CHAOS model, specifically CHAOS-7. The CHAOS model is one of the geomagnetic field models that consider external magnetic field measurements to enhance the precision with which the Earth's magnetic field is represented. This model was developed by employing magnetometer measurements obtained from satellite magnetometer measurements and ground observatory data [11]. Meanwhile, CHAOS-7 represents the latest iteration of the CHAOS model, offering enhancements that bolster the precision of forecast relative to preceding versions. These include more meticulous data selection during periods of minimal activity, incorporation of magnetic field data from the CryoSat-2 satellite, and relaxation of temporal constraints to facilitate the representation of complex temporal variations at finer spatial scales [12].

The data employed in this study, both measurement and model, are accessed through the Virtual Research Service from Swarm (VirES)[3]. In this study, we investigate the impact of geomagnetic storms on the magnetic field by calculating residuals between the observed magnetic field and the CHAOS model or carried out using Eq. (1).

$$H_{residual} = H_{observed} - H_{model} \tag{1}$$

It is hypothesized that geomagnetic storms will affect the residual values, as the model cannot accurately calculate the magnetic field during such events. The significant disturbance caused by the storm results in a greater discrepancy between the model and the data.

3 Result and Discussion

Based on the identification result of geomagnetic indices Kp and Dst, the geomagnetic storm events in May 2024 occurred from May 10 to 12, 2024. To determine the changes in the Earth's magnetic field, we used nine days consisting of a quiet day before the storm, a day when the storm occurred, and a calm day after the storm. The analysis that will be carried out in this study is divided into two parts. The first part is a time-based analysis to determine how much the Earth's magnetic field changes during quiet days and at the peak of a storm's activity. The second part is an analysis of the vector visualization and intensity of the Earth's magnetic field to determine changes based on the direction and intensity of the field.

3.1 Temporal Variations of Earth's Magnetic Field During the May Geomagnetic Storm

Figure 1(a) and 1(b) are the Kp and Dst indices showing the dynamics of geomagnetic activity over nine days. These graphs show that geomagnetic activity began to increase

[3] https://vires.services/.

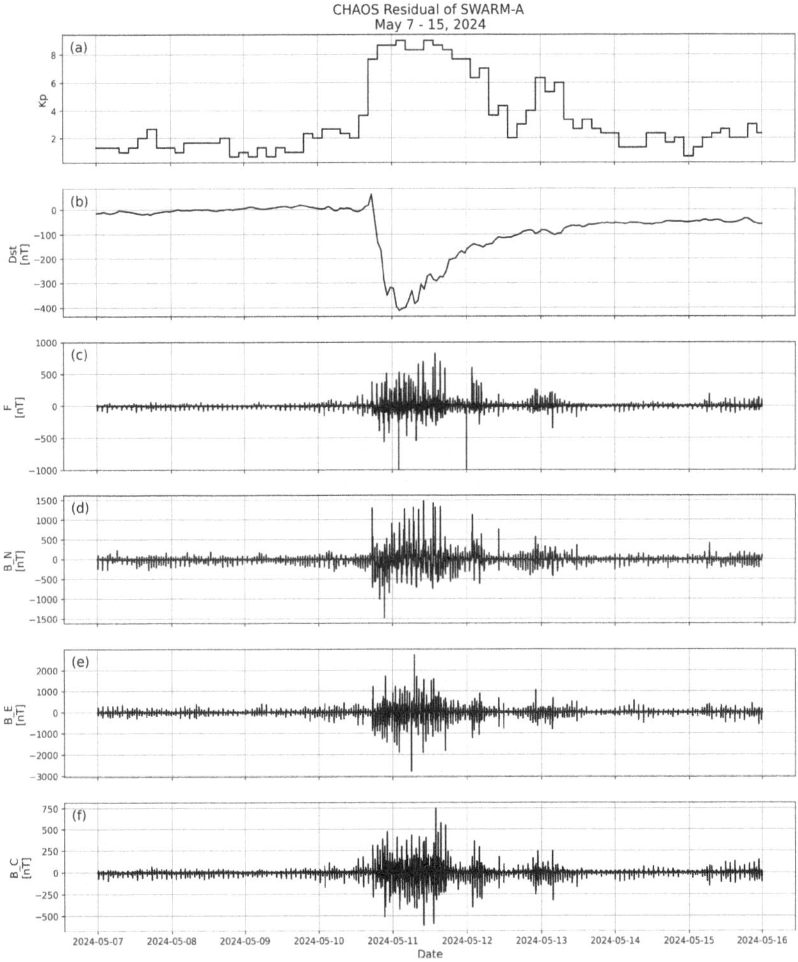

Fig. 1. Graph of the residual result of SWARM-A magnetometer data with CHAOS model; (a) Kp index, (b) Dst index, (c) Residual of Earth's magnetic field intensity, (d) Residual of Earth's magnetic field component in the direction of North-South, (e) Residual of Earth's magnetic field component in the direction of East-West, and (f) Residual of Earth's magnetic field component in the direction of center of the Earth on May 7–15, 2024.

on the evening of May 10 and lasted until around noon on May 12. The Kp index reached a maximum value of 9 during the peak of the geomagnetic storm on May 11, indicating that the storm was at the extreme storm level. Likewise, the Dst index shows the minimum value reaching -400 nT. The increase in geomagnetic activity over the three days led to fluctuations in the magnetic field residuals, resulting in variations in both the intensity and directional parameters of Earth's magnetic field, as shown in Fig. 1(c–f).

Figure 1(c) shows the residual of Earth's magnetic field intensity, with fluctuations reaching up to ±1,000 nT during the storm. Likewise, Fig. 1(d) and 1(e) show the residual of the Earth's magnetic field intensity in the direction of horizontal components of the

North-South and West-East directions, which provide changes in fluctuations between ±1,500 nT to ±3,000 nT. Figure 1(f) shows the residual of Earth's magnetic field intensity in the vertical component, pointing toward the center of the Earth. The fluctuations in this component are generally smaller than those in the horizontal component, with variations around ±700 nT. The differences between the horizontal and vertical components might be caused by the direction from which the solar storm approached.

The positive and negative values in the residual data is the result from the difference between the observed values of Earth's magnetic field and those predicted by the CHAOS model. Positive values indicate that the observed magnetic field is larger than the model's predicted value, while negative values indicate the opposite. The data from VirES show that the observed Earth's magnetic field exhibited values with a range between 18,000 and 50,000 nT over nine days.

3.2 Analysis of Changes in Earth's Magnetic Field Residual Based on Visualizations of Its Direction and Intensity

This section presents an analysis of changes in the Earth's magnetic field, focusing on the intensity and direction of the magnetic field during geomagnetic storm activity. The analysis uses contour and vector graphs to illustrate the variation of the residual magnetic field over time. The data presented here is based on residual data during geomagnetic storm events from May 10 to May 12, 2024. As evidenced by the visualization results presented in Figs. 2, 3 and 4, there were four instances when significant changes in the intensity and direction of the magnetic field occurred. These events happened in the second 12-h interval on May 10 (Fig. 2(b)), May 11 (Fig. 3(a) and (b)), and in the first 12-h interval on May 12 (Fig. 4(a)).

The most significant alteration in Earth's magnetic field occurred during the first 12 h of May 11 (Fig. 3(a)), which aligns with the peak time of the storm. This significance is characterized by more extreme changes in the intensity and direction of the magnetic field in all regions during this time interval. Figure 2(b) depicts a transitional period where normal conditions start to be influenced by geomagnetic storms. Figures 3(b) and 4(a) illustrate the period following the storm's peak, showing conditions as the storm subsides.

The changes in Earth's magnetic field residuals described earlier do not represent alterations in the actual orientation of Earth's magnetic field. Instead, they reflect the differences between observed data and the model's predictions. These differences may be caused by anomalies resulting from the influence of geomagnetic storms. During a geomagnetic storm, Earth's magnetic field is altered in a way that the model cannot accurately predict yet.

From Fig. 2, 3, 4(a) and (b), it is apparent that the intensity contour shows an opposite pattern between the first 12-h and second 12-h intervals. This phenomenon can be attributed to the local time when the satellite passes by, which determines when it measures the magnetic field. As illustrated in Figs. 2, 3, 4(c) and (d), due to its orbit, the Swarm satellite only passes over twilight regions on Earth, keeping the local time consistently around 6 PM and 6 AM. These two times divide the areas covered by the satellites into opposite patterns between the first and second intervals. During the first interval, the region between 0°–60° W and 0°–60° E is on the day side, while the area

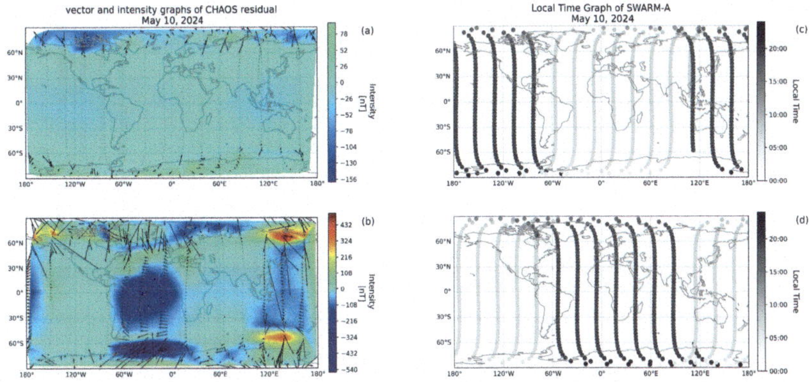

Fig. 2. Vector and intensity graphs of the Earth's magnetic field residual with the CHAOS model and the time chart of the SWARM-A satellite converted to local time on May 10, 2024; (a) vector and intensity graphs of the Earth's magnetic field residual at the first 12-h interval, (b) vector and intensity graphs of the Earth's magnetic field residual at the second 12-h interval, (c) Local time graph of SWARM-A satellite at the first 12-h interval, (d) Local time graph of SWARM-A satellite at the second 12-h interval.

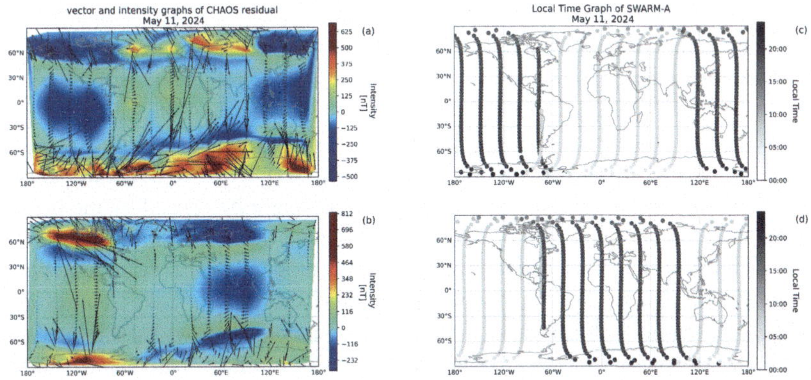

Fig. 3. Vector and intensity graphs of the Earth's magnetic field residual with the CHAOS model and the time chart of the SWARM-A satellite converted to local time on May 11, 2024; (a) vector and intensity graphs of the Earth's magnetic field at residual the first 12-h interval, (b) vector and intensity graphs of the Earth's magnetic field residual at the second 12-h interval, (c) Local time graph of SWARM-A satellite at the first 12-h interval, (d) Local time graph of SWARM-A satellite at the second 12-h interval.

between 60°–180° W and 120°–180° E is on the night side. In the second 12-h interval, this division reverses, with the previous day side becoming the night side, and vice versa.

Higher intensity changes tend to occur on Earth's day side, while lower intensity changes are observed on the night side. This is due to the distribution of solar radiation, which is more intense on the day side compared to the night side. Additionally, more significant changes occur as the Earth's magnetic field is compressed on the day side by the influence of solar wind [13]. This compression on the day side causes the formation

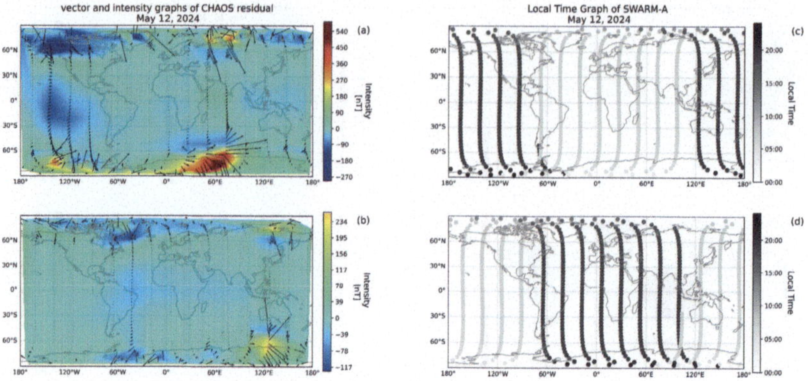

Fig. 4. Vector and intensity graphs of the Earth's magnetic field residual with the CHAOS model and the time chart of the SWARM-A satellite converted to local time on May 12, 2024; (a) vector and intensity graphs of the Earth's magnetic field residual at the first 12-h interval, (b) vector and intensity graphs of the Earth's magnetic field residual at the second 12-h interval, (c) Local time graph of SWARM-A satellite at the first 12-h interval, (d) Local time graph of SWARM-A satellite at the second 12-h interval.

of a magnetotail in the region on the Earth's night side [14]. The formation of the magnetotail results in a weaker magnetic field in this region, as indicated by the negative values in the residuals.

As illustrated in Figs. 2, 3 and 4, the intensity and direction of the Earth's magnetic field in the region around the poles always show more significant dynamics than other regions, even in the absence of geomagnetic storm activity. The dynamic state that causes disturbances in the region around the poles is might be caused by the polar cap phenomenon [15]. The polar caps cause open magnetic field flow in the polar region, allowing increased interaction between energetic particles and the Earth's magnetic field.

Although the residuals of Earth's magnetic field show significant changes during heightened geomagnetic storm activity, the CHAOS model still provided relatively accurate magnetic field predictions. The accuracy of the model could by calculated using Root Mean Square Error (RMSE) through Eq. 2 [16].

$$RMSE(\%) = \sqrt{\frac{\frac{1}{N}\sum_{i=1}^{N}(H_{obs} - H_{model})^2}{\frac{1}{N}\sum_{i=1}^{N}(H_{obs})}} \times 100 \tag{2}$$

Through the RMSE calculation, the CHAOS model gives an error of less than 1% on calm days, while on the peak day of the geomagnetic storm, the error value only reaches about 2.6%. Even during peak geomagnetic activity, this relatively small error indicates that the CHAOS model maintains good accuracy.

4 Conclusion

This study utilizes data from the Swarm A satellite and the CHAOS geomagnetic model to examine the Earth's magnetic field alteration during a geomagnetic storm in May 2024. The visualization results indicate a significant change in the intensity and direction of the residual magnetic field as the level of storm activity increases. The Earth's magnetic field around the North and South Poles is more dynamic, as indicated by the consistently larger discrepancies observed in the residual data, even when geomagnetic storm influences are absent.

The change of intensity and vector in the contour plot is attributable to the discrepancy between the measured value of the Earth's magnetic field and the CHAOS model, which is increasing. This demonstrates that the greater the activity of a geomagnetic storm, the more significant the discrepancy between the observed values and those produced by the model. This also validates the hypothesis regarding model predictions presented in the introduction. Furthermore, the asymmetry in the magnetic field lines resulting from the compression of the stronger solar wind also influences the transformation of the Earth's magnetic field. This phenomenon reduces the intensity and direction of the magnetic field on the night side of the Earth and increases on the day side.

Although the model increasingly shows inaccuracies in evaluating measured magnetic field values when storm activity increases, the CHAOS model still provides satisfactory evaluation results. The RMSE calculation during the storm yields an error value of less than 3%, indicating that the model continues to perform satisfactorily even during geomagnetic storms.

References

1. Martiningrum, D.R., Purnowo, A., Nuraeni, F., Muhamad, J.: Fenomena Cuaca Antariksa. Revisi edn. Pusat Riset Antariksa (2012)
2. Liu, J., Wang, W., Qian, L., Lotko, W., et al.: Solar flare effect in the earth's magnetosphere. Nat. Phys. **17**(7), 807–812 (2021)
3. Nasuddin, K.A., Abdullah, M., Abdul Hamid, N.S.: Characterization of the south atlantic anomaly. Nonlinear Process. Geophys. **26**(1), 25–35 (2019)
4. Telford, W.M., Geldart, L.P., Sheriff, R.E.: Applied Geophysics, 2nd edn. Cambridge University Press (1990)
5. Erdmann, W., Kmita, H., Kosicki, J.Z., Kaczmarek, Ł: How the geomagnetic field influences life on the earth - an integrated approach to geomagnetobiology. Origins Life Evol. Biospheres **51**, 231–257 (2021)
6. Tsurutani, B.T., Gonzalez, W.D., Gonzalez, A.L.C., Guarnieri, F.L., et al.: Corotating solar wind streams and recurrent geomagnetic activity: a review. J. Geophys. Res. **111**(A7) (2006)
7. Tsyganenko, N.A.: Data-based modeling of the earth's dynamic magnetosphere: a review. Ann. Geophys. **31**(10), 1745–1772 (2013)
8. Yamazaki, Y., Matzka, J., da Silva, M.V.S., Kervalishvili, G., et al.: Assessment of Geomagnetic Activity for the Kp=9 "Gannon Storm" in May 2024 Based on Version 3.0 Hpo Indices ESS Open Archive (2024)
9. Friis-Christensen, E., Luhr, H., Hulot, G.: Swarm: a constellation to study the earth's magnetic field. Earth Planets Space **58**(4), 351–358 (2006)
10. EoPortal: Swarm (Geomagnetic LEO Constellation). https://www.eoportal.org/satellite-missions/swarm. Accessed 29 July 2024
11. Olsen, N., Friis-Christensen, E., Floberghagen, R., Alken, P., et al.: The swarm satellite constellation and research facility (SCARF) and swarm data products. Earth Planets Space **65**(11), 1189–1200 (2013)
12. Finlay, C.C., Kloss, C., Olsen, N., Hammer, M.D., et al.: The CHAOS-7 geomagnetic field model and observed changes in the South Atlantic Anomaly. Earth Planets Space **72**(1) (2020)

13. Yu, J., Li, L.Y., Cao, J.B., Reeves, G.D., et al.: The influence of solar wind pressure and interplanetary magnetic field on global magnetic field and outer radiation belt electrons. Geophys. Res. Lett. **43**(14), 7319–7327 (2016)
14. Goodman, J.M.: Magnetosphere and Geomagnetic Storms. Space Weather & Telecommunications. The International Series in Engineering and Computer Science. Springer New York (2005)
15. Laundal, K.M., Cnossen, I., Milan, S.E., Haaland, S.E., et al.: North-south asymmetries in earth's magnetic field. Space Sci. Rev. **206**(1–4), 225–257 (2016)
16. Kambezidis, H.D.: Evaluation of models. Solar Resour. Compreh. Renew. Energy **3**, 27–84 (2012)

D-Region Ionospheric Variations Induced by a 2023 Fireball in Hokuriku, Japan, Using OCTAVE VLF/LF Transmitter Signals

Ryota Furuya[1], Hiroyo Ohya[1(✉)], Fuminori Tsuchiya[2], Masa-yuki Yamamoto[3], and Hiroyuki Nakata[1]

[1] Chiba University, Chiba 263-8522, Japan
ohya@faculty.chiba-u.jp
[2] Tohoku University, Sendai 980-8578, Japan
[3] Kochi University of Technology, Kami 782-8502, Japan

Abstract. Several studies have examined ionospheric variation associated with meteorites, meteoroids, or meteors based on Global Satellite Navigation System total electron content observations. However, there have been few quantitative studies of the D-region of the ionosphere (60–90 km), which is associated with meteoroids. We investigated variations in the D-region during the passage of a fireball over Hokuriku, Japan, at 14:33 UT on 23 April 2023 and a satellite re-entry at 12:38 UT on 26 December, 2023, using OCTAVE very low-frequency (VLF, 3–30 kHz) and low-frequency (LF, 30–300 kHz) transmitter signals. OCTAVE is an observation network of VLF/LF transmitter signals from low to high latitudes that we established. In this study, we used three transmitters, BPC (68.5 kHz), JJY (60.0 kHz), and JJI (22.2 kHz), observed at Rikubetsu observatory (RKB), Japan. Periodic variation of 200–500 s was observed in the amplitudes and phases of BPC–RKB path at the acoustic wave arrival times. Periodic variation of 300–500 s was observed in the amplitudes and phases of JJY60–RKB path at the arrival time of acoustic wave. For JJI–RKB path, there was no clear variation, because variation with a period of 200–300 s occurred before the arrival time of the acoustic waves. Based on the observed propagation time of the acoustic waves and ray tracing results, the acoustic waves or atmospheric gravity wave propagated obliquely from fireball trajectory up to a D-region height.

Keywords: Ionosphere · Fireball · VLF/LF Transmitter Signals

1 Introduction

Meteoroids enter the Earth's atmosphere from space at speeds of 11–73 km/s [1]. As meteoroids enter the atmosphere at speeds of several tens of km/s or more, they generate acoustic and atmospheric gravity waves that propagate into the ionosphere [2], causing traveling ionospheric disturbances (TIDs). During the Chelyabinsk meteoroid event on 15 February 2013, three types of TIDs were observed using total electron contents obtained by the GPS navigation system [2].

E. Fitrianingsih et al. (Eds.): ISAST 2024, SPPHY 416, pp. 205–212, 2025.
https://doi.org/10.1007/978-981-96-1344-1_22

The acoustic wave associated with the Chelyabinsk meteoroid event caused the phase of the LF transmitter signals (66.67 kHz) to vary by 4–7°, with a period of 200–230 s, in the D-region ionosphere [3–5]. The LF path was located ~1,575 km from the point where the Chelyabinsk meteoroid exploded. The amplitude of the variation in the D-region height was estimated to be 0.1–0.2 km. The corresponding election density disturbances were 2.7–4.6% at a height of 65–70 km. Moreover, during the Hokkaido fireball occurred at 11:55 UT on 18 October, 2018, D-region ionospheric variations with a period of 100–200 s due to acoustic waves were also reported [6]. However, D-region response due to fireballs of satellite re-entry has not been revealed. In this study, we investigated disturbances in the D-region associated with a meteoroid that passed over Hokuriku, Japan, at 14:33 UT on 23 April 2023 and satellite re-entry on 26 December 2023, using very-low-frequency (VLF) / low-frequency (LF) transmitter signals of the Observation of CondiTion of ionized Atmosphere by VLF Experiment (OCTAVE) network installed by our research group.

2 Observations

The fireball was observed at heights from 131.1 to 80.7 km, traveling over Hokuriku, Japan, at 14:33 UT on 23 April 2023, and the satellite was observed at heights from 89.9 to 89.1 km, traveling over Tokai, Japan, at 12:38 UT on 26 December 2023, based on optical network observations by the SonotaCo network, a Japanese amateur group that estimates the location and height of meteoroids using motion detection software. The fireball fell from east (37.79°N, 138.16°E) to west (37.63°N, 137.42°E), and the satellite fell from west to east (34.46°N, 139.48°E). Because the exact location of the beginning point of the satellite re-entry was unknown, we describe only the location of the end point in this study.

Figure 1 shows the location of trajectory of the fireball and the four VLF/LF paths used in this study, and Fig. 2 shows the trajectory of the satellite re-entry. The transmitters were located at Miyazaki, Japan (JJI, 22.2 kHz, 32.05°N, 130.82°E), Saga, Japan (JJY-60; 60 kHz, 33.47°N, 130.18°E), Fukushima, Japan (JJY-40; 40 kHz, 37.37°N, 140.85°E), and China (BPC, 68.5 kHz, 34.63°N, 115.83°E). The receivers were located at Rikubetsu, Hokkaido (RKB; 43.45°N, 143.77°E) and Kagoshima, Japan (KAG; 31.48°N, 130.72°E). The RKB receiver is one of the OCTAVE network, a worldwide one of VLF/LF transmitters that our research group installed to monitor the D-region ionosphere. We have observed the amplitude and phase of these signals. The amplitudes and phases have been recorded on hard disks with a time resolution of 0.1 s.

We also used vertical seismic velocity observations sampled 0.1 s intervals, measured near the fireball trajectory by Full Range Seismograph Network of Japan (F-net) operated by the National Research Institute for Earth Science and Disaster Resilience (NIED), and infrasound data sampled 1 s intervals, measured near the trajectory of satellite re-entry by a research group of Kochi University of Technology. The blue squares in Fig. 1(b) and Fig. 2 indicate locations of seismometer at KZK (37.30°N, 138.51°E) and infrasound sensor at TTY (34.98°N, 139.83°E), respectively.

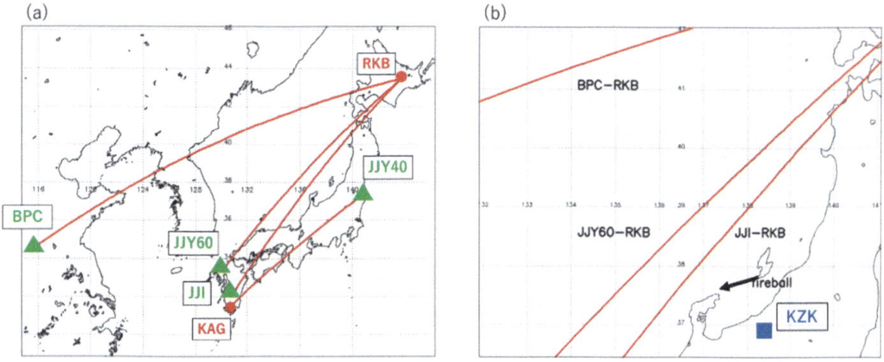

Fig. 1. (a) Location of VLF/LF observations. The green triangles and red circles indicate VLF/LF transmitters and two receivers. (b) Enlarged view of the fireball trajectory (the black arrow). The blue square indicates location of a seismometer at KZK.

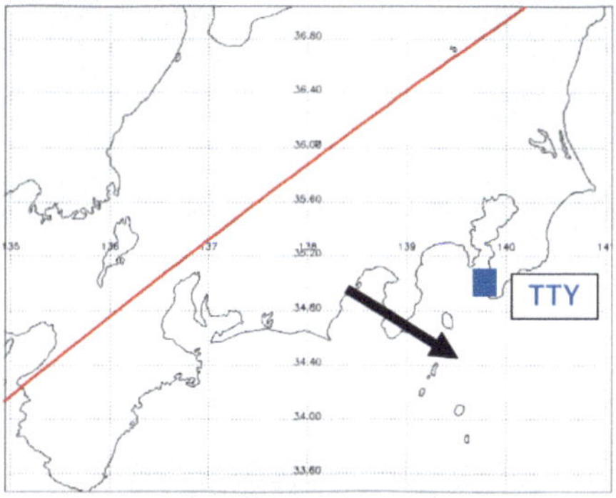

Fig. 2. Location of the satellite re-entry trajectory (the black arrow). The blue square indicates location of infrasound sensor at TTY.

3 Results and Discussion

3.1 Hokuriku Fireball

Figure 3 shows the waveforms of the VLF/LF amplitude and phase and the wavelet spectra on the BPC–RKB, JJY60–RKB, and JJI–RKB paths. The orange, blue, and green lines indicate occurrence time of the fireball, the arrival time of the acoustic wave generated at the beginning point, and the arrival time of the acoustic wave generated at the end point, respectively. The signals were filtered using a low-pass filter of >30 s and subtracted 5 min moving average.

The arrival time of the acoustic wave was calculated from the beginning point and end point of the fireball to the D-region ionospheric height, using a raytracing tool of GeoAc. Figure 4 shows the ray path of acoustic wave propagated from the end point to JJY60-RKB path using the GeoAc. Based on calculations using the Appleton–Hartree equation, the International Geomagnetic Reference Field 13, and the International Reference Ionosphere 2020 models [6], the estimated reflection heights for transmitted

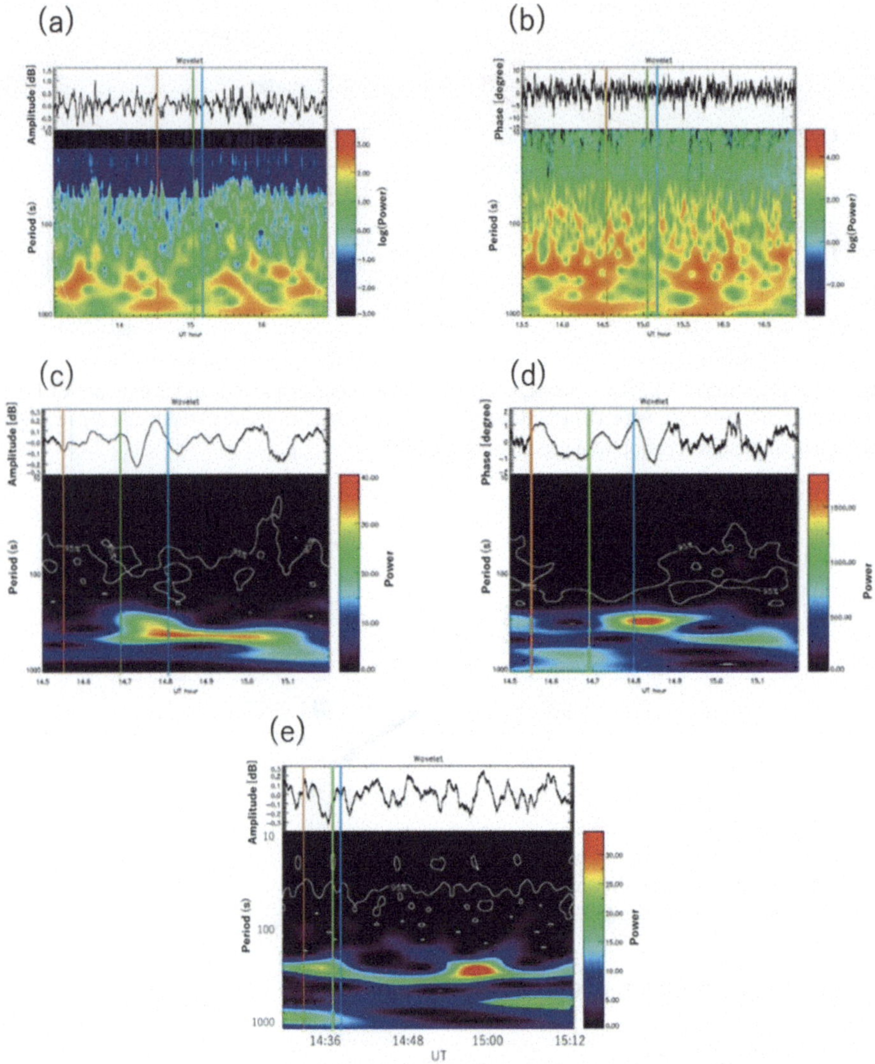

Fig. 3. (a) Amplitude for BPC–RKB path and the wavelet spectra, (b) phase for BPC–RKB path and the wavelet spectra. (c) amplitude for JJY60–RKB path and the wavelet spectra, (d) phase for JJY60–RKB path and the wavelet spectra, and (e) amplitude for JJI–RKB path and the wavelet spectra.

frequencies of 68.5 kHz, 60.0 kHz, and 22.2 kHz were 92.7 km, 92.3 km, and 86.5 km, respectively. We calculated that the arrival times of the acoustic wave from the beginning and end points to the BPC–RKB path were approximately 2197 s and 1782 s, to the JJY60-RKB path were approximately 904 s and 505s and to the JJI-RKB path were approximately 309 s and 235 s, respectively.

Periodic variation of 200–500 s was observed for the BPC-RKB paths after the arrival of the acoustic waves from the beginning point (Fig. 3). For the JJY60–RKB path, there was the variation with a period of 300–500 s after the arrival of the acoustic waves from end point. For the JJI–RKB path, there were no clear the variations, because variation with a period of 200–300 s continued before the arrival time of the acoustic waves.

Figure 5 shows coherence between the VLF/LF amplitudes and phases for BPC-RKB path after arrival time of acoustic wave excited from fireball, which was particularly significant at a period of 546 s.

Figure 6 shows the waveforms of vertical seismic wave at KZK and the wavelet spectra. The orange, blue, and green lines indicate occurrence time of the fireball, the arrival time of the acoustic wave generated at the beginning point, and the arrival time of the acoustic wave generated at the end point, respectively. No fluctuations in the vertical seismic wave were observed at arrival time of acoustic wave.

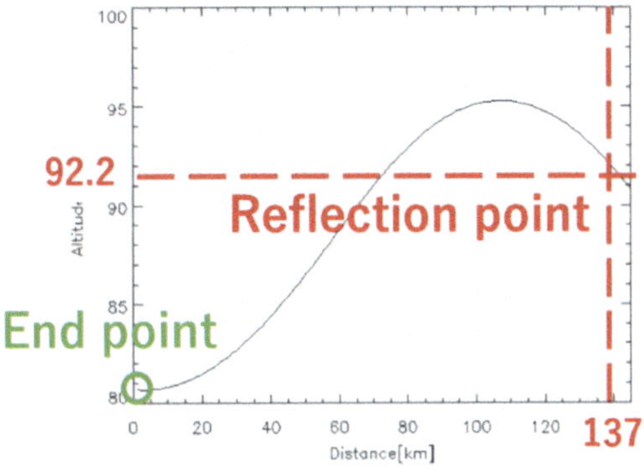

Fig. 4. Ray path of acoustic wave propagated from end point to JJY60-RKB path based on raytracing method.

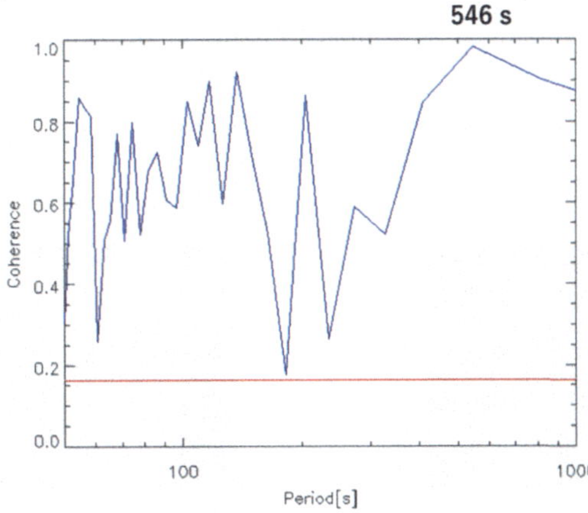

Fig. 5. Coherence between the LF amplitude and phase for the BPC-RKB path. The horizontal red line indicates the 95% significance level.

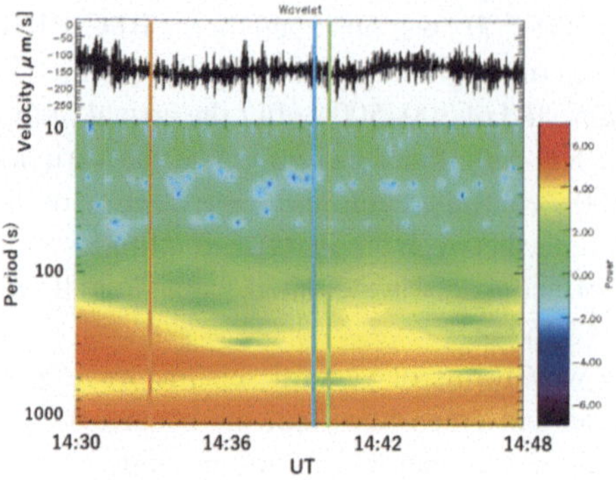

Fig. 6. Waveform of the F-net vertical seismic velocity at KZK and the wavelet spectra.

3.2 Satellite Re-entry

Figure 7 shows the waveforms of (a) the VLF/LF amplitude and (b) phase and the wavelet spectra on the JJY40-KAG path. The orange lines and green lines indicate the occurrence time of fire ball, and the arrival time of the acoustic wave generated from the end point, respectively. The estimated reflection height for the JJY40 signals was 92.4 km. We calculated that the arrival time of the acoustic wave from the end points to the JJY40-KAG path was approximately 754 s. Periodic variation of 200–400 s was observed for the JJY40-KAG path before the arrival of the acoustic waves from the end point. Because the beginning point is nearer JJY40-KAG path than end point, it is possible that observed variations were associated with the satellite re-entry.

Figure 8 shows coherence between the LF amplitude and phase for the JJY40-KAG path between observed time of satellite re-entry and arrival time of acoustic wave from end point, which is particularly significant at a period of 204 s.

Figure 9 shows the waveforms of the infrasound data and the wavelet spectra. The orange and green lines indicate the occurrence time of fire ball, and the arrival time of the acoustic wave generated from the end point, respectively. There are variations

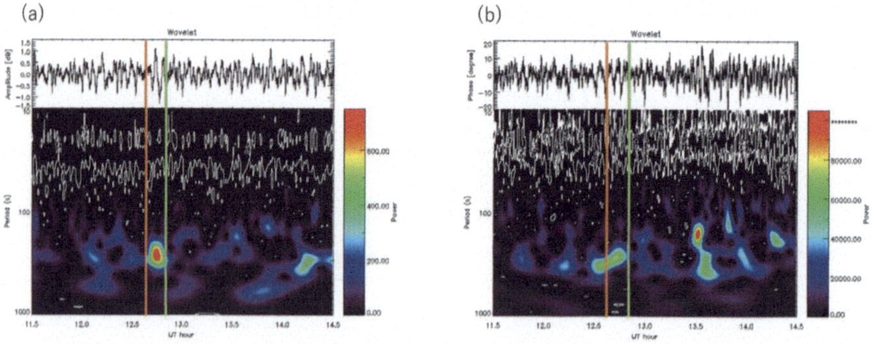

Fig. 7. (a) Amplitude for JJY40-KAG path and the wavelet spectra, (b) phase for BPC-RKB path and the wavelet spectra.

in the infrasound data at arrival time of acoustic wave, although it is unclear because variations with a period of 100–300 s continued before the time of the satellite re-entry. We investigated infrasound data at 8 other sites in Japan, although found no clear variations. This could be due to long horizontal distance between the end point of the fireball and each infrasound station (over 100 km).

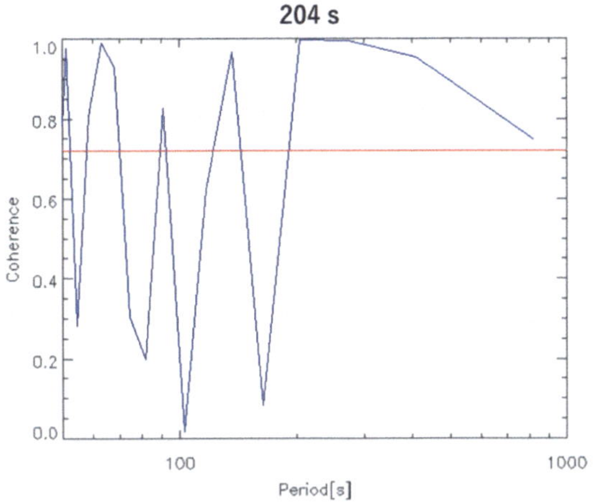

Fig. 8. Correspondence between the LF amplitude and phase at JJY40–KAG. The horizontal red line indicates the 95% significance level.

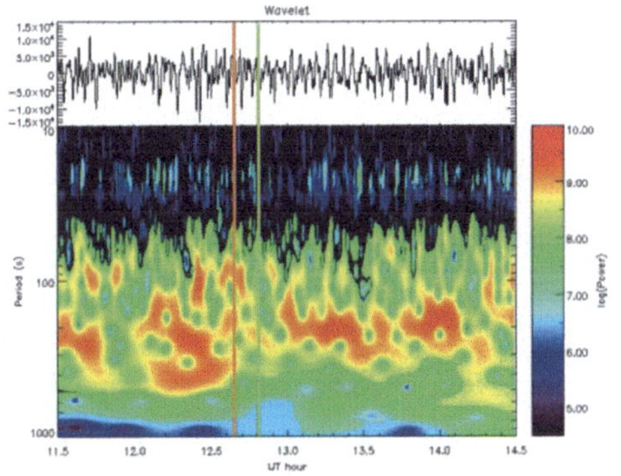

Fig. 9. Waveform of the infrasound data and wavelet spectra.

4 Conclusions

We investigated variation in the VLF/LF amplitudes and phases of the BPC–RKB, JJY60–RKB, and JJI–RKB paths after a fireball passed over Hokuriku, Japan, at 14:33 UT on 23 April 2023, and JJY40-KAG path after a satellite re-entry passed over Tokai,

Japan, at 12:38 UT on 26 December 2023.Periodic variations of 200–500 s and 300–500 s were observed in the amplitudes and phases of BPC–RKB and JJY60–RKB paths at the arrival time of acoustic waves, respectively. The variations might be caused by the acoustic waves or atmospheric gravity waves propagated from fireball/satellite re-entry trajectories upward to the D-region height.

References

1. Ceplecha, Z., et al.: Meteor phenomena and bodies. Space Sci. Rev. **84**(3/4), 327–471 (1998)
2. Yang, Y.-M., Komjathy, A., Langley, R.B., Vergados, P., Butala, M.D., Mannucci, A.J.: The 2013 Chelyabinsk Meteor ionospheric impact studied using GPS measurements. Radio Sci. **49**(5), 341–350 (2014)
3. Chernogor, L.F.: Ionospheric effects of the Chelyabinsk meteoroid. Geomag. Aeron. **55**(3), 353–368 (2015). https://doi.org/10.1134/S0016793215030044
4. Chernogor, L.F., Liashchuk, O.I., Shevelev, M.B.: Parameters of the infrasonic signal generated by the Kamchatka meteoroid. Kinematics Phys. Celestial Bodies **36**(5), 222–237 (2020). https://doi.org/10.3103/S0884591320050037
5. Ohya, H., Suzuki, T., Tsuchiya, F., Nakata, H., Shiokawa, K.: Variation in the reflection height of VLF/LF transmitter signals in the D-region ionosphere and the possible source: a 2018 meteoroid in Hokkaido, Japan. Radio Sci. **59**, e2023RS007801 (2018). https://doi.org/10.1029/2023RS007801
6. Ohya, H., Nishino, M., Murayama, Y., Igarashi, K., Saito, A.: Using tweek atmospherics to measure the response of the low-middle latitude D-region ionosphere to a magnetic storm. J. Atmos. Solar Terr. Phys. **68**(6), 697–709 (2006). https://doi.org/10.1016/j.jastp.2005.10.014

Can Milky Way's Rotation Curve Constrain Dark Matter Gravity-Matter Coupling?

Davendra S. Hassan[1(✉)], M. Dio Danarianto[2], and Anto Sulaksono[1]

[1] Department of Physics, FMIPA, Universitas Indonesia, Kampus UI, Depok 16424, West Java, Indonesia
davendra.shayna@ui.ac.id

[2] Research Center for Computing, National Research and Innovation Agency (BRIN), Jl. Sangkuriang, Bandung 40135, West Java, Indonesia

Abstract. We investigate the matter-gravity coupling of dark matter in the Milky Way by fitting the rotation curve of the Sérsic-based Galactic model with standard dark matter coupling and Yukawa-like effective coupling between DM and gravity. These models are fitted to different Milky Way rotation curve datasets, namely Sofue (2017 and 2020), Gaia DR3 derivations by Wang (2023) and Zhou (2023), and near-black hole data from the GRAVITY collaboration (2023). Our results show that there is a mixed statistical favour for the non-trivial DM-gravity coupling model based on the Bayes factor. We argue that more precise observational data covering a broader range of the galactic radii, such as outside of the Solar neighbourhood, are still required to gain a more exact insight into DM-gravity coupling in the Milky Way.

Keywords: Modified gravity · Dark matter · General relativity · Galaxy

1 Introduction

Dark matter (DM) has been proposed to explain various galactic or cosmological phenomena; however, its existence is still questioned. On the other hand, as a standard theory, general relativity itself must be modified to test for potential deviations and ensure its universal applicability. Numerous modified gravity (MG) theories proposed to replace GR. Theories such as $f(R)$ theory, scalar-tensor theories and massive gravity result in a simple deviation from the Newtonian inverse-square law into a Yukawa-correction potential

$$\Phi_{Tot}(r) = -G \int \frac{\rho(\mathbf{r}')}{|\mathbf{r} - \mathbf{r}'|} \left(1 + \beta e^{-\frac{1}{\lambda}|\mathbf{r}-\mathbf{r}'|}\right) d^3\mathbf{r}', \tag{1}$$

that is parameterized by β and λ, denote the strength and range of the interaction, respectively. Yukawa term could also arise from non-trivial DM-gravity

E. Fitrianingsih et al. (Eds.): ISAST 2024, SPPHY 416, pp. 213–222, 2025.
https://doi.org/10.1007/978-981-96-1344-1_23

coupling (NTDMC) [1], which assumes the gravitational interaction between DM and baryons vanishes at small scales, namely Ref. [2].

In this paper, we test the DM coupling with non-trivial models in the Milky Way by fitting to rotation curve (RC) datasets from Sofue (2017, 2020) [3,4], Gaia DR3-derived Wang (2023) [5] and Zhou (2023) [6], and GRAVITY collaboration data [7] using the nested sampling method. The result also compared to standard gravity to calculate data preference via Bayes Factor (BF).

2 Methodology

2.1 Rotational Velocity Calculation

The circular rotational velocity is derived from the gravitational potential as

$$V_{circ}^2(r) = r\frac{d\Phi(r)}{dr}. \tag{2}$$

We decompose the Milky Way rotation curve as a superposition of four components: a de Vaucouleurs bulge, an exponential disk, NFW (Navarro-Frenk-White) DM halo, and a point-mass supermassive black hole. For non-trivial DM coupling, the NFW profile is calculated along with the Yukawa potential (1) to obtain the modified circular velocity. In general, we derive the total rotational velocity of the Milky Way as

$$V_{Tot}^2(r) = V_{Disk}^2(r) + V_{Bul}^2(r) + V_{DM}^2(r) + V_{BH}^2(r) + V_{MG}^2(r). \tag{3}$$

Galactic Decomposition

Bulge. The bulge is represented by de Vaucouleurs profile (equivalent to Sérsic profile with $n = 4$), where the surface matter distribution (SMD) is modeled as

$$I(r) = I_e \exp\left(-b_n\left[\left(\frac{r}{a_b}\right)^{1/4} - 1\right]\right), \tag{4}$$

with $b_n = 7.6695$. I_e represents the surface matter distribution at the scale length radius $r = a_b$. The volume density of the bulge can then be reconstructed through an Abel transformation [8]. By substituting $M_b = \eta I_e a_b^2$, where $\eta = 22.665$ from Ref. [9,10], we obtain the Newtonian bulge velocity as

$$V_{Bul,N}^2(r) = \frac{b_n e^{b_n} M_b G}{\eta a_b^{9/4} r}\int_0^r \int_{r'}^{\infty} \frac{e^{-b_n(x/a_b)^{1/4}}}{x^{3/4}\sqrt{x^2 - r'^2}} r'^2 dx dr'. \tag{5}$$

Disk. Observations show that the Galactic disk can modeled as an exponential profile, or equivalently the Sérsic profile for $n = 1$, where the SMD is

$$\Sigma(r) = \Sigma_0 e^{-r/a_d}. \tag{6}$$

Since the necessity of a thin-thick dichotomy for the Galactic disk remains debated, we modeled the Galactic disk as a single thin disk for simplicity,

$$\rho_{Disk}(r, z) = \Sigma(r)\delta(z) = \Sigma_0 e^{-r/a_d}\delta(z), \tag{7}$$

where $\delta(z)$ represents the delta Dirac function, r is the radial distance from the GC, and z is the height measured from the Galactic plane. Following the derivation in Ref. [8], the Newtonian disk velocity is

$$V_{Disk}^2 = \frac{GM_d r^2}{2a_d^3}\left[I_0\left(\frac{r}{2a_d}\right)K_0\left(\frac{r}{2a_d}\right) - I_1\left(\frac{r}{2a_d}\right)K_1\left(\frac{r}{2a_d}\right)\right], \tag{8}$$

DM Halo. The distribution of DM is commonly modeled as a spherical halo surrounding galaxies to facilitate structure formation in the early universe. There have been various models for the halo density profile itself, and we follow the commonly used Navarro-Frenk-White (NFW) profile,

$$\rho_{DM}(r) = \frac{\rho_s}{\left(\frac{r}{r_s}\right)\left(1 + \frac{r}{r_s}\right)^2}. \tag{9}$$

where the parameter ρ_s denotes the scale density, while r_s denotes the scale radius. The Newtonian velocity can be easily derived as

$$V_{DM}^2 = \frac{4\pi G\rho_s r_s^3}{r}\left[\ln\left(1 + \frac{r}{r_s}\right) - \frac{r/r_s}{1 + \frac{r}{r_s}}\right], \tag{10}$$

while the Yukawa-correction velocity follows the same form as Ref. [1] as

$$\begin{aligned} V_{MG}^2 = -\frac{2\pi G\beta\rho_s r_s^3}{r}\Bigg[&\frac{2r}{r + r_s} + e^{-\frac{r+r_s}{\lambda}}\left(1 + \frac{r}{\lambda}\right)\left\{\mathrm{Ei}\left(\frac{r_s}{\lambda}\right) + e^{\frac{2r_s}{\lambda}}\mathrm{Ei}\left(-\frac{r_s}{\lambda}\right)\right. \\ &\left.-\mathrm{Ei}\left(\frac{r + r_s}{\lambda}\right)\right\} + e^{\frac{r+r_s}{\lambda}}\left(\frac{r}{\lambda} - 1\right)\mathrm{Ei}\left(-\frac{r + r_s}{\lambda}\right)\Bigg]. \end{aligned} \tag{11}$$

Central SMBH. The Galactic center harbors a SMBH known as Sagittarius A* (Sgr A*), whose density profile can be modeled to be a point mass at the nucleus of the Milky Way, as

$$\rho_{BH}(r) = \frac{M_{BH}\delta(r)}{4\pi r^2}. \tag{12}$$

We can immediately derive the velocity equation as

$$V_{BH}^2 = \frac{GM_{BH}}{r}. \tag{13}$$

2.2 Data Fitting

Our study primarily uses the grand rotation curve (GRC) data of the Milky Way from Ref. [3] and [4]. The Sofue (2017) dataset covers a range of 0.001 kpc $< r <$ 1000 kpc, while the Sofue (2020) dataset spans a smaller range but has more precise error margins ($\sigma_V \lessapprox 20$ km/s). We also incorporate the Wang (2023) and Zhou (2023) rotation curve data for the Solar neighborhood (5 kpc $< r <$ 30 kpc), which, albeit covering a more limited range, provides even smaller uncertainty ($\sigma_V \lessapprox 5$ km/s). We include the posterior mean fitting of the Sgr A* parameters from the GRAVITY collaboration (Table 2 in Ref. [7]). We create three distinct datasets for analysis by combining the available six datasets as follows:

1. Sofue (2017) and GRAVITY;
2. Sofue (2020) and GRAVITY; and
3. Sofue (2017), Wang (2023), Zhou (2023), and GRAVITY.
4. Sofue (2017), with averaged Wang (2023) & Zhou (2023) data, and GRAVITY.

We constrain the model's parameters using the static nested sampling method to estimate bayesian evidence $\mathcal{Z}$ and posterior distributions for individual parameters. The calculation was implemented using `dynesty` Python library. Our approach assumes a Gaussian form for the likelihood probability density

$$\mathcal{L}(\Theta) = \prod_i \frac{1}{\sqrt{2\pi}\sigma_i} \exp\left[-\frac{\left(v_{obs}(r_i) - v_{pred}(r_i)\right)^2}{2\sigma_i^2}\right]. \tag{14}$$

Here, $v_{obs}(r_i)$ represents the observed rotational velocity at radius r_i, while $v_{pred}(r_i)$ represents the rotational velocity predicted by a model $\mathcal{M}_i$ with the parameters Θ at r_i. Errors from the observed data is represented by σ_i. We use uniform prior distributions, presented in Table 1, and logarithmically transform the λ, ρ_s, and M_{BH} parameters to help localize the best-fit values.

Table 1. The prior choice for each parameter, along with their respective units.

Parameter	Range	Units
β	−150–150	–
$\ln(\lambda)$	−4–10	ln [kpc]
M_b	$10^8 - 10^{11}$	$M_\odot$
a_b	0.01–6	kpc
M_d	$10^7 - 3 \times 10^{13}$	$M_\odot$
a_d	1.5–20	kpc
$\ln(\rho_s)$	−5–20	ln [$M_\odot$/kpc^3]
r_s	0.1–1e4	kpc
$\ln(M_{BH})$	0.01–20	ln [$M_\odot$]

2.3 Model Comparison

In order to compare a model $\mathcal{M}_1$ to another model $\mathcal{M}_2$ that fits the observational data of the Milky Way, the previously computed evidences during the data fitting are used to calculate the Bayes Factor (BF), which is the ratio of each model's posterior probabilities,

$$BF_{12} = \frac{P(\mathcal{M}_1|\mathbf{D})}{P(\mathcal{M}_2|\mathbf{D})} = \frac{P(\mathbf{D}|\mathcal{M}_1)P(\mathcal{M}_1)/P(\mathbf{D})}{P(\mathbf{D})|\mathcal{M}_2)P(\mathcal{M}_2)/P(\mathbf{D})} = \frac{P(\mathbf{D}|\mathcal{M}_1)}{P(\mathbf{D}|\mathcal{M}_2)} = \frac{\mathcal{Z}_1}{\mathcal{Z}_2}. \quad (15)$$

The first model $\mathcal{M}_1$ is said to be preferred or supported by the data when $\log_{10} BF_{12} > 1$ [11].

3 Results and Discussion

3.1 Sofue (2017) and GRAVITY

Newtonian Model. We initially compare our Newtonian case results with Ref. [3], which featured a different fitting method that truncates the data for each component and fit the parameters using least-χ^2 separately [12]. Meanwhile, in our approach, the entire datasets are taken into account simultaneously to fit all 7 parameters. These difference in fitting methods led to the difference of best-fit results, where our posterior distribution results is displayed in Fig. 1. For $r \gtrsim 100$ kpc, the lower quantile of several data reaches $v_{obs} \leq 0$. This can be seen at the posterior predictive check, specifically the rotational velocity distribution curve in Fig. 1. Consequently, the posterior distribution of the DM parameters spread feature small values of ρ_s and arbitrarily large r_s to compensate zero to negative velocity.

NTDMC Model. The NTDMC model results can be distinguished by a double-peak feature of the DM halo velocity distribution caused by the extreme strength Yukawa-correction at the $r \sim 10$ kpc region. Our results suggest the need for more accurate data in the fitting process, using independent prior information on both the Galaxy and theoretical parameters, or – as suggested by Ref. [13] – to improve data accuracy at smaller radii to better constrain the NTDMC model. In this model we obtain $\log_{10} BF_{N/NTDMC} = 0.724$, which shows that the Newtonian model is slightly more preferred in this dataset.

3.2 Sofue (2020) and GRAVITY

Newtonian Model. One of the strengths of the reduced uncertainty of the Sofue (2020) dataset is that the posterior predictive check, or the decomposed rotation curve in Fig. 2, reveals much narrower CIs due to the smaller error margins. Nevertheless, the DM halo parameters still have uncertain distribution. This suggests the necessity of precise data at $r \gtrsim 100 kpc$ to fully constrain the DM halo.

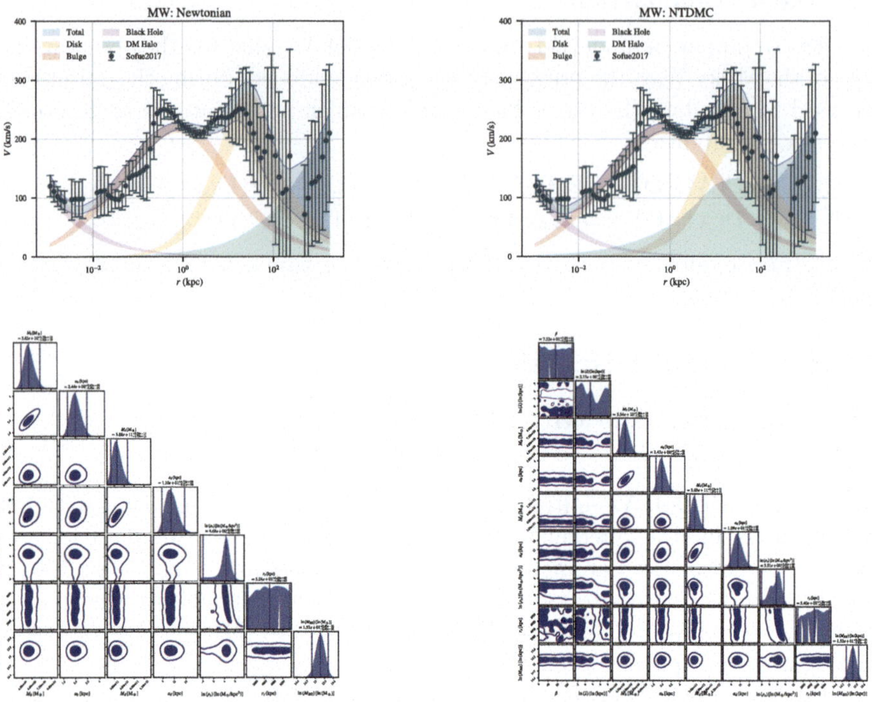

Fig. 1. Rotation curves (*top*) and marginal posterior distributions (*bottom*) for both models using the Sofue (2017) and GRAVITY dataset. The GRAVITY data has been omitted from the graph to conserve space on the axis of the top graphs, where the shaded regions show the 95% CIs. The navy areas and its white areas with navy outlines in the bottom graphs correspond to the 1σ and 2σ confidence regions, respectively. *Left*: Newtonian model. *Right*: NTDMC model.

NTDMC Model. The marginal posterior distributions in Fig. 2 now exhibit more constrained distributions for the DM halo and λ parameters; however, the bulge parameters' velocity distribution is replace by the double-peaked DM halo velocity distribution, which can be seen in the posterior predictive check. The Yukawa-correction now has an extreme strength at $\lambda \approx 0.1$ kpc, which although aids in constraining the DM halo parameters, supports our suggestion of more prior information needed to circumvent such cases. Nevertheless, the analysis of this dataset suggests that enhanced data precision has yielded improved parameter constraints. The NTDMC model is now strongly more supported statistically, shown by $\log_{10} BF_{N/NTDMC} = -345.043$.

3.3 Sofue (2017), Wang (2023), Zhou (2023) and GRAVITY

Newtonian Model. To test the necessity for further precise data which covers larger ranges, we combine the extensive Sofue (2017) dataset with the precise

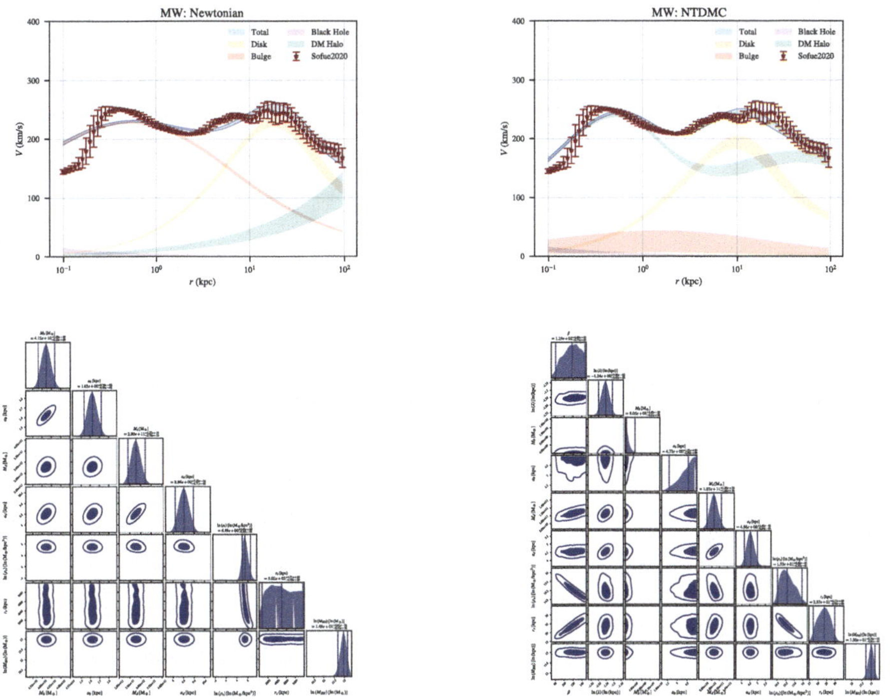

Fig. 2. Same as Fig. 1, but for the Sofue (2020) and GRAVITY dataset.

Gaia DR3-derived rotation curves. The posterior velocity distribution of the DM halo, seen in Fig. 3, is narrower, where the lower quantile doesn't tend to $v_{obs} \sim 0$ in the posterior predictive check. This is reflected in the more tightly constrained DM halo parameters compared to previous datasets. However, it's important to note that the Gaia DR3-derived data has more data points with high precision than those from Sofue (2017) hence it indirectly introduce a relatively more weight on the likelihood calculation, which leads to significant bias of the total rotation curve to the GAIA DR3-derived dataset, visible even in the Newtonian case.

NTDMC Model. The dataset's constrain on the DM halo parameters is now more apparent when examining the β parameter, which now features tighter constraints compared to the previous datasets. Nevertheless, there still remains the small probability of $\beta < -1$ values, indicating that the Yukawa-correction term can dominate the Newtonian term. We obtain a value of $\log_{10} BF_{N/NTDMC} = -10.882$, further showing that the NTDMC model is strongly more preferable in this dataset.

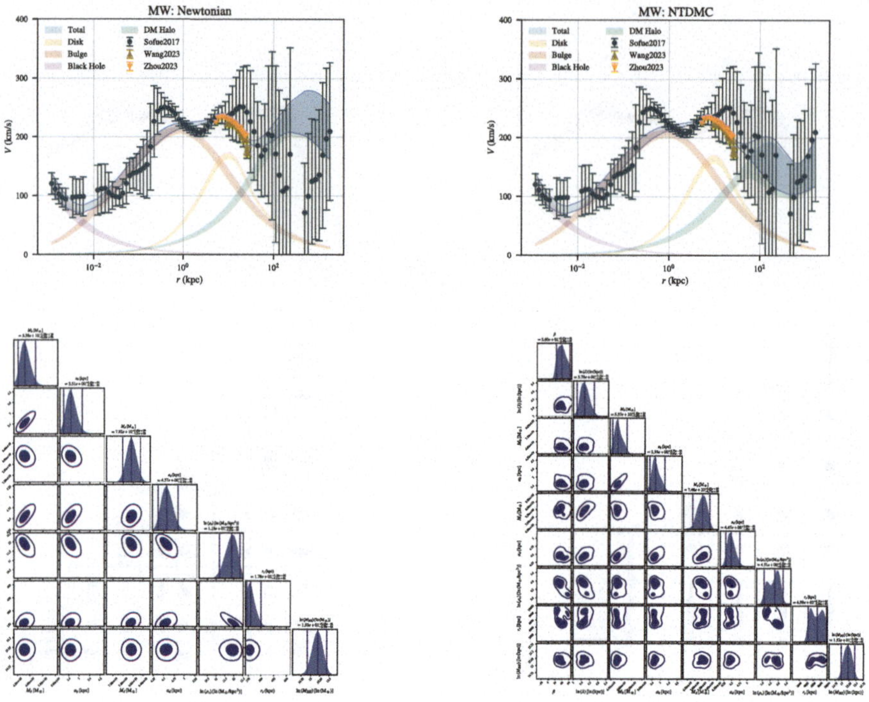

Fig. 3. Same as Fig. 1, but for the Sofue (2017), Wang (2023), Zhou (2023) and GRAVITY dataset.

3.4 Averaged Sofue (2017), Wang (2023), Zhou (2023) and GRAVITY

Newtonian Model. To forego the effects of the weight of the GAIA DR3-derived datasets, we calculated a weighted mean average and weighted standard deviation of the combined dataset within the 5–30 kpc region. This calculation corresponds to a 5-kpc window of each Sofue (2017) data point, from which every data point inside the window are taken into account into the weighted mean average and standard deviation. We then combine this averaged dataset in the region with the rest of the Sofue (2017) dataset. Although the data in the $1 < r < 100$ kpc region is indeed more precise than the inner and outer region data, it now has a more balanced statistical weight compared to the previous dataset. In Fig. 4, we observe that the DM halo posterior distribution still has the tendency to approach $V_{DM} \sim 0$, as evident by the small values of ρ_s and large values of r_s similar to the first dataset.

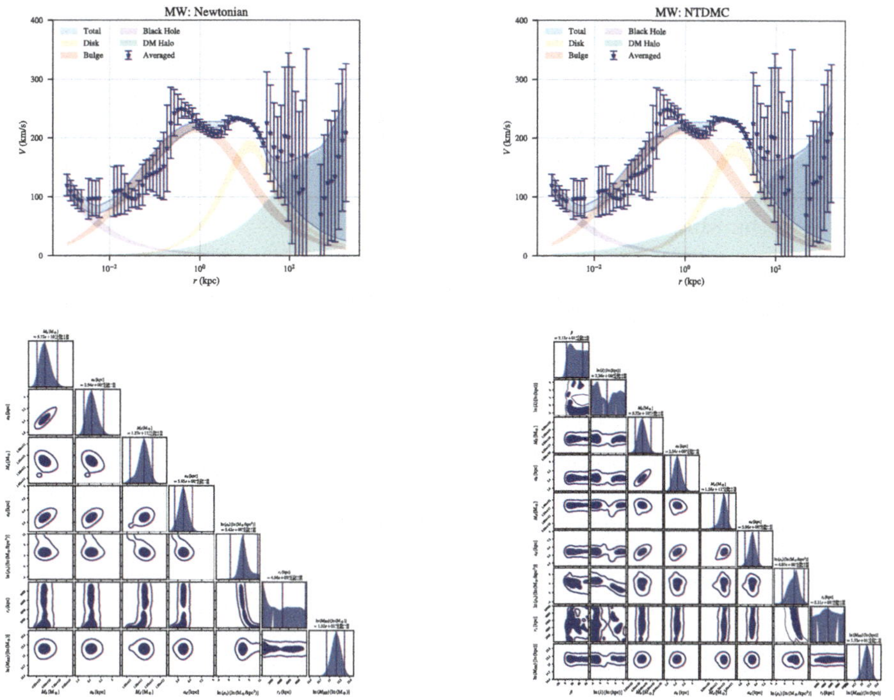

Fig. 4. Same as Fig. 1, but for the averaged Sofue (2017), Wang (2023), Zhou (2023) and GRAVITY dataset.

NTDMC Model. Similar to the previous, the lack of constraintment in the DM halo parameters lead to the same phenomenon in the Yukawa parameters. We obtain a value of $\log_{10} BF_{N/NTDMC} = 1.402$, which shows that the Newtonian model is more preferred in this dataset. Like the first dataset, this dataset features a more balanced absolute value of the BF, likely because of the less statistical weight the data has over the likelihood.

4 Conclusion

In this paper, we have shown the effects of Yukawa-correction on the NFW DM halo on the rotation curve of the Milky Way. The Yukawa-correction may enable a double-peaked DM halo rotation curve due to the additional 'fifth-force' introduced by $\beta > 0$. If λ is sufficiently small (i.e., $\sim 10^{-1}$ kpc), the effect may replace the bulge contribution. If we would like to retain bulge/disk value to be similar to the Newtonian case, we can either set a higher limit to the lower prior of λ (e.g., $\gtrsim$10 kpc) or choose tighter prior of bulge/disk parameters (e.g., from other independent measurements).

Our result shows that the quality of the data strongly affects the posterior distribution and data preference for the models considered. So far, our results do not show which gravity model is preferred with high significance. To better constrain all parameters, especially DM components (both on Newtonian or non-trivial coupling), we have shown that it is necessary to have precise data with error margins (i.e., $\sigma_V \lessapprox 5$ km/s) and covers both inner $r < 5$ kpc (especially up to sub-au scales) and outer $r > 30$ kpc regions of the Milky Way. Independent measurement of galactic morphology (i.e., Sérsic parameters measured from other than the rotation curve) could also enable better constraint of DM-gravity coupling in the Milky Way.

Acknowledgements. The computation in this work has been done using the facilities of MAHAMERU BRIN HPC, National Research and Innovation Agency of Indonesia (BRIN).

References

1. de Almeida, Á.O.F., Amendola, L., Niro, V.: Galaxy rotation curves in modified gravity models. J. Cosmol. Astropart. Phys. **2018**(8), 012–012 (2018)
2. Piazza, F., Marinoni, C.: Model for Gravitational Interaction between dark matter and baryons. Phys. Rev. Lett. **91**(14), 141301 (2003)
3. Sofue, Y.: Rotation and mass in the milky way and spiral galaxies. Publ. Astron. Soc. Jpn. **69**(1), R1 (2017)
4. Sofue, Y.: Rotation curve of the milky way and the dark matter density. Galaxies **8**(2), 37 (2020)
5. Wang, H.-F., Chrobáková, Ž, López-Corredoira, M., Sylos Labini, F.: Mapping the milky way disk with Gaia DR3: 3d extended kinematic maps and rotation curve to 30 kpc. Astrophys. J. **942**(1), 12 (2023)
6. Zhou, Y., Li, X., Huang, Y., Zhang, H: The circular velocity curve of the milky way from 5–25 kpc using luminous red giant branch stars. Am. Astron. Soc. Astrophys. J. **946**(73) (2023)
7. GRAVITY Collaboration: Polarimetry and astrometry of NIR flares as event horizon scale, dynamical probes for the mass of SGR A*. Astron. Astrophys. **677**, L10 (2023)
8. Mannheim, P.: Alternatives to dark matter and dark energy. Prog. Part. Nucl. Phys. **56**(2), 340–445 (2006)
9. Sofue, Y.: Dark halos of M31 and the milky way. Publ. Astron. Soc. Jpn. **67**(4), 75 (2015)
10. Vaccari, M.: Simulated observation of galaxies with ESA GAIA satellite. Master's thesis, University of Padova, Padova, Italy (2000)
11. Khelashvili, M., Rudakovskyi, A., Hossenfelder, S.: Dark matter profiles of SPARC galaxies: a challenge to fuzzy dark matter. Mon. Not. R. Astron. Soc. **523**(3), 3393–3405 (2023)
12. Sofue, Y.: Grand rotation curve and dark-matter halo in the milky way galaxy. Publ. Astron. Soc. Jpn. **64**(4), 75 (2012)
13. Henrichs, J., Lembo, M., Iocco, F., Amendola, L.: Testing gravity with the milky way: Yukawa potential. Phys. Rev. D **104**(4), 043009 (2021)

LSTM-Based Multivariate Time-Series Analysis for Ionospheric TEC Prediction of Single Station GNSS

Asnawi Husin(✉), Varuliantor Dear, Rizal Suryana, Faruk Afero, Adi Purwono, and Jiyo

Research Center for Space, National Research and Innovation Agency (BRIN), Bandung, Indonesia
asnawi@brin.go.id

Abstract. In this paper, we develop a predictive model for ionospheric Total Electron Content (TEC) using a straightforward multivariate Long Short-Term Memory (LSTM) machine learning algorithm. Time series data from a single GNSS station over eight years was used for training (90%). The model was trained on data from 2012–2019, covering both maximum solar activity from 2012–2016 and minimum solar activity from 2017 - 2019. The training process included solar and geomagnetic indices F10.7, Kp, and Dstas model inputs. The Ionosphere Weather Index (WTEC Index), an indicator of ionospheric changes due to geomagnetic disturbances, was also incorporated. The LSTM model was developed to capture seasonal variations on a monthly basis. The RMSE was approximately 4 TECU, calculated using (10%) of the test data year 2020, with the maximum observed TEC values ranging from 50 to 60 TECU. However, testing and validating the model using the 2022 observation data showed a maximum RMSE below 11 TECU (with maximum TEC observation data ranging from 100 to 110 TECU). The model's RMSE is reasonably accurate and can be adjusted with a 10% correction factor. The model's trend variations generally match. The data is observational, with a correlation coefficient of up to 0.8. A comparison of the model predictions for the IRI model revealed that it underestimates both the LSTM model and observation data. The evaluation demonstrates that the LSTM model effectively responds to geomagnetic storms, whether moderate or intense.

Keywords: Ionosphere · LSTM · TEC

1 Introduction

When traversing the ionosphere layers, the GNSS signal will experience a delay, quantified by total electron content (TEC). GNSS signals are used to obtain TEC measurements of the ionosphere. To mitigate the effects of ionospheric dispersion on GNSS signals, ionosphere models have been developed. Prediction models for TEC parameters have been extensively built and published for both global and regional applications. The ionospheric model commonly used as a reference standard is the International Reference Ionosphere (IRI) model. The IRI-2016 model and its latest version, IRI-2020, are

E. Fitrianingsih et al. (Eds.): ISAST 2024, SPPHY 416, pp. 223–232, 2025.
https://doi.org/10.1007/978-981-96-1344-1_24

empirical models that can predict daily, monthly, latitudinal, and longitudinal variations of various temporally and spatially ionospheric parameters [1, 2]. Based on observational data, several studies have been released assessing the performance, validation, and comparability of the IRI-2016 model. For example, [3] reported that the IRI-2016 TEC over the Equatorial Ionization Anomaly (EIA) crest regions is reasonably accurate as a climatologically model, with an accuracy of up to 90%. However, it shows relative weaknesses during periods of maximum solar activity in 2013 compared to the solar minimum 2008. [4] reported on the model's response during quiet and stormy days in Indian regions, finding that while the model simulation captured observed TEC on quiet days, it failed to simulate observed TEC with reasonable accuracy on stormy days.

Recently, machine learning techniques have been widely used for ionosphere modeling, such as Neural Networks (NN), Support Vector Machine (SVM), Deep Learning, Random Forest, Gaussian Process (GP) regression, Recurrent Neural Networks (RNN), particularly Long Short-Term Memory (LSTM) networks, Convolution Neural Networks (CNN) for ionosphere modeling. For instance, [5] used NN to construct a TEC-NN model in the crest region anomaly of Bhopal, India, based on one year of GNSS observations. Derived from the IRI model, the model included input parameters such as the sunspot number and NmF2. Other machine learning methods for TEC prediction are deep learning [6] and regression [7]. The LSTM is extensively utilized in TEC, such as [8] the hybrid LSTM and the CEEMDAN algorithm, which are used to reduce nonlinearity in the TEC signal by decomposing it into several intrinsic mode functions (IMFs), Multiple-attention Long Short-Term Memory, MA-LSTM [9], and piecewise LSTM [10]. The majority of input models for LSTM are based on the utilization of solar and geomagnetic indices. Because the variations in the TEC ionosphere are primarily driven by the influx of energetic particles from the sun and fluctuations in the magnetic field within the magnetosphere. These changes occur daily and seasonal, as well as during an 11-year cycle. Based on historical data on solar flux, geomagnetic index, and ionospheric weather (WTEC) index, this paper proposes an ionospheric TEC prediction using multivariate LSTM based on single GNSS data. We only use the LSTM tools/libraries with the simplest architecture options. We apply this model to predict ionospheric TEC and compare the predicted results with other predictive models, such as the IRI model.

2 Data Sources and Preparation

GNSS data from the Indonesia Continuous Operating References Station (INACORS) (http://inacors.big.go.id/) network was used in this study. Data from a single station, namely BAKO (−6.4911 106.848), was collected over nine years, from 2012 to 2020. The period includes the years of maximum solar activity (2012–2016) and minimum solar activity (2017–2020). Due to disruptions in observations, data collection faced multiple days of missing data. To solve this problem, data from the nearest stations, namely CTGR (−5.709 106.663) and CGON (−5.979 106.052), were used as a supplement. However, the data supplied by these two stations contributed just 10% of the overall contribution. Figure 1 shows the relative contribution of data used to develop the model, focusing on data from the central BAKO station. Figure 2 displays the dispersion of the CTGR and CGON relative to the BAKO central station.

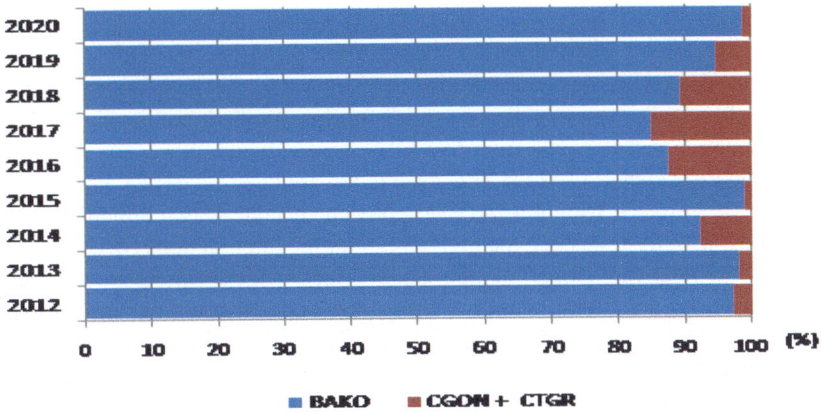

Fig. 1. Data availability for model construction. The primary station data (BAKO) exceeds 90%, while the additional data from CGON and CTGR contributes less than 10%.

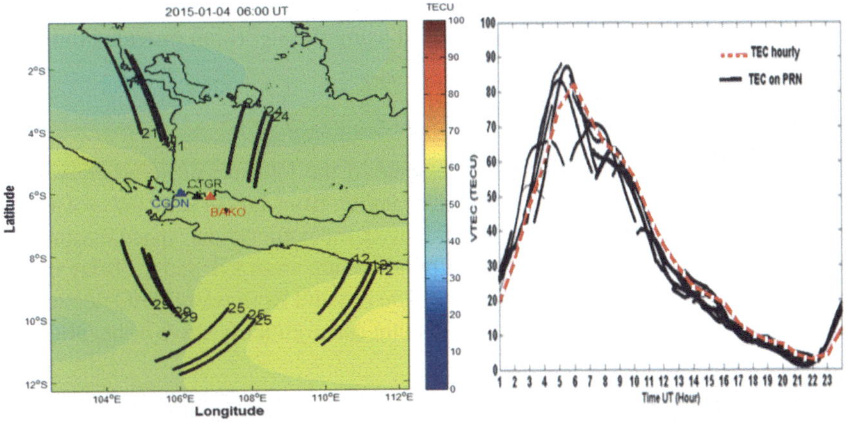

Fig. 2. The left panel: Map of the location of the central GNSS station BAKO and the neighboring stations CTGR and CGON. 2D TEC on the situation during 1-h observation (06:00UT) based on IPP of the five visible PRN: 12, 21, 24, 25, and 29 received by BAKO, CTGR and CGON station. Right panel: The daily TEC profile for 24 h displays TEC values corresponding to each PRN (black line), and the average TEC (red line) is computed to provide the daily TEC

TEC measurement is derived from RINEX (Receiver Independent Exchange) data of the GNSS INACORS with BAKO station as the main station for time series data analysis. TEC calculation is based on measurements of two different GNSS frequencies (L1 = 1.57542 GHz and L2 = 1.2276 GHz).The mathematical functions of TEC measurement are not discussed in detail. Several references can be referred to, for example, [11]. A software package that implemented these functions has been used for TEC measurements [12]. Boston College's Institute for Scientific Research created this program, which goes by the name GPSTEC. It applies a phase leveling process to obtain accurate TEC values. The data quality applied only considers data from stations with a duration of 24 h and elevations above 30 degrees. Figure 2 (right panel) presents a TEC calculation example with TEC's 24-h daily variation profile. The black line represents the TEC values based on PRN, while the red line represents the average TEC value used in this study. As shown in Fig. 2 (right panel), map TEC based on IPP for the specified satellite signal (PRN). The BAKO station receives data from five PRN signals, which it uses to calculate the TEC 2D. The PRN signals used for TEC calculation are PRN 12, 21, 24, 25, and 29.

The Inverse Distance Weighting (IDW) approach generates a two-dimensional TEC map. Figure 2 also illustrates that the PRN tracks for the three stations (BAKO, CTGR, and CGON) are relatively close, with the satellite track range spanning around 500 km from the center of the BAKO station. Hence, the TEC derived from measurements at BAKO station can effectively encompass an area with a radius of 500 km around BAKO. The TEC value, derived from the BAKO station, sufficiently represents these three stations. Figure 3 displays a dataset of TEC spanning nine years, namely from 2012 to 2020, with hourly resolution. According to the data, the daily TEC variations align with the F10.7 flux pattern. The maximum activity is observed between 2012 and 2016, while the minimum activity is observed between 2016 and 2020. Critical factors of space weather, as well as solar and geomagnetic activity, influence ionospheric electron density. The TEC deviation is then determined using the algorithm provided in [13].

$$\mathrm{DTEC} = \mathrm{Log}\left(\frac{TEC_{daily}}{TEC_{baseline}}\right) \tag{1}$$

DTEC in Eq. (1) is also known as the scale of the ionospheric weather W index. The DTEC as W index reveals TEC behavior ranging from a quiet state ($W = \pm 1$) to an intense storm ($W = \pm 4$), providing a helpful proxy index for driving space weather. The monthly data for DTEC, F10.7, Dst, and Kp indices are arranged, and a 27-day moving average is also used to get the baseline TEC. The Dst and Kpimpact DTEC fluctuations, especially during severe geomagnetic storms that cause either positive or negative TEC changes.

3 Model Architecture

LSTM is a type of RNN designed to handle time series problems by capturing long-term dependencies through memory cells that store information from previous time steps. However, using too much historical data can lead to overfitting, where the model remembers training patterns but fails to generalize to new data. Segmenting data by seasons can reduce overfitting, especially when the time series shows dynamic seasonal trends. Our approach segments data by month (January to December) to account for differing seasonal characteristics, such as equinoxes in March and September. The model learns monthly patterns to predict TEC deviations (DTEC) as described in Eq. 1. The machine learning algorithm used for the TEC prediction model is based on Eq. 2,

$$\mathrm{TEC}_{\mathrm{month}} = f(\mathrm{W},\ \mathrm{hod},\ \mathrm{F10.7},\ \mathrm{Kp},\ \mathrm{Dst}) \tag{2}$$

with month is group month January to December, and hod is the hour of the day; they present the seasonal and diurnal variation of TEC based on month groups; F10.7 represents the solar activity component. Kp and Dst indexes represent the geomagnetic

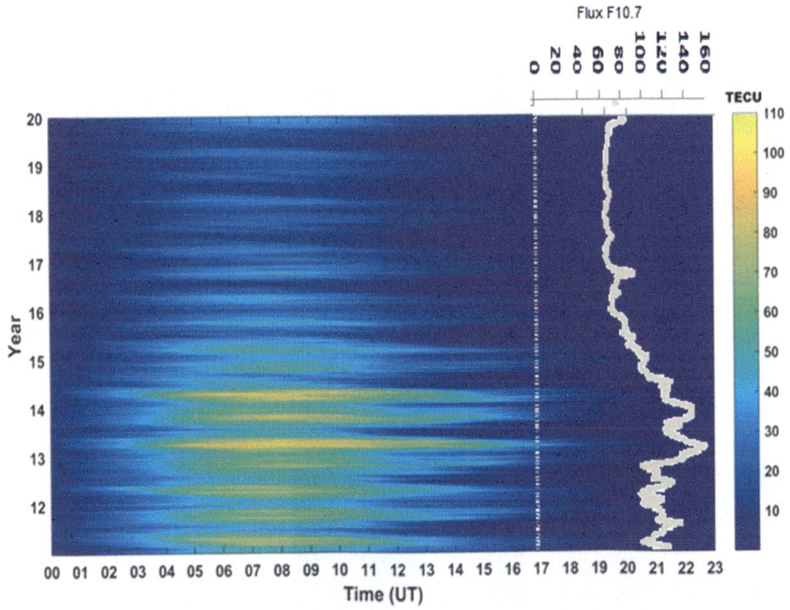

Fig. 3. Contour of the hourly resolution observational TEC data from 2012 to 2020 used in this study. Put in the image of the Flux F10.7 data representing solar activity.

activity component, and W (or DTEC) is the ionospheric weather W index (based on a 27-day moving average). Data division in the model learning process involves using 90% of the data for training and 10% for testing. As shown in the block diagram in Fig. 4, the LSTM architecture used in this study generally consists of several layers, as follows: Sequence Input layer: this layer accepts input with dimensions according to the feature dimension, which consists of historical data for month groups, January to December (2012–2020). LSTM Layer: The LSTM layer has 500 hidden units and outputs in sequence form. Fully connected layer with 500 units. Dropout layer, with a dropout rate of 0.1. In order to address the issue of overfitting, we incorporate two fully connected layers that receive input from both the LSTM layer and the dropout layer. The regression layer is responsible for the regression prediction task.

4 Results: Validation and Evaluation Model

Figure 5 shows the hourly variation of TEC contours, sampled from 10% of the testing data from the 2020 dataset. The analysis involves predicting results and calculating the Root Mean Square Error (RMSE) and correlation coefficients based on observation data. The trends in Figs. 5(a) and 5(b) show that the LSTM TEC model closely matches the observed data, with slight overestimation. The RMSE for each model from January to December is depicted in Fig. 5(c), with values primarily falling below 5 TECU. The March and November models, on the other hand, reach up to 7 TECU. With the TEC scale ranging from 50 to 60 TECU, an RMSE of 5–6 TECU suggests an appropriate level of reliability.

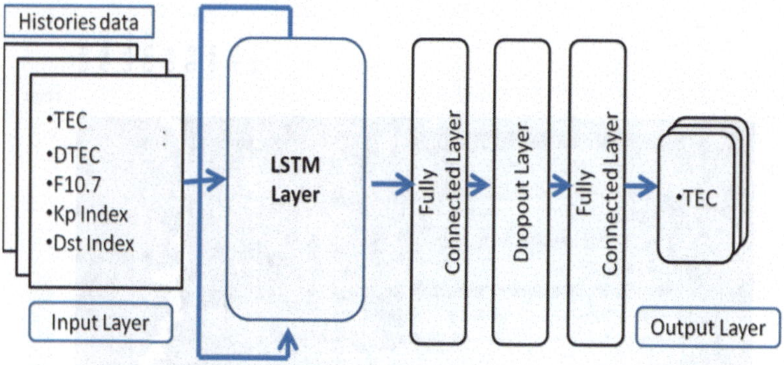

Fig. 4. LSTM model architecture for TEC prediction.

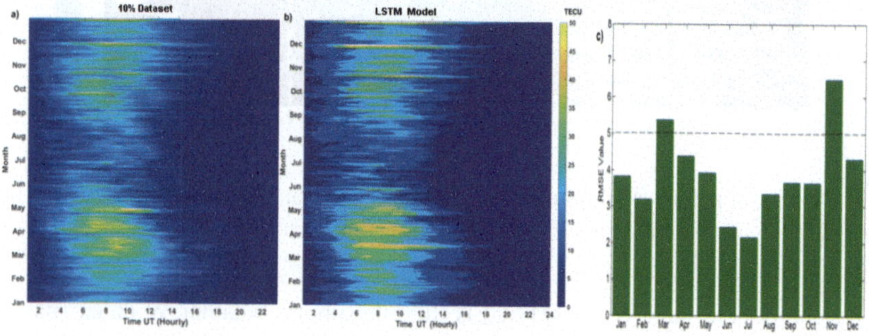

Fig. 5. The comparison between observation data (a) and model (b) for January to December of hourly variation in TEC (obtained from 10% datasets of the testing process). The models follow a similar pattern but slightly overestimate. (b) The models' root mean square error (RMSE) values are all below 5 TECU for each month, except for the March and November models.

Figure 6 presents an evaluation model based on observational data collected from January to December 2022. The figure displays the results of the LSTM and IRI2020 models in three panels, corresponding to the December solstice, equinox, and June solstice periods. The figure demonstrates that maximum TEC values typically occur during the equinoxes, as seen in observation data and predicted by both the LSTM and IRI2020 models.

The LSTM model closely matches the observations, with occasional overestimates or underestimates. In contrast, the IRI2020 model consistently underestimates and exhibits distinct trend patterns, including earlier predicted daily peaks (around 05:00–06:00 UT) compared to the observed peaks (around 09:00 UT). The IRI2020 model shows TEC increases starting at 00:00–01:00 UT, while both observations and the LSTM model start at 01:00–02:00 UT. Similar issues have been reported with previous IRI versions; for instance, [14] found that IRI-2016 underestimated GPS TEC in equatorial regions. The LSTM model's RMSE, ranging from 6 to 11 TECU (Table 2), is reliable given the 100–110 TECU TEC scale.

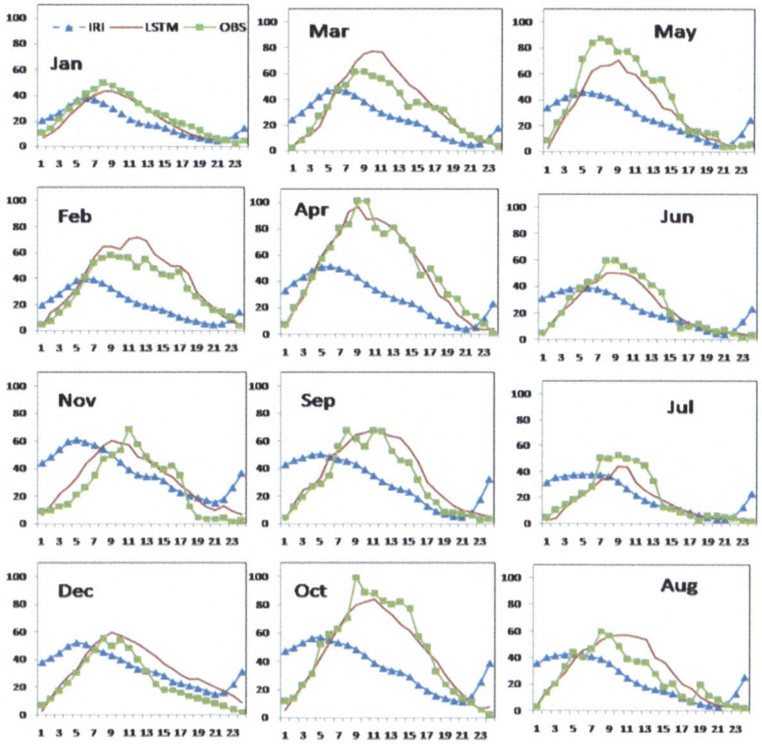

Fig. 6. Monthly validation of the LSTM model and IRI2020 model using 2022 observation data, demonstrating daily fluctuations

We assessed the model's effectiveness concerning geomagnetic storms during two specific events: November 4, 2021, and March 24, 2022. The models in Fig. 7 were tested during geomagnetic storms. Figure 7(a) shows a moderate storm on November 4, 2021, when DSt reached −100 nT, and Fig. 7(b) shows a strong geomagnetic storm on March 24, 2023, when DSt reached −150 nT. The red line represents observation data, the blue line represents the LSTM model, and the dashed line represents the IRI2020 model. During the main phase, the observed TEC and LSTM models demonstrate a geomagnetic storm response with a decrease (typical negative storm). However, the IRI model fails to show a response with unchanged variations pattern. The LSTM model performs better when reacting to geomagnetic storms than the IRI model, which is ineffective at simulating TEC during these disturbances [15, 16] (Table 1).

5 Conclusions

A-TEC prediction model has been developed using a machine learning, a multivariate LSTM, with a simple architecture. The model input is the solar activity parameter F10.7, the geomagnetic index Dst and Kp, and the ionospheric weather index. The model learns from historical data, which includes maximum solar activity (2012–2016) and minimum

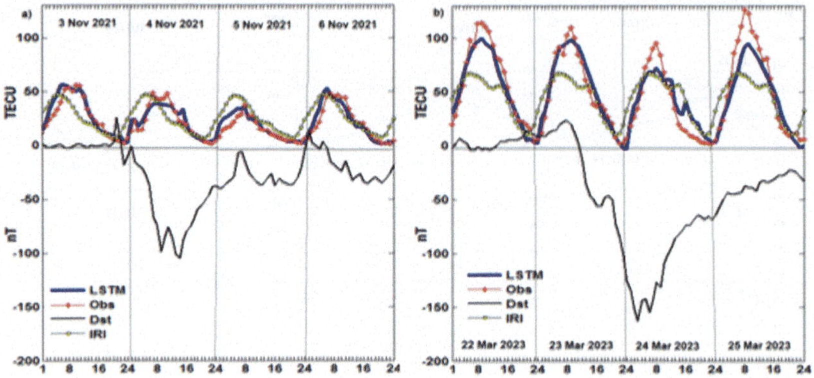

Fig. 7. Comparison response during geomagnetic storm between observation data, LSTM model, and IRI model a) is the case of a moderate storm that occurred on November 4, 2021, with Dst reaching −100 nT, and (b) is the case of an intense category storm with Dst reaching −150 nT on March 24, 2023.

Table 1. RMSE and R (coefficient correlation) validation models each month for 2022.

	RMSE	Correlation coefficients
January	7.18	0.902
February	10.49	0.850
March	12.41	0.917
April	10.34	0.931
May	8.145	0.951
June	6.16	0.934
July	7.18	0.908
Augustus	9.22	0.912
September	8.13	0.938
October	8.18	0.952
November	7.77	0.942
December	7.74	0.942

solar activity (2017–2020). Model validation using observation data for 2022 shows that RMSE varies monthly with a maximum of 11 TECUs based on maximum TEC observation data at a scale of 100–110 TECUs(confident 10%). The evaluation model shows that the LSTM model is capable and responsive in reacting to geomagnetic storms during moderate and intense ones. Based on the RMSE analysis and correlation coefficient, the results indicate that the model is quite feasible and can be used for predictions as a warning or mitigation for interference with satellite signals due to increased electron density in the ionosphere region near the single station during space weather activities.

Acknowledgments. We would like to express our gratitude to the Geospatial Information Agency (BIG), Indonesia, for access to INACORS, Gopi Seemala for GPSTEC,and the Pushkov Institute of Terrestrial Magnetism, Ionosphere and Radio Wave Propagation, Russian Academy of Sciences, especially Tamara Gulyaeva for providing the WTEC data for the BAKO station in 2022.

References

1. Bilitza, D.: International reference ionosphere 1990. Planet. Space Sci. **40**, 544–544.(1992). https://api.semanticscholar.org/CorpusID:120497099.1
2. Bilitza, D.: International reference Ionosphere 2000. Radio Sci. **36**(2), 261–275 (2021). https://doi.org/10.1029/2000RS002432.1
3. Mengistu Tsidu, G., Melaku Zegeye, M.: Comparison of quiet-time ionospheric total electron content from the IRI-2016 model and from gridded and station-level GPS observations. Ann. Geophys. **38**(3), 725–748 (2020). https://doi.org/10.5194/angeo-38-725-2020
4. Acharya, R., Majumdar, S.: Comparison of observed ionospheric vertical TEC over the sea in Indian region with IRI-2016 model. Adv. Space Res. **63**(6), 1892–1904 (2019). https://doi.org/10.1016/j.asr.2018.10.049.1
5. Sahu, S., Trivedi, R., Choudhary, R., Jain, A., Jain, S.: Prediction of total electron content (TEC) using neural network over anomaly crest region bhopal. Adv. Space Res. **68**(7), 2919–2929 (2021). https://doi.org/10.1016/j.asr.2021.05.027
6. Silva, A., Moraes, A., Sousasantos, J., Maximo, M., Vani, B., Faria, C .: Using deep learning to map ionospheric total electron content over Brazil. Remote Sens. **15**(2) (2023)
7. Iban, M.C., Şentürk,E.: Machine learning regression models for prediction of multiple ionospheric parameters. Adv. Space Res. **69**(3), 1319–1334 (2022). https://doi.org/10.1016/j.asr.2021.11.026,
8. Shaikh, M.M., Butt, R.A., Khawaja, A.: Forecasting total electron content (TEC) using CEEMDAN LSTM model. Adv. Space Res. **71**(10), 4361–4373 (2023). https://doi.org/10.1016/j.asr.2022.12.054
9. Liu, H., et al.: Ionospheric TEC prediction in china based on the multiple-attention LSTM model. Atmosphere **13**(11) (2022). https://doi.org/10.3390/atmos13111939
10. Xie, T., Dai, Z., Zhu, X., Chen, B., Ran, C.: LSTM-based short-term ionospheric TEC forecast model and positioning accuracy analysis. GPS Solut. **27**(2), 66 (2023). https://doi.org/10.1007/s10291-023-01406-8
11. Langley, R., Fedrizzi, M., de Paula, E., Santos, M., Komjathy, A.: Mapping the low latitude ionosphere with GPS. GPS World **13**, 41–46 (2002)
12. Rama Rao, P.V.S., Gopi Krishna, S., Niranjan, K., Prasad, D.S.V.V.D.: Temporal and spatial variations in TEC using simultaneous measurements from the Indian GPS network of receivers during the low solar activity period of 2004–2005. Ann. Geophys. 24(12), 3279–3292 (2006). https://doi.org/10.5194/angeo-24-3279-2006
13. Gulyaeva, T., Arikan, F., Hernandez-Pajares, M., Stanislawska, I. : GIM-TEC adaptive ionospheric weather assessment and forecast system. J. Atmos. Sol.-Terr. Phys. **102**, 329–340 (2013). https://doi.org/10.1016/j.jastp.2013.06.011
14. Wan, Q., et al.: Performance evaluation of IRI-2016 with GPS-derived TEC at the meridian of 110oE in China of 2014. J. Atmos. Solar-Terr. Phys. **201**, 105206 (2020). https://doi.org/10.1016/j.jastp.2020.105206

15. Asmare, Y., Kassa, T., Nigussie, M.: Validation of IRI-2012 TEC model over Ethiopia during solar minimum (2009) and solar maximum (2013) phases. Adv. Space Res. **53**(11), 1582–1594 (2014). https://doi.org/10.1016/j.asr.2014.02.017
16. Tariku, Y.A.: TEC prediction performance of IRI-2012 model during a very low and a high solar activity phase over equatorial regions. Uganda. J. Geophys. Res. Space Phys **120**(7), 5973–5982 (2015). https://doi.org/10.1002/2015JA021203

The Interplanetary Shocks Impact on the Electron Flux Dynamics

Siska Filawati(✉), Fitri Nuraeni, Silmie V. Fani, Dian Y. Risdianto, Rian P. Salasa, Mira Juangsih, Rizal Suryana, and Elvina A. Ratnasari

Research Center for Space, BRIN, Bandung 40135, Indonesia
sisk005@brin.go.id

Abstract. The interactions between disturbed solar wind from CME and CIR and magnetosphere generate interplanetary shock (IP shock) through steepening of fast magnetosonic waves. The IP shocks accelerate charged particle to high energies that can be hazardous to space-based technology and activities. This study aimed to understand the effect of gradient of IP shock to electron flux dynamics in energy of $\geq$0.8 MeV and $\geq$2 MeV. We determine the onset time of IP shock as the time when an abrupt change of Vp gradient and calculate the time lag between IP shocks arrival and electron flux anomalies. Our results show that CMEs and CIRs as the source of solar wind parameter have different affect to electron flux dynamics. Our results indicate that CMEs have a higher potential to generate IP shocks. However, CIRs have a greater potential to increase electron energy.

Keywords: Interplanetary Shocks · Solar Wind · Coronal Mass Ejection · Co-rotating Interaction Region · Electron Flux

1 Introduction

The sun is the nearest star to the Earth. The sun emits energy through the interplanetary medium that we call it as solar wind. Solar wind has characterization of the sun's activities. Some of the sun's activities that hazardous to the Earth is Coronal Mass Ejection (CME) and Co-rotating Interaction Region (CIR). CME and CIR has different physical process from the sun. CME is ejected plasma blast from solar corona during solar maxima, while CIR is generated at interplanetary medium due to compression between fast speed streams and slow speed streams that associated with coronal holes (CH) during solar minima [1, 2]. The interactions between disturbed solar wind from CME and CIR and magnetosphere produce interplanetary shock (IP shock) through steepening of fast magnetosonic waves. The shock is associated with abrupt increases or decreases both in the magnetic field magnitude and solar wind plasma parameters. The IP shock is known that it can accelerate charged particles to very high energies (several tens of MeV), posing a significant hazard to satellite technology and human activities in space [3]. Also, the IP shock compresses the magnetopause that face the sun and stretch the magnetotail become wider and longer. This leads to an increase of horizontal magnetic field measured by magnetometers on the ground [1], then further accelerate

E. Fitrianingsih et al. (Eds.): ISAST 2024, SPPHY 416, pp. 233–240, 2025.
https://doi.org/10.1007/978-981-96-1344-1_25

and transport the energetic electrons inward. It is also known that IP shock can greatly affect the radiation belt electron dynamics [3].

The outer radiation belt is dominated by electron within energy of >400 keV to 15 MeV at 4 R_E to 6 R_E (R_E, earth's radii 6387 km) [4]. These electrons come from solar wind plasma and ionosphere. Variation in electron flux occur as an accumulation of various processes, including source of disturbance, acceleration and loss processes [3, 4] which generate different geomagnetic storms. The shock triggers electron flux enhancements, characterized by either a dispersive or dispersionless energy within 1–2 min after the shock onset, which subsequently leads to a reduction in energetic electrons [3]. Generally, electron flux decreases during the main phase of a storm then returns to its pre-storm condition during the recovery phases. But in some cases, the electron flux increases higher than its pre-storm condition. The typical time scale for the electron flux enhancement of the outer belt is in 2 to 3 days [5]. The enhancement of the electrons flux of the outer radiation belt frequently occur not only in storms but also associated with solar wind disturbances with large scale of the interplanetary structures.

In this study, we examine the outer radiation belt dynamics by analyzing the 0.8 MeV and 2 MeV energy of electron flux following the interplanetary shock arrival. We calculate the gradient of proton bulk speed and time lag between onset time of IP shock and electron flux anomaly. Thus characterizing the impact of IP shock to the flux electron based on its solar disturbances.

2 Data and Method

We re-identified IP shock events database that listed in [6] from 2014 to 2019 by using solar wind and interplanetary magnetic field (IMF) data from ACE (Advanced Composition Explorer). The Level-2 solar wind and IMF data used are obtained from ACE database [7, 8]. The solar wind parameters included proton density (Np), proton bulk speed (Vp), and temperature (Tp), while the IMF data comprised the North-South component of magnetic field (Bz) and the magnitude of the magnetic field (Bt), sampled at 64 s interval. The electron flux in energy of $\geq$0.8 MeV and $\geq$2.0 MeV used in 1-min interval from GOES-15 [9].

We first visually inspected for abrupt changes in the solar wind parameters and IMF data. The onset times were set based on sudden enhancements in solar wind parameters and increases or decreases in the IMF. We observed that the onset times for the solar wind and IMF parameters varied in order of minutes, despite some data being either incomplete or missing, which prevented a complete observation of the changes. Thus, we consider the increase in Vp as the primary indicator of onset time of IP shock and apply this to other parameters. Using the onset time of Vp as IP shock onset time, we examined the gradient of Vp and also other parameters to characterize the IP shock impact on electron flux dynamics. Time lag between IP shocks and electron flux anomalies, whether indicated by either a decrease or an increase in the anomalies, is calculated as the delay time between the IP shocks onset time and the appearance of electron flux anomalies.

3 Result and Discussion

We identified 34 IP shock events during 2014–2019. Analyzing solar wind and interplanetary magnetic field parameters such as Vp, Np, Tp, Bt and Bz during each IP shock event to understand the impact of the shock on electron flux dynamics. The drivers of the interplanetary shocks are mainly coronal mass ejections (CMEs), although some are identified as originating from a combination of CMEs and co-rotating interaction regions (CIRs).

From the initial process, we found out that there are slight differences on onset time of IP shock occurrence between each solar wind parameter and interplanetary magnetic field parameter, as shown in Fig. 1 on 12 June 2014, an IP shock identified to occur on 12:43 UT, the Vp and Np recorded almost the same onset time, while the IMF rapid change occurred in a few seconds after the sudden increase in the solar wind parameters.

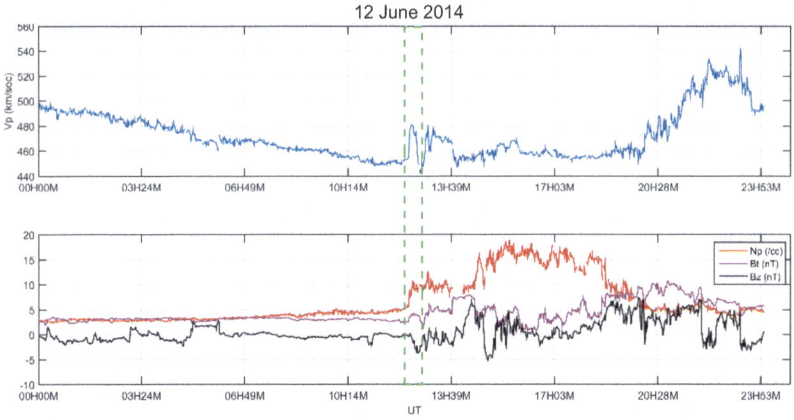

Fig. 1. IP shock on 12 June 2014 marks with the green stripes line. Upper panel is the proton speed (blue), while in the lower panel are proton density (red), total magnetic field (magenta) and North-South magnetic field component (black).

Positive gradient mostly observed from all parameters during the shock, means that every shock events generate a sudden increase in the interplanetary medium parameters. This leads to a confirmation that IP shocks as signature of the arrival of disturbances in the interplanetary medium. Except for Bz, almost all Bz data show a negative gradient. These indicate that during the occurrence of IP shocks, the IMF directed to southward, which is opposite to the magnetospheric magnetic field, that can leads a reconnection.

From the time lag correlation, the Vp, Np, Bt during shock event have negative correlation with electron flux, with time lag ranging from 3 min to 4 h. In contrast, Bz showed a positive correlation with electron flux with a time lag of 2 to 7 h after the onset time of the shock. It is showed that IP shock caused a decreased of electron flux sometime afterward the shock event.

We categorized IP shocks based on the driving solar storm and the channel energy of the electron flux. Out of these, 24 IP shock events were driven by CMEs, while 11 events were driven by CIRs during the period from 2014 to 2019. As depicted in Table 1 and 2, the IP shocks caused by CMEs exhibit a higher average Vp gradient compared to those caused by CIRs. This suggests that CMEs have a greater potential to generate IP shocks which means that IP shocks caused by CMEs caused a higher acceleration to solar wind particles compared to those caused by CIRs.

Additionally, we also calculated the time lag which determined from the onset time of IP shocks and electron flux anomalies. The identification of electron flux anomalies based on sudden increase or decrease in electron flux. The time lag calculations were divided based on the sources of disturbance (CMEs and CIRs) and the energy channels of electron $\geq$0.8 MeV as low-energy electron and $\geq$2 MeV as high-energy electron. For CME-driven IP shocks, the time lag for low-energy electron is shorter compared to high-energy electron. The time lag for low-energy electron ranges from 0 to 68 min, meanwhile for high-energy electron, it ranges from 37 to 168 min. This is reasonable because low-energy electron is more sensitive to energy enhancements, thus providing a clearer picture of electron flux in geostationary orbit. In contrast, for CIR-driven IP shocks, the time lag for low-energy electron is longer compared to high-energy electron. The time lag for low-energy electrons ranges from 30 to 148 min, while for high-energy electrons, it ranges from 42 to 101 min. In fact, the difference between low-energy electrons and high-energy electrons during CME-driven IP shocks and CIR-driven IP shocks is not very significant. Additionally, there is a considerable amount of 'not defined' for high-energy electrons during CIR-driven IP shocks.

It is important to note that 'not defined' occurs when there is bad or missing data. In Table 1 and 2 the high-energy electron channel has more instances of 'not defined' compared to the low-energy electron channel. This is due to the electron flux in the high-energy electron channel is contaminated by solar proton energy, which is strong enough to require filtering out these contaminated data. This also indicates the inability of the low-energy electron channel affected by solar proton energy [10].

Based on the energy channel categories, the time lag for CME-driven events are shorter compared to CIR-driven events for low-energy electron. On the other hand, for high-energy electron, the time lag for CME-driven events are longer compared to CIR-driven events. This indicates that CIR has a higher potential to increase electron energy compared to CME.

In general, there was a complex response in electron flux at low-energy and high-energy corresponding to each interplanetary shock driven by solar disturbances. However, through different physical processes, CIRs have a more significant effect on increasing electron energy, with shorter times in CIR-driven IP shocks compared to CME-driven IP shocks. CIRs increase electron energy at geostationary orbit (5–6 R_E), while CMEs do so at 4 R_E. GOES-15 is a satellite at geostationary orbit, thus able to observe electron energy enhancements during CIRs. In some previous studies, most electron flux dynamics have been attributed to non-adiabatic processes [11].

Vp gradient can be interpreted in relation to the time lag between the IP shock and electron flux anomaly. A high Vp gradient indicates a significant increase in bulk speed over a short time duration. Figure 2 and 3 show that a high Vp gradient tends to correspond with shorter time lag. In contrast, a low Vp gradient, which indicates a smaller increase in bulk speed over a longer time duration, corresponds with longer time

Table 1. IP shocks due to CMEs from 2014 to 2019, with Vp gradient, shock duration, and time lag categorized by electron energies of 0.8 MeV and 2.0 MeV

Date	Proton bulk speed (Vp) gradient (km/s^2)	IP shock duration (minutes)	Time lag between IP shock arrival and ≥0.8 MeV EF anomaly (minutes)	Time lag between IP shock arrival and ≥2 MeV EF anomaly (minutes)
20140107	2.751	4	36	112
20140109	30.323	4	40	Not defined
20140207	36.882	3	49	Not defined
20140213	16.043	4	Not defined	72
20140215	1.805	8	28	46
20140220	41.536	4	27	Not defined
20140419	13.622	3	45	Not defined
20140420	2.959	3	43	Not defined
20140507	12.596	3	53	Not defined
20140529	2.546	4	19	Not defined
20140607	21.695	5	42	Not defined
20140623	1.7474	5	67	Not defined
20150107	12.423	5	23	60
20150612	3.5693	10	24	56
20150622	12.453	9	21	Not defined
20151024	22.489	4	31	37
20151214	12.319	5	53	54
20161012	24.823	3	52	71
20161109	0.4345	13	68	65
20170527	2.4106	5	55	57
20170701	15.527	3	52	183
20170907	86.693	3	0	Not defined
20180215	14.402	3	36	Not defined
20180716	5.469	4	35	Not defined
Average	16.56	4.875	39.1	73.9

Not defined refers bad or missing data

lag. It is previously known that the average Vp gradient for CMEs are higher compared to CIRs, suggesting that CMEs have a higher potential to generate IP shocks. However, in relation to electron flux dynamics, CIRs are more potential to increase electron energy, which can lead a more significant changes in electron flux dynamics compared to CMEs.

Table 2. IP shocks due to CIRs from 2014 to 2019, with Vp gradient, shock duration, and time lag categorized by electron energies of 0.8 MeV and 2.0 MeV

Date	Proton bulk speed (Vp) gradient (km/s^2)	IP shock duration (minutes)	Time lag between IP shock arrival and ≥0.8 MeV EF anomaly (minutes)	Time lag between IP shock arrival and ≥2 MeV EF anomaly (minutes)
20150331	10.396	4	55	Not defined
20160105	5.4823	5	87	42
20160908	7.2232	9	42	101
20170831	14.959	3	57	Not defined
20180108	19.197	3	58	Not defined
20180505	0.3911	7	30	46
20190324	8.531	3	83	Not defined
20190708	22.567	3	54	Not defined
20190924	17.999	4	148	Not defined
20190927	0.6666	12	123	Not defined
Average	10.74	5.3	73.7	63

Not defined refers bad or missing data

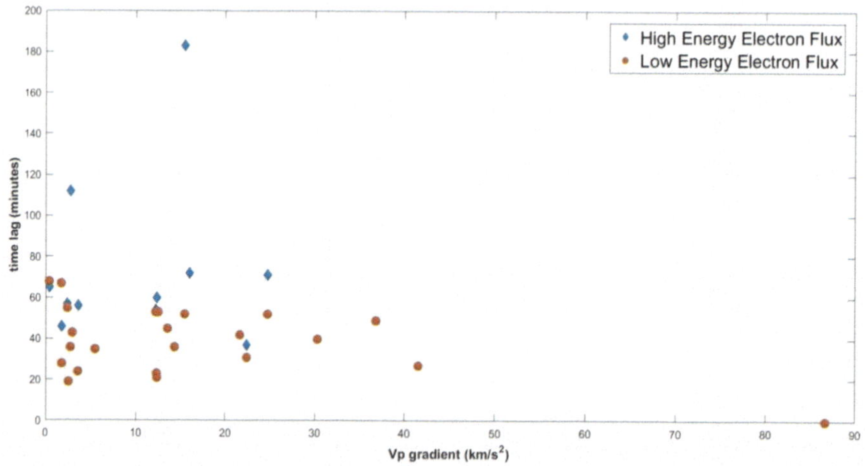

Fig. 2. Vp gradient and time lag in minutes for CME-driven IP shocks between high-energy electron flux and low-energy electron flux.

IP shocks are characterized by sudden increases in speed [12]. CMEs have higher speed enhancement compared to CIRs. Although CIRs have higher speed in the interplanetary medium, their speed enhancement are not as significant as CMEs. Vp gradient can illustrate the shock waves that occurred in the interplanetary medium by detailing the speed enhancement over a short duration. Therefore, as shown in Table 1 and 2, CME-driven shocks have a higher average Vp gradient. CME-driven shocks are regularly aligned with the Sun-Earth line, while CIR-driven shocks tend to propagate through the Sun-Earth line and the Parker spiral direction [12].

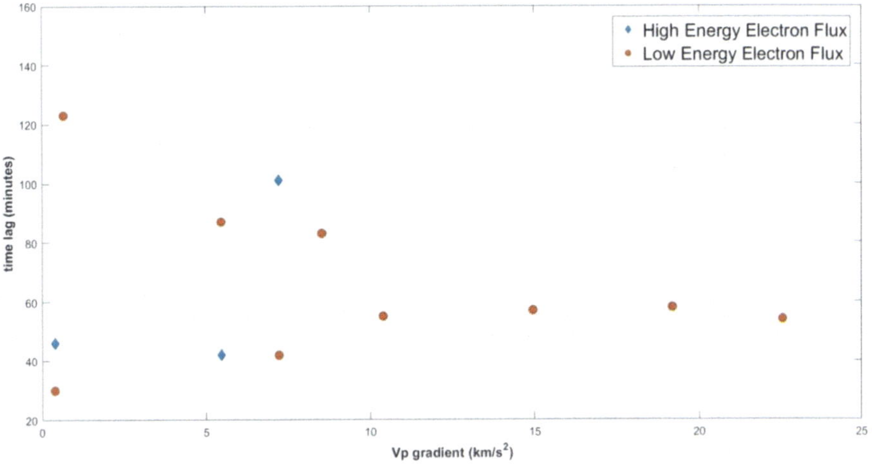

Fig. 3. Vp gradient and time lag in minutes for CIR-driven IP shocks between high-energy electron flux and low-energy electron flux.

4 Conclusion

The impact of IP shocks on electron flux dynamics was investigated by analyzing the time lag between the onset time of IP shocks and electron flux anomalies, categorized by energy channel. The Vp gradient is indicating IP shock that often associate with solar disturbances (CMEs and CIRs). Positive IP shock gradients were observed across almost all parameters, except for the Bz gradient. The time lag was used to assess the influence of IP shocks enhancement on electron flux anomalies. Anomalies in electron flux were detected from minutes to hours after the shock. Our results indicate that:

- CMEs have more significant impact on IP shocks, as evidenced by higher Vp gradients compared to CIRs. This suggests that CMEs have a greater potential to generate IP shocks.
- However, in terms to electron flux energy channels, CIRs exhibit a greater potential to increase high energy electron fluxes, leading to more substantial changes in electron flux dynamics compared to CMEs.

While time lag analysis does not fully explain the relationship between IP shocks and electron flux anomalies, it serves as an initial empirical parameter for understanding when electron flux anomalies occur.

Acknowledgments. We thank the ACE team for providing SWEPAM and IMF data. And we also thankful to the GOES team for providing electron flux data in low-energy and high-energy. We express our gratitude to the authors that we have cited in this manuscript, as well as the referees for valuable comments and suggestions that helped improve the manuscript.

References

1. Oliveira, D.M., Samsonov, A.A.: Geoeffectiveness of interplanetary shocks controlled by impact angles: a review. Adv. Space Res. **61**, 1–44 (2018)
2. Kallenrode, M.-B.: Space Physics: An Introduction to Plasmas and Particles in the Heliosphere and Magnetosphere. Springer (2004)
3. Liu, Y., Zong, Q.G., Zhou, X.Z., Hao, Y.X., Liu, Z.Y.: Understanding electron dropout echoes induced by interplanetary shocks: test particle simulations. J. Geophys. Res. Space Phys. **124**, 6759–6775 (2019)
4. Li, X., Temerin, M.A.: The electron radiation belt. Space Sci. Rev. **95**, 569–580 (2001)
5. Filawati, S.: Analysis of electron flux condition in outer electron radiation belt based on interplanetary magnetic field (Bz) and solar wind speed. J. Sains Dirgantara **15**(1), 39–50 (2018)
6. Oliveira, D.M.: Interplanetary shock data base. Front. Astron. Space Sci. **10**, 1240323 (2023)
7. Garrard, T., Davis, A., Hammond, J., et al.: The ACE science center. Space Sci. Rev. **86**, 649–663 (1998)
8. ACE Science Center. https://izw1.caltech.edu/ACE/ASC/index.html. Accessed 31 July 2024
9. GOES Data Access. https://www.ngdc.noaa.gov/stp/satellite/goes/dataaccess.html. Accessed 31 July 2024
10. Rodriguez, J. V. GOES EPEAD science-quality electron fluxes, algorithm theoretical basis document. NOAA NESDIS National Geophysical Data Center (2014)
11. Staples, F.A., Kellerman, A., Murphy, K.R., Rae, I.J., Sandhu, J.K., Forsyth, C.: Re-solving magnetopause shadowing using multimission measurements of phase space density. JGR: Space Phys. **127**, e2021JA029298 (2022)
12. Kilpua, E.K.J., Lumme, E., Andreeova, K., Isavnin, A., Koskinen, H.E.J.: Properties and drivers of fast interplanetary shocks near the orbit of the Earth (1995–2013). J. Geophys. Res. Space Phys. **120**, 4112–4125 (2015)

The Observation of Supernova Impostor AT2016blu Using Seimei 3.8 m Telescope

Chatief Kunjaya[1,2,3](✉), Keisuke Isogai[4,5], Daisaku Nogami[4], Rafa Nanda Akilah[1], Tian Ahmad Bayuadri[1], Sahabista Arkitanego Armantara[1], Kiki Vierdayanti[1], Fumihide Iwamuro[4], Anton T. Jaelani[1,2], and Putra Mahasena[1,3]

[1] Astronomy Research Division, FMIPA, Institut Teknologi Bandung, Jl. Ganesa 10, Bandung 40132, Indonesia
kunjaya@itb.ac.id
[2] Center for Space Science, Technology and Innovation, Institut Teknologi Bandung, Jl. Ganesa 10, Bandung 40132, Indonesia
[3] Bosscha Observatory, Jl. Peneropongan Bintang 1, Lembang 40391, Indonesia
[4] Department of Astronomy, Graduate School of Science, Kyoto University, Kitashirakawa Oiwake-Cho, Sakyo-Ku, Kyoto-Shi, Kyoto 606-8502, Japan
[5] Okayama Observatory, Kyoto University, 3037-5 Honjo, Kamogata-Cho, Asakuchi-Shi, Okayama 719-0232, Japan

Abstract. Supernova impostors are extragalactic object which have been discovered to brighten significantly like supernova, however after sometimes it become much brighter again which disprove supernova classification. Nevertheless, there are examples that such objects can finally explode as a true supernova. That makes supernova impostors interesting object to be monitored. Supernova impostor PSN J12355230+2755559 or AT2016blu was observed photometrically and spectroscopically using Seimei 3,8 m telescope at Okayama Astrophysical Observatory, Japan. Photometric observation was done using three color TRICCS camera with *g*, *r* and *I* filters, and spectroscopic observation using fiber-fed integral field KOOLS-IFU spectrograph. The object is situated in crowded area surrounded by HII region in a small satellite of galaxy NGC 4559, making the analysis quite difficult. The multi fiber spectrograph allowed us to choose the spectra of certain fibers to be integrated in order to minimize contamination from surrounding objects. The observation indicated that this object was found to be in outburst on March 9, 2024.

Keywords: Stars: individual: AT2016blu · Luminous Blue Variable · Stars: variables: S Doradus

1 Introduction

Supernova Impostors are objects in extra galaxies which once found to brighten significantly so that they may wrongly hypothesized as supernovae. However, in the other time they are rebrightening again which disprove the supernova hypothesis. This kind of objects are outbursting massive stars which remain intact after outburst. Most of the

E. Fitrianingsih et al. (Eds.): ISAST 2024, SPPHY 416, pp. 241–247, 2025.
https://doi.org/10.1007/978-981-96-1344-1_26

known supernova impostors are luminous blue variables (LBV) who destined to end their life as core collapse supernova. The prototype of this kind of objects is S Doradus, a blue supergiant in Large Magellanic Cloud.

Some of the known supernova impostors were truly exploded as supernova. One of the famous examples is SN 2006jc which was discovered as supernova impostor in 2004 by a Japanese amateur astronomer. It exploded as supernova two years later [1].

SN 2009ip is another example of a supernova impostor which finally exploded as true supernova [2–4]. In the Milky way, Eta Carinae dan P Cygni are examples of such objects. They showed great outburst in the past and are expected to become supernova in the future, based on the theory of stellar evolution which predict that massive stars will explode as supernova at the end of their life. Therefore, it is interesting to monitor such objects, to get some clues about their behavior and the critical condition triggering a supernova explosion.

There is a supernova impostor named AT2016blu or PSN J12355230+2755559. It is a dim and challenging object at a distance of about 8.9 Mpc, in a crowded area of a HII region near the galaxy NGC 4559 [5]. The region is possibly a satellite galaxy of NGC 4559 and called NGC 4559C or IC3550. The object was relatively quiet during 1999 to 2009 and became more actively outbursting from 2012 to 2022. Aghakhanloo (2023) [5] estimated that the object repeats its outburst every about 113 days based on long term photometric data and was expected to get bright again around February 2024, therefore we are encouraged to observe this object in that period.

2 Observation

AAVSO alert notice 847 [6] motivated us to try to observe the object using Seimei telescope [7] at the Okayama Astrophysical Observatory (OAO). Using Seimei telescope is particularly significant as Indonesia is set to have similar telescope at Mount Timau in Kupang, East Nusa Tenggara province in the near future. Observing such objects with comparable instrument at this new location will hopefully continue the research.

Competing with tight observation time allocation and not always good weather we finally could obtain data of the object from three nights observation. Photometric observation was done using a three-color camera, TRICCS, with g, r and i filters, simultaneously. Spectroscopy was done using fiber-fed integral field KOOLS-IFU low resolution spectrograph [8] with VPH-blue grism which has wavelength coverage 410 – 890 nm. We took also spectra of an empty field near the object to correct for sky emission due to nearby city light.

The KOOLS-IFU spectrograph, used for observing this object, has 110 usable optical fibers, each of which can provide an independent spectrum of every part of field of view. The pitch size of each fiber is 0.84 arc second and the total field of view is 8.4×8.0 (arc sec)^2. Table 1 provides details on the number images captured and the corresponding exposure times. For reduction procedure we took also images of dark, twilight and comparison star. For calibration we captured also images of Hg, Xe and Ne standard lamp.

Table 1. Observation log for AT2016blu in February and March 2024.

Date	Time (UTC)	Exp. Time (″)	No of frames	Instrument
February 12	14:16–14:21	60	15	TRICCS
February 17	18:07–19:57	60	20	TRICCS
	16:26–19:20	900	6	KOOLS-IFU
March 9	17:30–17:32	60	10	TRICCS
	17.43–18:45	900	3	KOOLS-IFU

3 Data Reduction

3.1 Photometry

The object's images obtained by TRICCS were then corrected using dark and flat images taken during twilight. From the clean image, in principle, we can extract the counts from the object using aperture photometry method. However, standard aperture photometry method with concentric aperture and sky may not be good as it is located in HII region and many other nearby objects. There are drawbacks to using a concentric annulus as the "sky" region, as it may include several other fainter objects within the sky area. To address this, we selected a nearby empty region for the sky subtraction procedure. The aperture size must be small to minimize contamination from the light of the surrounding objects.

The comparison star to be used, according to AAVSO identification number AUID, is 000-BPV-063 (RA = 12:36:07.80, Dec = +27:55:26.8) which have magnitude V = 17.532 and color (B-V) = 0.622 and check star 000-BPV-065 (RA = 12:35:32.36, Dec = +27:54:44.5, V = 18.342, (B-V) = 0.583). For calibration we transformed the V magnitude and (B-V) color into SDSS g, r, i magnitude system using a set of transformation formulae [9].

3.2 Spectroscopy

The raw image spectra of the object taken by KOOLS-IFU spectrograph [8] were then reduced using the reduction program made by one of the authors (FI). After dark subtraction dan flat fielding process we have to remove also the effect of optical distortion and cosmic ray. We did also correction for the sky emission due to the surrounding city light by observing empty field near the object. Wavelength calibration was done using the frames taken from the standard lamps.

From the 110 clean spectra derived from each fiber we choose eleven fibers around the center of field of view to extract the object's one-dimensional spectra. The effective field of view of the eleven fibers is about the same with the usual seeing size of Okayama sky in winter. This is expected to be the optimal compromise between minimizing contamination from nearby objects and maximizing light from the LBV itself.

The HST image [5] of the object revealed many other objects within the KOOLS-IFU spectrograph's field of view. Previous observations reported by other authors also indicated the existence of HII region features. To minimize the light contamination from other objects, it is better to use smaller field of view, which means using only small number of spectra from certain fibers around the center of field of view. However, fewer number of fibers used leads to lower signal to noise (S/N) ratio. So, we have to decide the number of fibers used for deriving one-dimensional spectra, to optimize between minimizing the surrounding objects contamination and maximizing S/N ratio.

4 Result and Discussion

Two clean *g* filtered images of the object in different dates are shown in Fig. 1 and Fig. 2 as examples of the observation result. The object is outside the big galaxy NGC 4559, in a small satellite galaxy NGC 4559C and marked by two straight lines in each frame.

Visual inspection of the images indicated that AT2016blu was not in outburst on February 17 but it significantly brighter on March 9, which indicated an outburst. One possible explanation of the outburst is the periastron passage of a companion star [5]. When the companion pass through the periastron, the distance of both components become closer and mass transfer will be enhanced, yielding more energy release. However, if that were true, another question may arise, why the object did not show outburst activities during 1999 to 2009?

Fig. 1. The region around the supernova impostor AT2016blu in *g* band, obtained on February 17, 2024. The object situated in a HII region called NGC4559C or IC3550. The nearby big galaxy is NGC 4559.

For spectroscopy, we did not use the spectra from all fibers, because they will include many other nearby objects. When we reviewed detail image obtained by Hubble Space Telescope, and if we want the light purely from the AT2016blu, ideally, we must use

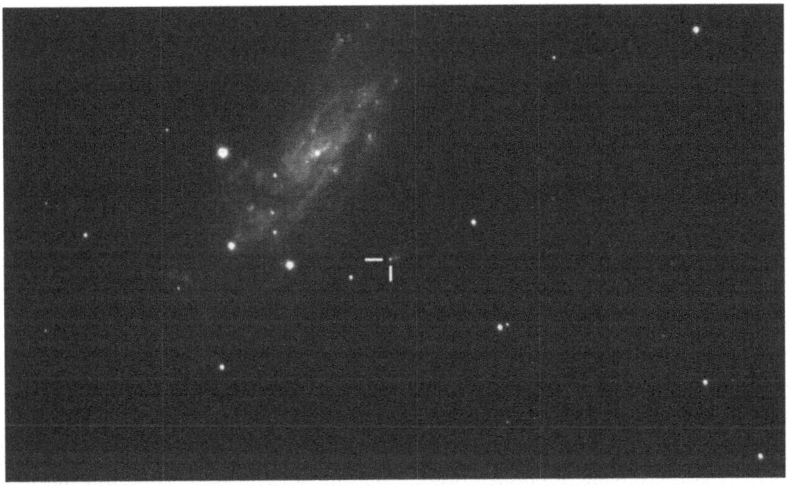

Fig. 2. AT2016blu on March 9, 2024. The object was significantly brighter compared to the one in Fig. 1, which indicate an outburst state.

only one spectrum from the fiber in the center, out of 110 spectra from all fibers, because many other fainter objects in the neighborhood of the object's 1″ region. But using only one fiber will yield a not so good S/N ratio spectrum. Therefore, we use 11 spectra from the central fibers as the compromise, to construct a one-dimensional spectrum which is shown in Fig. 3.

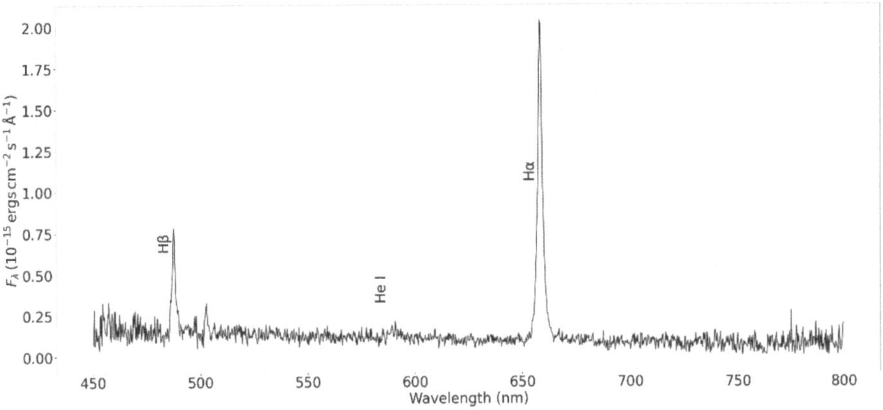

Fig. 3. Example of the spectrum of AT2016BLU, 17 February 2024. Two prominent emission lines are H_α and H_β. Helium I line is very weak.

The spectra obtained showed the existence of strong H_α and H_β emission. Both emission lines appeared also in the spectra from the fibers of surrounding region when we exclude the central fibers containing the LBV. When we subtract object spectrum with the surrounding's, the Balmer lines still exist. This indicate that both LBV and the

HII region contribute to the formation of the Balmer lines in the object spectrum. The HeI line which appeared in the spectrum of other author's observation [5] looks very weak in our spectra, comparable to noise. Did the HeI line disappear? Or just undetectable? It needs to be confirmed by other observation in the future.

5 Conclusion

The Seimei telescope with its imaging camera and fiber feed spectrograph is powerful enough for observing such faint and challenging object, provided data reduction and analysis are done in more sophisticated way. Observation of similarly challenging object with the same telescope size, better environment like in Timau mountain and with smaller sky seeing size is expected to yield better result.

The photometric images of the object showed significant increase of brightness of the object on March 9 compared to February 17. This indicate that the object was in outburst on March 9, which means the object continue to be active since 2012. Further analysis will be done to derive magnitudes and color of the object in quiescence and in outburst. It will provide a clue how the luminosity and the color change due to outburst. However, we must do the analysis more carefully to minimize the effect of the light contamination from surrounding objects. Standard photometric reduction procedure with concentric annulus for sky will not be suitable.

The spectra showed strong Balmer emission lines but Helium lines, which were visible in other observations, looks very weak. It needs to be investigated more to get a clue whether the Helium line is variable.

Prediction by Aghakhanloo [5] that in February 2024 AT2016blu would outburst again was quite accurate; this support the theory that the cause of the outburst is periastron passage of a small companion star. But its quiescence during 1999–2009 and multiple outbursts during February and March 2024 was not in line with the theory, and still need further explanation. More observation with big telescope equipped with adaptive optics will be necessary to separate the object from the surrounding objects.

Monitoring this object is advisable considering its potential to explode as core collapse supernova in near future, especially these years when this object has been active for 12 years. The monitoring can be continued in Mount Timau, Indonesia using similar 3.8 m telescope.

Acknowledgement. This work is supported by ITB International Research Program under the scheme of Staff Exchange and Research Grant with external funding.

References

1. Pastorello, A., Smartt, S.J., Mattila, S., et al.: A giant outburst two years before the core-collapse of a massive star. Nature **447**(7146), 829–832 (2007)
2. Drake, A.J., Howerton, S., McNaught, R., et al.: A new luminous outburst from SN 2009ip. ATel 4334 (2012)
3. Smith, N., Andrews, J.E., Filippenko, A.V., et al.: SN2009ip after a decade: the luminous blue variable progenitor is now gone. MNRAS **515**, 71–81 (2022)

4. Mauerhan, J.C., Smith, N., Filippenko, A.V., et al.: The unprecedented 2012 outburst of SN 2009ip: a luminous blue variable star becomes a true supernova. MNRAS **430**, 1801–1810 (2013)
5. Aghakhanloo, M., Smith, N., Milne, P., et al.: Recurring outburst of the supernova impostor AT 2016blu in NGC 4559. MNRAS **526**, 456–472 (2023)
6. Aghakhanloo, M.: AAVSO alert notice 847 (2024)
7. Kurita, M., Kino, M., Iwamuro, F., et al.: The Seimei telescope project and technical developments. PASJ **72**(3), 48:1–48:11 (2020)
8. Matsubayashi, K., Ohta, K., Iwamuro, F., et al.: KOOLS-IFU: Kyoto Okayama Optical low-dispersion spectrograph with optical-fiber integral field unit. PASJ **71**, 102 (2019)
9. Jester, S., Scheider, D.P., Richard, G.T., et al.: The sloan digital sky survey view of the palomar-green bright quasar survey. AJ **130**, 873–895 (2005)

Geomagnetic Field Effect Observed in Biak During the Solar Eclipse of April 20, 2023

Elvina Ayu Ratnasari[1], Mamat Ruhimat[1], Fitri Nuraeni[1](✉), Clara Y. Yatini[1], Moh Andi Aris Biantoro[1], and Jehunias Leonidas Tanesib[2]

[1] Research Center for Space, National Research and Innovation Agency (BRIN), Central Jakarta, Indonesia
fitr008@brin.go.id
[2] Nusa Cendana University, Kupang, Indonesia

Abstract. We analyze the effect of a solar eclipse on the geomagnetic field by processing geomagnetic observation data during the solar eclipse from the first contact until the moon's shadow leaves the Sun. During the eclipse, the moon blocks the radiation coming from the Sun, so the Earth receives less solar radiation. Then, the reduced radiation will affect the photoelectron process in the ionosphere and the magnetosphere. We conducted the geomagnetic observation in Biak during a total solar eclipse to determine its effect on the geomagnetic field. The total solar eclipse occurred on April 20, 2023. The coverage of the eclipse on Biak was 100%. We also compare the result in Biak with geomagnetic data obtained in Pontianak, which experienced a solar eclipse with only around 30% coverage. The results show that Biak's geomagnetic X component (north-south) has good responses to the solar eclipse. Similarly, the geomagnetic X component in Pontianak also decreased during the solar eclipse, but the amplitude is smaller than in Biak. As for the magnetic Y component, the response is not significant.

Keywords: Geomagnetic Field · Solar Eclipse

1 Introduction

The magnetosphere is a layer in the Earth's environment that contains a magnetic field originating from the superposition of several sources. The primary sources of the Earth's magnetic field come from (1) the main field originating from the Earth's core, (2) the plasma flow of the Earth's core, which is related to secular variations in the magnetic field, (3) the lithospheric field caused by magnetized crustal rocks, (4) induction due to ocean motion, (5) ionospheric magnetic field and (6) magnetospheric field. Measurements of the Earth's magnetic field activity can be carried out using magnetometers on the Earth's surface (ground-based magnetometer) or in space (space-based magnetometer), which will provide information related to studies of solar wind and the Earth's magnetosphere interaction. Long-term reliable geomagnetic recording is necessary to study the long-term behavior of the geomagnetic field, for example, its secular variations [1, 2]. High-resolution geomagnetic measurements provide beneficial information about changes in the geomagnetic field.

E. Fitrianingsih et al. (Eds.): ISAST 2024, SPPHY 416, pp. 248–255, 2025.
https://doi.org/10.1007/978-981-96-1344-1_27

On the significant date of April 20, 2023, a total solar eclipse passed parts of Indonesia, providing a unique opportunity for scientific observation. Although this eclipse is a Hybrid Solar Eclipse, it is seen as both a total and a partial solar eclipse in Indonesia. The first contact occurs at 02:36 UT (11:36 LT in Biak), and the last contact occurs at 05:56 UT (12:56 LT). This celestial event, where the moon shields the Earth from the Sun's radiation, is a spectacle and a catalyst for changes in the Earth's atmosphere, particularly the geomagnetic field. The disturbances from the Sun, carried by the solar wind, interact with the Earth's magnetic field in the atmosphere, making their way from high latitudes to low latitudes. During a solar eclipse, the decrease in solar radiation will affect ionization in the ionosphere [3, 4]. Several mathematical models [5, 6] show that the reduction of total electron content in the ionosphere in the area passed by a total solar eclipse can cause changes in current patterns and geomagnetic disturbances of several nanoTesla [7–10].

The task of detecting geomagnetic disturbances due to this eclipse is formidable, given the daily fluctuations in the geomagnetic field [11, 12]. The magnitude of the disturbance quantitatively depends on local geomagnetic conditions, latitude, longitude, and time [11, 13–15]. The influence of geomagnetism at different latitudes will provide an overview of the amplitude of geomagnetic disturbances that reach observation stations on the Earth's surface [16]. It is understood that the magnitude of the disturbance is intricately tied to local geomagnetic conditions, latitude, longitude, and time, further complicating our research. Nevertheless, these complexities make our study all the more intriguing and valuable.

In this comprehensive study, we meticulously analyzed the geomagnetic field data during the eclipse from Biak station (Geographic Coordinates 1.08° S, 136.05° E; Geomagnetic Coordinates 9.73° S, 207.39° E). The total eclipse with 100% coverage passed the Biak station. The eclipse commenced at 12:19:32 LT (03:19:32 UT) and concluded at 15:25:24 LT (06:25:24 UT). We then proceeded to compare the geomagnetic response in Biak with other stations at low latitudes in Indonesia that did not experience a total solar eclipse, namely the Pontianak observation station (Geographic coordinate 0.05° N, 134° E; Geomagnetic Coordinate 9.75° S, 181.96° E). The solar eclipse passed the Pontianak with a maximum coverage of only 30%. Figure 1 shows the path of the solar eclipse taken from https://www.timeanddate.com/eclipse/map/2023-april-20. Our goal is to determine whether there are differences in geomagnetic response to differences in eclipse coverage or differences in response between areas directly traversed by a total solar eclipse and areas further from the total eclipse path.

2 Data and Method

Geomagnetic disturbance shows the difference between the observed geomagnetic conditions and the pattern of quiet days. The pattern of quiet days is obtained from the average daily variation of 5 quiet days in the related month, obtained from the World Data Center Kyoto. We determined the geomagnetic field disturbances from the geomagnetic station by separating the geomagnetic field data recorded in the field from the geomagnetic quiet day pattern at that station. Geomagnetic disturbances during an eclipse must also be eliminated first from disturbances of the ring current, known as the

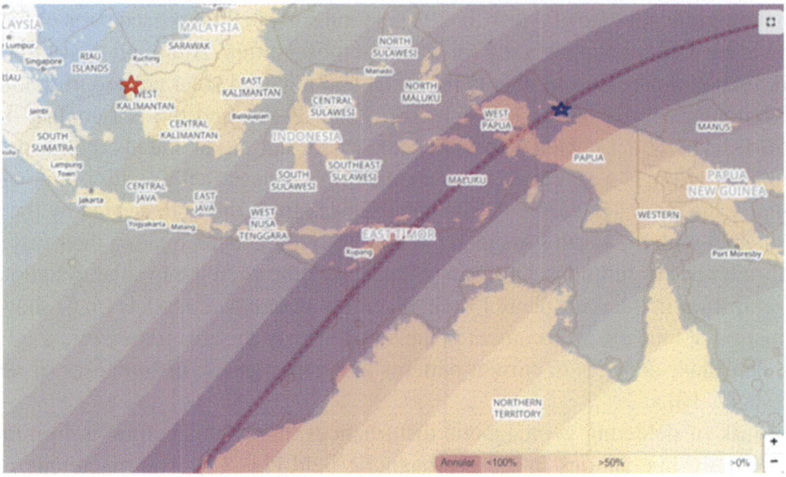

Fig. 1. The total solar eclipse path on April 20, 2023, crosses parts of Indonesia (https://www.timeanddate.com/eclipse/map/2023-april-20). The blue star indicates the location of Biak, and the red star indicates the location of Pontianak.

geomagnetic disturbance storm time index or Dst. We use the Dst index from the World Data Center Kyoto. To eliminate the influence of ring currents, we use the following Equations [17]:

$$X_D = X\text{-}Dst\ cos\ \lambda\ cos(-11^\circ) \tag{1}$$

$$Y_D = Y\text{-}Dst\ cos\ \lambda\ sin(-11^\circ) \tag{2}$$

where

X is the north-south component of geomagnetic disturbance
Y is the east-west component of geomagnetic disturbance
X_D and Y_D are the disturbances that have been eliminated from ring current interference
λ is the latitude of the observation station
11° represents the angular difference between the geographic and geomagnetic poles.

The geomagnetic field data used in this research is from the Geomagnetic Station in Biak, which experienced an eclipse with 100% coverage, and the Geomagnetic Station in Pontianak, which experienced an eclipse with around 31% coverage. The geomagnetic data and *Dst* index used start in March and April 2023.

The method employed in data processing is as follows:

1. Calculate the pattern of quiet days, which are days with minimal geomagnetic disturbances, from the two geomagnetic stations based on information on five quiet days from WDC Kyoto University.
2. Geomagnetic variations are obtained from observational data subtracted by the quiet-days pattern.

3. The geomagnetic variation obtained from step 2 is then adjusted by the *Dst* index value to eliminate the ring current effect, using Eqs. (1) and (2) to obtain a new set of X and Y component data, referred to as X_D and Y_D.
4. To determine the eclipse's effect on geomagnetic observations at the two geomagnetic stations, compare the geomagnetic disturbances from the two stations from step 3.

3 Result and Discussion

We analyzed geomagnetic data from the Biak and Pontianak Geomagnet stations to determine the effect of the solar eclipse on the geomagnetic field. The two places are also compared on a quiet day without any geomagnetic disturbances and on a disturbed day, apart from when an eclipse occurs. In initial processing, we find the geomagnetic quiet day pattern, which separates geomagnetic disturbances from the daily geomagnetic data for April 2023. This quiet day pattern is an average of the geomagnetic field data at each station, which corresponds to the quiet day information obtained from the WDC Kyoto University.

3.1 Geomagnetic Conditions on Quiet Day

To determine whether the solar eclipse affects the geomagnetic field, it is necessary first to know the magnetic field pattern when there is no eclipse and no disturbance, called a quiet day. The quiet day we chose was April 17, 2023, meticulously selected from the list of quiet days in April 2023 from WDC. Based on these data, the quiet days before the eclipse were April 13 and 17, 2023, and April 17, 2023, was the closest quiet day to the eclipse. Figure 2 shows the geomagnetic field pattern from the Biak and Pontianak stations on the quiet day of April 17, 2023.

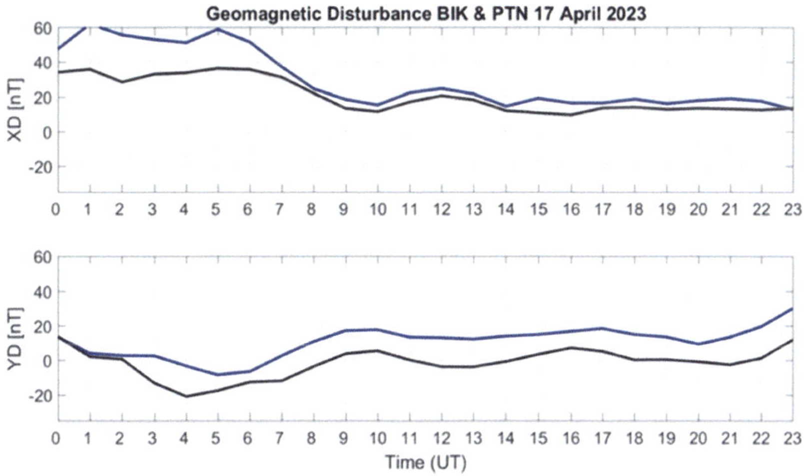

Fig. 2. Geomagnetic field pattern on April 17, 2023, for the X_D and Y_D components from Biak station (BIK, blue line) and Pontianak station (PTN, black line), respectively.

The daily variation of the geomagnetic field is intricately linked to the magnetic latitude position of the observation location. Despite Biak and Pontianak's almost identical magnetic latitudes (9.73° S and 9.75° S), the geomagnetic field variation exhibits a distinct pattern. The most significant difference is observed during the local daytime (UT + 7 for PTN and UT + 9 for BIK), a phenomenon likely caused by the influence of different geographical positions on the reception of solar radiation. This solar radiation is the primary generator of ionospheric currents, the leading cause of regular daily variations called Solar Quiet (Sq) [18]. Figure 2 illustrates that the geomagnetic variations of the X_D and Y_D components on quiet days from the Biak station are higher than those of the Pontianak station. Specifically, the X_D component variation is around 60 to 18 nT for the Biak station and 40 to 10 nT for the Pontianak station, while the Y_D component variation is 30 to −5 nT in the Biak station data and 10 to −25 nT for the Pontianak station data. These findings are consistent with previous studies on variations in the X component in Indonesia, indicating that for quiet days, data from Pontianak will be lower than stations with geomagnetic latitudes closer to the magnetic equator [19] and Biak station.

3.2 Geomagnetic Conditions During Disturbed Day

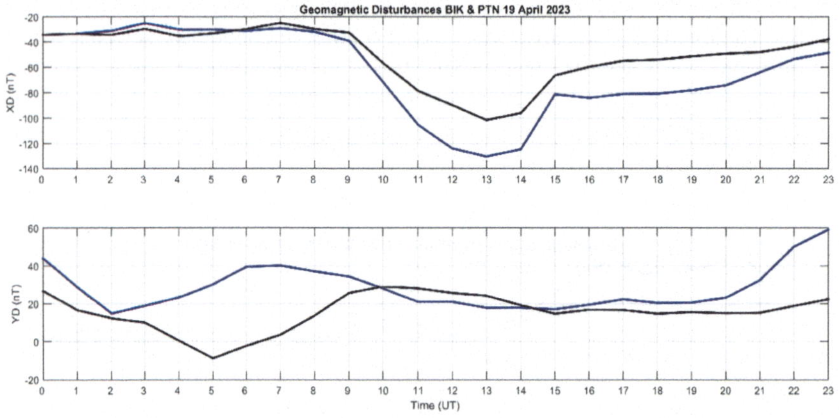

Fig. 3. Geomagnetic field pattern on disturbed day on April 19, 2023, for the X_D and Y_D components from Biak station (BIK, blue line) and Pontianak station (PTN, black line), respectively.

On April 19, 2023, a disturbance in the geomagnetic field caused a depression, with *Dst* reaching −36 nT at 15 UT. Under disturbed conditions, the variation patterns of the two stations still show similar patterns, as shown in Fig. 3. During the geomagnetic disturbance period from 09.00 UT to 15 UT, the depression at both stations has almost the same value. It is just that the depression in the X_D component is slightly more significant in the data from Biak station. This is in accordance with the results of [20] that the closer to the equator, the lower the magnetic response of the magnetic field compared to measurements at higher latitudes. The geomagnetic latitude of Biak station and Pontianak station is almost the same, namely 9.73° S and 9.75° S, so based on [19], Biak station will have nearly the same X_D component variation value, and this can be seen in Fig. 3 upper panel, with a slight standard deviation.

3.3 Geomagnetic Condition During April 20, 2023

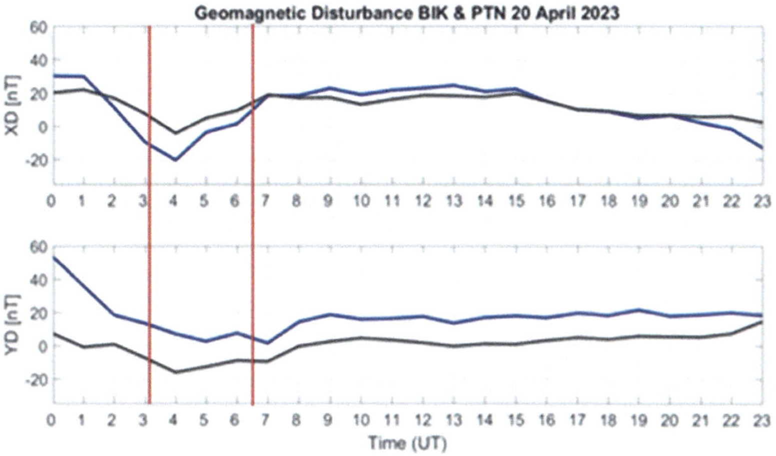

Fig. 4. Geomagnetic field profile on April 20, 2023, for components X_D and Y_D, respectively, from Biak station (blue line) and Pontianak station (black line). Red vertical lines show the start and end times of the eclipse.

The eclipse's effect is discernible in the X_D component data, with a noticeable decrease in value of up to −20 nT and a corresponding increase in the Y_D component, reaching 10 nT at Biak station compared to Pontianak station. The latter experienced a 30% partial eclipse, resulting in an X_D component value of −5 nT and a Y_D component value of −18 nT. The difference in the X_D value, as depicted in Fig. 4 of the upper panel, demonstrates that the eclipse coverage significantly influences the geomagnetic field response by 15 nT. Compared to [8], which reported a response value for the eclipse effect on the X_D component reaching 5 nT, our analysis revealed a more substantial value during the April 20, 2023 eclipse. This is likely due to the lingering effects of the previous geomagnetic disturbance on April 19, 2023.

4 Conclusion

The solar eclipse on April 20, 2023, affected the geomagnetic field in Indonesia, particularly at Biak station. The station experienced disturbances during the eclipse, notably in the X_D component, which was −20 nT. A key finding of our study was the comparison of the geomagnetic response at different locations in Indonesia, such as Pontianak station, which revealed a distinct influence of solar eclipse coverage on the geomagnetic field. This analysis of the eclipse event yielded a more significant value than previous studies, likely due to the effects of geomagnetic disturbances that occurred the day before. The magnetosphere is a layer in the Earth's environment that contains a magnetic field.

References

1. Cain, J.C., Mozzoni, D.T., Ferguson, B.B., Ajayi, O.: Geomagnetic secular variation 1995–2000. J. Geophys. Res. **108**(B3), 2161 (2003)
2. Yoshihara, A., Kondo, A., Ohno, M., Hamano, Y.: Secular variation of the geomagnetic field intensity during the past 2000 years in Japan. Earth Planet. Sci. Lett. **210**(1–2), 219–231 (2003)
3. Le, H., Liu, L., Yue, X., Wan, W.: The ionospheric behavior in conjugate hemispheres during the 3 October 2005 solar eclipse. Ann. Geophys. **27**, 179–184 (2009)
4. Liu, X., et al.: The response of geomagnetic daily variation and ionospheric currents to the annular solar eclipse on 21 June 2020. J. Geophys. Res. Space Phys. **127**, e2022JA030494 (2022)
5. Campbell, W.H.: An introduction to quiet daily geomagnetic fields. Pure Appl. Geophys. **131**, 315–331 (1989)
6. Yamazaki, Y., et al.: An empirical model of the quiet daily geomagnetic field variation. J. Geophys. Res. **116**, A10312 (2011)
7. Střeštík, J.: The response of the 11 August 1999 total solar eclipse in the geomagnetic field. Earth Moon Planet. **85**, 561–566 (2001)
8. Ruhimat, M., Winarko, A., Nuraeni, F., Bangkit, H., Aris, M.A., Suwardi, S.: Effect of March 9, 2016 total solar eclipse on geomagnetic field variation. J. Phys. Conf. Ser. **771**, 012036 (2016)
9. Vega-Jorquera, P., Lazzus, J.A., Tamblay, L., Palma-Chilla, L., Salfate, I., Pacheco, R.: Geomagnetic field variation during the total solar eclipse of 2 July 2019 in La Serena, Chile. Geomag. Aeron. **61**, 287–292 (2021)
10. Chernogor, L.F.: Geomagnetic effect of the solar eclipse of June 10, 2021. Kinematics Phys. Celestial Bodies **38**(1), 11–24 (2022)
11. Kim, J., Chang, H.: Statistical analysis of geomagnetic field variations during solar eclipses. Adv. Space Res. **61**(8), 2040–2049 (2018)
12. Chernogor, L.F., Holub, M.Y.: Geomagnetic effect of the solar eclipse of October 25, 2022, in Eurasia. Kinematics Phys. Celestial Bodies **40**(3), 117–137 (2024)
13. Baran, L.W., Ephishov, I.I., Shagimuratov, I.I., Ivanov, V.P., Lagovsky, A.F.: The response of the ionospheric total electron content to the solar eclipse on August 11, 1999. Adv. Space Res. **31**, 989–994 (2003)
14. Adekoya, B.J., Chukwuma, V.U., Reinisch, B.W.: Ionospheric vertical plasma drift and electron density response during total solar eclipses at equatorial/low latitude. J. Geophys. Res. **120**, 8066–8084 (2015)
15. Adekoya, B.J., Chukwuma, V.U.: Ionospheric F2 layer responses to total solar eclipses at low and mid-latitude. J. Atmos. Solar Terr. Phys. **138**, 136–160 (2016)
16. Yumoto, K.: The STEP 210° magnetic meridian network project. J. Geomagn. Geoelectr. **48**, 1297–1309 (1996)
17. Korte, M., Lühr, H., Förster, M., Haak, V.: Did the solar eclipse of August 11, 1999, show a geomagnetic effect? J. Geophys. Res. **106**(A9), 18563–18575 (2001)
18. Yamazaki, Y., Maute, A.: Sq and EEJ—a review on the daily variation of the geomagnetic field caused by ionospheric dynamo currents. Space Sci. Rev. **206**, 299–405 (2017)

19. Syafitri, N., et al.: Analysis of H-component of geomagnetic variation in Indonesian region. In: Proceeding of the 2022 International Conference on Computer, Control, Informatics and Its Applications the (IC3INA '22), pp. 341–344 (2023)
20. Woodroffe, J.R., Morley, S.K., Jordanova, V.K., Henderson, M.G., Cowee, M.M., Gjerloev, J.G.: The latitudinal variation of geoelectromagnetic disturbances during large ($Dst \leq -100$ nT) geomagnetic storms. Space Weather **14**, 668–681 (2016)

Literature Study of Satellite Payload Missions for Ionospheric Observations on the Next Generation BRIN Satellite

Rizal Suryana[1(✉)], Visca Wellyanita[1], Ferdiansjah[2], Agus Herawan[3], Patria Rachman Hakim[3], and Moh Andi Aris Biyantoro[1]

[1] Research Center for Space, National Research and Innovation Agency, Bandung, Indonesia
rizal.suryana@brin.go.id

[2] Department of Nuclear Engineering and Engineering Physics, Universitas Gajah Mada, Yogyakarta, Indonesia

[3] Research Center for Satellite Technology, National Research and Innovation Agency, Bogor, Indonesia

Abstract. Flares, coronal mass ejections (CME), and the solar wind, significantly influence geomagnetic and ionospheric activities, impacting space weather. The ionosphere is an important part of radio wave propagation. Instabilities or fluctuations in the ionosphere can lead to interference with high-frequency (HF) communication, absorption of radio waves, and disruptions in satellite communication. Additionally, changes in the ionosphere can significantly affect satellite navigation systems, particularly those relying on GPS and GNSS signals. Understanding ionospheric dynamics and space weather phenomena is crucial due to the significant impact of ionospheric changes on modern technology. Observational data from both ground-based remote sensing and in-situ observations are essential to gain insights into these phenomena. In-situ ionospheric observations employ diagnostic tools such as retarding potential analyzers, ion drift meters, and Langmuir probes for measuring ion density, ion temperature, ion velocity, ion composition, ion drift velocity, ion arrival angle, electron density, and electron temperature. These parameters can be used to understand the ionosphere's influence on radio wave propagation, communication systems, the thermal structure of the ionosphere, the distribution of electrons, interactions with other regions of space, and the impact on satellite orbits.

Keywords: Space Weather · Ionospheric Observations · Scientific Payload · Satellite Missions

1 Introduction

Space weather refers to the environment in interplanetary space, magnetosphere, ionosphere, and thermosphere, with the Sun as the main factor. It can affect the reliability and performance of ground-based and space-based technologies [1]. Solar activities

E. Fitrianingsih et al. (Eds.): ISAST 2024, SPPHY 416, pp. 256–267, 2025.
https://doi.org/10.1007/978-981-96-1344-1_28

such as flares, coronal mass ejections (CME), and solar wind are the main factors influencing the changes in space weather. These activities cause disturbances in geomagnetic and ionospheric conditions. The Sun continuously emits electromagnetic radiation, charged particles, and high-energy waves across the electromagnetic spectrum, including infrared, ultraviolet, visible light, and X-rays [2]. Space weather events can disrupt critical systems such as satellite operations, communications, navigation, power grids, and high-frequency (HF) radio communications. Given its significant potential impact, effectively disseminating space weather information and predictions is critical to minimizing risk to modern technology, which is essential for communications, navigation, energy, transportation, and various other of daily life.

The ionosphere is the upper part of the Earth's atmosphere that is influenced by solar activity, and it extends from 60 km to more than 1000 km above the Earth's surface. The solar radiation ionizes the atmosphere, generating free electrons and ions that can disrupt radio wave propagation, increase atmospheric drag, and affect navigation systems. Instabilities or fluctuations in the ionosphere can cause radio wave absorption, and disrupt satellite communication. Furthermore, changes in the ionosphere conditions also impact on global positioning systems (GPS) and global navigation satellite systems (GNSS) signals, resulting in measurement errors, signal attenuation, and potential navigation system failure.

For example, the UHF communication satellite (SATCOM) was disrupted in March 2002 by the equatorial plasma bubble, causing significant casualties among military personnel engaged in combat [3]. In early February 2022, 38 Starlink satellites have failed to reach their orbit due to geomagnetic storms [4] that increased particle density in the ionosphere and thermosphere, raising atmospheric drag. In January, 1994, the Intelsat K, Anik E1, and Anik E2 satellites lost attitude control due to surge in relativistic electron flux (greater than 2 meV), where the electron radiation belt triggered an internal electrostatic discharge (ESD) [5]. These incidents highlights the enormous impact of the ionosphere on modern technology and the importance of multi-point observations to understand the Magnetosphere-Ionosphere-Atmosphere (M-I-A) systems. A detailed understanding of ionospheric dynamics is essential for studying space weather as a certain subject in M-I-A systems [1]. The ionospheric bias is one of the main problems in positioning using GPS, where position accuracy levels can drop from centimeters to meters, and in the extreme conditions, errors can reach up to 150 m [6].

This paper will explain the satellite-based ionospheric observation missions that utilize the Ion Drift Meter (IDM), Retarding Potential Analyzer (RPA), and Langmuir Probe (LP). These sensors will measure key ionospheric parameters such as electron density (Ne) and ion density (Ni), electron temperature (Te), ion temperature (Ti), ion composition, ion energy, drift velocity, drift direction, and others. These parameters are invaluable for understanding ionospheric dynamics, the effect of electric and magnetic fields, and how those factors influence radio communication, GPS navigation, and satellite performance. In the context of space weather, these parameters are crucial for forecasting and mitigating disruptions caused by geomagnetic storms, ionospheric storms, or solar activity, all of which could potentially impact modern technology.

2 Observation Missions

Ionospheric observation from ground-based and space-based is crucial to understanding ionospheric dynamics. In-situ ionospheric measurements involve directly assessing the properties of ionospheric using instruments aboard satellites in low Earth orbit. These measurements have successfully collected ionospheric data across various satellite missions, including the Communications/Navigation Outage Forecasting System (C/NOFS), Ionospheric Connection Explorer (ICON), Challenging Minisatellite Payload (CHAMP), Swarm, Defense Meteorological Satellite Program (DMSP), Scintillation Observations and Response of the Ionosphere to Electrodynamics (SORTIE), HORYOU-IV, Ten-Koh, China Seismo-Electromagnetic Satellite (CSES), QB50, and Ionosphere Dynamics Explorer and Attitude Subsystem Satellite (IDEASSat/INSPIREsat-2). In-situ ionospheric observations are crucial for several reasons: the first, they enable the study of the dominant physical processes in the ionosphere [7]; the second, they provide real-time data on ionospheric conditions, helping to mitigate disruptions in these systems; and third, they allow for the direct measurement of ionospheric irregularities [8].

2.1 Scientific Objectives

The impact of ionosphere on modern technology has garnered significant attention from researchers worldwide, leading to increased use of satellite for ionospheric observation both now and in the future. Space-based ionospheric observations provide valuable data to address various scientific questions and practical challenges. The primary objective of satellite-based ionospheric observation missions is to investigate plasma phenomena that impact communication, navigation, ionospheric dynamics, and the geomagnetic environment. To achieve this goal, these missions will measure ionospheric parameters, as outlined in Table 1.

Table 1. Scientific objective of ionospheric observations.

Scientific objective	Ionospheric parameters
To understand ionospheric dynamics, plasma interactions, and satellite drag estimation	Ion density, ion temperature, ion velocity, and ion composition
To understand ionospheric and plasma phenomena that affect satellite operations, communication systems, and navigation accuracy	Ion drift velocity and ion arrival angle
To understand the space environment and study the structure, and dynamics of the ionosphere	Electron density and electron temperature

2.2 Objective Measurements

The charged particles in the ionosphere play a critical role in space weather, satellite communications, GPS navigation, and other technological systems. Understanding the ionospheric dynamics requires precise measurements of its physical properties. In-situ measurements provide the most direct and detailed data on ionospheric conditions, helping scientists unravel the complex behavior of ionized gases and their interactions with the Earth's magnetic field. The key measurement objectives for ionospheric observation using in-situ instruments highlight their importance for space weather research and the design and operation of spacecraft. To accomplish these goals, the satellite will carry the scientific instruments listed in Table 2.

Table 2. Objective measurement of satellite's payload.

Ionospheric parameters	Instrument
density, temperature, velocity, and composition of ions	RPA
drift velocity and arrival angle of ions	IDM
Electron density and electron temperature	LP

3 Ionospheric Sensors

The ionosphere contains charged particles electricity that depends on the changing magnetic and electric conditions in the space environment [9]. LEO satellites operate at altitudes between 250 km to 2000 km, with orbital periods ranging from 90–120 min, make them well-suited for ionospheric measurements. Ionospheric sensors are specialized instruments designed to measure various physical properties of the ionosphere. The measurements provide data that help scientists understand the dynamics, composition, and interactions of ionized particles in the ionosphere, particularly in the context of space weather, communication, and navigation systems. Here are some of the key ionospheric sensors:

3.1 Langmuir Probes

Irving Langmuir was a pioneer in plasma diagnostics. His work with electric probes, now famously known as Langmuir probes (LP), was pivotal in understanding and measuring fundamental plasma parameters. LP is a simple diagnostics tool to determine plasma parameters such as density electrons, temperature electrons, and electric potential [10]. This measurement method is suitable for space plasma, low-pressure electric discharges, and low-density cold plasma. The LP operates by inserting an electrically charged electrode into plasma and measuring the current-voltage (I-V) characteristics. The electrode of LP is applied to a sweeping voltage from a negative to positive voltage (-10 V to $+$

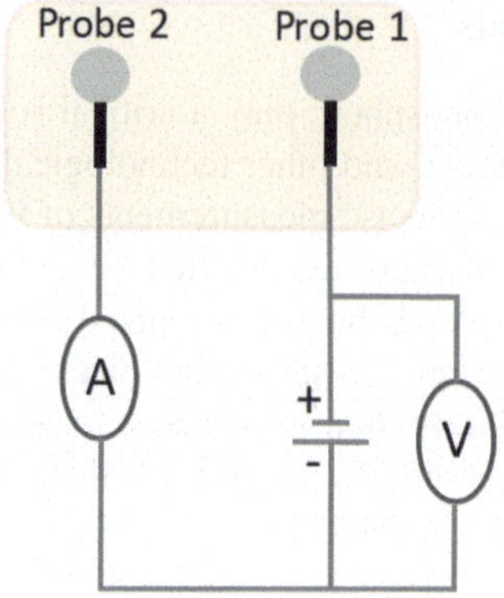

Fig. 1. A Basic schematic of Langmuir probe.

10 V), or only positive voltage, as shown in Fig. 1. The most commonly used geometries are spherical, cylindrical, and planar.

We assume that LP is supplied with a voltage ranging from -10 V to +10 V. Firstly, probe-1 is supplied with a voltage of -10 V, and probe-2 is supplied with a voltage of 0 V. The probe-1 will collect all the ions, and probe-2 will collect all the electrons. As a result, the ion from probe-1 and the electron from probe-2 will be attracted to each other, causing the current flow from probe-1 to probe-2 [11]. This condition is known as the ion saturation current as illustrated in the I-V curve shown in Fig. 2. The voltage on probe-1 will gradually change until 0 V. At this point, both probes will collect zero net current because there is no interaction between electrons and ions. Once the voltage on probe-1 equals 0 V, it will be increased incrementally until the maximum value (+10 V). In this condition, probe-1 will collect the entire electrons, and probe-2 will collect the entire ions. This there creates an interaction between the electron on probe-1 and the ion on probe-2, causing the current to flow from probe-1 to probe-2. This condition is known as the electron saturation current, as shown in Fig. 2.

In Region 1, the current remains nearly constant as the applied voltage becomes more negative. In Region 2, the probe collects fewer ions and starts to collect more electrons, resulting in an exponential increase in the electron current to increase as the voltage approaches zero. This condition is known as the retarding electron region. In Region 3, the probe collects entire electrons, and the region is referred to as the electron saturation current region. Initially, the current increases rapidly but eventually levels off as the probe reaches this saturation region.

Two essential parameters in the ionosphere are Ne and Te, which are critical for understanding ionosphere physics in scientific research [12]. The contamination and interference generated by the onboard satellite must always be a concern due to the limitation of the LP and the complex space plasma environment [12–14]. The current between the two probes is determined by the voltage difference, leading the probe current to reach the saturation region. The Te is calculated using Eq. (1).

$$T_e(eV) = \frac{I}{2\left(\frac{dI_+}{dV_p}\right)_{Vp=0} - \left(\frac{dI_+}{dV_p}\right)_{Vp=sat}} \quad (1)$$

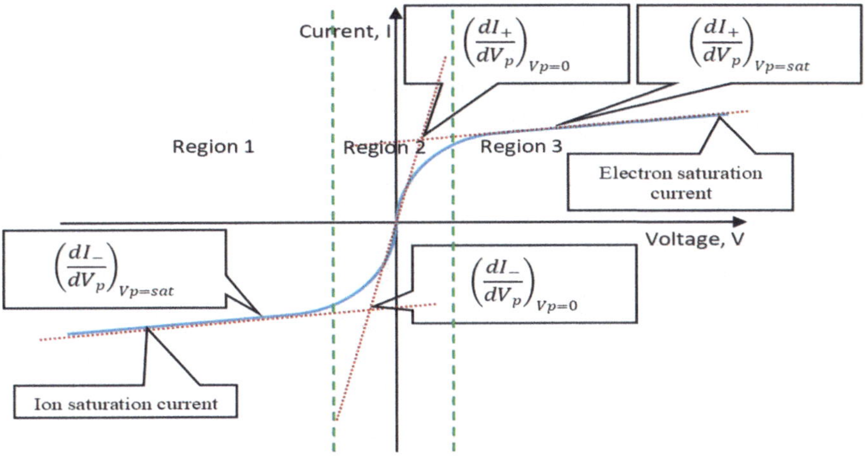

Fig. 2. The I-V characteristic curve of double Langmuir probe.

where $\left(\frac{dI_+}{dV_p}\right)_{Vp=0}$ and $\left(\frac{dI_+}{dV_p}\right)_{Vp=sat}$ are currents measured on two different points on the I-V curve, I is the slope of the natural logarithm of the electron current. When the electron temperature is known, electron density can be calculated using Eq. (2).

$$I = 0.61 N_e e A \sqrt{\frac{kTe}{m_i}} \tag{2}$$

where k is the Boltzmann constant, e is the electron charge, A is the surface probe area, and mi is the ion mass. The factor of 0.6 account for the reduction in ion density in the pre-sheath, which is the region where the ions are accelerated to the Bohm speed [10, 15]. Now, Eq. (2) can be modified as Eq. (3).

$$N_e = \frac{I}{0.61 Ae\sqrt{\frac{kT_e}{m_i}}} \tag{3}$$

In-situ measurement of plasma parameters using the LP offers several advantages, such as simplicity, local measurements, and cost-effectiveness. However, the LP has several disadvantages, including plasma perturbation, limited effectiveness primarily in low-density plasma, surface contamination, and sensitivity to magnetic fields [16].

3.2 Retarding Potential Analyzer

The RPA is an instrument used to measure the energy-charged particle (electron or ion) distribution function in a plasma and to calculate charged particle densities and temperatures [17]. It is used in space research to study spacecraft charging in a space environment, ionosphere, and other planetary atmospheres [18]. The RPA consists of several grids with progressively increasing retarding potentials, which are used to filter ions based on their energy. As ions pass through the grids, only those with sufficient energy can overcome the retarding potentials and reach the collector electrode [8]. The collector electrode measures the ion current, which is related to its energy distribution [19]. The measured current will produce a current vs voltage graph, as illustrated in Fig. 4. From this graph, we can extract information to determine the energy distribution of the ions.

The RPA works by using an electric field to selectively filter particles based on their kinetic energy. The current is measured as a function of the applied potential from the collector electrode. RPA consists of five electrodes in series: a floating electrode, two electron-repelling electrodes, an ion-retarding electrode, and a collector electrode, as shown in Fig. 3.

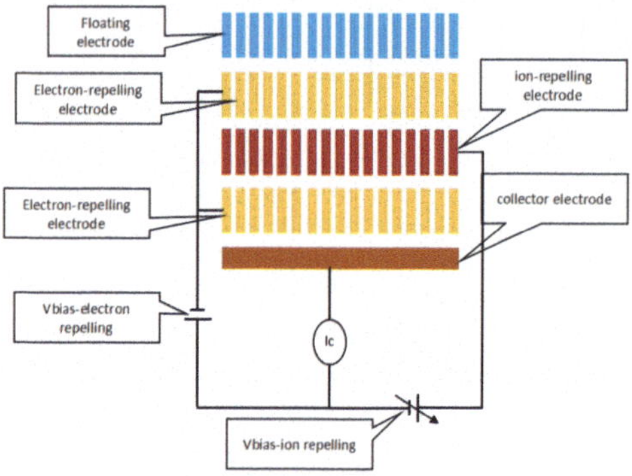

Fig. 3. The basic schematic of the retarding potential analyzer.

Each electrode serves a specific function in filtering and analyzing the energy of the incoming particles, and it plays a specific role in controlling the particles that enter the detector. The floating electrode is designed to allow charged particles from the external environment to enter the RPA, and it is often floating or slightly positive to minimize the impact of external electric fields [20]. The electron-repelling electrode prevents secondary electrons from reaching the collector and distorting the measurements. This electrode is biased negatively relative to the others to repel low-energy electrons that may have been produced by ion impacts or other processes, and ensure that only the primary charged particles are analyzed [21]. The ion-repelling electrode is the main component of the RPA, and its bias voltage determines which particles are allowed to enter the collector. The bias voltage is swept from 0 V to 15 V, allowing only high-energy particles to pass through to the collector electrode. The second electron-repelling electrode helps ensure that only high-energy particles pass through to the collector. As the retarding potential (voltage) increases, so the probe will repel positive ions that have low kinetic energy or are neutralized, and only ions with high kinetic energy can reach the collector [11]. The collector electrode measures the current of the particles that have successfully passed through all the grids, and the measured current produces a current vs. voltage curve, as shown in Fig. 4.

The I-V curve is used to determine the energy distribution of the particles, the ion energies, temperatures, and flow velocities. Knudsen gives a set of equations to explain

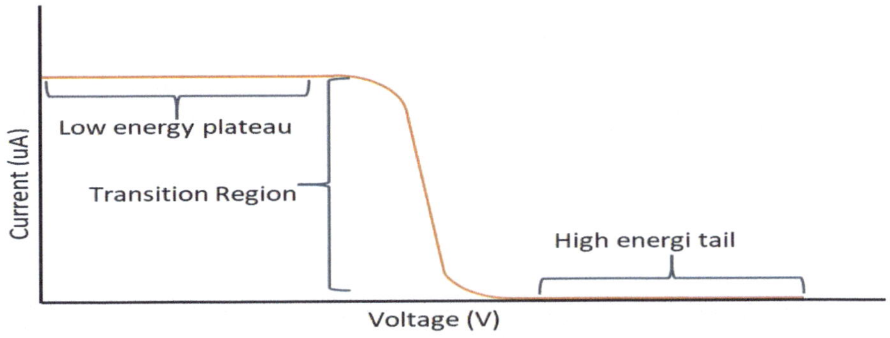

Fig. 4. The I-V curve of the retarding potential analyzer.

the I-V curve [22], as shown in Eq. (4).

$$I = T_r A e v_i \cos\theta \sum_{i=1}^{j} n_i \left(\frac{1}{2} + \frac{1}{2} \mathrm{erf}(k_i) + \frac{\exp(k_i^2)}{2a_i\sqrt{\pi}} \right) \tag{4}$$

where I is the ion current measured, Tr, A, e, and v are the fractional transparency of the grid stack, the aperture area, the charge of an ion, and the magnitude of the relative velocity, respectively. θ is an angle between the RPA aperture and the spacecraft velocity vector. I can be determined from the saturation current that corresponds to the current if the retarding potential is not significantly affecting the ion flux, as shown in Eq. (5).

$$I_s = n_i e A v_i \tag{5}$$

Vi is ion velocity that relates to ion temperature.

$$v_i = \sqrt{\frac{2k_B T_i}{m_i}} \tag{6}$$

where kB is the Boltzmann constant (1.38 × 10–23 J/K), ion density is calculated using Eq. (7).

$$n_i = \frac{I_s}{eAv_i} \tag{7}$$

The ion temperature is related to the shape of the I-V curve, and it is calculated using Eq. (8).

$$T_i = \frac{e}{k_B}\left(\frac{dV}{d(\ln I)}\right) \tag{8}$$

The slope $\frac{dV}{d(\ln I)}$ is determined from the linear portion of the retarding region. The RPA offers several advantages for measuring ionospheric and space plasma, including in-situ measurement of energy distribution, versatility, high sensitivity and compact design. The energy distribution of ions and electrons is crucial for understanding plasma dynamics, and it can be used to measure multiple parameters [17]. However, the RPA also has several disadvantages for in-situ measuring of ionospheric and space plasma, including complex data interpretation, limited to low-density plasma, and plasma perturbation. Extracting ionospheric or plasma parameters from the I-V curve can be complex and requires interpretation. Additionally, the RPA may collect excessive current in high-density environments [19].

3.3 Ion Drift Meter

An Ion Drift Meter (IDM) is a diagnostic tool used to measure the drift velocity of ions in the ionosphere, aiding in the study of the dynamics of the ionosphere and space weather phenomena. Its functions measure the components of ion velocity perpendicular to the spacecraft's motion, allowing scientists to infer the electric fields driving these ion drifts, which can be used to study ionospheric and magnetosphere physics. The RPA has limitations in characterizing ions drift vectors, so the IDM is necessary to measure the drift velocity of ions and ion arrival angle. By analyzing ion drift meter data, we can estimate the magnitude and the local electric field [23]. The IDM is similar to the RPA, consisting of an aperture, multiple bias grids, and a plate collector [24], but the IDM has the addition of a diaphragm and segmented collector electrodes. The IDM captures ions from the surrounding plasma through an aperture, which ensures that the incoming ions from a collimated beam, where the ions travel in nearly parallel trajectories. After passing through the aperture, the ion beam encounters a stack of grids. Each grid in the stack is either biased (set to a particular voltage) or grounded to control the movement of the ions. By applying different potentials, the IDM can filter out ions with certain energies or velocities, ensuring that only the desired ion population reaches the collector [23], as shown in Fig. 5.

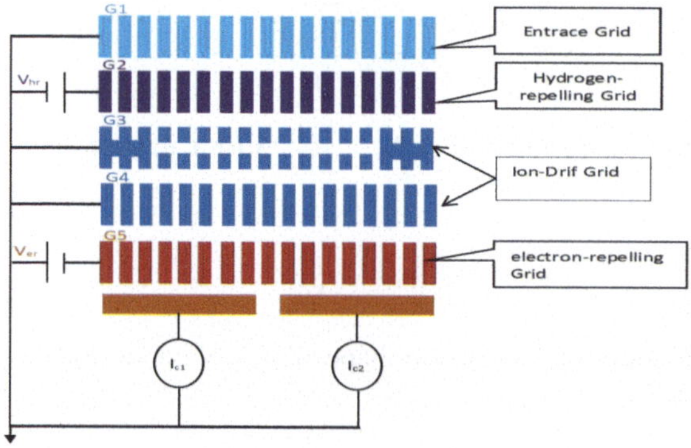

Fig. 5. The basic schematic of the ion drift meter.

G1 or entrance grid allows the ion from the surrounding plasma into the IDM sensors, and this grid ensures only ions traveling in specific directions enter the IDM. G2 or hydrogen-repelling electrode functions to filter low-mass ions like hydrogen. This is a critical function that ensures the IDM primarily measures the heavier ions by preventing lighter ions from reaching the collector. G3/G4 or ion drift electrode is designed to manipulate the ion trajectories, ensuring they are directed correctly toward the collector electrodes and narrowing their paths to hit specific segments of the collector. G5 or electron-repelling electrode functions as a stabilizer of the ion beam just before collection. Collector electrodes are segmented and positioned to collections that pass through all the previous grids. The distribution of the collected ions across the segments is used for determining both the drift velocity ions and the direction of the ion flow relative to the spacecraft.

Collectors are connected to individual ammeters, allowing the incoming ion arrival angles to be determined from the differences in ion current (I1 and I2) [11]. The ratio of the currents from two opposing segments of the collector grid is then calculated. This ratio is essential for determining the arrival angle. Mathematically, the ratio (R) can be expressed in Eq. (9).

$$R = \frac{I_1}{I_2} = \frac{\frac{W}{2} + D\tan\alpha}{\frac{W}{2} - D\tan\alpha} \tag{9}$$

where W is the length of the aperture, and D is the depth of the collector (from the aperture to the collector). From the current ratio, we can determine the tangent of the α.

$$\tan(\alpha) = \frac{I_{1-I_2}}{I_1 + I_2} \tag{10}$$

To determine the arrival angle (α), we can be inverted the tangent (arc-tan) of the Eq. (11).

$$\alpha = \arctan\left(\frac{I_1 - I_2}{I_1 + I_2}\right) \tag{11}$$

The IDM offers several advantages in the context of in-situ observations, including direct measurement of ion drift, high temporal resolution, global coverage, and detection of localized phenomena. The IDM is directly immersed in the plasma environment, it can allow for fine-scale temporal resolution, but it has to capture with high frequency. This is very useful to detect the occurrence of scintillation in the ionospheric structures [25]. Furthermore, it has disadvantages such as limited spatial coverage, dependence on the orbital path, complex calibration, and influence of satellite motion. The motion of the satellite can introduce noise in data extraction because the relative velocity of the satellite to the plasma must be taken into account, and sometimes the instrument receives signals from other sources.

4 Conclusion

Langmuir probes, retarding potential analyzers, and ion drift meters are simple diagnostic tools used to measure fundamental plasma parameters in the ionosphere to study the ionospheric dynamics and space weather phenomena. The power consumption of these sensors is relatively low, and their are small physical size, with dimensions approximately 10cm x 10cm x 10cm, makes them highly suitable to use as secondary payload on satellites. In addition, these sensors are passive, so they will be no interference with other sensors or the main satellite payload. To cover all areas of Indonesia and conduct long-term observations, the author suggests installing these sensors on constellation satellites. Electron density, electron temperature, ion density, ion temperature, ion velocity, and ion arrival angle are the fundamental plasma parameters that are crucial for understanding various space weather phenomena. When in-situ observations are combined with ground-based observations like GPS, ionosonde, all-sky imagers and others, these data can significantly contribute to our understanding of ionospheric processes and their impact on modern technology.

References

1. Cander, L.R.: Ionospheric Space Weather. Springer, Cham (2019)
2. Yifan, L., Qi, S., Honghao, Y., Fei, Y.: A review of the space environment effects on spacecraft in different orbits. IEEE Access **7**, 93473–93488 (2019)
3. Kelly, M.A., Comberiate, J.M., Miller, E.S., Paxton, L.J.: Progress toward forecasting of space weather effects on UHF SATCOM after operation Anaconda. Space Weather **12**, 601–611 (2014)
4. Tsyganenko, N.A., Andreeva, V.A., Sitnov, M.I., Stephens, G.K.: Magnetosphere distortions during the "satellite killer" storm of February 3–4, 2022, as derived from a hybrid empirical model and archived data mining. J. Geophys. Res. Space Phys. **127**, 1–18 (2022)
5. Nizam, A.: Analysis of potential anomaly for low orbiting satellites in solar cycle 23. Jurnal Sains Dirgantara **9**(2), 90–105 (2012)
6. Ririn, W.R.: Effect of ionospheric bias correction on single frequency GPS observation coordinate results using the Klobuchar model. Sepuluh Nopember Institute of Technology, Surabaya, Indonesia (2018)
7. Francisco, C., Henriques, R., Barbosa, S.: A review on CubeSat missions for ionospheric science. Aerospace **10**(7), 2–23 (2023)
8. Aol, S., Buchert, S., Jurua, E.: Ionospheric irregularities and scintillations: a direct comparison of in situ density observations with ground-based L-band receivers. Earth Planets Space **72**(164), 2–15 (2020)
9. Suneetha, E., Ratnam, D.V., Sivakrishna, K.: Review on space based ionospheric probes and payloads for monitoring ionospheric weather. In: 3rd International Conference on Advances in Computing, Communication Control and Networking (ICAC3N), Greater Noida, India, pp. 1844–1850 (2021)
10. Suryana, R., Tejumola, T.W., Kim, S., Cho, M.: Design on-board data processing for double Langmuir probe on lean satellite. In: 3rd International Conference on Research and Learning of Physics (ICRLP) on Journal of Physics: Conference Series, Padang, Indonesia, p. 012025. IOP (2021)
11. Lin, Z.W., et al.: Advanced ionospheric probe scientific mission onboard FORMOSAT-5 satellite. Terr. Atmos. Ocean. Sci. **28**(2), 99–110 (2017)
12. Yan, R., et al.: The regular features recorded by the Langmuir probe onboard the low earth polar orbit satellite CSES. J. Geophys. Res. Space Phys. **127**(1), 1–17 (2022)
13. Johansson, F.L., Eriksson, A.I., Gilet, N., Henri, P., Wattieaux, G., et al.: A charging model for the Rosetta spacecraft. Astron. Astrophys. **642**(A43), 1–12 (2020)

14. Wang, X., Samaniego, J.I., Hsu, H.W., Hornyi, M., Wahlund, J.E., et al.: Development of a double hemispherical probe for improved space plasma measurements. J. Geophys. Res. Space Phys. **123**, 2916–2925 (2018)
15. Merlino, R.L.: Understanding Langmuir probe current-voltage characteristics. Am. J. Phys. **75**(12), 1078–1085 (2007)
16. Li, P., Hershkowitz, N., Severn, G.: Building Langmuir probes and emissive probes for plasma potential measurements in low pressure, low temperature plasmas. J. Vis. Exp. **171**, 1–23 (2021)
17. Shu, T.L., Catherine, M.: Retarding potential analyzer: principles, designs, and space applications. AIP Adv. **10**(9), 095324 (2020)
18. Lai, S.T.: Fundamentals of Spacecraft Charging: Spacecraft Interactions with Space Plasmas. Princeton University Press, Princeton (2012)
19. Fanelli, L., Noel, S., Earle, G., Fish, C., Davidson, R., et al.: A versatile retarding potential analyzer for nano-satellite platforms. Rev. Sci. Instrum. **86**, 124501 (2015)
20. Hutchinson, I.H.: Principles of Plasma Diagnostics. Cambridge University Press, Cambridge (2002)
21. Miller, K.L., Knoebel, H.W.: Retarding potential analyzer for low-energy plasma measurements. Rev. Sci. Instrum. **41**(7), 1016–1020 (1970)
22. Knudsen, W.C.: Evaluation and demonstration of the use of retarding potential analyzers for measuring several ionospheric quantities. J. Geophys. Res. **71**, 4669–4678 (1966)
23. Luis, F.V.G., Javier, I.R., Hyeonseok, K.: Review of in-space plasma diagnostics for studying the Earth's ionosphere. J. Phys. D Appl. Phys. **55**(26), 263001 (2022)
24. Davidson, R.L., et al.: The gridded retarding ion drift sensor for the petitSat cubeSat mission. Rev. Sci. Instrum. **91**(6), 064502 (2020)
25. Kelley, M.C.: The Earth's Ionosphere: Plasma Physics and Electrodynamics, 2nd edn. Academic Press. Cornell University, Ithaca (2009)

Maximum Tides in West Coast of Sumatera Affected by Astronomical Constituents

Thomas Djamaluddin[1(✉)], Ibnu Nurul Huda[2,3], Edwards Taufiqurrahman[4], Lesi Mareta[5], Nailur Rahmi[6], Hendri[7], and Nanang Widodo[1]

[1] Research Center for Space, National Research and Innovation Agency (BRIN), Bandung, Indonesia
thom001@brin.go.id

[2] School of Astronomy and Space Science, Key Laboratory of Modern Astronomy and Astrophysics, Ministry of Education, Nanjing University, Nanjing, People's Republic of China

[3] Research Center for Computing, National Research and Innovation Agency (BRIN), Bogor, Indonesia

[4] Research Center for Oceanography, National Research and Innovation Agency (BRIN), Jakarta, Indonesia

[5] Institut Teknologi Sumatera (ITERA), Lampung, Indonesia

[6] Faculty of Sharia, State Islamic University (UIN) Mahmud Yunus, Batusangkar, West Sumatra, Indonesia

[7] Faculty of Sharia, State Islamic University (UIN) Sjech M Djamil Djambek, Bukittinggi, West Sumatra, Indonesia

Abstract. Maximum tides in coastal areas need to be a concern because they often cause tidal flooding. Analysis of maximum tides on the west coast of Sumatra uses long-term data from Sabang 13 years (2005–2018), Sibolga 28 years (1989–2004, 2005–2018), and Padang 25 years (1986–1998, 2005–2018). The analysis of maximum tides is associated with the analysis of the lunar and solar astronomical parameters that influence them. In general, the daily tide pattern has a semi-diurnal pattern which is related to the influence of the lunar phases which is clearly indicated by the monthly variation pattern. Maximum tides reach their peak during the new and full moon which is strengthened by lunar perigee conditions. Spectral analysis shows a semi-annual dominant period which is influenced by variations in solar declination. This is also shown by the annual variation of maximum tides which peak around June when the sun is most north (maximum declination 23.5°) and around December when the sun is furthest south (minimum declination −23.5°). There is no indication of the 18.6 years variation due to the lunar nodal cycle. Maximum tides are dominantly influenced by the moon phase, the nearest distance between the moon and the earth (perigee), and the sun's declination, which can be an early warning of potential flooding. The tidal flood on Padang coast on June 6 2024 is related to the new moon (June 6), perigee (June 2), and the northernmost sun (June).

Keywords: Maximum Tides · Astronomical Constituents · West Coast of Sumatera

E. Fitrianingsih et al. (Eds.): ISAST 2024, SPPHY 416, pp. 268–275, 2025.
https://doi.org/10.1007/978-981-96-1344-1_29

1 Introduction

Coastal areas are vulnerable to tidal flooding. Although tides are a daily phenomenon in coastal areas, maximum tides need to be watched out because they often cause tidal flooding. In extreme conditions, tidal flooding can cause damage to residential areas or other facilities. The results of a NASA study [1] obtained by the Sea Level Change Science Team and Thompson et al. [2] show the potential of tidal flooding on the coasts of the United States due to the nodal cycle effect. This constituent in the moon's orbit adds to the effects of rising sea levels due to global warming.

The results of a study of maximum tides on the north coast of Java with samples from Jakarta and Surabaya [3] show that maximum tides occur due to the influence of the position of the moon and the distance of the moon closest to the earth. In Jakarta, maximum tides occur during the northernmost or southernmost declination of the moon. Meanwhile, in Surabaya, maximum tides occur during the new moon or full moon. The solar constituent is seen in semi-annual variations. Maximum tides occur when the sun is most north (around June) and when the sun is most south (around December).

The results of a study of maximum tides on the south coast of Java [4] show that the influence of the moon is shown in the perigee and moon phase parameters. Maximum tides occur during the new and full moon phases. The influence of the sun can be seen from the maximum tides that occur when the sun is most south (annual period). In both [3] and [4] studies, there was no indication of the influence of the 18.6-year nodal cycle as reported by NASA [1] and Thompson et al. [2].

This study analyzes maximum tides on the west coast of Sumatra and compares them with astronomical parameters from the position of the moon and sun. The aim is to provide early warning of potential tidal flooding due to astronomical constituents. It needs to be explained here, coastal flooding is not only caused by tidal flooding, but also caused by meteorological flooding (high waves caused by strong winds and extreme weather on the coast). This study only limits on coastal flooding due to tides related to astronomical constituents.

2 Data and Method

Tide data were obtained from the University of Hawaii (UHSLC) tide gauge data-base https://uhslc.soest.hawaii.edu/data/?rq [5]. There are three data set on the coast of Sumatra, namely Sabang 13 years (2005–2018), Sibolga 28 years (1989–2004, 2005–2018) and Padang 25 years (1986–1998, 2005–2018). Astronomical parameters of the position of the moon and sun are obtained from NASA https://ssd.jpl.nasa.gov/horizons/app.html#/. Spectral analysis of tide data using the Lomb-Scargle method [6]. Tidal data and astronomical parameters were analyzed using Excel to obtain the correlation. To compare maximum tides with astronomical parameters, all values have been normalized.

3 Results

The general pattern of maximum tides is known, but the characteristics can vary on each coast. Understanding the characteristics of maximum tides on a coast can be used to anticipate potential tidal flooding. According to Parker [7], the main constituents that influences maximum tides is the differential gravitation of the moon and the sun. Monthly variations are generally influenced by the alignment of the moon and sun, namely when the new moon and full moon are. Tides will reach their maximum height when new and full moon conditions. Those are strengthened by lunar perigee conditions. The conditions of variations in the lunar nodal cycle also influence the long-term variations of 18.6 years.

Maximum tides reach their peak during the new and full moon which is strengthened by perigee conditions. Spectral analysis shows a dominant semi-annual period influenced by variations of solar declination (Fig. 1). Spectral analysis using the Lomb-Scargle method show a dominant 183 days period, half of a year or one period of solar declination variation.

To analyzed more detailed comparison, all value of maximum tides and astronomical parameters have been normalized. It is also shown by the semi-annual variation of maximum tides which reach a peak around June when the sun is most north (maximum declination 23.5°) and around December when the sun is most south (minimum declination −23.5°) (see lower left panel in Fig. 2, 3 and 4). There is no indication of 18.6 years variations due to lunar nodal cycle. In the lower left panel of Fig. 2, 3 and 4, it is shown the maximum tides (upper graphs) in line with variations in the absolute value of solar declination (lower graphs).

The semi-diurnal daily tide pattern (Fig. 2, 3 and 4, upper left) is also related to the variation pattern influenced by lunar phases (Fig. 2, 3 and 4, upper right), namely maximum during the new and full moon [8]. In the upper-right panel, it is shown that maximum tide variations (upper blue graphs) related to the moon phase pattern (lower grey graphs), i.e. maximum at new and full moon, and strengthened at moon's perigee (lower brown graphs). Maximum tide variations are not related to variations of declination of the northernmost or southernmost moon (Fig. 2, 3 and 4, lower-right panels) as is the case on the coast of Jakarta [3]. In the lower right panels, it is shown that maximum tides (upper blue graphs) are not in line with the variation of lunar declination (lower yellow graphs).

The semi-annual constituents are generally known for maximum tidal variations [8]. This causes the maximum tide to increase around June and December. Maximum tides are also influenced by the moon phase, the nearest distance between the moon and earth (perigee), and the sun's declination. It can be an early warning of potential tidal flooding. The highest maximum tides occur during June and December, strengthen by condition new Moon or full Moon and lunar perigee.

As an example, the tidal flood that occurred at Pasia Jambak Coast, Padang on June 6 2024 [9]. At this time the moon is in a new moon condition [10] and close to the event of perigee [11]. The high coastal flood occurred during maximum phase on June and related to the new Moon on 6 June 2024 and moon perigee on 2 June 2024. During the event there was no bad weather condition, so that the coastal flooding was caused by tidal flooding.

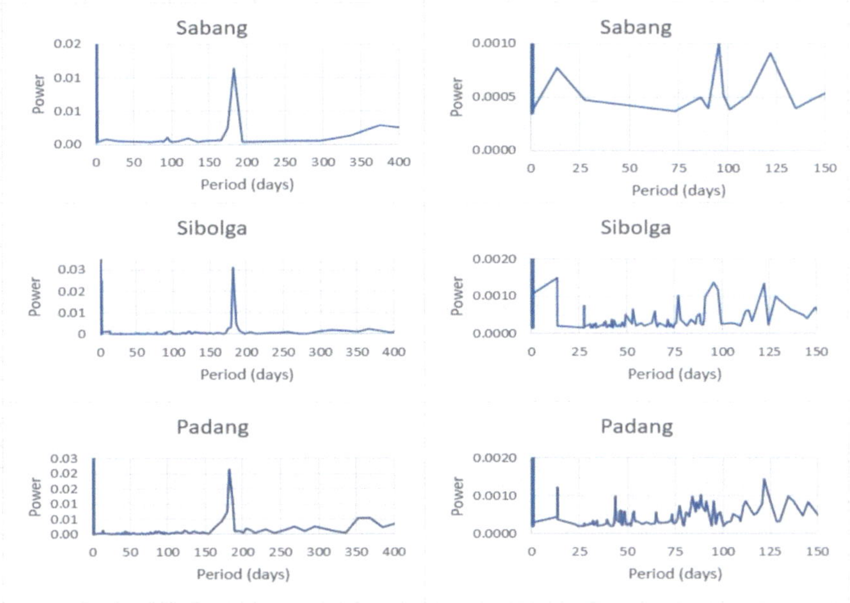

Fig. 1. Power spectral analysis of tides in Sabang, Sibolga and Padang. The graph shows the period (in days) and Lomb-Scargle Power. The main detected period is semi-annual period, around 183 days (half a year).

4 Conclusions

Spectral analysis of tides on the west coast of Sumatera and its empirical correlation with astronomical parameters shows that the maximum tides are related to the position of the moon and sun. The semi-diurnal tidal cycle on the west coast of Sumatra is related to lunar phases influence. During the new and full moons, maximum tides reach their highest every month. It is strengthened by perigee. The semi-annual period of the spectral analysis is also apparent in the annual variations. Maximum tides reach their highest conditions around June (when the sun is most-north) and around December (when the sun is most-south). As an example, the tidal flood event on the Padang coast on June 6, 2024 was related to a combination of the new moon which was strengthened by the perigee position and the peak of semi-annual variations in June.

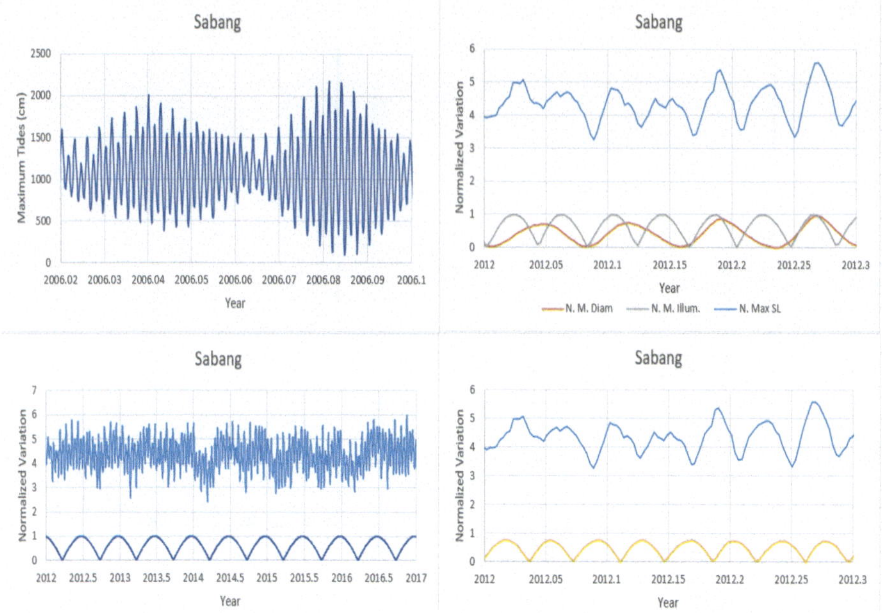

Fig. 2. Semi-diurnal variations of tides on Sabang coast (upper-left). Semi-annual variation of maximum tides is related to the absolute value of solar declination (lower-left). Maximum tides variations are related to lunar phase (grey line) and lunar perigee (brown line) (upper-right), not related to lunar declination (lower-right).

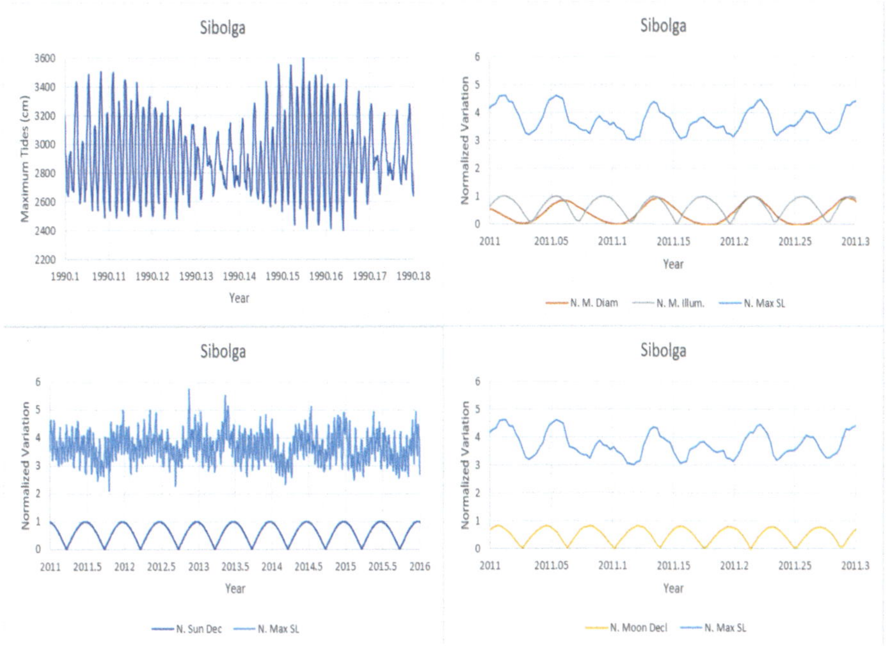

Fig. 3. Semi-diurnal variations of tides on Sibolga coast (upper-left). Semi-annual variation of maximum tides is related to the position of the sun (lower-left). Maximum tides variations are related to lunar phase (grey line) and lunar perigee (brown line) (upper-right), not related to lunar declination (lower-right).

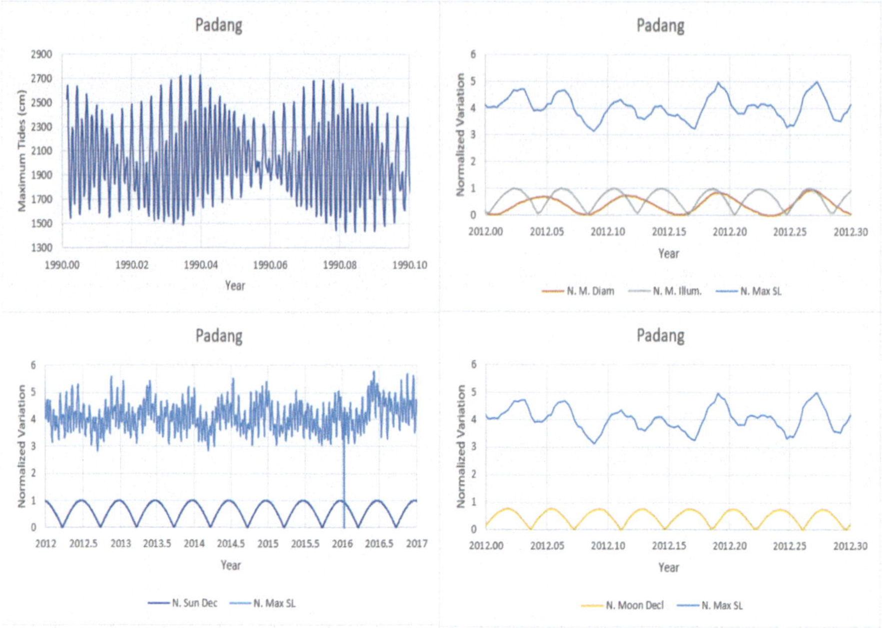

Fig. 4. Semi-diurnal variations of tides on Padang coast (upper-left). Semi-annual variation of maximum tides is related to the position of the sun (lower-left). Maximum tides variations are related to lunar phase (grey line) and lunar perigee (brown line) (upper-right), not related to lunar declination (lower-right).

References

1. NASA: Study Projects a Surge in Coastal Flooding, Starting in 2030s (2021). https://www.nasa.gov/feature/jpl/study-projects-a-surge-in-coastal-flooding-starting-in-2030s
2. Thompson, P.R., Widlansky, M.J., Hamlington, B.D., et al.: Rapid increases and extreme months in projections of United States high-tide flooding. Nat. Clim. Change **11**, 584–590 (2021). https://doi.org/10.1038/s41558-021-01077-8
3. Djamaluddin, T., et al.: Astronomical tidal constituents potentially trigger coastal floods on the north coast of Java. In: AIP Conference Proceedings, vol. 2941, p. 040004 (2023). https://doi.org/10.1063/5.0181475
4. Djamaluddinl, T., et al.: Differences in maximum tides responses of the North Coast and the South Coast of Java on the effects of the position of the moon and the sun. In: Journal of Physics: Conference Series, vol. 2773, p. 012010 (2024). https://doi.org/10.1088/1742-6596/2773/1/012010
5. Caldwell, P.C., Merrifield, M.A., Thompson, P.R.: Sea level measured by tide gauges from global oceans as part of the Joint Archive for Sea Level (JASL) since 1846. NOAA National Centers for Environmental Information (2001). https://uhslc.soest.hawaii.edu/data/?rq. Accessed 01 Mar 2023
6. VanderPlas, J.T.: Understanding the lomb–scargle periodogram. Astrophys. J. Suppl. Ser. **236**, 16 (28pp) (2018)
7. Parker, B.B.: Tidal analysis and prediction NOAA special publication NOS CO-OPS 3 (2007). https://tidesandcurrents.noaa.gov/publications/Tidal_Analysis_and_Predictions.pdf

8. Pugh, D., Woodworth, P.: Sea-Level Science: Understanding Tides, Surges, Tsunamis and Mean Sea-Level Changes. Cambridge University Press, Cambridge (2014)
9. Gelombang Pasang di Pantai Pasia Jambak Padang (2024). https://foto.bisnis.com/view/20240606/1771714/gelombang-pasang-di-pantai-pasia-jambak-padang#&gid=1&pid=1
10. Moon Phases (2024). https://www.timeanddate.com/moon/phases/
11. Moon Distances: Closest Approach (Perigee) (2024). https://www.timeanddate.com/astronomy/moon/distance.html

The Positioning Error of GNSS Receiver Associated with the Presence of an Equatorial Plasma Bubble: A Case Study over Biak

Sefria Anggarani[1(✉)], Asnawi[1], Adi Purwono[1], Agri Faturahman[1], and Rezy Pradipta[2]

[1] Research Center for Space, Research Organization for Aeronautics and Space, National Research and Innovation Agency (BRIN), Bandung 40135, Indonesia
sefria.anggarani@brin.go.id

[2] Institute for Scientific Research, Boston College, Chestnut Hill, MA 02467, USA

Abstract. Equatorial plasma bubbles (EPB), specifically anomalies in the ionosphere, have a substantial effect on the accuracy of Global Navigation Satellite System (GNSS) positioning. This study aims to examine the GNSS receiver signal's inaccuracy in the presence of EPB over Biak (1.02° S, 136° E, 9.11° dip lat.), Indonesia. We use GNSS data obtained from the Indonesian Continuously Operating Reference System (INACORS) to generate maps of the rate-of-total electron content (ROTI). These maps depict the occurrence of equatorial plasma bubbles (EPB). Likewise, we employ single-point positioning for a fixed station at Biak station, which is a part of INACORS, to determine the magnitude of positioning errors. Our study revealed that the GNSS signal exhibited a maximum positioning error of nearly 10 m with respect to the EPB, predominantly in the north-south direction.

Keywords: Equatorial Plasma Bubble · Positioning Error · GNSS

1 Introduction

Irregularities in the plasma density ionosphere can cause the Global Navigation Satellite System (GNSS) signal to fluctuate from satellites to receivers as it passes through the ionospheric layer. These irregularities can disrupt the carrier-to-noise ratio, cause cycle slips, or result in a loss of lock signal. It may contribute to accuracy and reliability [1–3]. Nowadays, many modern applications rely on GNSS signals, such as drones and aviation [4,5], precise agriculture [6,7] and autonomous vehicles [8,9]. However, these signals can be affected by variations due to space weather events and ionospheric dynamics.

In high latitude regions, ionosphere are greatly influenced by geomagnetic storm, where the ionospheric velocity increases, giving rise to irregularities [10,11]. Meanwhile, in equatorial and low latitude regions are more affected by

E. Fitrianingsih et al. (Eds.): ISAST 2024, SPPHY 416, pp. 276–282, 2025.
https://doi.org/10.1007/978-981-96-1344-1_30

equatorial plasma bubbles (EPB), which typically occur during specific seasons [12].

Researchers frequently conduct scintillation investigations, with an emphasis on indices [3,13,14]. However, due to ionospheric irregularities, there is a lack of widely reported evaluation of GNSS performance in terms of accuracy, particularly in the Indonesian region. Therefore, this research aims to provide evidence of the extent of error positioning that occurs during an EPB event. The scope of our research is located in Biak (1.02° S, 136° E, 9.11° dip lat.), a highly challenging environment within the Indonesian Equatorial Ionization Anomaly (EIA) region, with a data set of 20 April 2023, during a hybrid solar eclipse that occurred at that site. We investigate the prominent ionospheric irregularities in this region due to the positioning deviation that occurred at the reference point.

2 Data and Method

To determine EPB occurences, data from 181 GNSS receiver stations within the Indonesian Continuously Operating Reference Station (INACORS) network on 20 April 2023 were used for calculating the rate of TEC index (ROTI). The calculation of ROTI requires first determining the rate of TEC at every epoch (ROT) (1) and then computing the standard deviation of that rate over a 5-minute window throughout the calculation (2) [15]. This calculation generates ROTI values at 350 km from the ionospheric piercing point (IPP).

$$ROT = \frac{TEC_{t_2} - TEC_{t_1}}{t_2 - t_1} = \frac{\Delta TEC}{\Delta t} \tag{1}$$

$$ROTI = \sqrt{\langle ROT^2 \rangle - \langle ROT \rangle^2} \tag{2}$$

Furthermore, we utilize GNSS data from the Biak station, which is a component of INACORS, to ascertain positional errors associated with the EPB event using single-point positioning for one fixed station. The open-source software package RTKLIB, available at https://www.rtklib.com, is utilized for satellite navigation system positioning. The RTKLIB output provides the precise geographic coordinates of the received GNSS satellite signal, along with the time series of variations in the north, south, west, east, and vertical directions. Next, we compute the positioning error for each epoch in relation to the reference point. We determine the reference point by averaging position over a three-day period of observation, specifically the 19th–21st of April 2023, which are designated as quiet magnetic days. Furthermore, we generate a circular graph every hour to improve the visibility of the error in the north-south or east-west orientations, presenting it in a rounded form with each 5-meter dilution of precision.

3 Result and Discussion

In Fig. 1, we can see the distribution map of ROTI values from INACORS GNSS data receivers over the Indonesian region on 20 April 2023. The high values, with a color gradient ranging from green to yellow, indicate the presence of EPB. The EPB started to occur at 10:54 UTC in the eastern part of Indonesia, where subsequently the EPB also began to appear in several other places across the Indonesia region. In order to measure error in the GNSS receiver's position, we have specifically identified the single-point positioning for the fix station in Biak, which is part of INACORS. Figure 2 displays the hourly GNSS position solution's error over Biak. The horizontal axes depict the east-west position error from the central reference point (true position), whereas the vertical axes indicate the error along the north-south direction, with circles plotted every 5 m in radius.

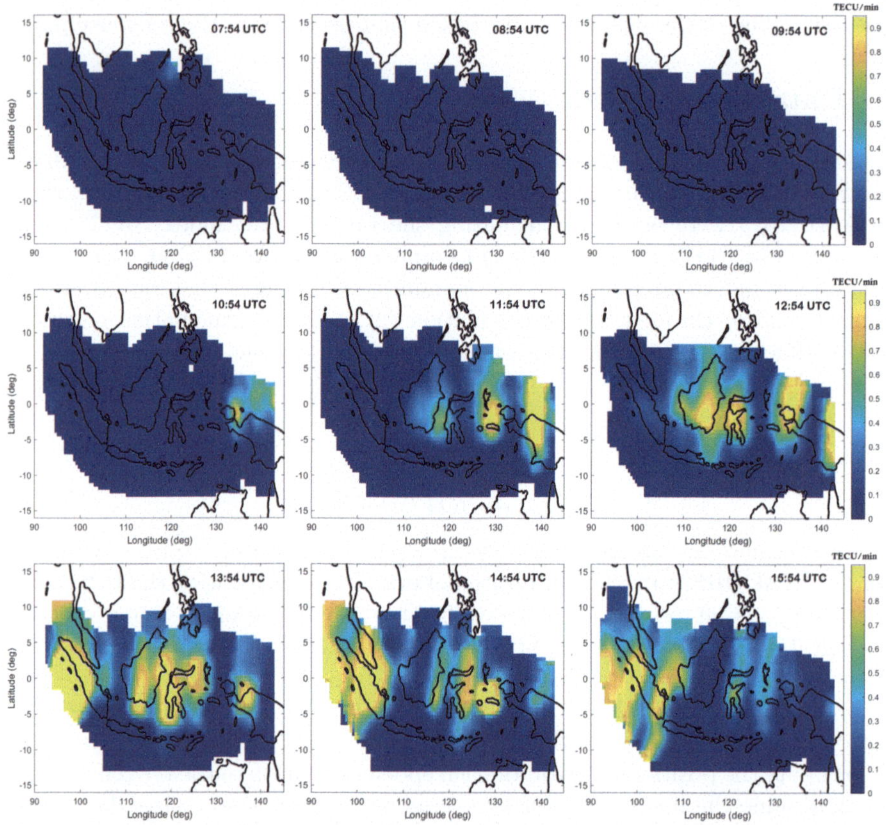

Fig. 1. The maps of ROTI values over Indonesia on 20 April 2023.

As can be seen in Fig. 2, from 7:00 to 9:00 UTC, the position error was within the first circle of a 5-m radius, stating that the accuracy and precision are in the nominal value. After 10:00 UTC, the EPB started to appear over Biak, and the GNSS position solution indicates the error magnitude within the second circle of the area with a radius of 10 m. This 10-meter position error persisted for approximately three hours. By 13:54 UTC, the EPB was still present over Biak; nevertheless, the position error began returning to around 5 m. However, there was still a systematic shift in the position error toward the northern area. After 14:00 UTC, the EPB over Biak disappears entirely, and the position solution tends to become more accurate and precise around the true position.

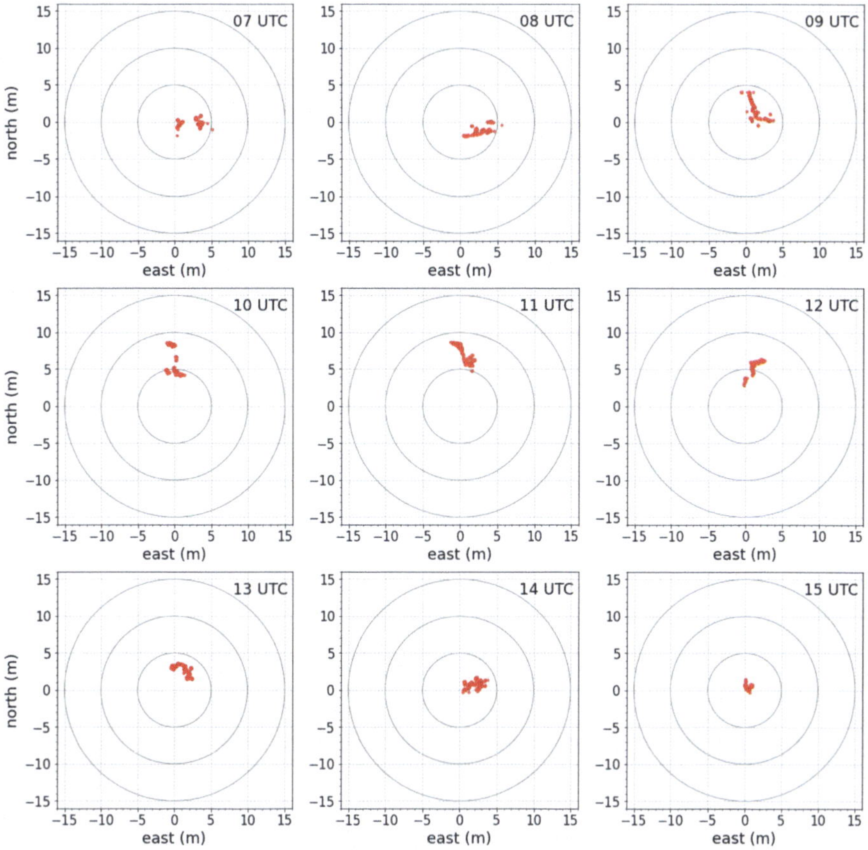

Fig. 2. The positioning error over Biak on 20 April 2023.

The positioning errors are more likely to occur in the north-south component than in the east-west direction. In order to be more convincing, we plotted the time series of position error for each component and computed the standard deviation, as illustrated in Fig. 3. The upper panel represents the north-south (n-s) direction, the middle panel represents the east-west (e-w) direction, and the lower panel represents the vertical direction, up-down (u-d). The north-south component's standard deviation value is 3.10, more than double that of the south component. The situation is potentially plausible due to the line of sight (LoS) satellite signal and the bubble structure. EPB structures, from a geometric perspective, typically extend in a north-south direction, in accordance with the magnetic field lines [16,17]. When EPB occurs, the ionosphere's unequal distribution of electron and ion density may cause rapid changes in the plasma's density. When GNSS signals traverse these EPB areas, the fluctuating plasma density can cause signal refraction.

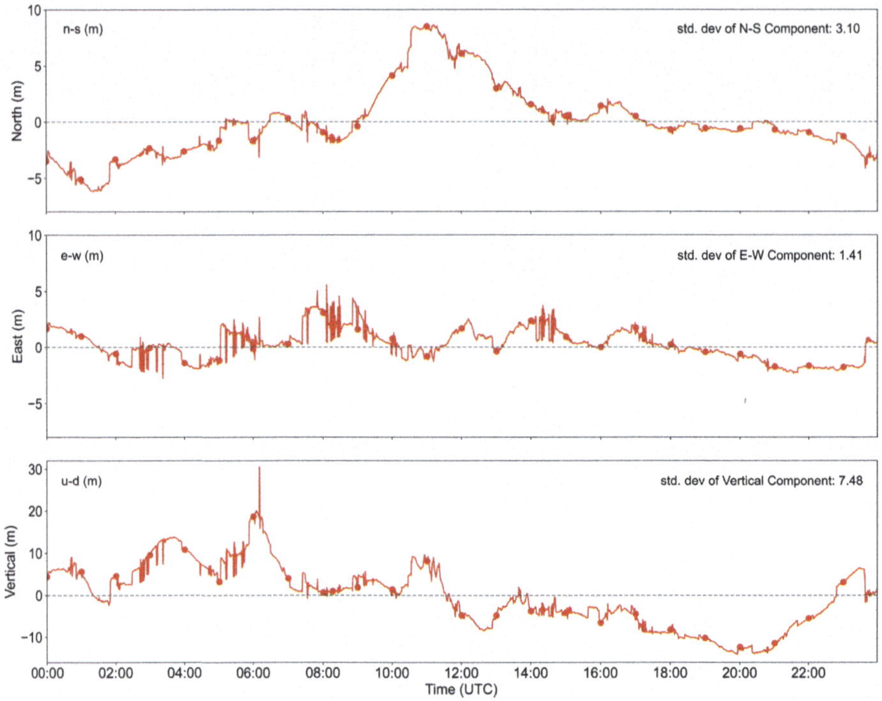

Fig. 3. The time series position error over Biak on 20 April 2023, includes each component's north-south (n-s), east-west (e-w), and up-down (u-d) components, along with their respective standard deviation values.

In satellite observation, signals coming from various satellites have different viewing angles for receivers on the earth's surface. The plasma bubble can disrupt many signals, particularly those from satellites, so that their signal traverses the ionospheric disruption area (including the plasma bubble region), resulting in variations in signal delay. The plasma bubble, which extends in the north-south direction, causes more significant fluctuations in signal paths that are parallel to or close to the north-south direction. As a result, the line of sight path that the satellite takes to the receiver is more disrupted in the north-south direction as compared to the east-west direction. In the Biak case, the error tends to be in the north direction.

The standard deviation of the north-south component is 3.10, twice as large as the east-west component. The vertical component has a standard deviation that is more than twice as high as the north-south component, while the west-east component has a standard deviation that is more than five times higher. Given the satellite's position above the Earth's surface with no signal coming from below, we can deduce that the vertical component's signal originates from a single side, resulting in a greater deviation. This results in greater variability for the vertical component compared to the other horizontal directions.

4 Conclusion

This study examines positioning errors in the GNSS receiver during EPB events over Biak. We observed positioning errors of up to 10 m, coinciding with the EPB occurrence. In this case study, plasma bubbles cause a north-south error tendency, particularly in the north direction, which is closely linked to how ionospheric disturbances affect satellite signals along their LoS path to the receiver. The north-south orientation of plasma bubbles leads to greater signal interference in that direction, resulting in larger positioning errors in the north-south component of the satellite positioning system. The result highlights the impact of ionospheric space weather on GNSS applications by demonstrating how plasma bubbles influence positioning accuracy.

Acknowledgment. The author is grateful to the Indonesian Geospatial Information Agency (Badan Informasi dan Geospasial, BIG) for providing the GNSS standard data (RINEX) from INACORS, which were used in this research.

References

1. Liu, Y., Fu, L., Wang, J., Zhang, C.: Study of GNSS loss of lock characteristics under ionosphere scintillation with GNSS data at Weipa (Australia) during solar maximum phase. Sensors **17**(10), 2205 (2017)
2. Ji, S., Chen, W., Weng, D., Wang, Z., Ding, X.: A study on cycle slip detection and correction in case of ionospheric scintillation. Adv. Space Res. **51**, 742–753 (2013)
3. Han, H., et al.: Effects of equatorial plasma bubbles on multi-GNSS signals: a case study over south China. Remote Sens. **16** (2024)
4. Rieke, M., Foerster, T., Geipel, J., Prinz, T.: High-precision positioning and real-time data processing of UAV-systems. Int. Arch. Photogramm. Remote Sens. Spatial Inf. Sci. **XXXVIII-1/C22**, 119–124 (2011)
5. Świerczynski, S., Felski, A.: Determination of the position using receivers installed in UAV. In: European Navigation Conference (ENC), pp. 1–4. IEEE, Poland (2019)
6. Radočaj, D., Plaščak, I., Jurišić, M.: Global navigation satellite systems as state-of-the-art solutions in precision agriculture: a review of studies indexed in the web of science. Agriculture **13** (2023)
7. Perez-Ruiz, M., Upadhyaya, S.: GNSS in precision agricultural operations. In: New Approach of Indoor and Outdoor Localization Systems. InTech (2012)
8. Abu-Shaban, Z., Seco-Granados, G., Benson, C., Wymeersch, H.: Performance analysis for autonomous vehicle 5g-assisted positioning in GNSS-challenged environments. In: 2020 IEEE/ION Position, Location And Navigation Symposium (PLANS), pp. 996–1003. IEEE, USA (2020)

9. He, Y., Li, J., Liu, J.: Research on GNSS INS & GNSS/INS integrated navigation method for autonomous vehicles: a survey. IEEE Access **11**, 79033–79055 (2023)
10. Shinbori, A., Otsuka, Y., Sori, T., Tsugawa, T., Nishioka, M.: Statistical behavior of large-scale ionospheric disturbances from high latitudes to mid-latitudes during geomagnetic storms using 20-yr GNSS-TEC data: dependence on season and storm intensity. J. Geophys. Res.: Space Phys. **127**(1) (2022)
11. De Abreu, A., et al.: Statistical analysis on the ionospheric response over South American mid- and near high-latitudes during 70 intense geomagnetic storms occurred in the period of two decades. J. Atmos. Solar-Terr. Phys. **245** (2023)
12. Aa, E., et al.: Statistical analysis of equatorial plasma bubbles climatology and multi-day periodicity using GOLD observations. Geophys. Res. Lett. **50** (2023)
13. Bumrungkit, A., Supnithi, P., Saito, S. et al.: A study of equatorial plasma bubble structure using VHF radar and GNSS scintillations over the low-latitude regions. GPS Solut. **26**(148) (2022)
14. Vankadara, R., Jamjareegulgarn, P., Seemala, G., Siddiqui, M., Panda, S: Trailing equatorial plasma bubble occurrences at a low-latitude location through multi-GNSS slant TEC depletions during the strong geomagnetic storms in the ascending phase of the 25th solar cycle. Remote Sens. (15) (2023)
15. Pi, X., Mannucci, A., Lindqwister, U., Ho, C.: Monitoring of global ionospheric irregularities using the Worldwide GPS Network. Geophys. Res. Lett. **24**(18), 2283–2286 (1997)
16. Shiokawa, K., Otsuka, Y., Ogawa, T., Wilkinson, P.: Time evolution of high-altitude plasma bubbles imaged at geomagnetic conjugate points. Ann. Geophys. **22**, 3137–3143 (2004)
17. Kil, H.: The morphology of equatorial plasma bubbles - a review. J. Astron. Space Sci. **32**(1), 13–19 (2015)

Orbital Dynamic of Temporary Captured Object (TCO): 2022 NX_1

Praza Gumilang Kembaren[1], Ridlo Wahyudi Wibowo[1,2], and Achmad Zainur Rozzykin[1,2(✉)]

[1] Department of Atmospheric and Planetary Sciences, Faculty of Science, Sumatera Institute of Technology, Bandar Lampung City, Lampung, Indonesia
achmad.rozzykin@sap.itera.ac.id

[2] Planetary and Space Sciences Research Group, Faculty of Science, Sumatera Institute of Technology, Bandar Lampung City, Lampung, Indonesia

Abstract. Temporary Captured Objects (TCO) are small Solar System objects temporarily captured by Earth's gravity, often referred to as Earth's temporary satellites. Five objects have been identified as TCO, one of which is 2022 NX_1. This research aims to investigate the orbital dynamics of 2022 NX_1 using N-body simulations conducted with the REBOUND module. The orbital elements data for the 2022 NX_1 used in this research were obtained from NASA's JPL Small-Body Database Lookup. During the simulation time span, 2022 NX_1's orbit was unstable when it was in close encounter with the Earth, and it was found that there were eight capture events of 2022 NX_1 by the Earth. The orbit of clone 2022 NX_1 is also unstable due to close encounter events, and this can be seen from its divergent orbit. No clones have crossed the orbit of any planet other than Earth.

Keywords: 2022 NX_1 · Dynamics · Simulations

1 Introduction

Temporary Captured Objects are small Solar System objects (such as asteroids and comets) that are temporarily captured by Planet's gravity. Temporary Captured Objects by the Earth (hereinafter abbreviated as TCO) are also known as Earth's temporary natural satellites. They are interesting to study because their orbits are close to Earth for a relatively long time, making them easily accessible to satellites in low Earth orbit. Studying TCO offers new opportunities for understanding the dynamics of the Earth-Moon system, testing models and evolutionary dynamics, evaluating techniques for utilizing asteroid material in space [1], and gaining insights into the origins and evolution of irregular satellite dynamics and short-period comets [2].

The term "captured" refers to the process in which an orbital transitions from a heliocentric orbit to a planetocentric orbit [3]. This transition occurs due to gravitational interactions between the object and the planet. According to Kary and Dones [4] and Granvik et al. [5], a captured object has a negative Keplerian planetocentric energy, and its distance from the planet is less than three times the planet's Hill radius. For Earth,

E. Fitrianingsih et al. (Eds.): ISAST 2024, SPPHY 416, pp. 283–294, 2025.
https://doi.org/10.1007/978-981-96-1344-1_31

this distance is approximately 0.03 astronomical units (au). In a three-body problem scenario, the capture of an object is only temporary [6].

Apart from the Earth, capture events of small Solar System objects, such as comets and asteroids, also occur on other planets, especially giant planets like Jupiter and Saturn. For instance, Comet Shoemaker-Levy 9 (C/1993 F2) was captured by Jupiter's gravity and collided with the planet [7]. Additionally, other comets, such as 39P/Oterma in (1936–1938), 147/Kushida-Muramatsu (May 1949–Jul. 1961), 82P/Gehrels 3 (Dec. 1966–Jul. 1974), 111P/Helin-Roman-Crockett (Jan. 1967–Jul. 1985), and P/1996 R2 Lagerkvist (Sep. 1983–Jun. 1993) [8] have been observed to be captured by Jupiter. Saturn's outer satellite Phoebe and Neptune's largest satellite, Triton, are thought to have become natural satellites after going through a capture process by their parent planet [9].

Identifying and observing TCO is quite challenging due to their small size, temporary visibility, faintness, and fast movement [1, 10]. Currently, five objects have been classified as TCO: 1991 VG, 2006 RH_{120}, 2020 CD_3, 2022 NX_1, and 2023 FY_3. 2022 NX_1 was discovered by D. Duszanowicz and J. Camarasa on July 2, 2022, at the Moonbase South Observatory in Namibia [7]. This asteroid was temporarily captured in January 1981 and June 2022. It is predicted to be captured again by Earth in 2051 [7, 11].

The orbital dynamics of planet-crossing asteroids, such as TCO, are significantly affected by close encounters with planets. This leads to variations in orbital elements, such as the semimajor axis elements [12, 13]. Asteroid orbits can also exhibit instability and transitions between orbital types (group movement) [12]. The evolution of asteroid orbits can be categorized into three groups: relatively stable orbits, group-shifting orbits, and chaotic orbits [14].

2 Data and Methods

2.1 Data

The orbital element data for the 2022 NX_1 used in this study was obtained from NASA's JPL Small-Body Database Lookup. JPL NASA SBDL is the primary reference database for small Solar System objects like asteroids and comets, developed by NASA and Caltech [14].

2.2 REBOUND

The REBOUND N-body integrator [15] was used to analyze the evolution of asteroid orbits over forward and backward integration timescales of 500 years each. REBOUND is a C programming language-based integrator that can also utilize Python for enhanced user-friendliness [16]. Furthermore, REBOUND has the capability to directly import data from NASA JPL Horizons to utilize ephemeris and masses of Solar System objects during the integration process [16]. The standard integrator used in REBOUND is Integrator with Adaptive Step-size control, 15th order (IAS15), an implicit integrator based on Gauß-Radau quadrature [17]. IAS15 can handle simulations showing orbital approaches and orbits with high eccentricity. It is also an adaptive integrator, capable of automatically selecting its time steps. The IAS15 integrator error is unaffected by nonconservative

Table 1. Optimal orbit element values and their uncertainties for asteroid 2022 NX_1 (Epoch 09-13-2023).

Orbital elements	Optimal value	Uncertainty ($\pm\sigma$)
e	0,02448362090275806	$5{,}8172 \times 10^{-8}$
$a(au)$	1,023216674464839	$8{,}2336 \times 10^{-8}$
$i(^\circ)$	1,0660778203	$2{,}8938 \times 10^{-6}$
$\Omega(^\circ)$	274,4822086593967	0,00010799
$\omega(^\circ)$	171,4168800639896	0,000132
$M(^\circ)$	254,1317360219585	0,00018472

forces like the Poynting-Robertson drag force [16] and has a better error value (in long-term simulations) compared to other integrators such as the leapfrog integrator, Wisdom-Holman integrator, and MVS integrator [17]. The default units in REBOUND are au for distance, $M_\odot$ for mass, and $year/2\pi$ for time.

2.3 Methods

The primary body in this simulation is the Sun, which also serves as the primary attractor. The other objects involved are the eight planets in the Solar System and the Moon [5]. Non-gravitational forces are not considered in the simulation. According to studies by Marcos and Marcos [18], radiation pressure has an insignificant impact on orbit calculations. Similarly, the Yarkovsky effect has been found to have an average change in the semimajor axis of 10^{-9} au per year.

This study also examines the orbital dynamics of the 2022 NX_1 clone. The clone is generated using a combination of the optimal value distribution and the uncertainty value of the 2022 NX_1 orbit elements, utilizing the NumPy normal random function. This function generates random numbers from a normal (Gaussian) distribution based on the uncertainty value (σ) in Table 1 [14]. Up to one hundred 2022 NX_1 clones were produced using this method. During the integration process, any clones that collide with Earth or the Moon are reviewed based on the collision criteria: geocentric distance $d_\oplus \leq 4,25 \times 10^{-5}$ au or selenocentric distance $d_☾ \leq 1{,}16 \times 10^{-5}$au [5]. This research also examines cases of clone collisions with other planets. The research will involve conducting the N-Object simulation twice: a forward-time simulation and a backward-time simulation.

Based on the capture criteria described in [4, 5], an asteroid is captured by the Earth if its geocentric energy is negative and is at a distance $<3R_{H\oplus}$. To calculate geocentric energy, the following equation is used:

$$\epsilon = \frac{1}{2}v^2 - \frac{\mu}{r} < 0 \tag{1}$$

where ϵ is the geocentric orbital energy, v is the relative speed of the asteroid to Earth, r is the distance of the asteroid to Earth, and μ is the gravity parameter ($\mu = GM_\oplus$).

Meanwhile, to calculate the influence of Earth's dominance, the Tisserand parameter value of the object relative to the Earth will be calculated:

$$T_E = \frac{a_E}{a} + 2\sqrt{\frac{a}{a_E}(1 - e^2)}\cos(i) \quad (2)$$

where T_E is the Tisserand parameter of the object relative to Earth, a_E is the semimajor axis of the Earth, a is the semimajor axis of the asteroid, e is the eccentricity of the asteroid and i is the inclination of the asteroid towards the Earth. The Tisserand parameter shows the orbital relationship between an asteroid and a planet. Objects that approach Earth will have a characteristic Tisserand value of $T_E \approx 3$ [19] and objects with a value of $T_E > 3$ indicate that the object is free from Earth's influence [14, 20].

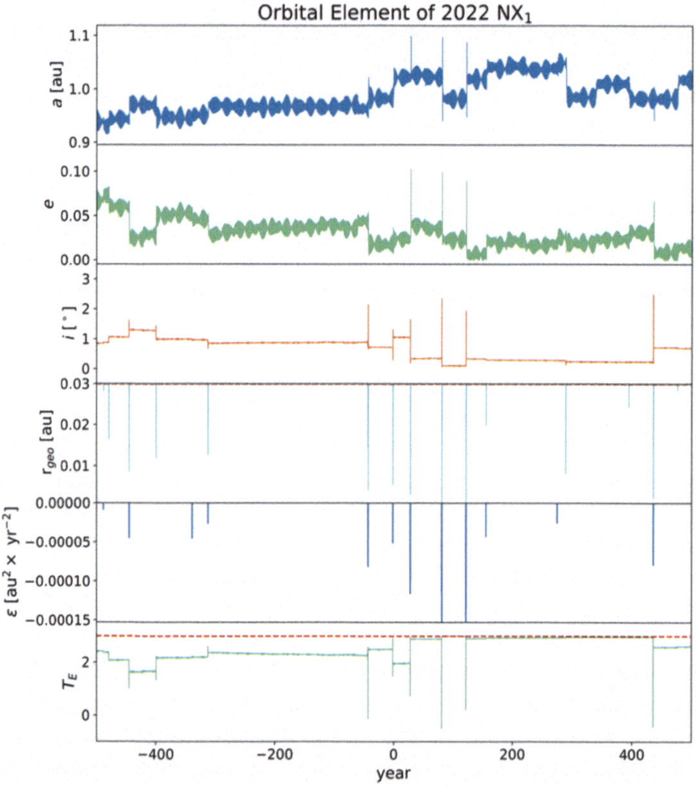

Fig. 1. Dynamics of the 2022 NX_1 orbital elements. The first panel is a plot of heliocentric orbit elements semimajor axis $\boldsymbol{a}$, eccentricity $\boldsymbol{e}$, inclination $\boldsymbol{i}$, geocentric distance $\boldsymbol{r_{geo}}$, geocentric energy ϵ, and Tisserand parameter $\boldsymbol{T_E}$.

3 Result and Discussion

3.1 Orbital Dynamic of 2022 NX$_1$

The dynamics of the orbital elements of 2022 NX$_1$ are illustrated in Fig. 1, which comprises six panels displaying the semimajor axis, eccentricity, inclination, geocentric energy, geocentric distance, and Tisserand parameters. Each element is plotted against the integration time (in years). The dynamics of the orbital elements of 2022 NX$_1$ are observed to be unstable due to its close encounters with Earth. Throughout the simulation period, there were 15 close encounter events (shown in the geocentric distance panel). Objects are considered to be in close encounter if the geocentric distance is <0.03 au. As discussed in [12, 13], the orbital elements dynamics are strongly influenced by close encounters events. Initially, the orbit of 2022 NX$_1$ was within the inner part of Earth's orbit before transitioning to an orbit around Earth (reference orbit image). Analyzing Fig. 1, it is apparent that the orbital elements exhibit stability over a certain period, which is limited by close encounter events (the orbit remains relatively stable until a close encounter event occurs). For instance, during the integration period from -312 (1771) to −42 (1981). On September 7, 1711, 2022 NX$_1$ passed close to Earth at a distance of 0.012617 au. Following this event, the orbital elements demonstrated a stable trend until 1981, before 2022 NX$_1$ once again passed close to Earth on January 16, 1981 at a distance of 0.00417 au.

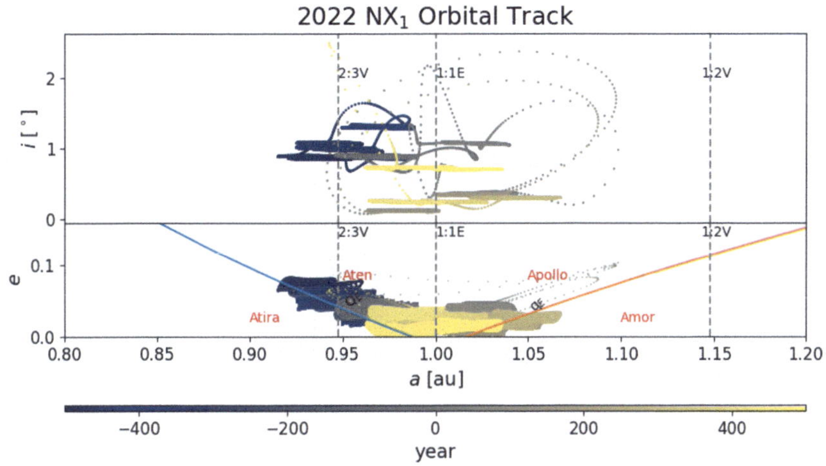

Fig. 2. The 2022 NX$_1$ orbital track.

In Fig. 2, the first panel depicts a plot of the semimajor axis (x-axis) and inclination (y-axis), while the second panel shows a plot of the semimajor axis against eccentricity. The color gradient indicates the year of integration. Furthermore, the lines for the aphelion distance Q_E (blue) and perihelion distance q_E (orange) illustrate the relationship between the values of a and e when they intersect the Earth's aphelion or perihelion distance values. It's evident that 2022 NX$_1$ has transitioned groups. Initially, it was located in the

2:3 mean motion resonance area with Venus for a considerable time before changing to the Apollo group. Subsequently, after becoming an Apollo-type asteroid, 2022 NX_1 once again experiences dynamics in the 1:1 mean motion resonance area with the Earth. Figure 2 also demonstrates that the orbital track of 2022 NX_1 seems to occupy more of the 2:3 mean motion resonance area with Venus.

3.2 NX_1 Capture Events by the Earth

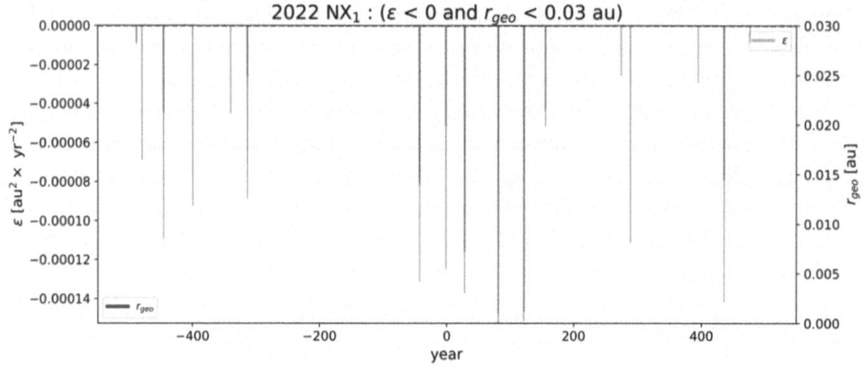

Fig. 3. Plot of geocentric energy ($\epsilon < 0$) and geocentric distance ($r_{geo} < 0.03$ au).

In Fig. 3, the plot shows the geocentric energy value when it is less than 0 (in blue) and the geocentric distance value when it is less than 0.03 au (in red). This plot was created to observe the dynamics of geocentric energy and geocentric distance based on the "captured" criteria described by Kary and Dones [4] and Granvik et al. [5]. The point

Table 2. 2022 NX_1 captures events by Earth and its duration.

No	Start	End	Duration
1	17 October 1578	17 November 1578	31 days
2	15 October 1711	12 November 1711	28 days
3*	11 October 1980	7 January 1981	88 days
	26 January 1981	23 February 1981	28 days
4	12 June 2022	02 July 2022	20 days
5*	06 September 2051	6 November 2051	61 days
	3 January 2052	23 February 2052	51 days
6	15 September 2104	13 February 2105	150 days
7	19 May 2145	28 October 2145	162 days
8	27 March 2459	24 May 2459	58 days

where geocentric energy and geocentric distance intersect, according to these criteria, is shown in a black plot, indicating the possibility of a capture event. To confirm a capture event, data exploration and selection are performed according to the existing criteria. During the simulation time, it was observed that 2022 NX_1 was captured by Earth eight times (shown in the Table 2 below).

The table provides the times when 2022 NX_1 was captured and escaped capture with Earth. It also indicates the duration of 2022 NX_1's capture. Notably, during the capture event, there are instances when the geocentric energy value changes to positive while 2022 NX_1 is within Earth's Hill sphere, marked with an asterisk (*) in the table. These instances occurred during the capture events in 1980–1981 and 2051–2052. The longest capture event during the simulation occurred from 2104 to 2105, lasting about 162 days.

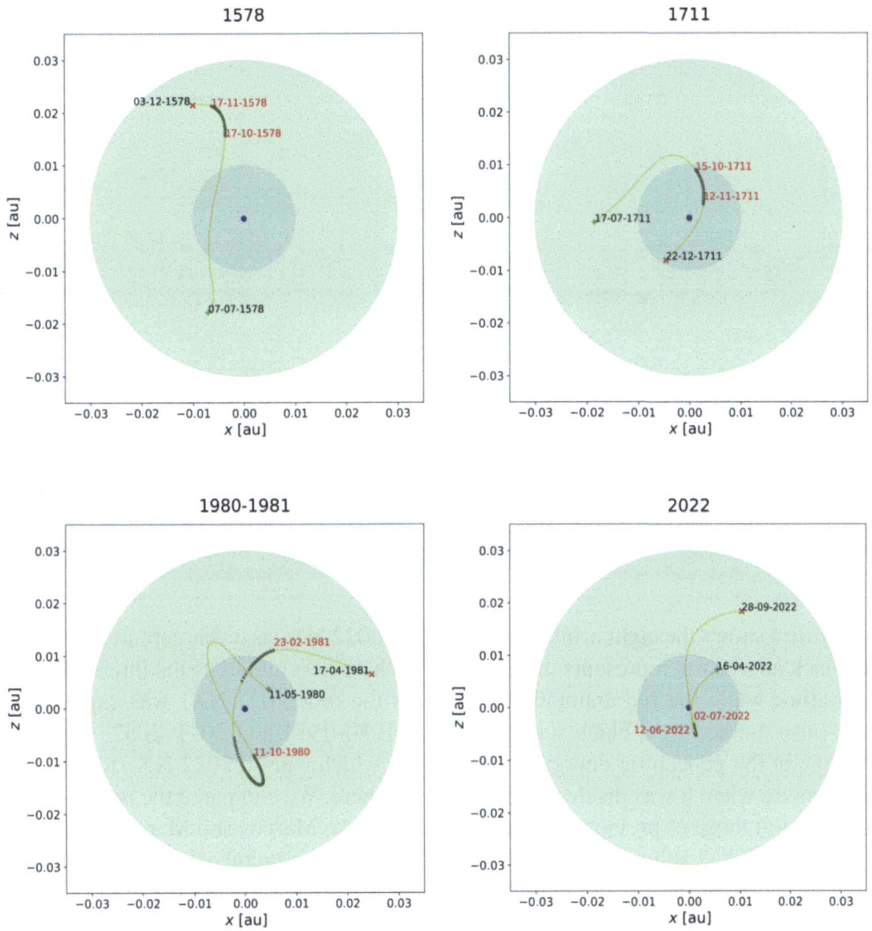

Fig. 4. 2022 NX_1 trajectory during a capture event with the Earth over the simulated period. The light blue color illustrates the Earth's Hill radius and Earth represented by a blue dot.

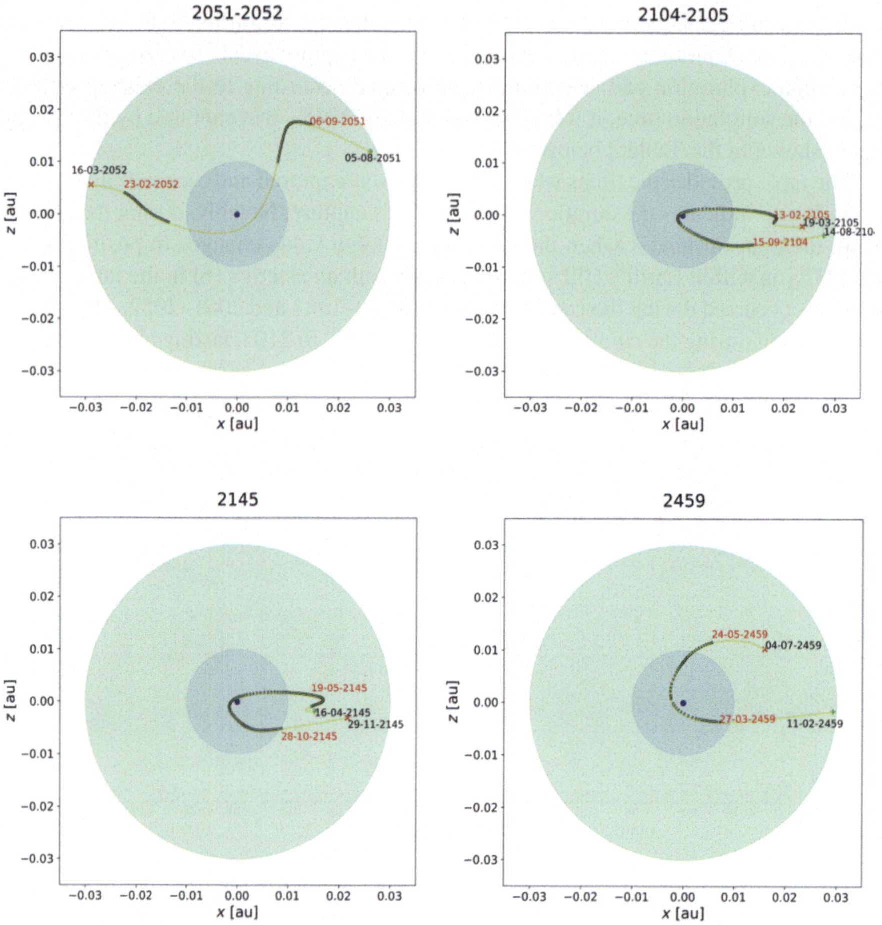

Fig. 4. *(continued)*

Figure 4 shows the eight orbital trajectories of 2022 NX_1 as it was captured by Earth. The black annotation represents the time 2022 NX_1 enters and exits the three times the Hill radius, while the red annotation represents the time 2022 NX_1 was captured by Earth (also marked by a black scatter plot). In 1980–1981 and 2051–2052, there was a change in the geocentric energy of 2022 NX_1, which caused 2022 NX_1 to "escape" from capture when it was inside the Earth's Hill sphere. We compared the results of this research with those of previous research. In their study, Marcos and Marcos [11] found that the 2022 NX_1 capture event by Earth was divided into several episodes: 1981, 2022, and 2051. We attempted to compare them in the Table 3 below:

Based on the data provided in the table, it appears that the timing of 2022 NX_1 being captured by Earth is consistent with previous research (as indicated in bold). We believe the slight variance may be attributed to differences in the use of orbital elements epoch and the objects included in the simulation. Notably, our study did not incorporate the three large asteroids: Ceres, Pallas, Vesta, or the Pluto-Charon barycenter.

Table 3. Comparison of the captured date and duration of the 2022 NX_1 with previous research conducted by Marcos and Marcos [11].

Episode		Start	End	Duration
1981	First Capture	1 October 1980	7 January 1981	98 days
		11 October 1980	**7 January 1981**	**88 days**
	Second Capture	26 January 1981	24 February 1981	29 days
		26 January 1981	**23 February 1981**	**28 days**
2022		11 June 2022	2 July 2022	21 days
		12 June 2022	**2 July 2022**	**20 days**
2051	First Capture	4 September 2051	6 November 2051	63 days
		6 September 2051	**6 November 2051**	**61 days**
	Second Capture	2 January 2052	23 February 2052	52 days
		3 January 2052	**23 February 2052**	**51 days**

We have identified the potential for 2022 NX_1 to become a natural satellite of the Moon. This finding originated from the 2022 NX_1 capture event in 2104–2105. Specifically, the selenocentric energy of 2022 NX_1 was found to be negative on November 19, 2104, at a distance of 0.0062 au. However, it's important to note that this was only detected in one data snippet. Furthermore, the selenocentric distance value when the selenocentric energy of 2022 NX_1 is negative does not meet the capture criteria. On another related note, Fig. 1 illustrates 15 close encounter events of 2022 NX_1 towards Earth, with the closest encounter occurring on November 12, 2104, at a geocentric distance of 0.0009715 au. Lastly, we compare our results with those of previous studies, as shown in Table 4.

Table 4. Date and distance of the close encounter of 2022 NX_1 to Earth.

No	Date	Distance
1	16 January 1981	0.00417 au
	16 January 1981	**0.00414 au**
2	26 June 2022	0.00543 au
	27 June 2022	**0.00543 au**
3	4 December 2051	0.00303 au
	4 December 2051	**0.00303 au**

From Table 4, it can be observed that the comparison of the date and distance of the 2022 NX_1 close encounter event in previous research is very similar to that in this research.

3.3 Orbital Dynamic of 2022 NX_1 Clones

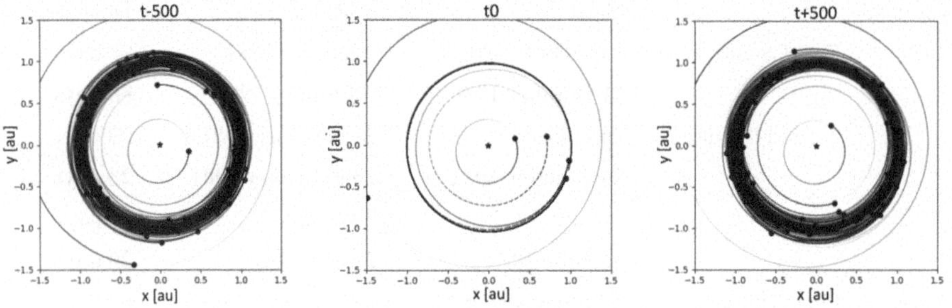

Fig. 5. The evolution of the TCO 2022 NX_1 clone orbit is depicted in three stages: the orbit on the left shows the integration at year −500, the middle orbit shows the initial integration time, and the orbit on the right shows the integration at year +500. The Sun is represented by the star marker, followed by the planets Mercury, Venus, Earth, the Moon, and Mars.

In the simulation results, none of the clones became Venus-crossers or Mars-crossers in either the 500-year backward or forward time simulations. The orbits of the clones are mainly around Earth's orbit. It is suspected that the spread of these clones is due to a close encounter event related to the main object 2022 NX_1. Additionally, during the simulation, no clones were found to be ejected, but some clones were found to collide with planet Earth (Fig. 5).

The Fig. 6 shows the distribution of orbits. The blue dots indicate the initial positions of 100 clones which then evolve over 500 years, resulting in the appearance of orange dots. This indicates that the clones are scattered. Over the -500 years of evolution, the orbital clones have spread across Earth with semimajor axis ranging from 0.9 to 1.1 au, eccentricity from 0 to 0.1, and inclinations from 0° to 1.5°. When compared with the distribution area of near-Earth objects as described by Granvik, et al. [5], all clones are still within areas that could be captured by the Earth-Moon system.

In the advanced time simulation, it is found that the clones mostly occupy the regions where $a > a_E$ and $q < q_E$, which are the orbital regions of the Apollo-type near-Earth asteroid group. The orbital clones are distributed around the Earth with a semimajor axis from 0.9 to 1.5 au, eccentricity from 0 to 0.15, and inclination from 0° to 1.5°. Similar to the exploration with backward simulation times, all clones are still within areas that could potentially be captured by the Earth-Moon system.

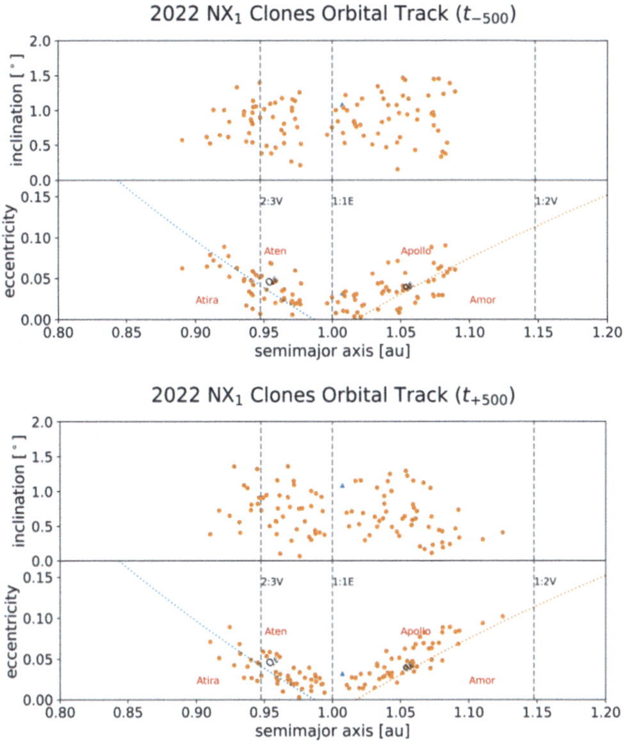

Fig. 6. 2022 NX$_1$ clone distribution for backward time simulation (top) and forward time simulation (bottom).

4 Conclusion

2022 NX$_1$'s orbit became unstable due to a close encounter event with the Earth. There have been 15 close events with the Earth. During the simulation time, it was found that there were eight capture events of 2022 NX$_1$ by the Earth. In the backward time simulation, it was found that 2022 NX$_1$ was predominantly in the 2:3 mean motion resonance region with Venus before moving to the NEO group. The dynamics of the clone orbit elements also show instability, both in backward and forward time simulations. Clones generally inhabit areas around Earth, dominant in the 2:3 mean motion resonance area with Venus and the Apollo group type area. During the simulation time, no clones were found that crossed the orbit of planets other than Earth, and clones were ejected. It was discovered that there was a clone that collided with planet Earth (in the backward time simulation).

References

1. Jedicke, R., et al.: Earth's minimoons: opportunities for science and technology. Front. Astron. Space Sci. **5**, 1–31 (2018). https://doi.org/10.3389/fspas.2018.00013

2. Iwasaki, K., Ohtsuki, K.: Dynamical behaviour of planetesimals temporarily captured by a planet from heliocentric orbits: basic formulation and the case of low random velocity. Mon. Not. R. Astron. Soc. **377**(4), 1763–1771 (2007). https://doi.org/10.1111/j.1365-2966.2007.11750.x
3. Heppenheimer, T.A., Porco, C.: New contributions to the problem of capture. Icarus **30**(2), 385–401 (1977). https://doi.org/10.1016/0019-1035(77)90173-7
4. Kary, D.M., Dones, L.: Capture statistics of short-period comets: implications for comet D/Shoemaker – Levy 9 comet D/Shoemaker – Levy 9 orbited Jupiter for decades. Icarus **224**(121), 207–224 (1996)
5. Granvik, M., Vaubaillon, J., Jedicke, R.: The population of natural Earth satellites. Icarus **218**(1), 262–277 (2012). https://doi.org/10.1016/j.icarus.2011.12.003
6. Urrutxua, H., Bombardelli, C.: A look at the capture mechanisms of the 'temporarily captured asteroids' of the earth. In: 26th International Symposium on Spaceflight Dynamics, number ISSFD-2017-074, Matsuyama, pp. 1–7 (2017)
7. de la Fuente Marcos, C., de la Fuente Marcos, R.: How to become a mini-moon: some hints from 2022 NX_1. Res. Notes AAS **6**(8), 160 (2022). https://doi.org/10.3847/2515-5172/ac8809
8. Ohtsuka, K., Ito, T., Yoshikawa, M., Asher, J., Arakida, H.: Quasi-Hilda comet 147P/Kushida-Muramatsu: another long temporary satellite capture by Jupiter. Astron. Astrophys. **489**(3), 1355–1362 (2008). https://doi.org/10.1051/0004-6361:200810321
9. Huang, T.-Y., Innanen, K.A.: The gravitational escape/capture of planetary satellites. Astron. J. **88**(10), 1537 (1983). https://doi.org/10.1086/113444
10. Jedicke, R., et al.: Minimoons & drifters. In: Asteroids, Comets, Meteors 2014 (2014)
11. De La Fuente Marcos, R., De León, J., De La Fuente Marcos, C., Licandro, J., Serra-Ricart, M., Cabrera-Lavers, A.: Mini-moons from horseshoes: a physical characterization of 2022 NX1with OSIRIS at the 10.4 m Gran Telescopio Canarias. Astron. Astrophys. **670** (2023). https://doi.org/10.1051/0004-6361/202245514
12. Milani, A.: Dynamics of planet-crossing asteroid. In: The Dynamics of Small Bodies in the Solar System, pp. 95–126 (1999)
13. Michel, P., Morbidelli, A., Bottke, W.F.: Origin and dynamics of Near Earth Objects. Comptes Rendus Phys. **6**(3), 291–301 (2005). https://doi.org/10.1016/j.crhy.2004.12.013
14. Dermawan, B., et al.: Dinamika orbit asteroid yang analog dengan orbit bumi. J. Sains Dirgant. **7**(2), 164–177 (2010)
15. Rein, H., Liu, S.F.: REBOUND: an open-source multi-purpose N-body code for collisional dynamics. Astron. Astrophys. **537**, A128 (2012)
16. Andersson, J.: Determination of the Orbit and Dynamic Origin of Meteoroids. Uppsala Universitet (2018)
17. Rein, H., Spiegel, D.S.: IAS15: a fast, adaptive, high-order integrator for gravitational dynamics, accurate to machine precision over a billion orbits. Mon. Not. R. Astron. Soc. **446**(2), 1424–1437 (2014)
18. de la Fuente Marcos, C., de la Fuente Marcos, R.: Dynamical evolution of near-Earth asteroid 1991 VG. Mon. Not. R. Astron. Soc. **473**(3), 2939–2948 (2018). https://doi.org/10.1093/mnras/stx2545
19. Siregar, S.: Fisika Tata Surya, 1st edn. Penerbit Fakultas Matematika dan Ilmu Pengetahuan Alam ITB, Bandung (2017)
20. Higuchi, A., Ida, S.: Temporary capture of asteroids by a planet: dependence of prograde/retrograde capture on asteroids' semimajor axes. Astron. J. **151**(1), 16 (2016). https://doi.org/10.3847/0004-6256/151/1/16

Characterizing Radiation Belt Electron Flux Behavior During Geomagnetic Storms Driven by CMEs and CIRs

Anwar Santoso[1,2], Sismanto Sismanto[3](✉), Eddy Hartantyo[3], Rhorom Priyatikanto[1], La Ode Muhammad Musafar Kilowasit[1], Mamat Ruhimat[1], Fitri Nuraeni[1], Mira Juangsih[1], Setyanto Cahyo Pranoto[1], Siska Filawati[1], Dadang Nurmali[1], and Iskandar Bakri[1]

[1] Research Center for Space, National Research and Innovation Agency (BRIN), Bandung, Indonesia

[2] Physics Department, Faculty of Mathematics and Natural Sciences, Gadjah Mada University, Yogyakarta, Indonesia

[3] Geophysics Laboratory, Physics Department, Faculty of Mathematics and Natural Sciences, Gadjah Mada University, Yogyakarta, Indonesia

sismanto@ugm.ac.id

Abstract. The behavior of electron flux was studied by examining 42 geomagnetic storms driven by Co-rotating Interaction Regions (CIR) and 48 geomagnetic storms driven by Coronal Mass Ejections (CME). This study used solar wind parameters (Nsw, Vsw, Psw), the Dst index, electron flux, and the interplanetary magnetic field southward-northward (Bz IMF) during solar cycle 24. The results indicate that the decrease in electron flux before (PRE) and during (POST) both CME-driven and CIR-driven storms is likely related to an increase in solar wind speed. For both types of storms, the decrease in electron flux generally begins almost simultaneously with the onset of the geomagnetic storm. The electron flux then starts to rise after the storm returns to normal conditions. The behavior of electron flux following both CME-driven and CIR-driven geomagnetic storms involves complex processes.

Keywords: Coronal Mass Ejection · Co-rotating Interaction Region · Geomagnetic Storm · Radiation Belt · Solar Wind · Interplanetary Magnetic Field

1 Introduction

Understanding the behavior of electron flux in the outer radiation belt during geomagnetic storms is a highly interesting topic in space weather research because enhancements in electron flux can cause deep dielectric charging and damage to spacecraft, such as satellites and space-based stations [1, 2]. Among the various phenomena in the Sun's atmosphere, Coronal Mass Ejections (CMEs) are major causes of geomagnetic storms, along with Co-rotating Interaction Regions (CIRs) in interplanetary space, where slow solar wind is compressed by fast wind [3, 4]. Based on their sources, geomagnetic storms

E. Fitrianingsih et al. (Eds.): ISAST 2024, SPPHY 416, pp. 295–302, 2025.
https://doi.org/10.1007/978-981-96-1344-1_32

can be categorized into CME-driven storms and CIR-driven storms [5, 6]. Interplanetary CMEs (ICMEs) have been shown to be responsible for storms with characteristics such as storm sudden commencement (SSC), the most intense events measured by Dst depression, saturation of the cross-polar cap potential, and the largest geomagnetically induced currents (GICs). CIR-driven storms typically occur gradually, are measured by Dst depression, and have longer-lasting impacts [6–8]. Both CME-driven storms and CIR-driven storms can lead to enhancements in electron flux [9, 10]. In addition to CME-driven and CIR-driven storms, High-Intensity, Long-Duration, Continuous AE Activity (HILDCAA) also causes electron flux enhancements. HILDCAA is a phenomenon that occurs in polar regions, characterized by an increase of more than 1,000 nT in the AE index and lasting for 2 days. It is generally associated with geomagnetic storms and typically occurs during solar minimum [9, 11]. Poloidal-mode ULF waves, which have a sufficiently strong electric field in the direction of electron motion, also play a role in the dynamics of the radiation belt [12, 13].

Among these sources of electron flux changes, CMEs have the greatest impact on Earth's environment compared to other solar phenomena. However, their effect on the electron population in the outer radiation belt is not as significant as their impact on the inner radiation belt [14]. Meanwhile, CIRs, which involve the interaction of high-speed solar wind with slower wind, can cause a more prolonged increase in electron flux compared to CMEs [8].

It is generally recognized that during a geomagnetic storm, the electron flux decreases in the main phase of the storm and recovers afterward [15, 16]. However, the processes governing the growth and loss of radiation belt electrons are not yet fully understood, even with theories such as adiabatic transport and magnetopause shadowing [2, 4, 5, 12]. It is known that the temporal and spatial growth of radiation belt electrons varies with geomagnetic storms. [4] found that the electron phase space density correlates well with the Interplanetary Magnetic Field in the z-direction (IMF Bz), the Symmetric H component (Sym-H), solar wind bulk speed (Vsw), and Auroral Low (AL) indices. Regarding the type of triggering storm, [17] found that the increase in electron flux associated with CIR-driven storms is generally higher than that associated with CME-driven storms. Additionally, the trend of electron flux variation decreases with increasing energy for all geomagnetic storms. During CIR-driven storms, the electron flux response is related to Vsw and Sym-H. [18] found that the electron flux response occurs not only during geomagnetic storms but also during non-storm periods. The magnitude of relativistic electron fluxes is higher during storms than in non-storm periods. They also found that relativistic electron fluxes are correlated with solar wind speed, although high solar wind speed is not always a necessary condition for the appearance of relativistic electron fluxes.

The purpose of this research is to understand the interplanetary space conditions that influence the increase in electron flux during CME-driven storms and CIR-driven storms, which can enhance the capability to develop preliminary early warning systems for future activities. This study is based on statistical analysis.

2 Data and Method

We identified geomagnetic storms with a Dst index ≤ -50 nT during solar cycle 24 (December 2008 to December 2019). We used 1-h Dst index data obtained from the Geomagnetic Data Center in Kyoto (https://wdc.kugi.kyoto-u.ac.jp/). Solar wind and plasma data, including density (Nsw), pressure (Psw), speed (Vsw), and the interplanetary magnetic field in the z-direction (Bz), were obtained from the ACE satellite, which orbits in interplanetary space. These data, provided as 1-h averages, can be accessed at https://omniweb.gsfc.nasa.gov/form/dx1.html. CME-driven storms were further classified using CME event lists from [14, 16, 19–21], and we also verified them against the CME events list from the Solar LASCO CME Catalog Version 2 (https://cdaw.gsfc.nasa.gov/CME_list/). CIR-driven geomagnetic storms were identified based on CIR event lists from [16] and [5]. Electron flux data were obtained from the GOES satellite, which orbits at geostationary altitude (6.6 RE, where RE = 6378 km). The electron flux data, originally in 5-min intervals, were processed into 1-h averages to align with the other data in this study.

Using Dst data, we identified 48 CME-driven storms and 42 CIR-driven storms during the period from December 2008 to December 2019. We calculated the correlation between Bz and solar wind parameters (Nsw, Vsw, Psw), as well as the Dst index, with the electron flux (EF). These calculations were categorized based on the onset of the minimum Dst index (storm). 'PRE' refers to the 3 days before the day of the storm, 'POST' refers to the 6 days after the day of the storm, and 'FULL' refers to the entire 10-day period (3 days before and 6 days after the day of the storm). Additionally, we calculated the highest correlations between Bz-EF, Nsw-EF, Vsw-EF, Psw-EF, and Dst-EF for each occurrence. Tables 1 and 2 present the correlations, with bold text indicating correlations of 50% or higher.

3 Result and Discussion

For 48 CME-driven storms, 44 are associated with the presence of electron flux data, while 4 geomagnetic storms have no electron flux data, as detailed in Table 1. For the 42 CIR-driven geomagnetic storms, 35 correspond to the presence of electron flux data, and 7 show no relation to electron flux, as shown in Table 2. Table 1 also presents the correlation between Bz and solar wind parameters (Nsw, Vsw, Psw), as well as the Dst index with electron flux (EF) during CME-driven storms. Meanwhile, Table 2 presents the correlation between Bz and solar wind parameters (Nsw, Vsw, Psw), as well as the Dst index with electron flux (EF) during CIR-driven storms.

Table 1 shows the correlation coefficients between the electron flux and Bz, the solar wind parameters (Nsw, Vsw, Psw) and Dst indices during 48 CME-driven storms. The findings are as follows:

a. Four of the CME-driven storms had no electron flux data.
b. During the pre-storm phase up to the maximum of the CME-driven storms, 19 electron flux events correlated with Vsw, with 7 of them having correlation values of 50% or higher, consistent with the results of [18] and [22].

Table 1. Correlation values for the entire period, as well as for the pre-, post-, and full phases of CME-driven storms. Values in bold indicate correlations of 50% or higher.

CME-Driven Geomagnetic Storm				CORRELATION														
No	Date of Magnetic Storms	Intensity (nT)	Time of Minimum Dst (UT)	PRE					POST					Full				
				Bz-EF	Nsw-EF	Vsw-EF	Psw-EF	Dst-EF	Bz-EF	Nsw-EF	Vsw-EF	Psw-EF	Dst-EF	Bz-EF	Nsw-EF	Vsw-EF	Psw-EF	Dst-EF
1	15-Feb-10	-58	23.00	No data EF														
2	6-Apr-10	-81	15.00	No data EF														
3	29-May-10	-80	12.00	-0,20	-0,04	-0,03	-0,04	-0,22	-0,38	0,28	-0,15	0,30	-0,49	-0,20	-0,03	0,15	0,12	-0,41
4	4-Aug-10	-74	01.00	-0,11	0,12	-0,23	0,03	0,01	-0,20	-0,15	0,16	-0,10	-0,27	-0,15	-0,07	0,05	-0,06	-0,25
5	4-Feb-11	-63	21.00	-0,30	-0,11	0,46	0,02	-0,28	0,00	-0,18	-0,32	-0,25	0,16	-0,08	-0,18	-0,02	-0,22	-0,06
6	1-Mar-11	-88	14.00	0,04	-0,34	-0,35	-0,42	0,00	0,02	-0,07	0,04	-0,08	0,04	-0,01	-0,09	-0,38	-0,12	0,27
7	11-Mar-11	-83	05.00	0,24	0,24	-0,20	0,23	0,35	0,40	-0,11	-0,28	-0,21	**0,57**	0,46	0,10	0,07	0,09	**0,50**
8	28-May-11	-80	11.00	0,03	-0,23	-0,24	-0,34	0,34	0,13	0,34	0,09	0,41	-0,36	0,04	0,17	-0,16	0,11	0,11
9	6-Aug-11	-115	03.00	0,26	-0,46	**-0,70**	-0,45	0,39	-0,24	-0,18	0,08	-0,13	-0,17	0,10	-0,16	-0,47	-0,22	0,30
10	10-Sep-11	-75	04.00	0,10	-0,21	-0,34	-0,22	0,27	0,31	0,27	0,32	0,29	-0,11	0,16	0,07	-0,04	0,07	0,14
11	17-Sep-11	-72	15.00	0,09	0,34	**0,62**	0,41	0,05	0,07	-0,43	0,28	-0,32	-0,43	0,17	0,00	0,08	0,06	-0,23
12	26-Sep-11	-118	23.00	0,04	-0,06	-0,09	-0,07	0,04	-0,04	-0,05	-0,20	-0,07	0,23	-0,08	-0,02	0,33	-0,03	-0,21
13	25-Oct-11	-147	01.00	-0,16	0,16	-0,18	-0,04	0,08	-0,02	-0,01	-0,09	-0,02	0,03	-0,10	-0,01	-0,15	-0,02	0,12
14	1-Nov-11	-66	15.00	-0,22	-0,06	0,19	0,02	-0,26	0,30	0,27	-0,19	0,20	0,41	0,25	-0,05	-0,16	-0,13	0,25
15	9-Mar-12	-145	08.00	0,05	-0,20	**-0,63**	**-0,50**	**-0,50**	-0,03	0,31	0,40	0,41	0,41	0,02	0,11	-0,07	0,03	0,23
16	15-Jul-12	-139	16.00	0,10	-0,06	0,04	-0,05	-0,03	-0,03	0,15	0,03	0,19	-0,12	0,03	0,03	0,03	0,02	-0,08
17	1-Oct-12	-122	04.00	0,25	-0,13	-0,15	-0,19	0,30	-0,03	-0,01	0,10	-0,02	0,21	0,00	0,02	-0,07	-0,04	0,37
18	9-Oct-12	-109	08.00	0,12	-0,15	0,04	-0,18	0,11	-0,04	0,25	0,41	0,37	-0,35	0,05	0,03	0,16	0,17	-0,09
19	14-Nov-12	-108	07.00	0,02	0,28	-0,20	0,01	0,17	-0,17	-0,33	0,18	-0,32	-0,09	-0,04	-0,37	0,38	-0,31	-0,15
20	17-Mar-13	-132	20.00	**-0,56**	**0,57**	0,39	**0,58**	-0,27	-0,01	-0,02	0,06	0,02	-0,07	-0,19	0,19	-0,15	0,16	0,12
21	1-Jun-13	-124	08.00	0,10	-0,04	-0,36	-0,07	0,24	-0,44	-0,05	-0,34	-0,24	0,00	-0,19	-0,03	-0,31	-0,14	0,13
22	29-Jun-13	-111	06.00	-0,20	0,21	-0,12	0,19	0,05	-0,04	0,16	0,41	0,32	-0,29	-0,04	0,16	0,34	0,24	-0,18
23	6-Jul-13	-87	18.00	0,28	0,40	-0,08	0,36	0,28	0,13	0,27	-0,26	0,19	0,27	0,19	0,32	-0,18	0,25	0,23
24	2-Oct-13	-72	07.00	-0,08	-0,06	-0,09	-0,06	0,07	0,10	-0,04	-0,40	-0,08	0,45	0,00	0,01	-0,08	-0,07	-0,09
25	19-Feb-14	-119	08.00	-0,24	-0,03	0,27	-0,03	-0,10	0,08	-0,05	-0,01	-0,07	0,14	0,05	-0,12	0,32	-0,07	-0,06
26	27-Feb-14	-101	23.24	0,29	-0,39	**0,63**	-0,22	-0,21	0,01	-0,06	0,14	-0,08	0,46	0,14	-0,23	0,41	-0,14	0,21
27	12-Apr-14	-92	08.00	0,32	0,10	-0,06	0,10	0,19	-0,08	-0,06	-0,24	-0,21	0,28	-0,01	-0,03	-0,32	-0,03	0,15
28	30-Apr-14	-76	09.00	0,31	0,04	0,03	-0,01	0,38	0,12	0,02	-0,32	-0,08	**0,51**	0,05	0,00	-0,40	-0,14	0,48
29	12-Sep-14	-97	23.00	No data EF														
30	22-Dec-14	-71	05.00	-0,20	-0,33	-0,22	-0,32	0,23	-0,28	-0,24	-0,02	-0,27	-0,01	-0,30	-0,26	-0,13	-0,31	0,06
31	23-Dec-14	-53	22.00	-0,17	0,01	0,33	0,17	0,11	**0,52**	-0,24	-0,05	-0,26	-0,05	-0,28	-0,26	-0,12	-0,29	0,01
32	7-Jan-15	-107	11.00	0,32	-0,14	**0,59**	-0,05	0,39	-0,18	-0,10	0,02	-0,17	**0,50**	0,09	-0,08	0,26	-0,13	0,45
33	17-Mar-15	-222	23:00	0,06	-0,02	-0,17	-0,12	0,11	-0,10	0,25	0,10	0,22	-0,19	-0,03	-0,15	0,18	-0,07	-0,18
34	23-Jun-15	-204	5:00	0,08	-0,25	-0,38	-0,30	0,25	-0,01	-0,14	0,11	-0,01	-0,35	0,04	-0,07	-0,20	-0,15	0,04
35	23-Jul-15	-83	07.00	0,21	0,25	-0,05	0,27	0,24	-0,16	0,04	-0,21	0,02	0,22	0,01	0,04	0,02	0,03	0,06
36	16-Aug-15	-94	07.00	0,32	-0,26	**-0,55**	-0,39	0,48	0,10	0,16	0,11	0,23	-0,07	0,13	-0,04	-0,06	-0,07	0,07
37	27-Aug-15	-101	20.00	0,25	-0,15	0,21	-0,07	0,27	0,00	-0,05	0,22	-0,04	-0,10	0,04	-0,03	0,24	0,05	0,02
38	9-Sep-15	-105	08.00	0,39	-0,16	0,10	-0,13	0,37	-0,08	0,09	0,06	0,11	-0,26	0,17	0,03	0,08	0,05	0,01
39	7-Nov-15	-89	06.00	-0,06	0,12	**0,51**	0,26	-0,22	0,40	0,04	-0,07	0,13	0,14	0,27	0,05	0,11	0,20	0,05
40	20-Dec-15	-155	23:00	-0,11	0,22	-0,10	0,13	-0,07	-0,02	0,15	-0,02	0,11	-0,03	-0,06	0,18	-0,06	0,11	-0,06
41	1-Jan-16	-110	1:00	0,12	0,27	0,02	0,25	0,10	-0,24	-0,26	0,05	-0,21	-0,45	-0,15	0,09	-0,11	0,07	-0,17
42	6-Mar-16	-98	21.00	-0,19	-0,23	-0,20	-0,24	0,04	-0,02	0,09	-0,20	-0,01	0,09	-0,07	-0,02	-0,22	-0,12	0,14
43	13-Oct-16	-103	18:00	0,31	-0,25	-0,19	-0,24	0,31	0,04	0,29	-0,19	0,16	-0,12	0,08	0,21	-0,38	0,04	0,16
44	22-Apr-17	-50	15.00	0,06	**-0,59**	0,22	-0,48	0,04	-0,13	0,31	0,14	0,29	-0,15	0,10	-0,05	0,07	0,04	0,13
45	28-May-17	-125	8:00	0,05	**0,79**	0,20	**0,79**	0,33	0,11	0,35	-0,32	0,27	0,29	0,30	0,20	0,10	0,20	0,09
46	16-Jul-17	-72	16.00	0,40	-0,27	-0,32	-0,38	0,33	-0,09	-0,09	0,46	0,14	-0,11	0,06	0,05	-0,25	-0,12	0,34
47	8-Sep-17	-124	2:00	0,08	-0,10	0,33	0,02	0,05	-0,08	-0,04	-0,05	-0,02	-0,12	-0,08	-0,05	-0,04	-0,03	-0,11
48	11-Sep-18	-60	10.00	No data EF														
Number of highest correlation values in the groups: PRE, POST, and FULL for each CME-Driven storm event				4	8	19	7	8	5	7	8	5	21	6	4	15	4	17

c. During the post-storm and recovery phases of the CME-driven storms, 21 electron flux events correlated with the Dst index, but only 3 had correlation values of 50% or higher, consistent with the findings of [23]. Almost all of the electron flux events showed low correlation with other parameters.
d. For the full duration of the CME-driven storms, 17 electron flux events correlated with the Dst index, but almost all of the electron flux events showed low correlation with other parameters.

Table 2 shows the correlation coefficients between the electron flux and Bz, the solar wind parameters (Nsw, Vsw, Psw) and Dst indices during 42 CIR-driven storms. The findings are as follows:

a. Seven of the CIR-driven storms had no electron flux data.
b. During the pre-storm phase up to the minimum of the CIR-driven storms, 13 electron flux events correlated with Vsw, with 5 of them having correlation values of 50% or higher, consistent with the results of [18] and [22].

Table 2. Correlation values for the entire period, as well as for the pre-, post-, and full phases of CIR-driven storms. Values in bold indicate correlations of 50% or higher.

CIR-Driven Geomagnetic Storm				Correlation														
No	Date of Magnetic Storms	Intensity (nT)	Time of Minimum Dst (UT)	PRE					POST					Full				
				Bz-EF	Nsw-EF	Vsw-EF	Psw-EF	Dst-EF	Bz-EF	Nsw-EF	Vsw-EF	Psw-EF	Dst-EF	Bz-EF	Nsw-EF	Vsw-EF	Psw-EF	Dst-EF
1	22-Jul-09	-83	06.00								No data EF							
2	12-Apr-10	-67	01.00								No data EF							
3	2-May-10	-71	18.00	0,03	-0,11	-0,29	-0,19	0,11	0,00	0,14	0,16	0,15	-0,23	0,04	0,04	-0,18	0,03	0,10
4	11-Oct-10	-75	19.00	-0,07	-0,02	-0,36	-0,06	0,20	0,33	0,34	-0,26	0,26	0,37	0,19	0,19	-0,14	0,09	0,16
5	6-Apr-11	-60	18.00	-0,12	**0,57**	0,25	**0,61**	-0,10	0,15	0,12	-0,22	0,08	0,24	0,09	0,25	-0,27	0,26	0,18
6	5-Jul-11	-59	00.00	-0,25	-0,19	0,14	-0,19	0,02	-0,01	-0,20	0,08	-0,16	0,16	-0,08	-0,05	0,13	0,00	0,08
7	24-Apr-12	-120	04.00	0,12	0,25	-0,20	0,22	0,24	-0,27	0,21	0,23	0,30	-0,38	-0,05	0,12	0,17	0,14	-0,14
8	13-Oct-12	-90	07.00	0,26	0,41	0,46	0,48	-0,44	-0,34	0,21	0,27	0,44	**-0,59**	-0,09	0,25	0,29	0,45	**-0,50**
9	26-Jan-13	-51	22.00	-0,02	-0,10	**-0,63**	-0,51	0,47	0,05	-0,10	-0,12	-0,13	0,08	0,01	-0,12	-0,30	-0,31	0,22
10	1-Mar-13	-55	10.00	-0,26	-0,18	-0,21	-0,21	-0,11	0,04	-0,07	0,13	-0,04	-0,16	-0,04	-0,14	0,12	-0,11	-0,18
11	29-Mar-13	-59	16.00	0,15	-0,15	0,16	-0,04	0,02	0,01	0,12	0,26	0,42	-0,49	0,05	0,26	0,23	0,38	-0,17
12	27-Aug-13	-59	21.00	0,25	-0,18	**-0,63**	-0,41	**0,53**	-0,21	-0,46	-0,17	-0,48	0,30	0,03	-0,21	-0,32	-0,36	0,39
13	30-Oct-13	-56	23.00	0,08	0,08	0,05	0,10	0,36	0,12	-0,29	0,12	-0,35	0,25	0,06	-0,10	0,05	-0,14	0,30
14	8-Dec-13	-66	08.00	0,28	-0,15	-0,35	-0,24	0,29	0,05	-0,02	-0,47	-0,28	0,29	0,12	-0,06	-0,42	-0,25	0,29
15	27-Aug-14	-79	18.00	0,33	-0,31	-0,14	-0,34	0,37	0,25	-0,03	0,15	0,14	0,26	0,27	-0,13	0,06	-0,06	0,24
16	17-Feb-15	-64	23.00	-0,11	0,12	0,02	0,08	-0,06	-0,01	0,37	0,06	0,38	-0,01	-0,06	0,28	0,04	0,28	-0,04
17	24-Feb-15	-56	03.00	-0,13	-0,23	-0,32	-0,25	0,13	0,12	-0,01	0,20	0,12	-0,14	0,02	0,03	-0,05	-0,01	0,07
18	2-Mar-15	-54	08.00	0,45	-0,12	-0,43	**-0,60**	**0,50**	0,04	-0,02	0,03	-0,02	0,06	0,22	-0,12	-0,10	-0,30	0,25
19	2-Apr-15	-54	08.00	0,15	-0,40	**-0,52**	-0,41	0,72	-0,13	-0,27	**0,50**	-0,07	-0,30	0,01	-0,30	0,27	-0,22	0,13
20	16-Apr-15	-79	23.00	-0,26	-0,48	**0,52**	-0,33	-0,32	0,15	-0,13	0,09	-0,15	0,05	0,03	-0,24	0,16	-0,25	0,06
21	13-May-15	-76	06.00	-0,16	-0,09	-0,11	-0,18	-0,02	-0,12	-0,15	0,37	0,08	-0,28	-0,01	-0,22	0,35	-0,11	-0,17
22	8-Jun-15	-73	07.00	-0,06	-0,18	-0,15	-0,21	-0,02	0,27	-0,09	0,18	-0,02	0,07	0,06	-0,31	0,43	-0,05	-0,26
23	5-Jul-15	-67	04.00	0,05	-0,09	-0,29	-0,17	0,14	0,12	-0,29	0,26	-0,19	-0,27	0,06	-0,33	0,34	-0,20	-0,29
24	13-Jul-15	-61	10.00	0,08	-0,21	-0,08	-0,29	0,00	0,00	-0,12	0,40	0,31	-0,36	0,01	-0,11	0,21	-0,06	-0,23
25	7-Oct-15	-124	23:00	-0,08	-0,02	0,25	0,12	-0,38	-0,18	-0,18	0,18	-0,17	-0,35	-0,15	-0,02	0,29	0,13	-0,45
26	20-Jan-16	-93	16.00	0,03	-0,28	-0,48	-0,33	0,18	0,15	-0,23	0,16	-0,19	0,14	0,08	-0,26	0,10	-0,23	0,12
27	3-Feb-16	-53	02.00	0,15	-0,29	-0,02	-0,30	-0,16	-0,29	-0,17	-0,25	-0,29	-0,30	-0,18	-0,20	-0,01	-0,30	-0,20
28	18-Feb-16	-57	00.00	0,04	-0,42	0,39	-0,24	-0,29	-0,14	0,34	-0,23	0,15	0,06	-0,07	0,07	0,04	0,11	-0,14
29	8-May-16	-88	08.00	0,01	0,22	0,00	0,22	0,01	-0,32	0,23	0,12	0,40	-0,30	-0,10	0,20	0,04	0,21	-0,11
30	3-Aug-16	-52	06.00	0,08	-0,14	-0,34	-0,27	0,15	0,05	0,24	-0,17	0,24	-0,06	0,14	0,15	-0,36	0,15	0,24
31	23-Aug-16	-74	21.00	0,12	0,13	-0,18	0,07	0,24	-0,06	-0,11	-0,04	-0,25	0,09	0,05	-0,07	0,06	-0,13	0,07
32	2-Sep-16	-58	01.00	-0,12	0,04	-0,02	0,09	**-0,58**	-0,11	0,42	-0,32	0,04	0,10	-0,11	0,13	-0,18	0,04	-0,14
33	29-Sep-16	-66	09.00	-0,07	-0,24	0,35	0,15	-0,01	0,04	0,09	-0,16	-0,01	0,08	-0,01	-0,02	0,06	0,07	0,04
34	1-Mar-17	-61	22.00	0,07	-0,18	-0,30	-0,38	0,26	-0,18	0,04	0,28	0,22	-0,48	-0,01	0,17	-0,43	-0,16	0,30
35	27-Mar-17	-74	14.00	0,21	0,14	0,15	0,18	0,09	-0,05	0,29	0,21	0,39	-0,29	0,05	0,09	0,18	0,24	-0,14
36	28-Sep-17	-75	06.00	0,06	-0,11	-0,14	-0,20	0,03	0,05	0,12	0,10	0,32	-0,26	0,09	0,07	-0,09	0,05	-0,02
37	8-Nov-17	-72	01.00	0,39	-0,29	**-0,54**	-0,47	0,44	-0,11	0,01	0,22	0,30	-0,38	0,14	-0,06	-0,14	-0,16	0,07
38	20-Apr-18	-66	09.00								No data EF							
39	6-May-18	-56	02.00								No data EF							
40	26-Aug-18	-174	07.00								No data EF							
41	7-Oct-18	-53	21.00								No data EF							
42	5-Nov-18	-53	05.00								No data EF							
Number of highest correlation values in the groups: PRE, POST, and FULL for each CIR-Driven storm event				3	4	13	10	6	1	6	6	9	13	3	4	12	11	8

c. During the post-storm and recovery phases of the CIR-driven storms, 13 electron flux events correlated with the Dst index, but only 1 had a correlation value of 50% or higher, consistent with the findings of [23]. Almost all of the electron flux events showed low correlation with other parameters.
d. For the full duration of the CIR-driven storms, 17 electron flux events correlated with Vsw, but almost all of the electron flux events showed low correlation with other parameters.

For the case studies, we examine the behavior of electron flux before, during the course of, and after both CME-driven and CIR-driven geomagnetic storms. Examples are shown in Fig. 1 and Fig. 2.

From Fig. 1 and Fig. 2, it can be observed that the electron flux tends to decrease from the onset until the recovery phase of both CME-driven and CIR-driven storms. In both types of storms, the decrease in electron flux typically begins almost simultaneously with the onset of the geomagnetic storm. The electron flux starts to rise again after the geomagnetic storm returns to normal conditions [15, 24]. The behavior of the electron flux following both CME-driven and CIR-driven storms is complex, sometimes showing

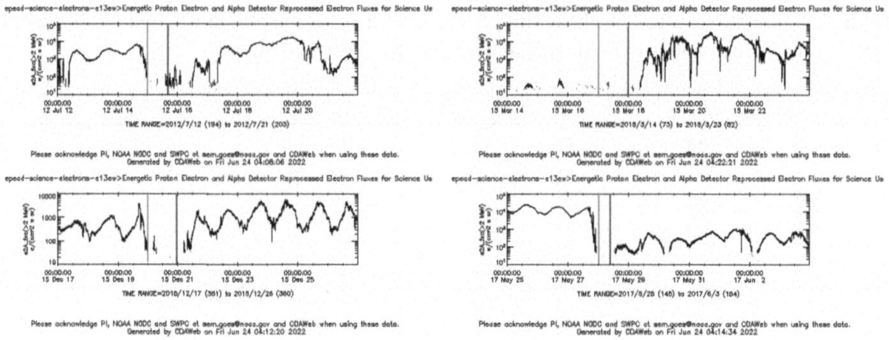

Fig. 1. Electron flux behavior during a CME-driven storm (blue line: onset; red line: peak of the geomagnetic storm) (https://cdaweb.gsfc.nasa.gov/cgi-bin/eval2.cgi).

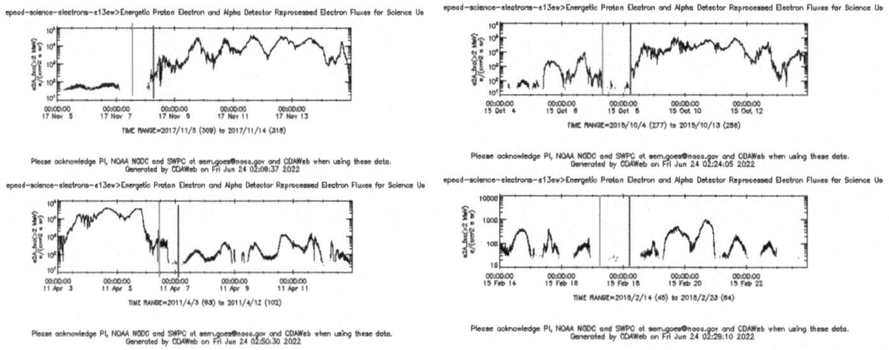

Fig. 2. Electron flux behavior during a CIR-driven storm (blue line: onset; red line: peak of the geomagnetic storm) (https://cdaweb.gsfc.nasa.gov/cgi-bin/eval2.cgi).

an increasing profile, a decreasing profile, or even no response at all after the geomagnetic storm [23].

4 Conclusion

The electron flux anomaly is related to both CME-driven and CIR-driven storms. The PRE, POST, and FULL periods of the geomagnetic storm were analyzed to identify which parameters most significantly affect electron flux. In the PRE condition, the highest correlation was found between solar wind speed (Vsw) and electron flux. In the POST condition, the highest correlation was between the Dst index and electron flux. This indicates that solar wind speed can serve as a precursor for understanding electron flux conditions during the main phase of the geomagnetic storm, while the Dst index is a better indicator of electron flux conditions during the recovery phase of the storm.

Acknowledgments. The author would like to thank the anonymous reviewer for their assistance in evaluating, supervising, processing data, and analyzing the results of this manuscript. The author also wishes to acknowledge the management teams:

(i) Kyoto Geomagnetic Data Center for the availability of Dst index data (http://wdc.kugi.kyoto-u.ac.jp).

(ii) Space Physics Data Facility, GODDARD Space Flight Center, NASA for the availability of solar wind data (https://omniweb.gsfc.nasa.gov/cgi/nx1.cgi).

(iii) Space Physics Data Facility, GODDARD Space Flight Center, NASA for the availability of electron flux data (https://cdaweb.gsfc.nasa.gov/cgi-bin/eval2.cgi).

References

1. Baker, D.N., Kanekal, S.G.: Solar cycle changes, geomagnetic variations, and energetic particle properties in the inner magnetosphere. J. Atmos. Sol. Terr. Phys. **70**(2–4), 195–206 (2008)
2. Miyoshi, Y., Kataoka, R.: Probabilistic space weather forecast of the relativistic electron flux enhancement at geosynchronous orbit. J. Geophys. Res. Space Phys. **113** (2008)
3. Webb, D.F., Howard, T.A.: Coronal mass ejections: observations. Living Rev. Sol. Phys. **9**, 1 (2012)
4. Temmer, M.: Space weather: the solar perspective. Living Rev. Sol. Phys. **18**, 1 (2021)
5. Pandya, M., Bhaskara, V., Ebihara, Y., Kanekal, S.G., Baker, D.N.: Variation of radiation Belt electron flux during CME-and CIR-driven geomagnetic storm: Van Allen probes observations. JGR Space Phys. (2019)
6. Hajra, R., Sunny, J.V.: Corotating Interaction Regions during Solar Cycle 24: A Study on Characteristics and Geoeffectiveness. Solar Phys. **297**, 30 (2022)
7. Borovsky, J.E., Denton, M.H.: Diiferences between CME-driven storms and CIR-driven storms. J. Geophys. Res. **111**, A07S08 (2006)
8. Heber, B., Sanderson, T.R., Zhang, M.: Corotating interaction regions. Adv. Space Res. **23**(3), 567–579 (1999)
9. Burns, A., et al.: The effects of corotating interaction region/high-speed stream storms on the thermosphere and ionosphere during the last solar minimum. J. Atmos. Solar Terr. Phys. **83**, 79–87 (2012)
10. Filawati, S., Nuraeni, F.: Effect of CIR and CME on electron flux in 2011. In: Prosiding Seminar Nasional Sains Antariksa (2017)
11. Hajra, R., Echer, E., Tsurutani, B.T., Gonzalez, W.D.: Solar cycle dependence of High-Intensity Long-Duration Continuous AE Activity (HILDCAA) events, relativistic electron predictors? J. Geophys. Res. Space Phys. **118** (2013)
12. Zong, Q.G., Rankin, R., Zhou, X.Z.: The interaction of ultra-low-frequency pc3–5 waves with charged particles in earth's magnetosphere. Rev. Mod. Plasma Phys. **1**(10) (2017)
13. Zong, Q.G., et al.: Energetic electron response to ULF waves induced by interplanetary shocks in the outer radiation belt. J. Geophys. Res. **114**, A10204 (2009)
14. Kataoka, R., Miyoshi, Y.: Flux enhancement of radiation belt electrons during geomagnetic storms driven by coronal mass ejections and corotating interaction regions. Space Weather **4**(9), 1542–7390 (2006)
15. Friedel, R.H.W., Reeves, G.D., Obara, T.: Relativistic electron dynamics in the inner magnetosphere—a review. J. Atmos. Sol. Terr. Phys. **64**(2), 265–282 (2002)
16. Shen, X.-C., et al.: Statistical study of the storm time radiation belt evolution during Van Allen era: CME-versus CIR-driven storms. J. Geophys. Res. Space Physics **122**, 8327 (2017)
17. Zhao, H., et al.: On the relation between radiation belt electrons and solar wind parameters/geomagnetic indices: dependence on the first adiabatic invariant and L^*, JGR: Space Phys. (2016)

18. Belakhovsky, V.B., et al.: Relativistic electron flux growth during storm and non-storm periods as observed by ARASE and GOES satellites. Earth Planets Space (2023)
19. Turner, N.E., Mitchell, E.J., Knipp, D.J., Emery, B.A.: Energetics of magnetic storms driven by corotating interaction regions: a study of geoeffectiveness. In: Tsurutani, B., McPherron, R., Lu, G., Sobral, J.H.A., Gopalswamy, N. (eds.) Recurrent Magnetic Storms: Corotating Solar Wind Streams, vol. 167, pp. 113–124 (2006)
20. Chi, Y., Shen, C., Wang, Y., et al.: Statistical study of interplanetary coronal mass ejections from 1995 to 2015. Sol. Phys. **291**, 2419 (2016)
21. Chi, Y., Shen, C., Luo, B., Wang, Y., Xu, M.: Geoeffectiveness of stream interaction region from 1995 to 2016. Space Weather **16**, 1960 (2018)
22. Hao, Y.X., et al.: Electron dropout echoes induced by interplanetary shock: Van Allen Probes observations. Geophys. Res. Lett. **43**(11), 5597–5605 (2016)
23. Turner, D.N., Erickson, P.J., Fennell, J.F., Foster, J.C., Jaynes, A.N., Verronen, P.T.: The response of Earth's electron radiation belts to geomagnetic storms: statistics from the Van Allen Probes era including effects from different storm drivers. Space Sci. Rev. **214**, 17 (2018)
24. Miyoshi, Y., Morioka, A., Obara, T., Misawa, H., Nagai, T., Kasahara, Y.: Rebuilding process of the outer radiation belt during the 3 November 1993 magnetic storm: NOAA and Exos-D observations. J. Geophys. Res. **108**(A1), 1004 (2003)

Observations of Solar Radio Bursts Using Callisto at Timau National Observatory for SWIFtS

Timbul Manik[1](✉), Peberlin Sitompul[1], Mario Batubara[1], Musthofa Lathif[1], Farahhati Mumtahana[1], Abdul Rachman[2], Radial Anwar[3], Taufiq Hidayat[4], and Christian Monstein[5]

[1] Research Center for Space, Indonesia National Research and Innovation Agency (BRIN), Bandung, Indonesia
timbul.manik@brin.go.id

[2] Timau National Observatory of Indonesia (TNOI), Indonesia National Research and Innovation Agency (BRIN), Kupang, Indonesia

[3] School of Applied Science, Telkom University, Bandung, Indonesia

[4] Department of Astronomy, Faculty of Mathematics and Natural Sciences, Institute Technology of Bandung, Bandung, Indonesia

[5] Istituto Ricerche Solari (IRSOL), Università della Svizzera italiana (USI), 6605 Locarno-Monti, Switzerland

Abstract. The Timau National Observatory of Indonesia (TNOI) is established at Mount Timau, Kupang, East Nusa Tenggara, Indonesia, to host the development of multi-wavelength astronomy, especially at visible and radio wavelengths. The construction of a 3.8-m optical telescope as the main facility is being undertaken, and it is expected to be completed in 2024. In addition to constructing this optical telescope, a plan is to build radio telescopes to work in frequencies below 1 GHz, and around 1–50 GHz is also being considered. Three Callisto (Compound Astronomical Low-cost Low-frequency Instrument for Spectroscopy and Transportable Observatory) Spectrometers will be set up at TNOI to observe solar radio bursts for astronomical science and education and RFI monitoring. The observation system contains an HF Log Periodic dipole antenna equipped with a suitable pre-amplifier and an up-converter to observe in the frequency range of 10–45 MHz and a high gain Log Periodic antenna equipped with a pre-amplifier and a signal splitter to observe in frequency range of 45–180 MHz as well as in frequency range of 180–870 MHz, and three Callisto receivers working simultaneously, that will be able to observe several type of solar radio bursts. This research will describe and discuss the three Callisto observation system set-up plans, the expected results, and the observation system data archive. The observation system daily data will then be available at the e-Callisto International Network central data server at Fachhochscule Nordwestschweiz (FHNW) operated by ETH Zurich to be open to the international community.

Keywords: Callisto · TNOI · Solar Radio Bursts

E. Fitrianingsih et al. (Eds.): ISAST 2024, SPPHY 416, pp. 303–312, 2025.
https://doi.org/10.1007/978-981-96-1344-1_33

1 Introduction

The Sun is an active star that serves as a source of energy and disturbances, emits radiation in the form of mass ejections such as solar wind, coronal mass ejections (CMEs), and solar energetic particles (SEPs), as well as electromagnetic emissions like solar flares and solar radiation. These solar emissions affect space and Earth, including the ionosphere and Earth's atmosphere [1]. The flash of light from the solar flare takes about 8 min to reach Earth, while the effects of a CME are estimated to reach Earth within several hours to days after a solar radio burst [2]. Both phenomena can be observed across the electromagnetic spectrum, from gamma rays at hundreds of MeV to radio waves with wavelengths of tens of meters. Solar flares and CMEs can stimulate plasma oscillations that emit radiation at metric and decametric wavelengths. Solar radio bursts are classified into five types: Type I bursts are narrowband features of short duration associated with active regions; Type II bursts are thought to be triggered by magnetohydrodynamic (MHD) shock waves associated with CMEs; Type III bursts result from energetic particles traveling along open magnetic field lines; while Type IV bursts show broad continuum emission with rapidly varying fine structures. The smooth, short-lived continuum following a Type III burst is called Type V [3].

The establishment of the Timau National Observatory of Indonesia (TNOI) on Mount Timau, Kupang, East Nusa Tenggara, Indonesia, aims to support astrophysical research as a host for the development of multi-wavelength astronomical observations, particularly in visible and radio wavelengths. It also serves as part of space weather and space environmental research. As indicated in the Master Plan, the main observing facility at TNOI is a 3.8-m optical telescope, which is expected to be completed in 2024. Plans are being considered to develop a radio telescope that will be operating below 1 GHz and in the 1–50 GHz range [4]. As an initial step towards constructing a radio telescope operating below 1 GHz, Callisto will be installed at the TNOI to observe solar radio bursts for astronomical scientific and educational purposes and for Radio Frequency Interference (RFI) monitoring. Three Callisto observation systems are installed to observe solar radio bursts with different characteristics. Initial RFI measurements in the vicinity of TNOI are also conducted. The results of these preliminary RFI measurements are used to determine Callisto's operating frequencies in configuring the radio frequency measurement to achieve the best frequency gap with the lowest radio interference. This paper describes and discusses the Callisto observation system, the initial RFI measurements, expected and preliminary results, and the archive of observation system data.

2 Methods and Results

Figure 1 shows the planning for the placement of several instrumentations at Timau, as mentioned in the master plan of Timau National Observatory of Indonesia (TNOI). It indicates the locations of the optical telescope (D), geomagnetic sensor (I), including radio telescopes in the high-frequency single-dish parabolic antenna (B), and low-frequency array radio antennas (H), and so on. Callisto instrumentations were installed next to geomagnetic sensors (J) and used in the same building for its control room.

The set-up plan of three Callisto observation systems at TNOI is shown in Fig. 2(a). Each observation frequency range group is divided into three groups to see the different

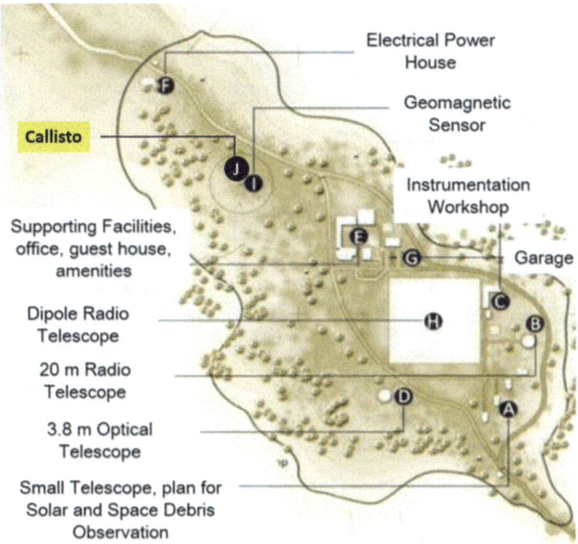

Fig. 1. The placement of the instrumentation as mentioned in the Master Plan of TNOI, shows the locations of optical and radio telescopes, etc., including Callisto (J).

characteristics of the radio burst type as follows: 1) an HF Log Periodic dipole antenna equipped with an up-converter to observe in the frequency range of 10–45 MHz, 2) a high gain Log Periodic antenna equipped with a pre-amplifier to observe in frequency range of 45–180 MHz as well as in 3) frequency range of 180–870 MHz. Three Callisto receivers were expected to work simultaneously, allowing the observation of several types of solar radio bursts, as mentioned before. In the current implementation of equipment installation, we report that from the original plan to install three Callisto observation systems, as mentioned above, the installation that could be carried out was for two observation systems, specifically for the frequency ranges of 45–180 MHz and 180–870 MHz, while the observation for the third frequency range of 10–45 MHz is not implemented yet, and will be conducted on a subsequent occasion so that the radio burst observation plan for the frequency of 10–870 MHz could be implemented shortly.

Figure 2(b) illustrates the schematic diagram of a Callisto spectrometer and the system configuration for observing solar activity. It consists of an antenna that receives radio waves from various frequencies and is related to solar activity, such as solar flares, coronal mass ejections (CMEs), and other types of activity. Callisto also uses frequency dividers and reducers, so the received radio signals are divided or reduced to various frequency ranges. Callisto also performs filtering to eliminate noise and then amplifies the received signals. GPS is required to synchronize observation times between Callisto systems. The timing of Callisto is controlled by a GPS clock, ensuring that its timing is very accurate to within <1 ms. In the algorithm we built for data processing, GPS timing is also selected so as not to exceed 1 ms to overcome the problem of delay differences and observation accuracy.

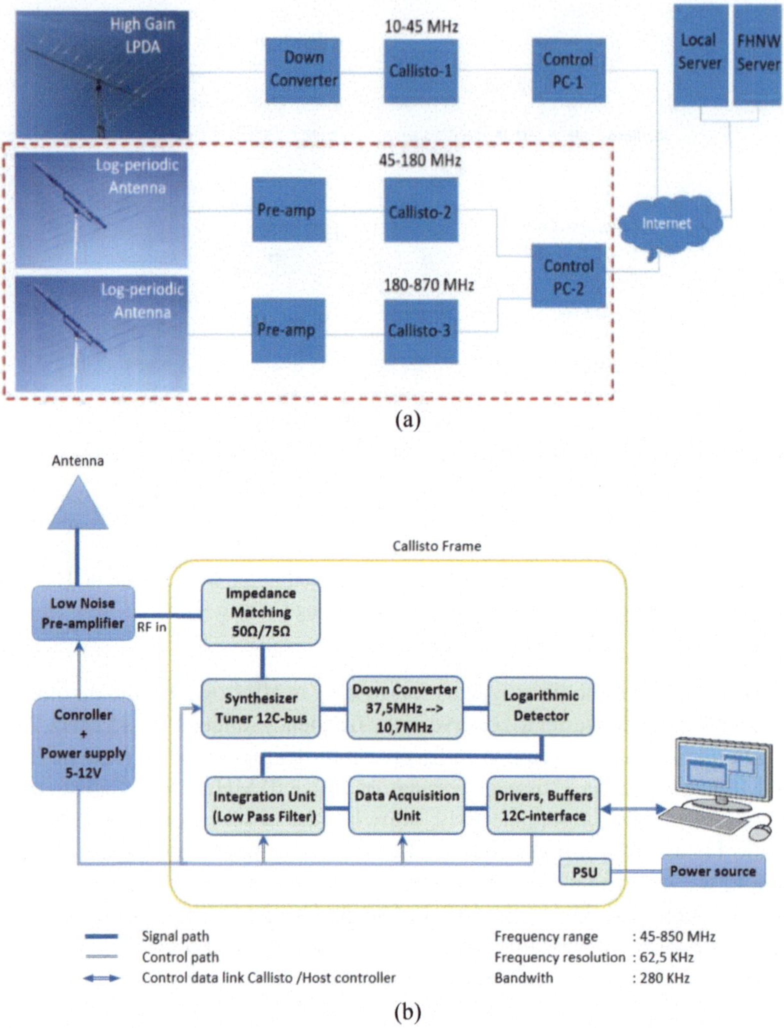

Fig. 2. (a) Set-up plan of Three Callisto at Timau National Observatory, and (b) Basic Diagram Scheme of the single Callisto spectrometer.

The expected data from Callisto operation at Timau will be in FITS file format, achieving 400 frequencies per sweep with a time resolution of 0.25 s per 200 channels/spectrum (800 pixels per second), with time integration of 1 ms and bandwidth of approximately 300 kHz. Overall dynamic range will be greater than 50 dB. Operational subroutines are built using SSW-IDL, a software based on the IDL programming language integrated with software and database libraries related to solar physics, incorporating inputs from various solar observation libraries, including astronomy and the Python programming language. Callisto can continuously observe radio spectra 24 h/day, 365 days/year, with 13,200 channels/spectrum [1, 5].

Callisto can capture solar radio frequencies constituting a single solar radio burst event. The data retrieval algorithm is built to arrange files automatically. The data storage period is 15 min following the buffer memory capacity owned by the Callisto system. The file type is selected in the FITS format to standardize the format of the Callisto system already operating on the international Callisto network. FIT format data stored per 15 min can be combined using SSW IDL software for further data analysis.

The collection, transfer, and processing schemes of the Timau Callisto data are planned as shown in Fig. 3. As per the planning, Timau Callisto data is collected locally and will sent to the FHNW server in Switzerland operated by ETH Zurich for open access to the international community on the International e-Callisto Network. Data will transmitted to the BRIN local server for real-time data processing purposes. From the BRIN local server, the data is further transmitted to separate PCs for subsequent processing, as depicted in Fig. 3 [6]. To test the Callisto system in Timau, the observation data is stored in the system's internal memory as an HDD. Data for further processing is sent periodically based on the high solar activity status to the researcher for clarification and confirmation of radio burst detection results. We have built an open source-based Callisto data processing system for data processing and automatic data transfer in the future where the spectrograms of the frequency radio-waves intensity generated by using Gaussian Filter and image enhancement, and the Numpy Python modules will be used as a script, which run automatically by task-scheduler. The image of the spectrogram produced in this process then will be displayed on the Space Weather Information and Forecast Services (SWIFtS) Homepage on (https://swifts.brin.go.id) for real-time support for the daily services.

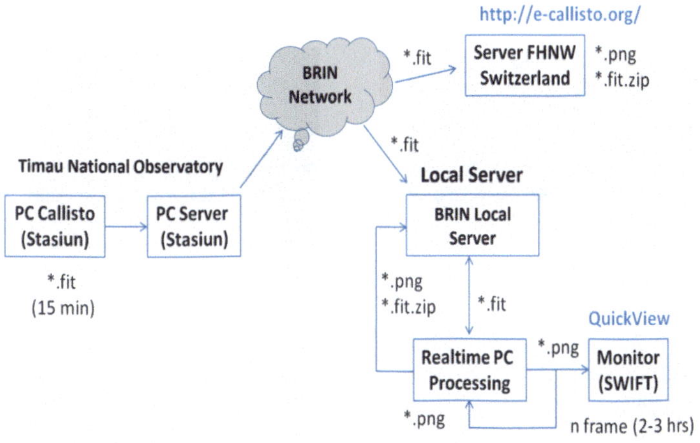

Fig. 3. Planning of the collection, transfer, and data processing of Timau Callisto.

After the recent installation, the operation of Callisto at Timau has not been functioning properly due to frequent power outages affecting continuous operation, internet network issues for data transfer, and technical operator constraints in the field, which need to monitor and maintain the observation system to keep it running. Therefore, it is necessary to develop an automated observation system since the Timau site is in a

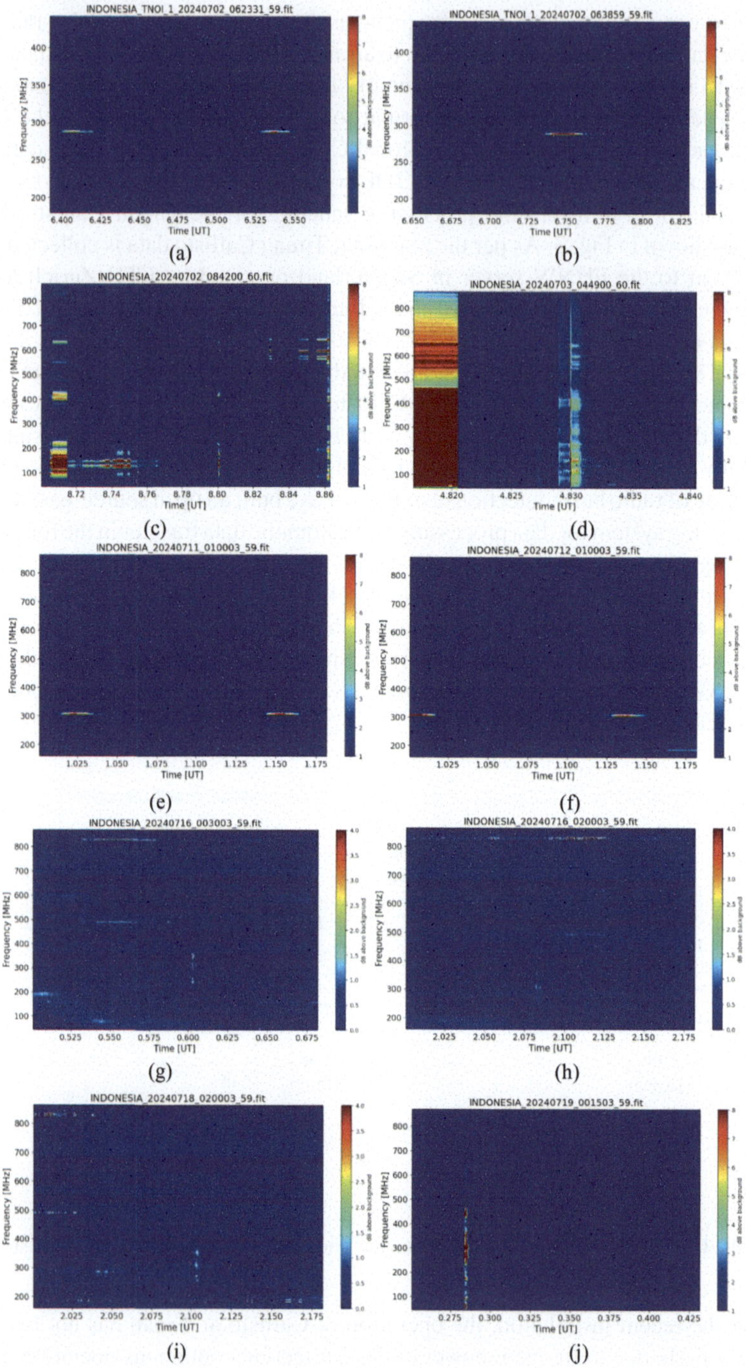

Fig. 4. Receptions of radio waves like bursts by Callisto1 and Callisto2 at the TNOI.

remote area. The observations obtained some signals on July 2 and 3, 2024; July 11, 12, 16, 18, and 19 July 2024, from both Callisto systems in operation as shown sequentially in Fig. 4(a) to (j).

3 Discussion

From the un-continuous reception of Callisto on 2, 11, and 12 July 2024, as shown in Fig. 4(a), (b), (e), and (f), visually there, no solar radio bursts were detected in the obtained data. However, there were recurring signals at around 300 MHz lasting about 1–2 min, repeating several times over certain intervals for the signal received. Sitompul et al. [7] also reported receiving short signals at around 300 MHz, indicating their use for radio navigation for flight and maritime applications. On the other hand, the reception of Callisto at a wide frequency range of up to 870 MHz on July 2, 2024, at 08:50 UT and a frequency range of 45–80 MHz on July 3, 2024, at 04:52 UT as shown on Fig. 4(c) and (d), still require clarification of the solar activity occurring at those times.

On July 16, 2024, there was a recording of a radio signal showing a type III burst at a frequency between 200–350 MHz, at 00:37 UT as shown in Fig. 4(g), and at 02.06UT as shown in Fig. 4(h). Refer to Space Weather Live and NOAA's Space Weather Center report [8, 9] as shown in Fig. 5(a) and (b) mentioned that the reception of the radio burst signal at 00:37 UT and 02:06 UT correlated with flares C8.8 and flare C5.8 that occurred at the same time, respectively. Solar activity has been at a high level over the past 24 h. NOAA also reported that the largest solar flare activity on that day was an X2-class flare that occurred at 13:26 UT from active region 3738 (N09W83).

On July 18, 2024, Timau Callisto obtained a radio signal record showing a type III burst at a frequency between 250–350 MHz at 02.08UT, as shown in Fig. 4(i). NOAA's Space Weather Center report, as shown in Fig. 6(a), mentions that the reception of the radio burst signal at 02.08UT correlates with solar activity of the C3.1 solar flare that occurred simultaneously. Solar activity has been at a moderate level throughout the last 24 h. NOAA also reported that the largest solar flare activity event on that day was an M2.2 class flare that occurred at 10.14 UT from active region 3751 (S08E14). Then, on July 19, 2024, Timau Callisto also received a radio signal showing a type III burst at a wide frequency of up to 650 MHz at 00.17 UT, as shown in Fig. 4(i). NOAA's Space Weather Center report, as shown in Fig. 6(b), mentions that the reception of the radio burst signal at 00:17 UT did not correlate with solar flare activity simultaneously and that solar activity was at an intermediate level over the past 24 h. NOAA also reported that the largest solar flare activity event on that day was an M3.1 class flare at 08.23 UT from active region 3753 (N12W76).

From radio frequency interference measurements utilizing Callisto on 2, 3, and 11 July 2024, we obtained day and night time, 04.00 UT, 10.00 UT, 16.00 UT, and 22.00 UT as shown in Fig. 7(a), (b), and (c). It can be explained that the horizontal axis represents radio frequencies from 45–870 MHz, and the vertical axis represents RFI intensity in mV~ dB. FM radio transmitters at 80–108 MHz frequencies saw the largest interference class. Around a frequency of 150 MHz, significant activity from amateur radio frequency users was observed. At approximately 250 MHz, signals from US military satellite downlinks were detected, while receiving signals from satellite communications to Earth

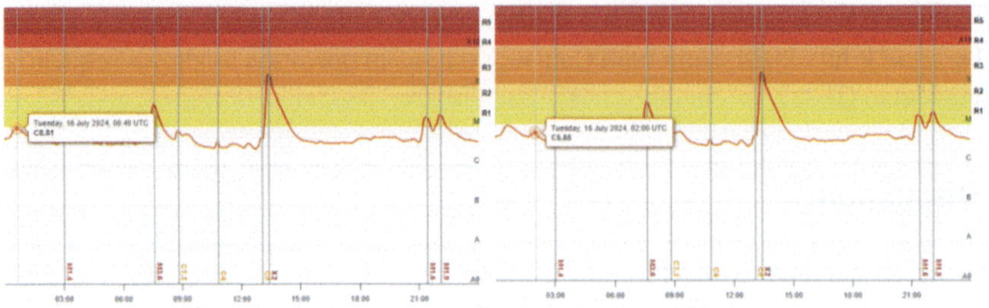

(a) (b)

Fig. 5. NOAA's Space Weather Center report 16 July 2024.

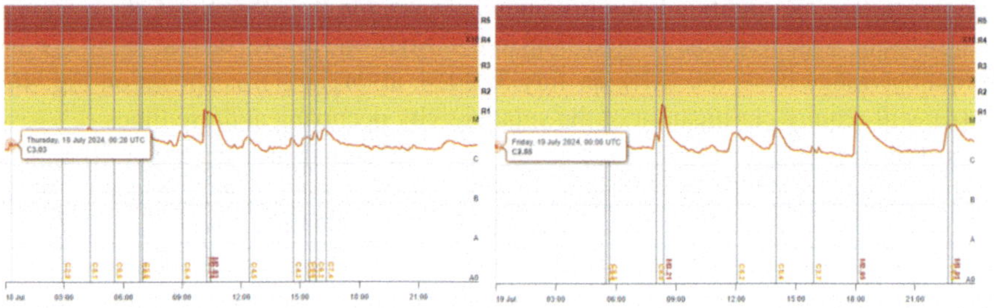

(a) (b)

Fig. 6. NOAA's Space Weather Center report: (a) 18 July 2024; (b) 19 July 2024.

at frequencies around 240–260 MHz indicates good sensitivity to measure solar radio emission signals. At around 380–400 MHz, the continuous use of frequencies can be seen both during the day and at night. The allocation of radio spectrum use based on regulations is for satellite communications from the Earth into space [10]. Digital Video Broadcasting and Terrestrial Broadcast (DVB-T) usage was shown at frequencies above 450–800 MHz, while signals above 850 MHz are usually used as cellular phone signals. There was stable radio interference below 800 MHz. Based on this obtained measurement, effective radio emission observations can be conducted up to frequencies of 430 MHz, considering the existence of DVB-T signals.

Frequency measurements of Callisto can be performed in the following options: a) frequencies from 45 MHz to 430 MHz, excluding the FM band, 80–110 MHz, b) frequencies from 110 MHz to 430 MHz, and c) then frequencies from 430–870 MHz. These initial measurement results are then used to determine Callisto's operating frequencies in frequency measurement configurations to achieve the best frequency gap with the lowest radio interference. Callisto operating frequencies are determined by excluding frequencies with the most significant interference. Based on these frequency choices, specific frequency configuration files are prepared for implementation in Callisto operations.

RFI in solar radio observation is something that needs to be overcome. We also consider these problems. As a first step, we do two things regarding software and hardware. On the software side, we create a separate file for the scheduling of observations and the range of observation frequencies implemented in the scheduler configuration file. Then,

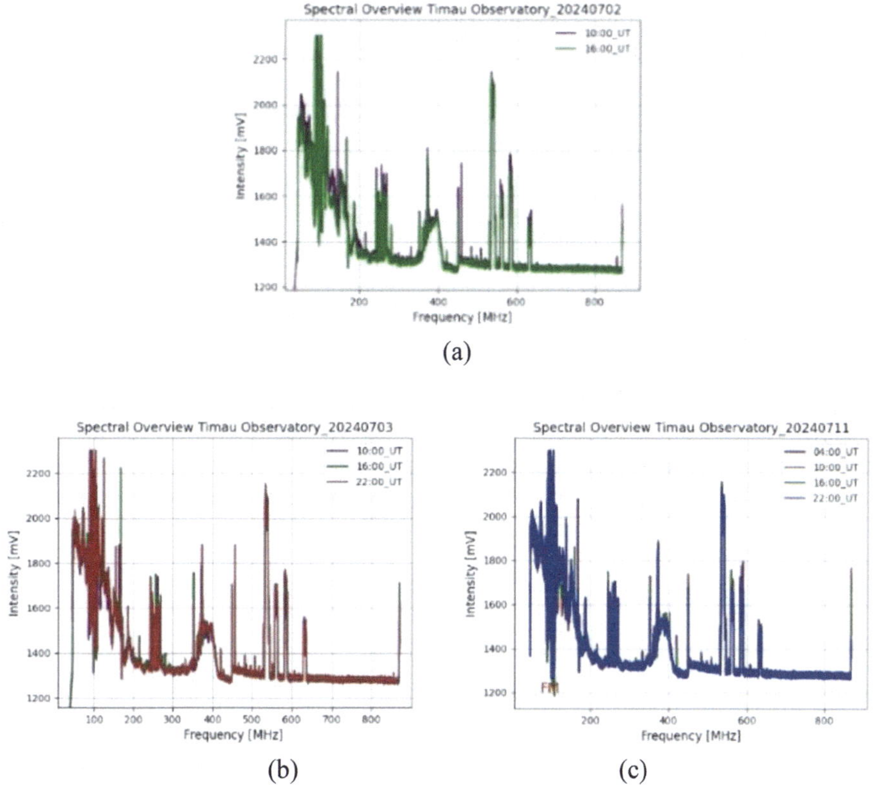

Fig. 7. Spectral Overviews Plots on 2, 3, and 11 July 2024 show Radio Frequency Interference measurements at Timau National Observatory through Callisto.

on the hardware side, devices that emit interference are also not activated around the observation location.

On the other hand, to mitigate RFI, the Radio Quiet Zone of the TNOI will be determined by consulting with the authorities on using radio frequencies to protect. However, particular signals will be difficult to avoid due to the increase in development. However, at these frequencies, radio burst characteristics will still be obtainable.

4 Summary

Operation of Callisto at Timau National Observatory can detect solar radio bursts and determine their characteristics, such as type I, II, III, IV, and V. Further data processing can be performed according to the research targets.

Reception of Callisto on 2, 11, and 12 July 2024 show recurring signals at a frequency of around 300 MHz lasting about 1–2 min and repeating several times, indicating their use for radio navigation for flight and maritime applications. The reception of Callisto at a wide frequency range up to 870 MHz on July 2, 2024, and on July 3, 2024, requires clarification of the solar activity. Moreover, the receptions of type III bursts on 16 and 18

July 2024 correlated with solar activity flares matched with the NOAA's Space Weather report, while the reception on 19 July 2024 did not correlate to solar activity.

RFI measurements up to 870 MHz at TNOI were dominantly affected by strong FM signals, some spectrums used for data uplink and downlink of satellite communications and satellite communications from the Earth into space, while RFI due to digital broadcasting and cellular phones were less and devoid of radio interference. Initial measurements of radio frequency interference were used to determine Callisto's operating frequencies in frequency measurement configurations to achieve the best frequency gap with the lowest radio interference.

Observations of current solar radio bursts using Callisto at Timau could be conducted manually. These observations will be further refined for an automated observation system.

A data processing for homogenous noise removal, unless background signal removals from the original data have been developed, includes an open source-based Callisto data processing and automatic data transfer system in the future. The spectrograms of the frequency of radio wave intensity generated using the Gaussian filter and image enhancement, as well as the Numpy Python modules, will be used as a script run automatically by the task scheduler.

References

1. Gopalswamy, N.: Coronal mass ejections and space weather. In: Tsuda, T., Fujii, R., Shibata, K., Geller, M.A. (eds.) Climate and Weather of the Sun-Earth System (CAWSES): Selected Papers from the 2007 Kyoto Symposium, pp. 77–120. TERRAPUB, Tokyo (2009)
2. Benz, A.O., Monstein, C., Meyer, H.: Callisto–a new concept for solar radio spectrometers. Solar Phys. **226**, 143–151 (2005)
3. Wild, J.P., McCready, L.L.: Observations of the spectrum of high-intensity solar radiation at metre wavelengths-I. The apparatus and spectral types of solar burst observed. Aust. J. Sci. Res. Ser. A **3**, 387–398 (1950)
4. Mumpuni, E.S., et al.: A Brief Overview of the Timau National Observatory, LAPAN Space Science Center, Bandung (2017). (in Indonesian)
5. e-Callisto. https://e-callisto.org. Accessed 31 July 2024
6. Manik, T., Sitompul, P., Batubara, M., Harjana, T., Yatini, C.Y., Monstein, C.: Solar radio observation using Callisto spectrometer at Sumedang West Java Indonesia: current status and future development plan in Indonesia. In: Dorotovic, I., Fischer, C.E., Temmer, M. (eds.) Proceedings of Ground-based Solar Observations in the Space Instrumentation Era: ASP Conference Series, vol. 504, pp. 331–337 (2016)
7. Sitompul, P.P., et al.: A study for a radio telescope in Indonesia: parabolic design, simulation of a horn antenna, and radio frequency survey in frequency of 0.045–18 GHz. Aerospace **11**, 52 (2024). https://doi.org/10.3390/aerospace11010052
8. Space Weather Live. www.spaceweatherlive.com. Accessed 31 July 2024
9. NOAA's Space Weather Prediction Center. www.swpc.noaa.gov. Accessed 31 July 2024
10. Regulation of Indonesia Ministry of Communications and Informatics No. 12, year of 2022 (2022). (in Indonesian)

The foF2 Depression over Indo-Australian Region Following the May 2024 Super Storm

Jiyo Harjosuwito[1](✉), Annis Siradj Mardiani[1], Varuliantor Dear[1], and Rezy Pradipta[2]

[1] Research Center for Space, Research Organization for Aeronautics and Space (ORPA), National Research and Innovation Agency (BRIN), Bandung, Indonesia
jiyo@brin.go.id

[2] Institute for Scientific Research, Boston College, Chestnut Hill, MA, USA

Abstract. In this paper we discuss the foF2 depression over the Indo-Australian region, which occurred several hours after the geomagnetic super storm of 10 May 2024. Observations by using ionosondes in Perth, Learmonth, Darwin, Kupang, Cocos Island and Pontianak to obtain ionospheric parameters, foF2. The results show that the foF2 depression has occurred over Perth, Learmonth, Darwin, Kupang and Cocos Island during daytime on 11 May 2024. However, over there is no response over Pontianak. Another result is time delay between storm onset and onset time of depression with the Moderate-Severe level, which is between 5 h (Perth) to 7.5 h (Kupang & Cocos Island). The difference in time delay indicates that the impact of magnetic storms propagates from high latitude to low latitude areas. In this super storm event, there was no foF2 increasing before depression.

Keywords: foF2 Depression · Super Storm · Time Delay

1 Introduction

Research on foF2 depression in Indonesia caused by geomagnetic storms is still lacking. Previous studies have conducted by [1] which examined the effects of geomagnetic storms on the decrease in TEC over Pontianak on March 17 and June 23, 2015; and then by [2] conducted a study on the effects of the October 29–November 3, 2003 storm to foF2 over Sumedang.

A geomagnetic storm is an incidental event that has been known to have an influence on changes in the ionosphere's circulation, which is inherently one of the sources of foF2 depression [3]. Moreover, the research on foF2 depression has important meaning in the development of models or prediction methods for MUF depression as an early warning information of HF radio communications disturbance.

On 10 May 2024 there was a geomagnetic storm with a Severe level with a Dst index decreasing to −412 nT [4] This storm is called a Super Storm because the decrease in Dst values is categorized as very extreme. One of the impacts of a Super Storm is that it will change the circulation in the ionosphere layer which will reduce the electron density of that layer.

E. Fitrianingsih et al. (Eds.): ISAST 2024, SPPHY 416, pp. 313–320, 2025.
https://doi.org/10.1007/978-981-96-1344-1_34

The foF2 is the maximum frequency of radio waves that propagate vertically and reflected by the ionosphere layer. The foF2 value is proportional to the Maximum Usable Frequency (MUF), which can be used in HF (High Frequency) radio communications. Therefore, the ratio of foF2 to its median for last 30 days will be proportional to the MUF depression; that is one of the space weather information used in civil aviation navigation [5].

Research regarding the impact of the 10 May 2024 Super Storm to foF2 change over Indonesia and Australia region is important to improve the understanding of the relation between MUF-Depression and Geomagnetic Storm. This result will improve the references in developing MUF Depression models or forecasting methods for civil aviation navigation information services.

2 Data and Method

In order to identify geomagnetic storms, we examine data from geomagnetic indices with a resolution of one minute [6]. Therefore, we utilize the SYM-H as a proxy for the high resolution for Dst index, acquired from NASA's Space Physics Data Facility (SPDF) at the Goddard Space Flight Centre [7].

We used foF2 data obtained from observations at ionosonde stations in Indonesia, Australia's west coast, and the Indian Ocean [8]. The observation stations are Pontianak (0.03°S, 109.33°E), Kupang (10.16°S, 123.73°E), Cocos Island (12.16°S, 96.88°E), Darwin (12.55°S, 130.85°E), Learmonth (22.25°S, 114.08°E), and Perth (31.94°S, 115.95°E). The Ionosonde generates an ionogram every 15 min.

The ionograms were scaled based on the rules in UAG Report 23A [9], to obtain the foF2 value. The foF2 depression is measured based on the ratio (r) between foF2 and the median for the last 30 days (30d-med). For this reason, we use foF2 data obtained from observations at these 6 stations since April 10 2024 until 10 May 2024. For each station, 30d-med is determined and then we obtained r for May 10th, 11th, and 12th.

The graph of r on the date of the Super Storm and several days after is used to examine the impact of the Super Storm on foF2 depression. The important parameter that will be obtained is the time delay between Storm Onset (SO) and the starting time of depression foF2 with Moderate ($0.5 < r \leq 0.7$) and Severe ($r < 0.5$) levels. Ionospheric storms occur when foF2 depression reaches *Moderate* to *Severe* levels [5, 10]. The starting time for post storm depression (PSD) [11] is determined based on the time when the foF2 depression reaches the Moderate level.

3 Result and Discussion

Based on the SYM-H on 10 May 2024 (see Fig. 1), the onset of the main phase of geomagnetic storm (SO) is 17:15 UT. The main phase of the storm occurred during 9 h and reached a minimum value of −518 nT. The z-component of the planetary magnetic field, Bz, decreased below −10 nT at 17:59 UT and fluctuated with negative values until 16:16 UT on 11 May 2024. The solar wind speed increased and exceeded 500 km/s at 17:50 UT, and reaches its maximum at 01:24 UT on 12 May 2024, with a value of 1026 km/s. Then, the solar wind speed decreased and reached a value below 500 km/s on May 14 at 17:15 UT. Based on SYM-H, the main phase occurs for 9 h, from May 10 at 17:15 UT to May 11 at 02:16 UT. Then, the recovery phase occurs approximately 123 h, from the end of the main phase until May 16 at 05:28 UT.

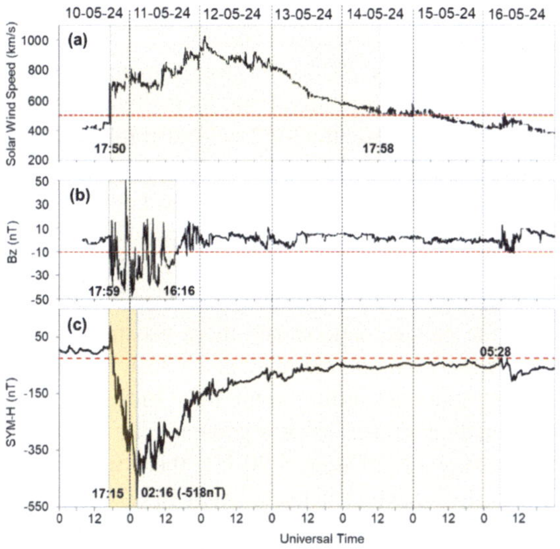

Fig. 1. Solar wind speed (a), z-component of IMF (b), and SYM-H (c) during 10–16 May 2024 (Source: https://omniweb.gsfc.nasa.gov/form/omni_min.html).

Figure 2 shows the decrease in foF2 from May 10th to May 12th, 2024. Figure 2 shows the observational findings obtained over Pontianak. There was no observation at this station on 10 May due to technical issues. The data collection commenced on May

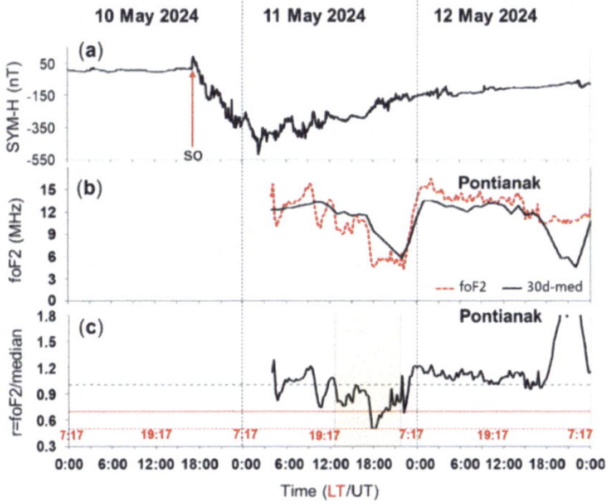

Fig. 2. SYM-H (a), foF2 (dash-red) and 30-days median (solid-black) (b), and *r* over Pontianak (c) during 10–12 May 2024.

11, 2024, at 4:00 Universal Time (UT). The foF2 depression occurred from 17:35 UT to 19:35 UT, aligning with the early morning hours in the local time zone (UT+7). Based on this, and the relatively significant time lag between the onset of the main phase of the geomagnetic storm (SO), the foF2 depression was driven by other effects rather than the geomagnetic storm. This finding is comparable to [12] for foF2 over Phu Thuy and Vanimo, which are closer to the geomagnetic equator.

Figure 3 shows that on 10 May 2024 foF2 over Kupang has been depressed since 17:35 UT. However, based on the foF2 trend and its median, this depression is not the effect of a geomagnetic storm because the foF2 trend still follows its median and occurs in the early morning local time. The foF2 depression, which is the result of a geomagnetic storm, began at 23:50 UT and entered the moderate level at 0:45 UT on May 11, 2024. At 0:45 UT, this is the start time of the post-storm depression (PSD) in the ionosphere above Kupang, so the delay time between SO and PSD is 7 h 30 min.

Figure 4 depicts the foF2 values observed over Cocos Island from May 10th to May 12th, 2024. Prior the geomagnetic storm, there was a decrease in the foF2 level at this station, which reached a moderate level at 17:35 UT. The depression occurred during the early morning hours, according to the local time. The resemblance between the foF2 trend and its median suggests that a geomagnetic storm was not responsible for it. Subsequently, on May 11, 2024, at 0:20 UT, there was a decline in the critical frequency of the F2 layer (foF2), which persisted until 18:00 UT. The depression reached a moderate level at 0:45 UT. According to the foF2 trend and its median, this indicated

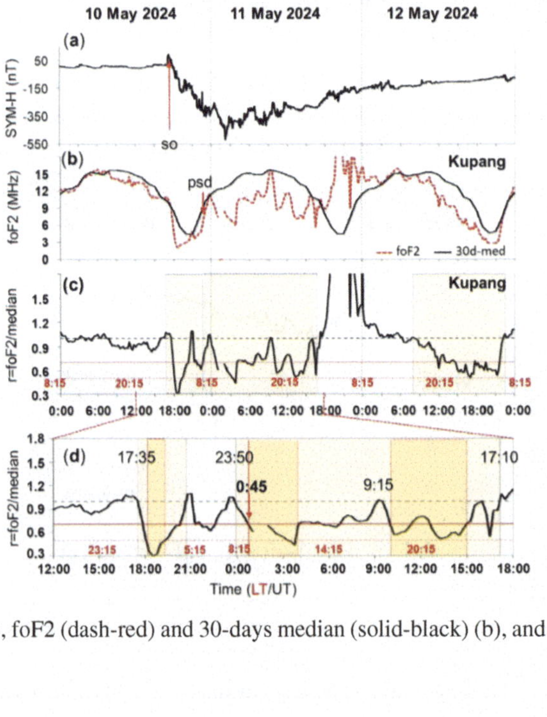

Fig. 3. SYM-H (a), foF2 (dash-red) and 30-days median (solid-black) (b), and *r* over Kupang for 10–12 May 2024.

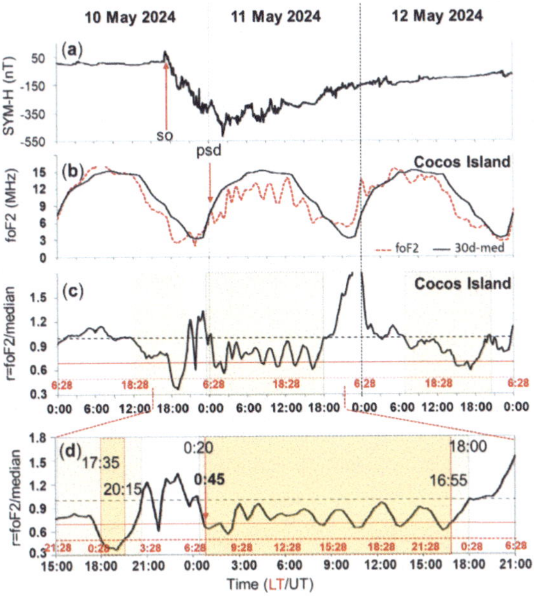

Fig. 4. SYM-H (a), foF2 (dash-red) and 30-days median (solid-black) (b), and *r* over Cocos Island for 10–12 May 2024.

the initiation of PSD over this station. The time difference between SO and PSD is approximately 7 h and 30 min.

Figure 5 shows the foF2 values for the period of 10–12 May 2024 in Darwin. The foF2 value over this station on May 10 has been below its median at 13:40 UT and reaches a moderate level around 17:50 UT. This depression is not attributed to a geomagnetic storm as it takes place during the early morning and the foF2 trend remains consistent with its median. At 21:10 UT, the foF2 decreases and remains low until 8:40 UT the following day. The severity of the depression reaches a moderate level precisely at 0:00 UT on 11 May 2024. According to the foF2 graph pattern and its median, the starting time for PSD over Darwin can be determined. The time delay between SO and PSD is approximately 6 h and 45 min.

Figure 6 depicts the foF2 levels during the period of 10 to 12 May 2024 at Learmonth. There was no decrease in the foF2 value before 20:40 on May 10. Subsequently, the foF2 recognized a period of decreased activity until 10:10 UT on the next day. The foF2 depression at 22:55 UT was of moderate level, causing a time delay of 5 h and 40 min between SO and PSD at Learmonth.

Finally, Fig. 7 depicts the foF2 observations in Perth during May 10 to 12, 2024. The foF2 depression started on May 10, 2024, at 17:35 UT and continued until 10:10 UT the next day. The foF2 depression reached the Moderate level at 18:50 UT or 1:50 LT. The cause of foF2 depression in the early morning is complex so that 18:50 cannot be set as the starting time of PSD. Therefore, we use the foF2 trend and its median to determine the starting time of PSD. The median foF2 has increased rapidly at 22:15 UT, but there

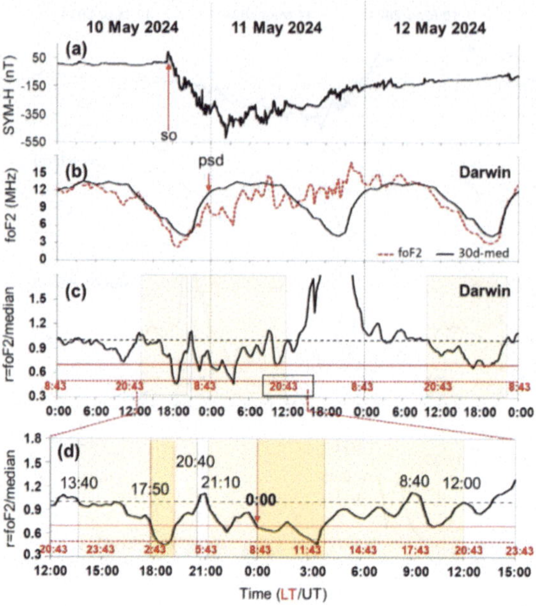

Fig. 5. SYM-H (a), foF2 (dash-red) and 30-days median (solid-black) (b), and *r* over Darwin for 10–12 May 2024.

is no increase in foF2. Therefore, 22:15 UT is determined as the starting time of PSD and then we obtain the time delay between SO and PSD is 5 h.

From this discussion we conclude delay between SO and the start time of Moderate-Severe level depression for 5 stations, as follows: ~5 h for Perth, ~5 h and 40 min for Learmonth, ~6 h and 45 min for Darwin, ~7 h and 30 min for Cocos Island and Kupang, meanwhile above Pontianak there was no foF2 depression. Difference in time delay indicates that the effects of geomagnetic storms propagate equatorward [12–15]. This is a general understanding of the effects of geomagnetic storms. An interesting finding from this superstorm is, foF2 value does not increase first but is directly depressed. This phenomenon is different from geomagnetic storms in general, where foF2 increases and then decreases. This contrast is an open question and requires further study.

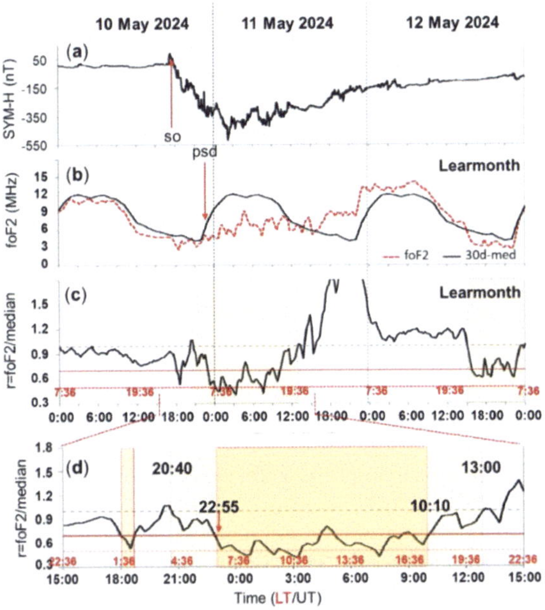

Fig. 6. SYM-H (a), foF2 (dash-red) and 30-days median (solid-black) (b), and *r* over Learmonth for 10–12 May 2024.

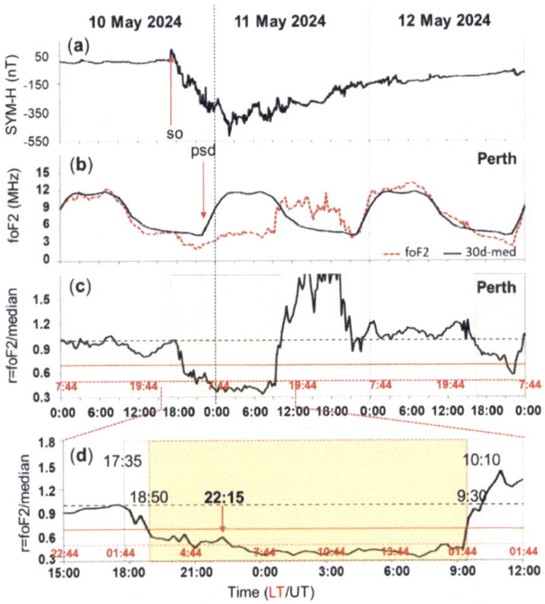

Fig. 7. SYM-H (a), foF2 (dash-red) and 30-days median (solid-black) (b), and *r* over Perth for 10–12 May 2024.

4 Conclusion

The foF2 depression occurred hours after the onset of a geomagnetic storm. The time delay between the onset of the main phase of magnetic storms is 5 h, 5 h and 40 min, 6 h and 45 min, 7 h and 30 min, and 7 h and 30 min, respectively, for Perth, Learmonth, Darwin, Cocos Island, and Kupang. The ionospheric response at higher latitudes (Perth & Learmonth) is earlier than its response at lower latitudes (Darwin, Cocos Island, Kupang). This result is agreed with previous common results that the propagation of ionospheric disturbances due to geomagnetic storms is equatorward.

References

1. Ekawati, S., Santoso, A.: Comparison between the 17 March and 23 June 2015 geomagnetic storm effect on different response of ionospheric TEC. Jurnal Sains Dirgantara **16**, 67–78 (2019)
2. Santoso, A., et al.: Indonesia ionospheric storm model related to geomagnetic storm. Jurnal Sains Dirgantara **15**, 25–38 (2017)
3. Timocin, R.: The effect of different phases of severe geomagnetic storms on the low latitude ionospheric critical frequencies. Adv. Space Res. **64**, 2280–2289 (2019). www.elsevier.com/locate/asr
4. World Data Center for Geomagnetism, Kyoto (2024). https://wdc.kugi.kyoto-u.ac.jp/dst_realtime/202405/index.html
5. Kauristie, K., et al.: SpaceWeather services for civil aviation—challenges and solutions. Remote Sens. **13**, 3685 (2021). https://doi.org/10.3390/rs13183685
6. Wanliss, J.A., Showalter, K.M.: High-resolution global storm index: DST versus SYM-H. J. Geophys. Res. **111**, A02202 (2006). https://doi.org/10.1029/2005JA011034
7. SPDF, Goddard Space Flight Center. https://omniweb.gsfc.nasa.gov/form/omni_min.html. Accessed 29 June 2024
8. Commonwealth of Australia 2024, Bureau of Meteorology. https://sws.bom.gov.au/HF_Systems/1/3. Accessed 29 June 2024
9. Piggott, W.R., Rawer, K.: The URSI Handbook of Ionogram Interpretation and Reduction (WDC-A report UAG-23, World Data Center A for Solar-Terrestrial Physics, NOAA (1978). https://www.ngdc.noaa.gov/stp/space-weather/online-publications/stp_uag/documentation/stp_uag-list.pdf
10. International Civil Aviation Organization, Manual on Space Weather Information in Support of International Air Navigation (2018)
11. Fiori, R.A.D., Kumar, V.V., Boteler, D.H., Terkildsen, M.B.: Occurrence rate and duration of space weather impacts on high-frequency radio communication used by aviation. J. Space Weather Space Clim. **12**, 21 (2022). https://doi.org/10.1051/swsc/2022017
12. Lynn, K.J.W., Sjarifudin, M., Harris, T.J., Le Huy, M.: Combined TOPEX/Poseidon TEC and ionosonde observations of negative low-latitude ionospheric storms. Annales Geophysicae **22**, 2837–2847 (2004). SRef-ID: 1432–0576/ag/2004-22-2837
13. Cherniak, I., Zakharenkova, I.: Large-scale traveling ionospheric disturbances origin and propagation: case study of the December 2015 geomagnetic storm. Space Weather **16**, 1377–1395 (2018). https://doi.org/10.1029/2018SW001869
14. Danilov, A.D.: Ionospheric F-region response to geomagnetic disturbances. Adv. Space Res. **52**(2013), 343–366 (2013). https://doi.org/10.1016/j.asr.2013.04.019
15. Kishore, A., Kumar, S.: Large scale traveling ionospheric disturbances during geomagnetic storms of 17 March and 23 June 2015 in the Australian Region. J. Geophys. Res. Space Phys. **128**(11), e2023JA031740 (2023). https://doi.org/10.1029/2023JA031740

Innovative Aviation and Aeronautics Technology Toward a Green and Sustainable Environment

Motor-Propeller Matching Analysis of Axial Flux Brushless DC Motor

Najwa Rashika Az-Zahra Raharema(✉), Muhammad Harist Herti Ramadhani, Muhammad Faiz Kamal, Kresno Bayu Aditya Seno, and Firman Hartono

Institut Teknologi Bandung, Bandung, Indonesia
{13622098,13722019,13722059,13123005}@mahasiswa.itb.ac.id,
firman.hartono@itb.ac.id

Abstract. Optimizing the flight performance of an Unmanned Aerial Vehicle (UAV) is essential, especially for electric-driven propulsion systems, thereby it requires a proper motor-propeller matching process. This process analyzes the motor and propeller torque in search of the optimum motor speed. This method can help determine the optimal configuration between the propeller and electric motor to ensure the motor produces maximum power with minimal energy consumption, i.e., improving the motor's efficiency. This study aims to analyze the motor-propeller matching problem on an axial flux brushless direct current motor configuration. The result of this analysis is a graph showing intersections between the motor and propeller torque, which indicates the optimum operating conditions for the motor. This means that the motor's output power will produce sufficient torque required by the propeller. Insufficient torque will lead to more current consumption, and excessive torque and rpm will cause the motor to accelerate. A coupled motor-propeller power curve describing the propulsion system's available power is also presented. A quantitative approach to find the motor's available power and matching torque is done through a simulation using MATLAB. The produced thrust and required power for the propeller data is obtained from a propeller performance map.

Keywords: Motor-propeller matching · Axial flux motor · Motor performance · Propeller performance

1 Introduction

Recently, electric motors have shown vast growth for UAV propulsion systems due to their great practicability and cost-efficiency [1]. In addition, constructing and operating an electric motor is relatively easy, making it one of the best options for UAV propulsion systems. Generally, electric motors are divided into two categories, namely brushed direct current (BDC) and brushless direct current (BLDC). The main difference is that there is no brush to deliver electric current to the coil on a BLDC motor. Instead, it uses an electric speed controller (ESC) to power each coil in sequence [2]. Since it works without brushes,

E. Fitrianingsih et al. (Eds.): ISAST 2024, SPPHY 416, pp. 323–332, 2025.
https://doi.org/10.1007/978-981-96-1344-1_35

it has many advantages—not requiring periodic treatment, minimizing carbon snarling, and no sparks arising [3].

There are two types of BLDC motors based on their magnetic flux configuration: axial and radial flux. Radial flux motors produce magnetic flux perpendicular to the shaft, whereas the axial flux motors produce magnetic flux parallel to the shaft and the winding arrangements are in a planar fashion. It has been quite proven that the axial flux BLDC motor is capable to compete with the radial flux motor, as it has better heat dissipation, flexible air gap arrangement, and produces higher torque and efficiency [4]. Axial flux BLDC motor' high-power density, efficiency, wide speed range, robustness, low cost, and less maintenance also make them popular in many applications [5], including UAVs.

The propeller and motor are both main parts of the UAV propulsion system. The main task of a propeller is to generate thrust, a result of the propeller's interaction with the fluid in which it operates. This motion generates aerodynamic forces with a distinct direction towards its axis [6]. The propeller will not generate any thrust without the required power supply from the motor through the propeller's axis. If both motor and propeller are properly matched, the flight vehicle will perform remarkably, and thus, the research on matching problems between motor and propeller is important [7].

There are several previous researches regarding the engine/motor-propeller matching method. Most of them are for marine vehicle applications, considering a propulsion overload and excessive fuel consumption problems may happen without proper engine-propeller matching [8]. Moreover, the ship's propeller differs from the ones used in aerial vehicles and is more expensive [9]. Other studies on engine-propeller matching in aerial vehicles have also been conducted. For instance, Zheng et al. (2015) developed the propeller matching characteristics on quadrotor aircraft [7], Gur and Feldman (2023) discuss the various aspects of reciprocating, turboshaft, and electric engine to propeller matching process [6], and Anggraeni et al. (2014) explore the engine and propeller selection for surveillance UAV [10].

However, research in axial flux BLDC motors regarding the motor-propeller matching still needs to be done. The superior feature of the axial flux BLDC motor, which has been mentioned before, will only result in an inefficiency if a motor-propeller matching is not done. In this research, a static testing displayed a significant difference in loaded and unloaded torque and RPM, indicating that improper motor-propeller matching may be the reason. Therefore, a study on the motor-propeller matching analysis of the axial flux BLDC motor is needed.

The purpose of this research is to analyze the motor-propeller matching of an axial flux BLDC motor by identifying the torque curve intersections across different RPMs and throttle settings. The motor's optimal operating conditions, which are essential to maximize its performance and efficiency, are represented by these intersections. Based on the understanding of these matching points, this study offers important insights that can guide towards the selection of propellers for a given axial flux BLDC motor to ensure the motor's optimal performance in various situations. The motor and propeller parameters can also be adjusted to meet those intersections. The motor used in this study is a single-rotor, single-stator axial flux BLDC motor with a 9-pole 12-slot configuration. For consistency, this paper will use the term "motor" to refer to the corresponding axial flux BLDC motor to suit the discussion better. The motor design is illustrated in Fig. 1.

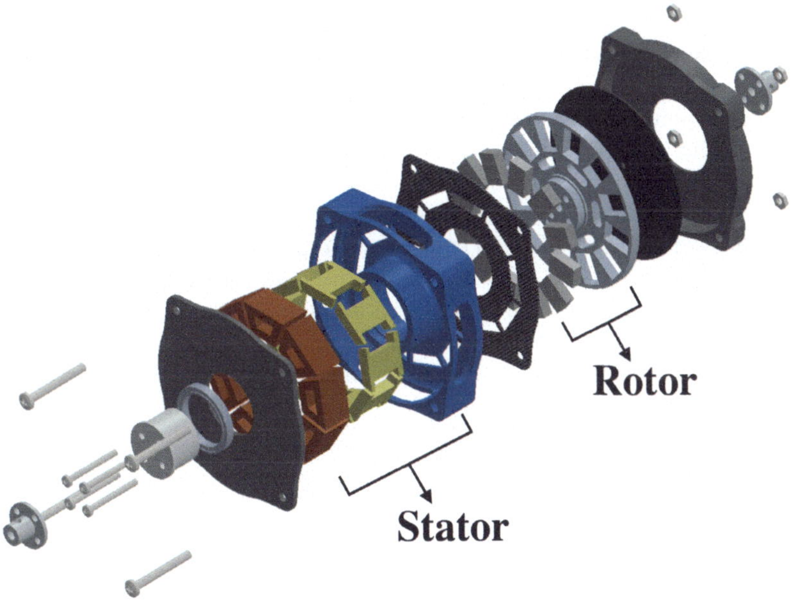

Fig. 1. Single-stator single-rotor axial flux BLDC motor design.

2 Method

To find the motor-propeller matching, the performance of both motor and propellers, in the forms of torque in newton meter (Nm) and the rotational speed of the motor in revolution per minute (rpm), is needed. Propeller performances are usually represented on propeller performance maps, which show the propeller thrust coefficient C_T and coefficient of power C_P as a function of advance ratio J [11]. The advance ratio J, thrust T, and shaft power P can be found using Eq. 1, 2, and 3.

$$C_T(J) = \frac{T}{\rho_a n^2 D^4} \tag{1}$$

$$C_P(J) = \frac{P}{\rho_a n^3 D^5} \tag{2}$$

$$J = \frac{V_\infty}{nD} \tag{3}$$

where ρ_a is the air density, V_∞ is the freestream velocity, n is the rotational speed (in rpm), and D is the propeller's diameter. Both propeller and motor performance models should be available based on flight conditions variables, such as altitude, atmospheric conditions, airspeed, and temperature. If the predefined conditions are considered, i.e., flying at those known variables, the propeller performance model will only depend on its rotational speed and pitch angle. Likewise, the motor performance model will depend only on its speed and throttle settings [6].

The corresponding propeller data used in this research is fixed APC propellers in four different dimensions: 10×4, 11×7, 14×10, and 16×10. The propeller's size is represented by two numbers separated by an 'x', where the first number indicates the propeller's diameter in inches and the second number denotes the pitch in inches per revolution. To construct the propeller performance map, the C_T and C_P data are taken from the APC Propeller Performance Data website [12] and plotted against J for various rotational speeds. To visualize this relation, the resulting C_T and C_P vs J graphs for 10×4 propeller dimensions are presented in Fig. 2. The propeller torque Q_p is obtained using the relation between the propeller torque coefficient C_Q and the propeller coefficient of power C_P, as expressed in Eq. 4 and 5 [13].

$$C_Q = \frac{C_P}{2\pi} \tag{4}$$

$$C_Q = \frac{Q_p}{\rho_a n^2 D^5} \tag{5}$$

To find the matching point, the motor torque Q_m and propeller torque Q_p are plotted against the rotational speed to find the correct rotational speed to match the output torque of the motor and propeller's required torque. It was done using MATLAB in which the source code was obtained from [14] and [15]. The original code, which was made for a simple BLDC motor, was modified to specifically account for the torque calculation of a single-rotor, single-stator axial flux BLDC motor, as expressed in Eq. 6 [16].

$$Q_{out} = \frac{\sqrt{2}}{240}\pi^2 \frac{N_p}{N_s} k_d k_f k_{io}(1 - k_{io}^2) A_e B_g \alpha_i D_{out}^3 \gamma \tag{6}$$

where N_p and N_s, represents the number of poles and slots respectively. The distribution factor and form factor are denoted respectively as k_d and k_f. Additionally, k_{io} indicates the ratio of the rotor's inner diameter to its outer diameter. The electrical loading A_e measures the current per unit area in the windings, and B_g denotes the air gap flux density. The ratio of the pole arc to the pole pitch is represented by α_i, and D_{out} denotes the rotor's outer diameter. Finally, γ signifies the efficiency of the motor.

Furthermore, the power available provided by the propulsion system (motor and propeller) will be plotted against the airspeed for various throttle settings.

This graph is referred to as the "power curve". The systematic flowchart to conduct the motor-propeller matching analysis is illustrated in Fig. 3.

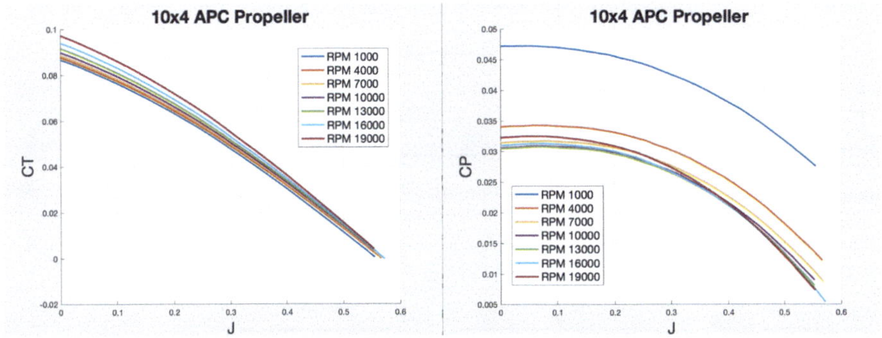

(a) (b)

Fig. 2. (a) Thrust coefficient vs advance ratio graph for 10 × 4 APC propeller (b) power coefficient vs advance ratio graph for 10 × 4 APC propeller.

3 Results and Discussion

3.1 Static Test Result

The specifications of the motor and the static test result using a 4S lithium-polymer battery were obtained and tabulated in Table 1. The air-gap flux density and electrical loading were obtained using simulation through ANSYS Maxwell software.

Table 1. Motor specifications and static test result.

Specifications	Value	Specifications	Value
Motor Diameter	90 mm	KV	924 RPM/V
Motor Thickness	40 mm	Peak Current (180 s)	22 A
Shaft Diameter	5 mm	Idle Current (16.8 V)	0.19 A
Weight (Cable included)	420 g	Maximum Power	550 W
Configuration	9N12P	Internal Resistance	0.2 Ω
Air-gap Flux Density	0.901 T	Electrical Loading	10.794 A/mm

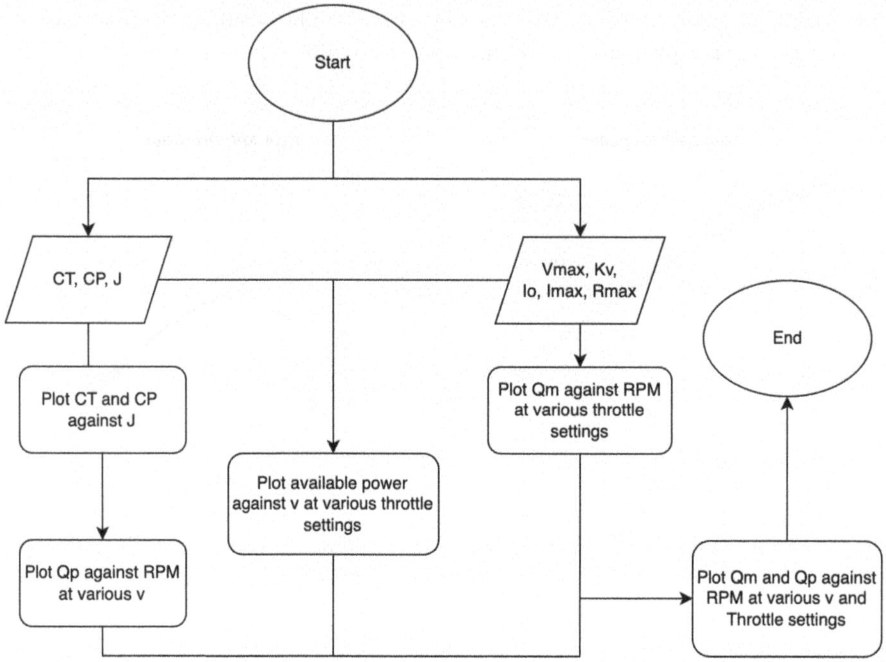

Fig. 3. Motor-propeller matching process.

3.2 Motor-Propeller Matching

The matching process for all propeller dimensions was done. Apparently, not all graphs intersect due to mismatches between the motor's torque and power output and the propeller's required torque. At a particular RPM, the torque curves will not align if the motor produces less power than what the propeller needs. Besides, the motor might be optimized for different load conditions than the propeller. A high-speed or heavy-load applications propeller, for instance, might need higher torque levels than what the motor can produce. The motor-propeller matching for the 10×4 APC propeller is depicted in Fig. 4. For comparison, the number of intersection points of the motor for various propeller dimensions is tabulated in Table 2.

From Table 2, there are interesting trends with the propeller dimensions: as the dimensions increase, it becomes more difficult for the motor to match the required torque, possibly owing to increased load and required torque at higher RPMs. A higher number of intersections implies better compatibility, permitting more flexible operation, whereas fewer intersections indicate inefficiencies or possible overload risks. The motor's optimal operating zones are highlighted by the distribution of these intersections within the RPM range, where the motor and propeller are well-matched. To ensure effective and consistent performance, choosing the best propeller requires finding a balance between the number of matching points and the motor's safe operating range.

Table 2. Number of intersections of various propeller dimensions.

Propeller Dimension	Number of Intersections
10×4	67
11×7	53
14×10	36
16×10	27

From Fig. 4, notice that the motor curve is discontinued, i.e., stops at a certain point. Theoretically, the motor's torque curve should go all the way up, as illustrated in Fig. 5 [17], where v_r is the required voltage. This issue arises as a consequence of the limitations of the motor. These limitations prevent the motor from supplying torque beyond certain points. Such limitations are affected by current limit, internal resistance, and back EMF. Once the motor reaches the current limit, the motor is unable to generate more torque, especially at higher RPMs. Moreover, due to internal resistance and increasing back EMF, the motor could not draw enough current to increase the torque.

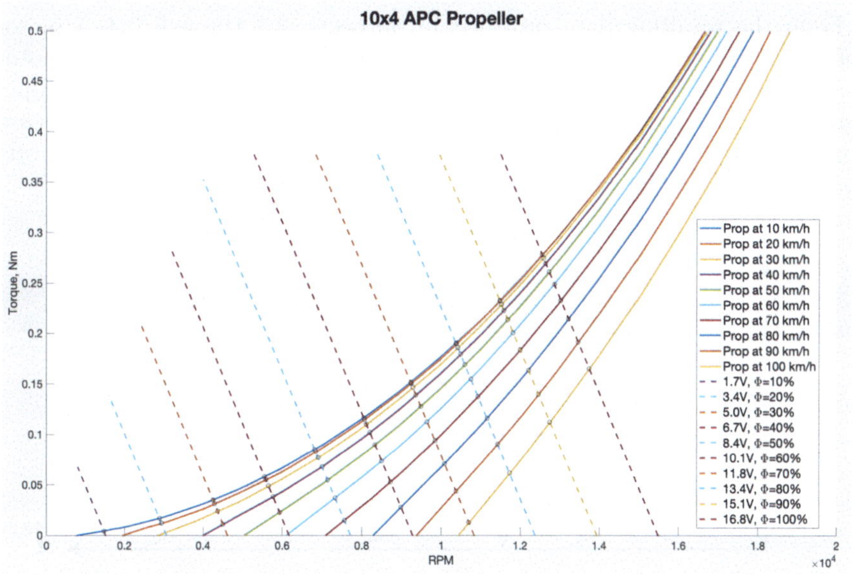

Fig. 4. Motor torque vs RPM graph of 10×4 APC propeller for various velocity and motor throttle settings (Φ).

3.3 Power Curves

To show the power provided by the propulsion system, the motor and propeller are coupled, i.e., working together as a single unit. Here, the motor and propeller performance characteristics are considered together to show how the motor and propeller operate as an integrated system. This integrated power curve can be useful to identify the operational limit of the combined propulsion system. Figure 6 shows the power curve for the coupled motor and 10 × 4 APC propeller at various throttle settings. From the graph, it can be seen that the motor is capable of supplying more power as the throttle increases, which supports the general hypothesis that a higher throttle will result in a higher power output. It is also worth noticing that at higher velocities, the propulsion system might not be able to provide sufficient power at lower throttle settings.

The C_T and C_P versus J graphs in Fig. 2 offer important information regarding the propeller's efficiency under various operating conditions. The propeller's most efficient operating range is marked by the point where the C_T and C_P curves maintain a good balance. A propeller with higher C_T at a given J generates thrust more efficiently, whereas a propeller with lower C_P at the same J is more power-efficient. Therefore, the graph can be used to determine the operational range in which the propeller produces the optimum thrust with minimal power consumption.

From the resulting data provided in Subsects. 3.1 through 3.3, a motor-propeller matching analysis can be performed in a short time. This rapid motor-propeller matching analysis can help to save time in the development of a propulsion system by shortening the design and testing phase. Moreover, it helps to identify the optimal motor-propeller combination, configuration, and throttle control to enhance the overall efficiency of the propulsion system.

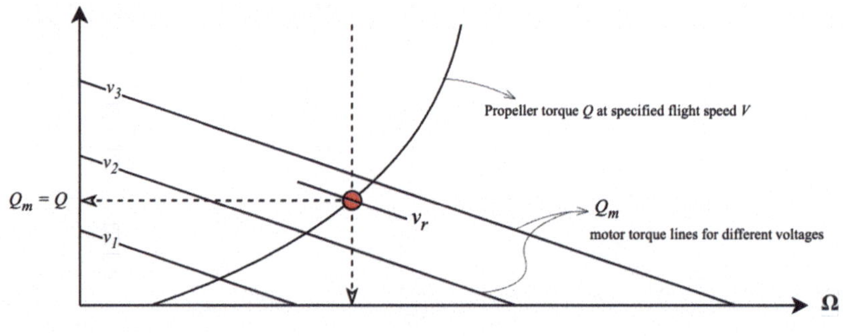

Fig. 5. Typical motor-propeller torque matching curve.

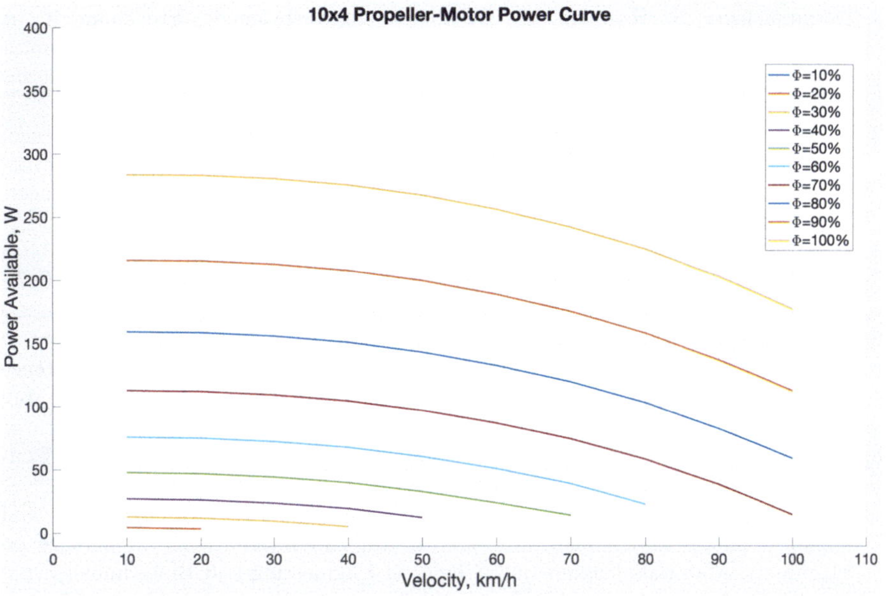

Fig. 6. Motor-10 × 4 APC propeller available power at various throttle settings (Φ).

4 Conclusion

The motor-propeller matching analysis has been done for various propeller dimensions, resulting in intersections between motor and propeller torque curves vs rpm for various airspeed and throttle settings. These intersections indicate the correct rpm to match the torque of the motor and propeller. Moreover, the motor-propeller matching analysis produced an integrated power curve of the motor and propeller which expresses the power available provided by the propulsion system. The torque intersection graph and the power curves provide important information that emphasizes the significance of choosing propellers that assure effective motor operation under various flight conditions. Further research can be done to make a motor-propeller analysis which also integrates the aerodynamic and other parameters from the flight vehicle. With this, a power curve that compares the power available from the propulsion system and the power required from the flight vehicle at various airspeeds to maintain a level flight can be constructed.

References

1. Joshi, D., Deb, D., Muyeen, S.M.: Comprehensive review on electric propulsion system of unmanned aerial vehicles. Front. Energy Res. **10**, 752012 (2022). https://doi.org/10.3389/fenrg.2022.752012

2. Derammelaere, S., Haemers, M., De Viaene, J., Verbelen, F., Stockman, K.: A quantitative comparison between BLDC, PMSM, brushed DC and stepping motor technologies. In: 19th International Conference on Electrical Machines and Systems (ICEMS), pp. 1–5. IEEE (2016)
3. Nizam, M., Waloyo, H.T., Inayati, I.: Brushless direct current electric motor design with minimum cogging torque. In: 1st International Conference on Electrical Engineering, Computer Science and Informatics 2014. Institute of Advanced Engineering and Science (2014)
4. Prasetio, K.D., Yuniarto, M.N.: Design and performance test of axial Halbach brushless DC motor with power density 1.5 Kw/Kg. Jurnal Teknik ITS **5**(2)(2017). https://doi.org/10.12962/j23373539.v5i2.20658
5. Awari, P., Sawarkar, P., Agarwal, R., Khergade, A., Bodkhe, S.: Speed control and electrical braking of axial flux BLDC motor. In: 6th International Conference on Computer Applications, Electrical Engineering-Recent Advances (CERA), pp. 297–302. IEEE (2017). https://doi.org/10.1109/CERA.2017.8343344
6. Gur, O., Feldman, J.: Engine-propeller matching. In: The 62nd Israel Annual Conference on Aerospace Sciences (2023)
7. Zheng, F., Wang, W., Li, S., Zhang, J.: Matching characteristics of propeller and engine of fuel-powered quadrotor aircraft. In: 2nd International Conference on Machinery, Materials Engineering, Chemical Engineering and Biotechnology, pp. 352–357. Atlantis Press (2015). https://doi.org/10.2991/mmeceb-15.2016.70
8. Stapersma, D., Woud, H.K.: Matching propulsion engine with propulsor. J. Mar. Eng. Technol. **4**(2), 25–32 (2005). https://doi.org/10.1080/20464177.2005.11020189
9. Carlton, J.: Marine Propellers and Propulsion. Butterworth-Heinemann (2018). https://doi.org/10.1016/B978-0-7506-8150-6.X5000-1
10. Anggraeni, D., Sumaryanto, A.R., Sumarna, E.: Engine and propeller selection for propulsion system LAPAN Surveillance UAV-05 (LSU-05) using analytic and experimental test. In: International Seminar of Aerospace Science and Technology (ISAST) II-2014, vol. II, pp. 41–50. Pustekbang LAPAN (2014)
11. Giannakakis, P., Laskaridis, P., Nikolaidis, T., Kalfas, A.I.: Toward a scalable propeller performance map. J. Propul. Power **31**(4), 1073–1082 (2015)
12. APC Propeller Performance data (2024). www.apcprop.com/technical-information/performance-data/. Accessed 27 Apr 2024
13. McCormick, B.W.: Aerodynamics, Aeronautics, and Flight Mechanics, 2nd edn. Wiley, Hoboken (1995)
14. Engdancili. Read APC propellers performance data (2024). https://github.com/dciliberti/read-APC-prop-perfo-data/releases/tag/v1.0.0,GitHub. Accessed 29 Apr 2024
15. Engdancili. simple-BLDC-model. GitHub (2024). https://github.com/dciliberti/simple-BLDC-model/releases/tag/v1.0.3. Accessed 29 Apr 2024
16. Usman, H., Ikram, J., Alimgeer, K.S., Yousuf, M., Bukhari, S.S.H., Ro, J.S.: Analysis and optimization of axial flux permanent magnet machine for cogging torque reduction. Mathematics **9**(15), 1738 (2021). https://doi.org/10.3390/math9151738
17. Drela, M.: DC Motor/Propeller Matching. Lab 5 lecture notes. Technical report. MIT (2005)

Computer Vision Implementation in a Hand Gesture-Based Indoor Drone Guidance System

Javensius Sembiring(✉), Adhitya Rizky Andhira Kusumah, Lalu Muhamad Alhadad, Patrisius Bagus Alvito Baylon, Yazdi Ibrahim Jenie, and Hari Muhammad

Faculty of Mechanical and Aerospace Engineering, Institut Teknologi Bandung, Bandung 40116, Indonesia
javen.sembiring@itb.ac.id

Abstract. This research aimed to develop an indoor drone guidance system that relies on human hand gestures for control, as Global Positioning System (GPS) is unavailable indoors. A computer vision system was implemented using Mediapipe and a Convolutional Neural Network (CNN) to recognize and interpret hand gestures captured by a drone camera. The system was trained on a dataset of 10 distinct hand gestures, achieving a 95% accuracy rate with rapid processing time. Real-world tests were conducted using a DJI Tello drone. The developed system provides a user-friendly, accessible method for controlling drones indoors, requiring no prior experience.

Keywords: Human-Drone Interaction · Computer Vision · Gesture Recognition · Mediapipe · Hybrid Model · UAV

1 Introduction

Unmanned Aerial Vehicles (UAVs), commonly known as drones, were initially developed for military applications [1]. However, the demand for civil drones has increased during recent years, with applications ranging from warehouse inspection to film-making [2]. In recent years, the aircraft maintenance, repair, and overhaul (MRO) industry has started using drones in their activities, particularly for the visual inspection of aircraft airframes [3].

Human-Drone Interaction (HDI) is a multidisciplinary field of study derived from Human-Robot Interaction (HRI) that aims to understand, design, and evaluate drone systems for use by or with humans [4]. In particular, the use of hand-gesture-based control systems offers a significant advantage over traditional remote controllers, as gestures are easier to learn, more intuitive, and require less technical expertise. A key application of this technology is in the aircraft maintenance, repair, and overhaul (MRO) industry, where drones are increasingly used for visual inspections of aircraft. These inspections often take place inside enclosed hangars, where GPS signals are unavailable, making precise control of drones a critical challenge.

Compared to remote controls, a hand-gesture-based system simplifies the control interface, reducing the need for complex equipment. The concept of operation for the

E. Fitrianingsih et al. (Eds.): ISAST 2024, SPPHY 416, pp. 333–343, 2025.
https://doi.org/10.1007/978-981-96-1344-1_36

hand-gesture-based system is shown in Fig. 1. The gestures can be captured by a camera, either mounted on the drone or placed in the ground control system. This setup requires only a camera as the hardware, eliminating the need for complicated control systems. The drone interprets hand gestures via a Convolutional Neural Network (CNN)-based algorithm, translating them into control commands that enable the drone to navigate around the aircraft. Given the GPS-denied environment inside hangars, the drone relies on its internal sensors, such as the Inertial Measurement Unit (IMU) and optical flow sensors, to maintain accurate navigation and avoid collisions.

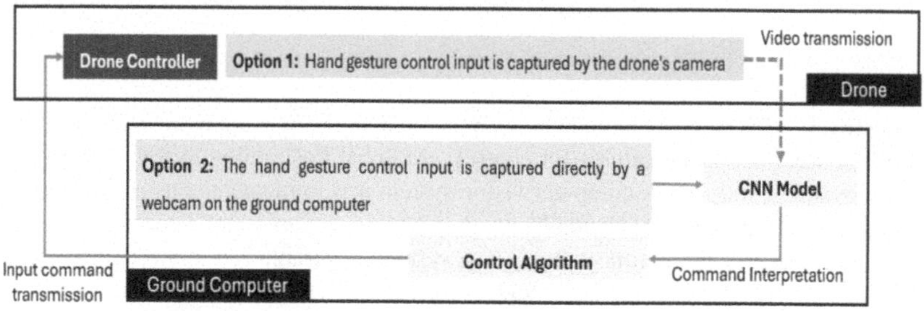

Fig. 1. Hand-gesture-control drone concept of operation.

This research focuses on advancing hand-gesture-based control for drone manoeuvres that can be implemented in the inspection process. The system will be tested to ensure it can command the drone to a user-defined position while relying solely on its internal sensors (IMU and optical flow) in a GPS-denied environment, simulating real-world conditions inside an enclosed aircraft hangar. The precision of the navigation process and the response time of the hand-gesture control will affect the ability to operate the drone safely in close proximity to human personnel and inspected aircraft, which is a critical factor in such environments.

2 Related Work

Numerous studies have been conducted in the field of Human-Drone Interaction, particularly focusing on interactions facilitated through hand gestures using vision-based gesture recognition. Vision-based gesture recognition allows an intuitive interaction between the user and the drone by analyzing visual information through feature extraction [5]. Several feature extraction techniques, including landmarks, skeletonization, hand and finger features, and machine learning-based approaches, have been explored. Landmarks and skeletonization represent hand shape and movement, while hand and finger features emphasize specific attributes. In contrast, machine learning-based features allow automatic discovery of features to improve gesture discrimination.

In [6], the authors have successfully developed a non-wearable drone control system by implementing computer vision. The system utilized Mediapipe, TensorFlow, and OpenCV for hand recognition and drone control. Mediapipe identified 21 hand landmarks [7], which were fed to a TensorFlow-based gesture recognition model. The system recognized takeoff and landing gestures, with a fail-safe mode included. Experimental results using a Parrot Mambo drone showed 95% hand gesture recognition accuracy and 90% command execution accuracy. However, this research can be further extended to allow more control commands such as directional movements.

In another research titled "Hand Gesture Recognition using DenseNet201-Mediapipe Hybrid Modelling," the authors introduced a hybrid framework for hand gesture recognition that integrates Mediapipe for hand detection and tracking with various neural network architectures for gesture classification [8]. The system was designed to recognize 18 gesture classes and a 'no gesture' class. The hybrid model detects hand gestures using Mediapipe and creates a boundary box around the hand gesture for image cropping, the cropped image was then passed into the neural network model for classification. The neural network model compared in this research are Xception, ResNet152V2, ResNetV2, and DenseNet201. Training result shows that DenseNet201 model had the best performance and was able to achieve a validation accuracy of 97.55% with very little loss.

3 Dataset Generation and Preprocessing

The hand gesture dataset was generated from video recordings of volunteers performing predefined gestures in an indoor setting. The videos were collected from 5 volunteers using both mobile devices and webcams, resulting in videos saved in MP4 format. Figure 2 shows the 10 predefined hand gestures along with its associated control commands including takeoff, land, yaw, and translational movements.

Fig. 2. Predefined gestures and associated control commands.

Python programming language, in conjunction with OpenCV and MediaPipe, was employed for hand gesture dataset preprocessing. MediaPipe, a Google-developed framework specializing in high-speed machine learning solutions, provided the MediaPipe Hand Landmarker model for accurate hand landmark detection within input images.

To prepare the dataset for model training, a preprocessing pipeline was implemented. OpenCV handled initial video frame processing, converting frames from BGR to RGB format for subsequent MediaPipe analysis. This ensured consistent and suitable input for the hand landmark detection process. The MediaPipe framework was employed to precisely locate and extract hand landmarks, representing key hand points. To ensure high-quality data, a minimum detection confidence threshold of 95% was implemented.

To mitigate bias from skin tone and lighting, landmarks were drawn on a blank image using NumPy library. A boundary box was created by identifying the minimum and maximum landmark coordinates, isolating the hand gesture. The cropped image was saved in JPG format for dataset creation. Examples of extracted images before and after preprocessing are shown in Fig. 3.

(a) (b)

Fig. 3. Extracted image for training data before preprocessing (a) and after preprocessing (b).

Preprocessing results in approximately 4000 images of hand landmarks on a blank image, which were then manually classified into 10 gesture classes. Each gesture class contained 100 images, resulting in a final hand gesture dataset of 1000 images. This dataset was then split into training and validation sets at an eight-to-two ratio. The training set was used to train the model, while validation set was used to evaluate the classification ability of the model.

4 CNN Model Development

4.1 Model Development

A custom CNN model was developed for real-time drone control based on hand gestures captured by the drone camera. The architecture, comprising five convolutional and three dense layers, aimed to balance classification performance and computational efficiency. Convolutional layers, using 32 filters each, extracted features from basic edges to complex gesture patterns. The `ReLU` activation function was employed to enhance feature extraction and mitigate the vanishing gradient problem. Batch normalization layers were inserted before convolutional and specific dense layers to standardize input data. This technique accelerated training, enhanced convergence, and improved overall model performance by mitigating internal covariate shift.

Max pooling layers were added after each convolutional layer to reduce the dimensions of the feature maps, reducing the computational load of the model and prevent overfitting. These layers identify the maximum value within a specified region (window) of the feature map, effectively capturing the most significant features. Max pooling layers not only reduced the size of the feature map, but also increased the model's robustness to variations in hand gesture positions.

Following the convolutional layers, the CNN model introduced three dense layers to integrate the characteristics identified by the convolutional layers and classify the hand gesture. Each neuron in the dense layers was interconnected with all neurons in the preceding layer, allowing the model to combine learned features in various ways to make predictions. The initial two dense layers utilized `ReLU` activation functions to introduce non-linearity, while the final output layer utilized a `softmax` activation function to transform raw output scores of the CNN model into probability distributions across the predefined hand gesture classes.

In order to improve the model's generalization capability and prevent overfitting, a dropout layer was introduced after the first dense layer. This layer serves as a regularization technique to ensure the model performs well on unseen data by reducing the likelihood of memorizing the training data. During each training iteration, the dropout layer randomly deactivated 60% of the input nodes, forcing the model to learn more robust features by preventing over-reliance on any single node. Table 1 below shows the parameters used in the proposed CNN model.

Table 1. Parameters in model development.

Parameter	Value	Parameter	Value
Optimizer	rmsprop	Max Pooling	(2 × 2)
Learning Rate	0.0001	Activation	ReLU
Loss	Sparse categorical cross-entropy	2D Convolutional Layers	32, (3 × 3)
Input Shape	(256, 256, 3)	Dropout	0.6
		Softmax	10 × 1

4.2 Model Training

The model was trained to recognize 10 distinct hand gestures using the generated dataset, comprised of 1000 images across 10 gesture classes. Model training process was conducted on the Google Collaboratory platform to reduce the training duration by utilizing its GPU and TPU support [9]. The model was trained for 500 epochs to ensure the model had sufficient exposure to the dataset to learn and refine its capability to recognize the gestures. To prevent overfitting, an early stopping regularization method was implemented, monitoring the model's performance on a validation set and stopped training when performance was no longer improving after a certain number of epochs.

The evaluation of the CNN model's performance was based on two primary metrics: loss and accuracy. The loss metric measured the difference between the model's predicted labels and the actual labels, with lower values indicating better performance. Accuracy, on the other hand, measured the proportion of correctly classified gestures out of the total predictions, serving as a straightforward indicator of the model's effectiveness. Figure 4 and Fig. 5 show the trend of each metric throughout the training process.

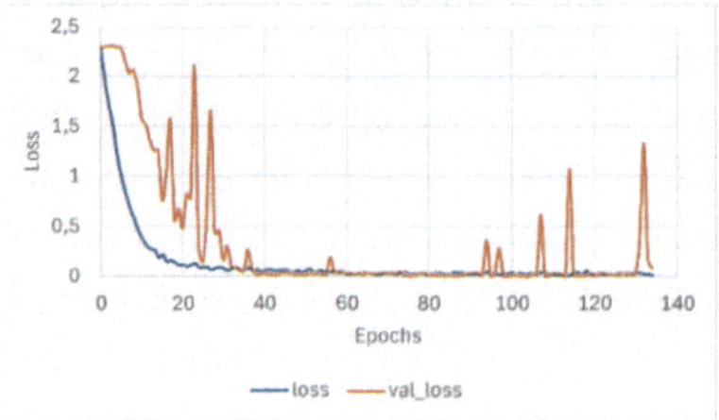

Fig. 4. Training and validation loss comparison.

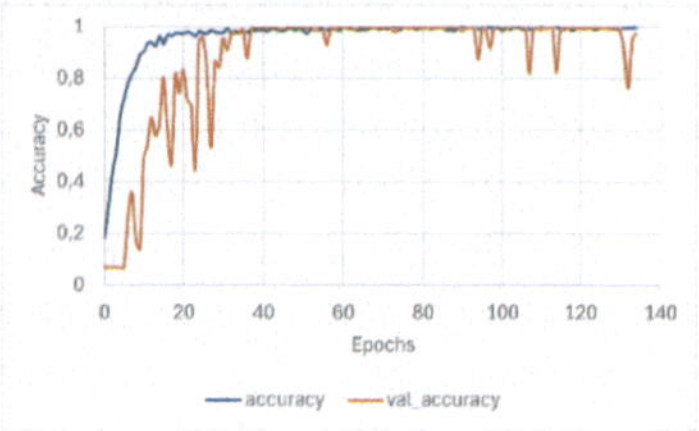

Fig. 5. Training and validation accuracy comparison.

The results indicated that the proposed CNN model achieved optimal performance after 135 epochs, successfully recognizing all 10 hand gestures with 100% validation accuracy. This highlights the model's ability to generalize effectively and maintain performance on unseen data. The final validation loss was notably low, recorded at 2.45×10^{-5}, further validating the model's precision and capability to distinguish between different hand gestures.

4.3 Model Evaluation

The prediction and classification capabilities of the CNN model proposed in this research were tested to evaluate its effectiveness in real-time applications. The evaluation involved performing predefined hand gestures in a live video stream captured from a laptop webcam, replicating real-world conditions where the model needs to recognize gestures dynamically. The model processed the continuous feed of hand gestures in real time from the live video stream. The model demonstrated high accuracy, correctly identifying over 95% of hand gestures in real time, as shown in Table 2.

Precision, measuring true positive classifications, averaged 90% for the custom CNN model. However, Move Backward and Move Left classes exhibited lower precision, indicating potential misclassifications. Recall, measuring correctly identified positive instances, reached an average of 94%, highlighting the model's ability to accurately classify hand gestures. Overall, the model demonstrated robustness and reliability with high precision and recall values.

5 Implementation

5.1 Guidance System Overview

A Python-based indoor drone guidance system incorporating a CNN model was developed for real-time operation. The system prioritizes minimal latency for precise control, particularly during visual inspections. After establishing communication with the drone, the system processes the live video stream to identify and classify hand gestures, generating control commands. The guidance system utilized OpenCV for image processing,

Table 2. Prediction results of the proposed CNN model.

Class	Precision	Recall
Takeoff	94%	100%
Land	88%	88%
Move Forward	100%	100%
Move Backward	63%	87%
Move Right	100%	100%
Move Left	68%	74%
Move Up	100%	100%
Move Down	100%	100%
Yaw Right	95%	100%
Yaw Left	96%	94%

Mediapipe for hand landmarking, and TensorFlow for CNN model execution, enabling real-time performance. A laptop with an 11th Gen Intel Core i7 processor and 16 GB RAM served as the ground control station, supporting simultaneous process execution.

5.2 Drone Communication

The drone used in this research was a DJI Tello Drone, establishing communication with the ground control station (GCS) using User Datagram Protocol (UDP) over a wi-fi network. UDP allows faster communication by eliminating the necessity for a stable connection with the GCS prior to data transmission [10]. However, this protocol may lead to noisy data and potential data loss. To mitigate these issues, a modification on the communication protocol such as adding acknowledgement of received packets [11] can be implemented to enhance the data transmission reliability.

UDP communication was executed via a Python script utilizing DJITelloPy, a dedicated Python library for Tello drone. Activation of the UDP client requires sending the "command" to Tello drone's IP address 192.168.10.1 on UDP port 8889. Responses to commands are relayed to the UDP server of the GCS, with responses indicating "ok" or "error" according to the validity of the command [12]. For example, if the user type in `move_up(x)` command in the Tello Drone interface command line then the drone will fly up with a movement range between 20–500 cm according to x value provided by user. All API commands are listed in [12]. In this research, control commands for translational movements were configured to move 30 cm up or down and 40 cm in other directions, while yaw movements were set to 90 degrees both clockwise and counterclockwise.

5.3 Image Processing and Recognition

Video data was acquired from a Tello drone's 5 MP camera at 1280 × 720 pixels and 30 FPS. Preprocessing involved resizing frames to 720 × 480 pixels and converting color format to RGB for MediaPipe analysis. A relaxed detection confidence threshold of 35% accommodated the drone's video quality and potential noise. Subsequent image processing steps included landmark drawing, bounding box creation, cropping, and resizing to 256×256 pixels for CNN input.

A Convolutional Neural Network (CNN) was employed for image recognition, processing 256×256 RGB images. Five convolutional layers extracted image features, generating feature maps that were subsequently classified by three dense layers. To ensure reliable control commands, a classification threshold of 95% was implemented. The final command was determined using NumPy's `argmax` function to select the class with the highest probability.

5.4 Real-World Application

One of the primary applications for the hand gesture-based guidance system developed in this research is the aircraft maintenance (MRO) industry, particularly in the visual inspection of aircraft airframes. Traditional methods are labor-intensive, time-consuming, and prone to human errors, resulting in operational inefficiencies.

Effective implementation of the guidance system requires a ground control station with sufficient computational power for real-time processing. The drone should also be equipped with a high-resolution camera for accurate gesture recognition. Customization may be necessary to accommodate different drone models and communication protocols. Integrating the guidance system into drones allows for safer and more efficient inspections through hand gesture control, reducing the risk of accidents.

Overall, the implementation of the hand-gesture based guidance system in the MRO industry contributes towards a more efficient and safer inspection process, subsequently enhancing the reliability and safety of aircraft operation.

6 Result and Analysis

6.1 CNN Model Comparison

In this research, two CNN models were considered: DenseNet201 and a custom CNN model developed specifically for the research. The primary objective was to evaluate and compare the performance of these models in real-time hand gesture recognition. The evaluation focused on key performance metrics such as accuracy, inference time, and robustness. Both models were evaluated in an indoor environment using the same hardware setup. The test was conducted by performing predefined hand gestures in a live video stream captured by the laptop's webcam and the respected results are presented in Table 3.

The results indicate that DenseNet201 has an average accuracy of 93% with an average inference time of 110 ms. In contrast, the custom CNN model achieved an average accuracy of 95% with an average inference time of 23 ms. This significant difference in inference time suggests that the custom CNN model proposed in this research is more efficient and better suited to be implemented in a real-time guidance system where responsiveness is essential.

Table 3. CNN model performance comparison.

Model	DenseNet201	Custom CNN
Average Accuracy	93%	95%
Average Inference Time	110 ms	23 ms
Average Precision	43%	90%
Average Recall	38%	94%

Inference time directly affects the system's ability to respond to user commands in real time. According to [13], 100 ms delay commonly leads to measurable degradation of performance. Hence, the reduced inference time minimized the delay between user command and drone response, creating a more seamless and intuitive control experience. This responsiveness is important to ensure precise control in dynamic environments, especially when drones are operated in close proximity to human personnel.

6.2 Guidance System Implementation

The custom CNN model was integrated into the indoor drone guidance system developed in Python. This system was designed to operate in real-time with minimal latency to ensure responsiveness to the control commands. The commands were given by performing predefined hand gestures in a live video stream captured by Tello drone's camera. The evaluation was focused on key metrics such as accuracy and consistency of the drone's responses. The guidance system underwent 50 tests, with the results summarized in Table 4. The test results show that despite the absence of GPS and the navigation process relying solely on the drone's internal sensors, IMU, and optical flow, the drone can navigate in an indoor setting. However, there is a deviation between the drone's actual movement and the user's intended position when conducting manoeuvres using only internal sensors.

While overall accuracy was high, translational commands like Move Forward and Move Backward exhibited superior performance compared to other commands. This variation can be attributed to the Tello drone's optical flow sensor, which is sensitive to ambient lighting conditions. Fluctuations in lighting during testing might have affected sensor accuracy, leading to velocity miscalculations and less precise movements.

As previously stated in Dataset Generation and Preprocessing section, the custom CNN model's robustness under varying lighting conditions was ensured by drawing hand landmarks on a blank image. Additionally, several data augmentation methods such as random translation and random brightness was implemented to further improve the model's robustness. However, the custom CNN model showed a decrease in gesture recognition accuracy in different camera angles, particularly over 15° rotation. This finding highlighted the importance of a larger dataset which includes variations in camera angles and hand rotation.

Table 4. Desired and actual drone movement comparison.

Command	Desired	Actual	Standard Deviation
Move Forward	40 cm	37.31 cm	5 cm
Move Backward	40 cm	38.37 cm	5 cm
Move Right	40 cm	53.37 cm	15 cm
Move Left	40 cm	46.02 cm	10 cm
Move Up	30 cm	20 cm	10 cm
Move Down	30 cm	20 cm	10 cm
Yaw Right	90°	88°	3°
Yaw Left	90°	88°	3°

7 Conclusion

A hand gesture-based guidance system was successfully developed utilizing a custom CNN model. The model achieved 95% accuracy in classifying predefined hand gestures and was able to interpret them as control commands for real-time drone guidance, with an inference time of 23 ms. This approach allows responsive and seamless interaction between the user and the drone. The developed guidance system was evaluated as an intuitive interaction method, capable of navigating the drone with decent accuracy and minimal delay. Its responsiveness is crucial for real-time applications, ensuring accurate drone guidance without significant lag. The system's intuitive nature allows accessible and user-friendly operation even for individuals without prior experience in drone operations.

In real-life implementation, the significant deviation observed when relying solely on internal sensors such as IMU and optical flow makes this approach unsuitable for precise navigation. The risk of collision with inspection personnel or the inspected aircraft is high, which could compromise safety during the inspection process. Therefore, relying only on internal sensors is not adequate for complex indoor environments. To enhance navigation accuracy and ensure safe operation, additional sensors are required. One potential solution is the integration of a localizer beacon system, which can be installed to assist the drone in accurately determining its position within the indoor setting, thereby improving the overall navigation performance and reducing the risk of collisions.

Several recommendations for future research can be explored. Redefining similar hand gestures and expanding the training dataset by implementing data augmentation methods such as random rotation can improve model performance and robustness. Enhancing image processing techniques using adaptive thresholding methods or gaussian filtering algorithm to reduce noise can further increase accuracy. Evaluating the system with different drones and advanced equipment can provide valuable insights for future development.

Acknowledgements. The authors would like to acknowledge the financial support provided by Institute of Research and Community Services (LPPM) and Institute for Innovation and

Entrepreneurship Development (LPIK) of Institut Teknologi Bandung (ITB). Their funding enabled the execution of this research project.

References

1. Mátyás, P., Máté, N.: Brief history of UAV development. Repüléstudományi Közlemények **XXXI**(1), 155–166 (2019)
2. Commercial UAV News. Commercial Drones in 2023: Trends and Growth Prospects. https://www.commercialuavnews.com/international/commercial-drones-in-2023-trends-and-growth-prospects. Accessed 06 Oct 2023
3. Novák, A., Sedláčková, A.N., Bugaj, M., Kandera, B., Lusiak, T.: Use of unmanned aerial vehicles in aircraft maintenance. Transp. Res. Procedia **51**, 160–170 (2020)
4. Introduction | Human-Robot Interaction, 08 February 2012. https://humanrobotinteraction.org/1-introduction/. Accessed 13 Oct 2023
5. Trigueiros, P., Ribeiro, F., Reis, L.P.: Vision-based gesture recognition system for human-computer interaction. In: IV ECCOMAS Thematic Conference on Computational Vision and Medical Image Processing, Madeira (2013)
6. Naseer, F., Ullah, G., Siddiqui, M.A., Khan, M.J., Hong, K.-S., Naseer, N.: Deep learning-based unmanned aerial vehicle control with hand gesture and computer vision. In: 2022 13th Asian Control Conference, Jeju (2022)
7. Google. Hand Landmarks Detection Guide. https://developers.google.com/mediapipe/solutions/vision/hand_landmarker. Accessed 16 Nov 2023
8. Padhi, P., Das, M.: Hand gesture recognition using DenseNet201-mediapipe hybrid modelling. In: 2022 International Conference on Automation, Computing and Renewable Systems (ICACRS 2022), Bhubaneswar (2002)
9. Mandal, S.: How to use Google Colab - GeeksforGeeks, 25 January 2024. https://www.geeksforgeeks.org/how-to-use-google-colab/. Accessed 04 June 2024
10. GeeksforGeeks. Reliable User Datagram Protocol (RUDP), 07 December 2022. https://www.geeksforgeeks.org/reliable-user-datagram-protocol-rudp/. Accessed 05 Sept 2024
11. Fortiner. What is User Datagram Protocol (UDP)? Fortinet. https://www.fortinet.com/resources/cyberglossary/user-datagram-protocol-udp. Accessed 02 June 2024
12. Löw, J.: Tello - DJITelloPy API Reference. https://djitellopy.readthedocs.io/en/latest/tello/. Accessed 05 June 2024
13. de Vries, S.C.: UAVs and Control Delays (2005)

Initial Development of Swarming Drone in Normal DJI Tello Drone with ESP8266

Javensius Sembiring(✉), Lalu Muhamad Alhadad, Adhitya Rizky Andhira Kusumah, Patrisius Bagus Alvito Baylon, Yazdi Ibrahim Jenie, and Hari Muhammad

Faculty of Mechanical and Aerospace Engineering, Institut Teknologi Bandung, Bandung 40116, Indonesia
javen.sembiring@itb.ac.id

Abstract. This study introduces an initial setup for swarming control algorithm designed for DJI Tello drones, which employs ESP8266 microcontrollers as communication bridges in a laboratory scale environment. The research focuses on overcoming the challenge of coordinating multiple Tello drones that typically have the same IP address, complicating conventional control methods through Python scripts. By integrating ESP8266 modules, each linked to a drone's Wi-Fi network and connected to a central computer via USB serial communication, the proposed methodology enables individualized control of each drone within the swarm. Control commands are inputted through a keyboard interface, with instructions transmitted through the ESP8266 modules using the UDP protocol. Experimental procedures involved calibration and precise adjustments to optimize flight performance, followed by command execution and movement verification. The findings illustrate successful swarming behavior, albeit with minor discrepancies between actual and intended movements attributed to sensor noise and environmental variables. Based on the results obtained, drone 1 had an average error of 1.16 cm for translational movement and 0.35° for rotational movement. While drone 2, has an average error of 1.45 cm for translational movement and 1.05° for rotation. Despite these variations, the approach was deemed viable, with errors manageable through command modifications. Future improvements may entail the integration of additional sensors, feedback mechanisms, and advanced networking protocols to enhance precision and scalability. This research establishes a foundation for effective and dependable indoor drone swarming utilizing readily available and cost-effective hardware components.

Keywords: Swarming Drone · UAV Development · Alternative Mission · Normal DJI Tello

1 Introduction

The growing interest in autonomous systems has resulted in significant advancements in drone technology. The concept of swarming drones, inspired by the collective behavior of animals such as ants, fish, bees, and birds, has emerged as a promising technology with applications in various domains including search and rescue, environmental monitoring,

E. Fitrianingsih et al. (Eds.): ISAST 2024, SPPHY 416, pp. 344–353, 2025.
https://doi.org/10.1007/978-981-96-1344-1_37

agriculture, and surveillance [1]. Swarming drones involve multiple drones working together autonomously to perform complex tasks that would be challenging for a single drone to accomplish within a reasonable timeframe.

However, achieving effective swarming behavior in drone technology remains a challenging task due to the requirements of robust communication, precise coordination, and real-time decision-making capabilities among the drones. Developing efficient communication systems to meet these requirements is crucial for the success of swarming drones [2]. This paper introduces a new communication method for swarming drones using a microcontroller to enhance communication and control capabilities.

Fig. 1. Picture of a normal DJI Tello drone.

The DJI Tello drone, shown in Fig. 1 known for its affordability, compact size, ease of use, and ability to receive control commands using programmable languages, is a popular choice for drone enthusiasts [3]. While the DJI Tello EDU version offers swarming capabilities, it comes at a higher price and requires an additional router to connect the drone to a computer [4]. In contrast, the normal DJI Tello drone faces limitations in network configuration, as it cannot switch to receiving hotspot Wi-Fi like the DJI Tello EDU version and can only broadcast Wi-Fi as a hotspot device. Based on the dji discussion forum, it is stated that the normal communication system for swarming only applies to DJI Tello EDU [5].

One significant challenge in using the normal DJI Tello drone for swarming is its limitation in network configuration, where drones with the same IP address cannot be controlled simultaneously even with the addition of another Wi-Fi receiver module [6]. This limitation necessitates the development of an alternative communication method to enable effective swarming behavior.

To address this challenge, this paper proposes an alternative communication method for the normal DJI Tello drone for swarming, utilizing the ESP8266 microcontroller as a repeater to facilitate communication between the computer and multiple drones simultaneously. The ESP8266 acts as an intermediary, receiving data from the computer via USB serial communication and transmitting it to the drones via Wi-Fi. This setup allows for unique addressing of each drone and enables effective swarming capabilities, overcoming the network configuration limitations of the normal DJI Tello drones.

2 System Architecture

The system architecture designed to facilitate swarm behavior in DJI Tello drones involves a blend of hardware and software elements aimed at overcoming the inherent constraints posed by the drones' network configuration and processing capabilities. The DJI Tello drone, renowned for its cost effectiveness and compact design, is a favored option for educational and research applications due to its standard quality camera, stable flight performance, and user-friendly interface, rendering it suitable for indoor swarming experiments. However, managing multiple DJI Tello drones concurrently presents notable difficulties primarily attributable to their default network setup. Each DJI Tello drone connects through Wi-Fi, sharing a common IP address, which complicates simultaneous control using conventional methods and results in IP address conflicts and communication issues.

Fig. 2. Picture of an ESP8266 microcontroller.

2.1 To Overcoming Network Configuration Challenges

To tackle these challenges, the ESP8266 microcontroller shown in Fig. 2, was selected for its adaptability and cost efficiency in facilitating dependable communication and control within robotic systems [7]. The ESP8266 is an economical Wi-Fi microchip with comprehensive TCP/IP capabilities and microcontroller functionality, extensively utilized in Internet of Things (IoT) contexts. Its capacity to establish Wi-Fi connections renders it suitable for establishing ad-hoc networks and enabling communication among multiple devices. In this framework, the ESP8266 microcontrollers serve as intermediaries, managing data transmission between the computer and the drones. Each ESP8266 module is linked to the computer through USB serial communication and to the drones via Wi-Fi, surmounting the network configuration constraints of the DJI Tello drones.

2.2 Communication Setup

The architecture of the communication setup is shown in Fig. 3. The communication arrangement encompasses both USB serial communication and Wi-Fi communication. The computer dispatches data to the ESP8266 microcontrollers via USB serial communication, with each ESP8266 module connected to the computer through a USB-to-serial converter. This configuration facilitates the transmission of control directives and data from the computer to the ESP8266. Each ESP8266 module is programmed to connect to a specific DJI Tello drone by storing the drone's Wi-Fi SSID and password in it is code, ensuring that each ESP8266 module links to the appropriate drone's Wi-Fi network. Upon connection, the ESP8266 establishes a Wi-Fi connection with its designated drone, enabling it to relay commands and receive status updates.

The network setup addresses the issue of managing multiple drones with a shared IP address by establishing unique addressing and efficient communication protocols. Each ESP8266 microcontroller is assigned a distinct identifier to enable the computer to differentiate between various modules based on their COM serial numbers. This allows the computer to send specific commands to individual drones, ensuring accurate routing of instructions. The ESP8266 modules function as Wi-Fi access points, connecting to the broadcast Wi-Fi network of each designated DJI Tello drone. To prevent interference and maintain reliable communication, the drones are configured to operate on separate Wi-Fi channels. Communication involves transmitting control commands from the computer to the ESP8266 modules, which then relay these commands to the drones. Feedback and status updates from the drones are relayed back to the ESP8266 modules and subsequently forwarded to the computer for monitoring and further processing.

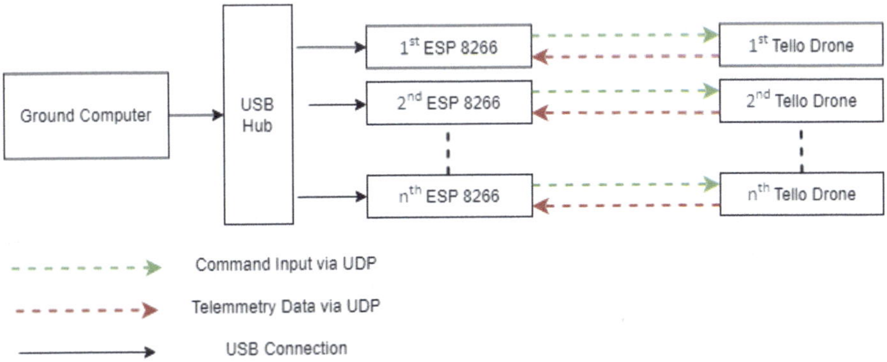

Fig. 3. Communication setup of DJI Tello drone.

2.3 Communication Protocols

For the packet control from the ESP8266 to the drone will be packet in UDP (User Datagram Protocol). The User Datagram Protocol (UDP) is a key component of the Internet Protocol suite, intended to facilitate swift and efficient communication between applications across a network [8]. In contrast to the Transmission Control Protocol (TCP), UDP functions in a connectionless manner, eliminating the need to establish a session between the sender and receiver prior to data transmission. This streamlined approach results in minimal header overhead, encompassing crucial elements such as source and destination ports, length, and checksum, thereby enhancing the speed of data transfer

[9]. Nevertheless, UDP does not guarantee the reliable delivery of packets or incorporate mechanisms for flow control or error correction. Applications that utilize UDP, such as real-time multimedia streaming, online gaming, and Domain Name System (DNS) queries, prioritize rapid responsiveness over reliability. Additionally, UDP supports broad-cast and multicast communication, enabling efficient data dissemination to multiple recipients. While UDP excels in minimizing latency and is particularly suitable for time sensitive applications, it lacks inherent features to address packet loss or ensure sequential delivery, necessitating independent management by applications if required.

The integration of hardware and communication elements entails configuring the ESP8266 microcontrollers to establish Wi-Fi connections with the drones and communicate with the computer through USB serial connections. Tailored scripts are developed to manage data transmission and control commands effectively. The DJI Tello drones are programmed to connect to their respective ESP8266 microcontrollers, and initial testing is conducted to verify stable connections and dependable communication. The computer implements a swarming algorithm comprising modules for path planning, obstacle avoidance, and coordination. This algorithm generates control commands that are transmitted to the drones via the ESP8266 modules. Testing and calibration procedures are carried out to refine communication parameters and ensure optimal system performance.

3 Swarming Control Algorithm

The developed swarming control algorithm in this research aims to facilitate synchronized movement and management of multiple DJI Tello drones through a centralized control system. Utilizing the Software Development Kit (SDK) provided by DJI Tello, users can transmit commands and receive telemetry data via a Wi-Fi connection [10]. The Tello SDK offers a range of commands for functions like takeoff, landing, multidirectional movement, aerial flips, and other maneuvers, enhancing it is adaptability for implementing swarm behaviors.

In this setup, the computer serves as the central control hub, executing the swarming algorithm and establishing communication with the drones via ESP8266 microcontrollers. The algorithm is structured to interpret high-level commands, such as those inputted from a keyboard, and convert them into precise directives for each drone within the swarm. This process encompasses several essential elements, including initialization, command interpretation, and swarm synchronization.

3.1 Initialization

In the initialization stage, the computer initiates a USB serial link with individual ESP8266 microcontrollers. These microcontrollers come preset with the Wi-Fi network name and password of the corresponding Tello drone, enabling them to connect to the drone's network automatically upon activation. Each ESP8266 is allocated a distinct identifier according to its COM serial number by the computer, guaranteeing that subsequent directives are accurately directed to the specific drone.

3.2 Command Parsing

The algorithm's command parsing module decodes keyboard inputs provided by the user and converts them into Tello SDK commands. For instance, the act of pressing arrow keys may trigger movement directives such as "forward," "back," "left," or "right," whereas other keys could be assigned functions like takeoff, landing, or executing aerial flips. This real-time processing of inputs by the algorithm guarantees prompt and effective control over the drones. The mapping between SDK commands and corresponding keyboard inputs is detailed in Table 1. List of SDK commands and keyboard inputs.

Table 1. List of SDK commands and keyboard inputs.

Command	Description	Keyboard Assign
command	Enter SDK mode	c
takeoff	Auto takeoff	f
land	Auto land	g
up x	Ascend "x" cm	Up arrow
down x	Descend "x" cm	Down arrow
left x	Move left "x" cm	a
right x	Move right "x" cm	d
forward x	Move forward "x" cm	w
backward x	Move backward "x" cm	s
cw x	Rotate "x" degrees clockwise	l
ccw x	Rotate "x" degrees counterclockwise	k

3.3 Swarm Coordination

The primary coordination mechanism of a swarm involves the use of simple movement commands rather than additional sensors. The Tello Software Development Kit (SDK) provides specific instructions for controlling the drone to move a set distance in any direction (e.g., "move forward x cm", "move left x cm"), which are utilized to synchronize the drones' movements within the swarm. An algorithm calculates the necessary movements for each drone based on user input and ensures that all drones carry out these movements simultaneously.

For example, when the user presses the "W" key to move the swarm forward by a specific distance, the algorithm generates the "move forward x cm" command for each drone, with x representing the desired distance. These commands are sent via USB serial to the ESP8266 modules, which then transmit them to the drones through Wi-Fi. To verify the accuracy of the swarm's movement, the algorithm compares the actual movement with the intended distance. The drones' internal sensors, such as the Inertial Measurement Unit (IMU), provide basic feedback on the movement. Following the execution of a movement command, the drones send telemetry data indicating their new positions. This data is used by the algorithm to confirm if the drones have traveled the expected distance.

For instance, after sending the command "move forward 50 cm," the algorithm checks if each drone has moved approximately 50 cm forward based on the telemetry data received. This straightforward validation process ensures that the drones are following the prescribed movement commands accurately and allows for easy adjustments if needed.

4 Experimental Setup and Results

The suggested alternative approach for swarming with the utilization of ESP8266 in the standard DJI Tello model will be assessed for its capabilities and precision through experimental validation. The drone will be directed to fly using a previously described keyboard interface.

Fig. 4. ESP8266 module setup.

The experimental procedure involves analyzing the response of the drones following a control command input via the computer keyboard for both drones. Initially, the drones will undergo calibration in all axes and center of gravity adjustments to ensure optimal flight performance and prevent any performance degradation resulting from skipped calibration procedures. The experiment will commence by configuring the C++ code in the ESP8266, including the insertion of the drone's Wi-Fi SSID and password. Subsequently, each ESP8266 module will be connected to the computer and the COM port of each module will be set. The setup of the ESP8266 module is shown in Fig. 4. Upon activation of the drone, the ESP8266 module will automatically establish a connection with the drone, initiating the transmission of packets from the computer to the drone through the UDP algorithm specified in the C++ code. For validation, the drone's travel distance is measured using a ruler to determine the distance the drone moves from the initial and final command conditions.

The experimental process and outcomes will be presented in the Fig. 5 and Table 2.

Fig. 5. Normal DJI Tello drone in swarm flight testing.

Table 2. Results of flight testing.

Command	Desired Movement	Actual Movement Drone 1	Actual Movement Drone 2
Ascend [cm]	35	37.3	36.9
Descend [cm]	35	33.5	34.1
Move left [cm]	35	34.6	33.2
Move right [cm]	35	33.9	32.9
Forward [cm]	35	36.2	36.5
Backward [cm]	35	34.5	35.5
Yaw left [degrees]	90	89.8	88.9
Yaw right [degrees]	90	89.5	89.0

The outcomes presented above illustrate the motion of drone 1 and drone 2 in accordance with the prescribed movement commands outlined in the Python program. The program specifies a translation movement of 35 cm and a rotational movement of 90°. However, in practice, the actual movement of the drones may deviate from the programmed commands. This discrepancy could be attributed to sensor noise, leading to drifting phenomena during flight. Additionally, environmental factors play a crucial role in influencing drone movement, as certain sensors like optical flow perform optimally under adequate lighting conditions. To mitigate errors arising from sensor readings, manual measurements using a ruler are employed to ascertain the precise location of movement in translation or rotation.

Nevertheless, the findings suggest that alternative approaches for conventional DJI Tello swarming utilizing ESP8266 and Python are feasible. The drones can be maneuvered based on user-defined movement parameters. Moreover, the afore-mentioned errors are deemed insignificant as adjustments can be made to the movement distances to compensate for any discrepancies, thereby enabling the desired distance movement to be achieved by modifying the Python commands in response to the observed errors.

5 Conclusions and Future Work

Based on the findings presented in the study, the use of ESP8266 modules for coordinating swarms of DJI Tello drones through UDP communication exhibits promising capabilities, albeit with observed discrepancies between intended and actual movements. The experiments illustrated that the drones successfully executed translational movements of approximately 35 cm and rotational movements of 90° as instructed via the Python interface. However, minor variations were detected in the recorded movements, slightly deviating from the programmed values. These disparities are likely attributable to sensor imprecisions, environmental influences on sensor functionality, and the inherent constraints of the UDP communication protocol in ensuring accurate command execution without mechanisms for error correction feedback.

To enhance the accuracy and dependability of future swarming endeavors involving DJI Tello drones and ESP8266 modules, several recommendations can be proposed. Firstly, enhancing sensor calibration procedures or integrating supplementary sensors could enhance the precision of drone movements, particularly in settings where environmental conditions fluctuate. Implementing feedback mechanisms within the communication protocol, such as acknowledgment signals or telemetry data exchange, could help mitigate errors and improve synchronization among drones within the swarm. Furthermore, refining the algorithm to enhance error management and real-time adjustment capabilities could contribute to more seamless and predictable swarm behavior.

Looking forward, forthcoming research could explore the integration of ma-chine learning algorithms or computer vision techniques to autonomously adjust drone movements based on environmental feedback and sensor data. Investigating advanced networking protocols beyond UDP, such as TCP/IP or MQTT, could offer improved reliability and resilience for swarm coordination applications. Additionally, expanding the scale of the experiment to involve more drones and conducting tests in diverse environments would further validate the scalability and adaptability of the proposed swarming methodology.

Acknowledgements. The authors would like to acknowledge the financial support provided by Institute of Research and Community Services (LPPM) and Institute for Innovation and Entrepreneurship Development (LPIK) of Institut Teknologi Bandung (ITB). Their funding enabled the execution of this research project.

References

1. Cuevas, E., Fausto, F., Gonzalez, A.: A swarm algorithm inspired by the collective animal behavior. In: Intelligent Systems Reference Library, pp. 161–188 (2020). https://doi.org/10.1007/978-3-030-16339-6_6
2. Chen, W., Liu, J., Guo, H., Kato, N.: Toward robust and intelligent drone swarm: challenges and future directions. IEEE Netw. **34**, 278–283 (2020). https://doi.org/10.1109/mnet.001.1900521
3. Tello | Ryze Tech – Feel the Fun. https://store.dji.com/product/tello. Accessed 13 June 2024
4. Tello Official Website-Shenzhen Ryze Technology Co., Ltd. https://www.ryzerobotics.com/tello-edu. Accessed 4 Sept 2024
5. FORUM D Controlling multiple Tello with one computer. https://forum.dji.com/thread-146899-1-1.html. Accessed 04 Sept 2024
6. Two devices using the same IP Address - Session Hijacking (2018). https://community.spiceworks.com/t/two-devices-using-the-same-ip-address-session-hijacking/680423. Accessed 04 Sept 2024
7. Samsugi, S., Ardiansyah, A., Kastutara, D.: Arduino dan Modul Wifi ESP8266 sebagai Media Kendali Jarak Jauh dengan antarmuka Berbasis Android. Jurnal Teknoinfo **12**(1), Article no. 1 (2018). https://doi.org/10.33365/jti.v12i1.42
8. User Datagram Protocol (UDP). https://www.geeksforgeeks.org/user-datagram-protocol-udp/. Accessed 14 June 2024
9. Differences Between TCP and UDP. https://www.spiceworks.com/tech/networking/articles/tcp-vs-udp/. Accessed 04 Sept 2024
10. SDK | Tello Dji. https://alfredo-reyes-montero.gitbook.io/tello-dji/sdk. Accessed 14 June 2024

SPICS: A Novel High Lift Devices Mechanism Based on 3D-Printed Morphing Wing for Small UAV

Mutaqin Aryawijaya(✉) and Yazdi Ibrahim Jenie

Faculty of Mechanical and Aerospace Engineering, Institut Teknologi, Bandung, Indonesia
13620046@mahasiswa.itb.ac.id

Abstract. Preventing stall is a crucial challenge in fixed-wing aviation as it significantly impacts flight safety and performance. This research focuses on the development of morphing wing technologies to improve high-lift devices in small Unmanned Aerial Vehicles (UAVs). The main aim is to design, develop, and test a morphing wing that enhances lift and provides better resistance to stalling. This research proposed the SPICS (Slotted-Pillar Internal Compliant System) morphing mechanism through several iterations to achieve a parabolic deflection. Using Fused Deposition Modeling (FDM) 3D printing technology, a test section with a 40 cm span and 20 cm chord length was manufactured. The SPICS design was subjected to testing in a controlled wind tunnel environment, where it demonstrated the ability to effectively adjust its camber. This adjustment led to an improvement in the lift coefficient (C_L), achieving a $C_{L_{max}}$ of up to 1.65. Moreover, the stall angle increased from 10° to 14°, indicating better stall resistance. During the stall tests, the SPICS successfully recovered from stall conditions and maintained its C_L at higher angles of attack. These findings suggest that the SPICS mechanism could be a promising alternative to conventional articulated flaps for small-scale UAVs.

Keywords: Morphing Mechanism · Parabolic Deflection · Flaps · 3D Print

1 Introduction

1.1 Background

Stall prevention remains a significant challenge in modern fixed-wing aviation, especially for small Unmanned Aerial Vehicles (UAVs). Stall occurs when an aircraft loses lift, either due to inadequate airspeed or steep angles of attack, which can greatly impact flight safety and performance [1]. In these cases, high-lift devices like flaps are commonly used, effectively increasing the aircraft maximum lift coefficient (C_L) at low-speed flight, especially when taking off or landing.

However, conventional flaps also lower the limit of stall angle, making the aircraft more prone to stall and harder to recover once it hits the condition. Recent studies highlight the problems in the design of flap mechanism, i.e. plain, single slotted, double

E. Fitrianingsih et al. (Eds.): ISAST 2024, SPPHY 416, pp. 354–363, 2025.
https://doi.org/10.1007/978-981-96-1344-1_38

slotted, or other, that typically result in gaps in the whole wing airfoil, resulting to premature flow separation and stall [2].

With advancement of aircraft material technology, the concept of morphing wing starting to gain interest as a potential solution for a better high lift device mechanism. Morphing wing relies on the wing structural flexibility to change the wing shape without resulting in any gaps. Furthermore, the concept also allows the design of parabolic flap mechanism, which not only eliminates any gaps on the wing, but the camber discontinuities as well. As a result, parabolic flap morphing wing designs can potentially improve aircraft's maximum lift at low airspeed and stall angle limitation [3].

Several studies highlighted the implementation of morphing wing, resulting in many concepts of morphing mechanism, especially for the airfoil camber manipulation. Some of the mechanisms including the Variable Camber Compliant Wing (VCCW) design [4], the FishBAC [5], the Kinetic Internal Nexus Compliant System (KINCS) concept [6], the Airfoil Recambering Compliant System (ARCS) [6], and Rear Spar Articulated Wing Camber (RSAWC) [7]. Their practical application for small UAVs, however, has not been adequately studied. This practicallity incluides the use of low-cost 3D printing technology for manufacture and the use of servo for the mechanism powerdrive.

Therefore, this research aims to develop a novel morphing wing mechanism for camber manipulation on UAVs and proposed it as the Slotted-Pillar Internal Compliant System (SPICS) concept. The process involves creating a parabolic flap deflection to boost lift and delay stall. This design is then manufactured using Fused Deposition Modeling (FDM) 3D printing technology. The performance of this innovative morphing wing will be tested in a wind tunnel to verify its aerodynamic improvements. This approach aims to significantly enhance the aerodynamic capabilities of UAVs.

2 Methodology

2.1 Design and Development of SPICS Morphing Mechanism

Design Requirements and Objectives (DRO). The SPICS morphing wing was designed to meet specific requirements ensuring suitability for small UAV applications and effective aerodynamic lift performance. The wing span should limited to 40 cm to fit within the FTMD wind tunnel. The morphing mechanism should powered by four mini servos, each with a maximum torque of 3.5 kgf.cm, using a 6-V power supply, and all servos were mounted inside the wing. The wing was designed to prevent stall at a minimum angle of attack of 12° at a Reynolds Number of 76630 [8], and it could morph into at least five different equivalent airfoil shapes. Materials chosen were suitable for FDM 3D printing to ensure structural integrity. The design objectives included manufacturing the wing using 3D printing technology, ensuring the design was suitable for wind tunnel testing, achieving parabolic deflection for both leading and trailing edges, comparing various morphing wing designs to find the most effective mechanism, and demonstrating the wing's ability to adjust its shape to enhance lift and increase the stall angle.

Airfoil Selection. The NACA 0012 airfoil was selected as the base model for this research due to its zero camber, which allows for a clear analysis of the effects of camber on aerodynamic lift and stall characteristics. By introducing camber through the morphing mechanism, the study aimed to assess its impact on these aerodynamic properties. This approach contrasts with previous research by B. C. Moulton [6], which used the NACA 2412 airfoil and focused on performance at low angles of attack and servo integration.

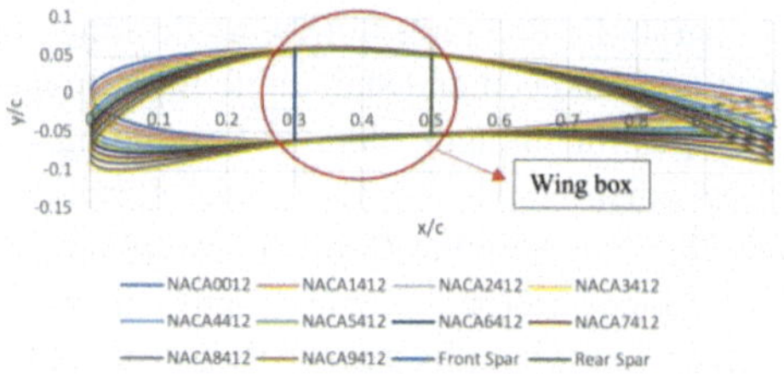

Fig. 1. Placement of the structural wing box on the airfoil designs

As in Fig. 1, the wing section design involved stacking airfoil series from NACA 0012 to NACA 9412, incrementally increasing camber to study its aerodynamic effects. A region from $x/c = 0.3$ to $x/c = 0.5$, where curvature changes are minimal, was identified as ideal for the wing box. This area provides structural integrity and houses the servo and wiring components, ensuring a stable base for the mechanisms and a secure connection to the UAV fuselage. By varying the camber from NACA 0012 to NACA 9412, this study aimed to analyze how increased camber affects aerodynamic lift and stall characteristics. This approach sought to determine how the morphing mechanism could adjust the airfoil shape to delay or prevent stall. Previous simulation data comparing these airfoils and plain flaps provided a solid foundation for this experimental analysis [8].

Morphing Mechanisms. Several morphing mechanisms were considered using CAD software Fusion 360 shown in Fig. 2, including KINCS, ARCS, LINCS, and PICS. KINCS (Kinetic Internal Nexus Compliant System) utilizing internal linkages and elastic elements for smooth and continuous deformation. However, it did not provide the required parabolic deflection as strain stopped after the kinks, leading to linear deflection [6]. ARCS (Airfoil Recambering Compliant System) employing interconnected arcs for significant camber changes. The model was thick due to TPU material limitations, resulting in higher weight and failure to achieve parabolic deflection as strain did not propagate through the arcs [9].

In other hand, this research also developing several morphing mechanisms, such as LINCS (Linear Internal Compliant System), using straight internal members to alter the airfoil shape. Showed better strain distribution and maintained local thickness, but required further refinement. PICS (Pillar Internal Compliant System) featuring internal pillars providing good strain distribution and parabolic deflection. Although promising, it faced issues with maintaining local thickness.

Using Finite Element Analysis (FEA) in Fusion 360, we ran static loading simulations to evaluate these mechanisms. The results as shown in Fig. 3 indicated that LINCS

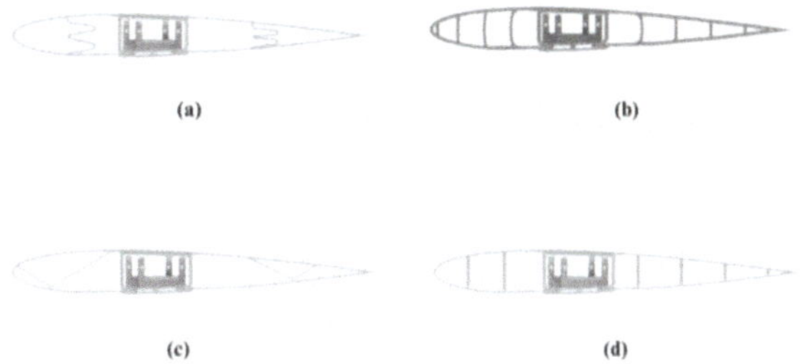

Fig. 2. KINCS mechanism (a), ARCS mechanism (b), Self developed LINCS mechanism (c), Self developed PICS mechanism (d).

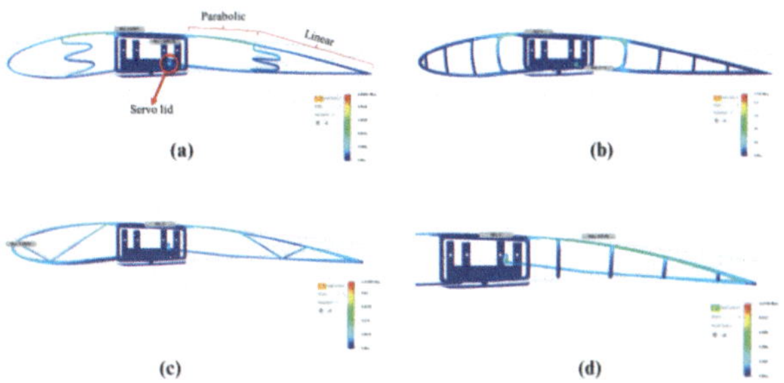

Fig. 3. KINCS static strain test (a), ARCS static strain test (b), LINCS static strain test (c), PICS static strain test (d).

showed promise but there still a linear region which we want to avoid, hence required further refinement. PICS demonstrated the best parabolic deflection along the trailing edge, despite some changes in local airfoil thickness. Therefore, we will focus on testing the PICS design.

The Novel Morphing Mechanism: SPICS. To address the limitations of the previous PICS design, we developed the SPICS (Slotted-Pillar Internal Compliant System) model. By adding several placed slots, the lower surfaces of both the leading and trailing edges could slide, achieving parabolic deflection while maintaining fixed thickness between the upper and lower surface (Fig. 4).

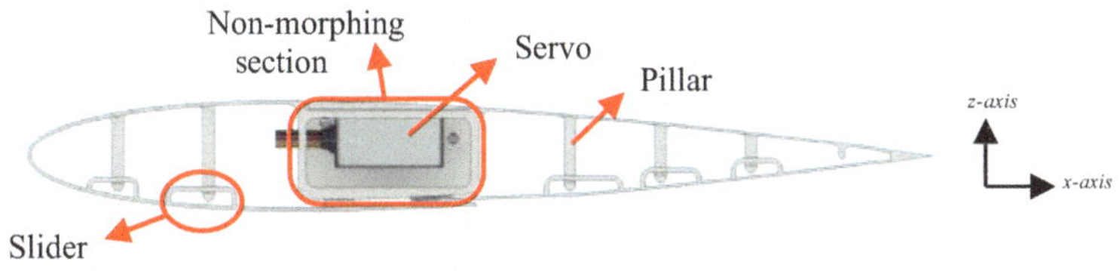

Fig. 4. SPICS mechanism.

SPICS consists of three main parts: the non-morphing section, pillar, and slider. The non-morphing section maintains structural integrity and serves as the wing's structural box. The pillar preserves the wing's local thickness, while the slider enables the pillar to move along the x-axis, allowing for parabolic deflection. We constructed a wing test section with a 20 cm chord length using FDM 3D printing with PLA material to compare the LINCS, PICS, and SPICS mechanisms. The printing settings used are detailed in Table 1.

Table 1. 3D printer settings.

Setting	LINCS Model	PICS Model	SPICS Model
Layer Height, mm	0.2	0.2	0.2
Line Width, mm	0.4	0.4	0.4
Wall Thickness, mm	0.8	0.8	0.8
Infill Density	15%	15%	15%
Built Plate Temperature, °C	60	60	60
Ext Material	PLA +	PLA +	PLA +
Ext Print Temperature, °C	210	210	210
Ext Flow Rate	100%	100%	100%
Support Structure	Normal	Normal	Tree
Support Placement	Everywhere	Everywhere	Everywhere
Build Plate Adhesion Type	Brim	Brim	Brim
Brim Width, mm	3	3	3

All three mechanisms (LINCS, PICS, and SPICS) were successfully printed using the FDM 3D printing settings. These models were evaluated for their ability to achieve parabolic deflection and maintain local thickness, which are crucial for enhancing aerodynamic efficiency and preventing stall.

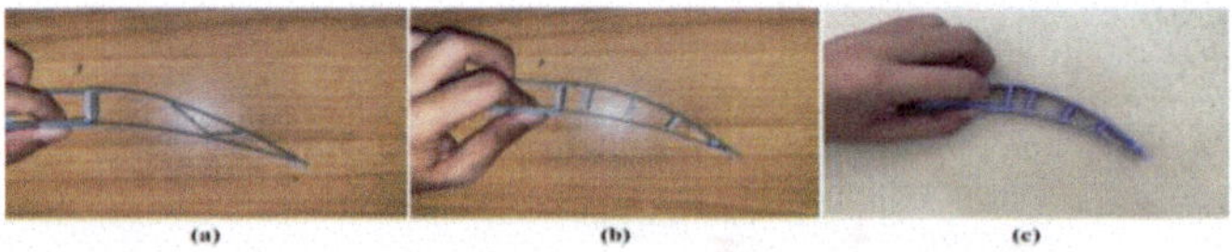

Fig. 5. Mechanism deflection test (a) LINCS mechanism, (b) PICS mechanism and (c) SPICS mechanism deflection test.

As in Fig. 5, LINCS mechanism unable to perform a full parabolic deflection as we have simulated before. PICS mechanism demonstrated impressive parabolic deflection capabilities but encountered difficulties in maintaining local thickness.

In other hands, developed the SPICS mechanism allowed the lower surfaces of the trailing edge (TE) and leading edge (LE) to slide, achieving parabolic deflection while maintaining a fixed local thickness. Hence, SPICS model emerged as the most promising candidate. Based on this design, we constructed a morphing wing test section with a 40 cm span and 20 cm chord length, fully 3D printed with PLA material. This setup was then prepared for wind tunnel testing to validate its aerodynamic performance.

2.2 Experimental Setup

Wind Tunnel Setup. To validate the aerodynamic lift of the SPICS morphing wing, we conducted tests in a subsonic wind tunnel at the Faculty of Mechanical and Aerospace Engineering, Institut Teknologi Bandung (FTMD ITB). Figure 6 shows the wind tunnel setup included an Arduino UNO R3 for data acquisition, a 5 kg load cell for force measurements, and a laptop for data processing. Testing was performed at a Reynolds number of approximately 76,630, aligning with reference values from previous studies [8] to ensure comparability with existing simulation data. In terms of the blockage ratio, it's important to clarify that the wind tunnel test section measures 40 × 40 cm, with the wing's frontal area facing the airflow being 2 × 40 cm, resulting a blockage ratio of 5%, which is within acceptable limits for subsonic wind tunnel testing. The dimensions were influenced by the morphing system and servo thickness. Our focus was on observing changes in C_L and $C_{L_{max}}$, rather than their absolute values, comparing each configuration against itself. Therefore, the effects of the blockage ratio are negligible in this testing, and no significant corrections were needed.

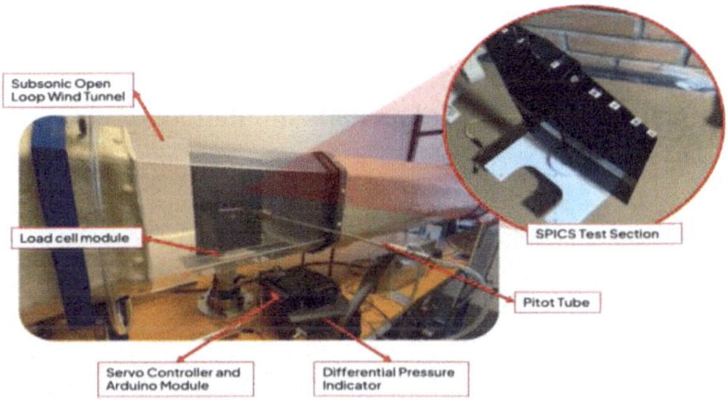

Fig. 6. Wind tunnel test setup.

Calibration and Data Acquisition. Load cell calibration, a 5 kg load cell was calibrated using a known mass (249 g DJI Mini 3 drone) to ensure accurate force measurements. An Arduino code adjusted the calibration factor until the readings matched the known mass value consistently. Pitot tube calibration, the pitot tube was calibrated using a Differential Pressure Indicator (DPI) by recording pressure readings at various known airspeeds to create a calibration curve for accurate airspeed conversion. Arduino and servo tester setup, an Arduino UNO R3 was connected to a laptop for real-time data recording, while a digital servo ESC tester module controlled the servo mechanisms, connected to a 6 V power supply.

Test Procedures. Two main tests were conducted to evaluate the performance of the SPICS morphing wing: the regular aerodynamic lift test and the stall test. The regular aerodynamic lift test aimed to assess the wing's lift characteristics across different camber configurations and angles of attack. The wing section was secured in the wind tunnel, and the main shaft was manually rotated to set the initial angle of attack to 0°, then locked in place. The Arduino and servo tester were connected for data acquisition and control. Aerodynamic lift data was collected with the wing in a neutral position to establish a baseline. The wing's camber was then adjusted using servo motors controlled by PWM signals, with values ranging from 800 to 1700 in increments of 100. Data was collected for each camber setting at angles of attack from 0° to 18°, incrementing by 2°. For each configuration, lift force and angle of attack were recorded.

The stall test aimed to evaluate the wing's performance under stall conditions and its ability to recover from stall. The wing section was secured in the wind tunnel, and the initial angle of attack was set to 0°. The Arduino and servo tester were connected for data acquisition and control. To identify the stall condition, the angle of attack was incremented by 4° from 0° to 12° while maintaining the wing at PWM 800 (baseline state). After the wing reached stall condition, the PWM was increased to 1700 to observe the morphing wing's stall mitigation capabilities. Lift forces and angle of attack were recorded during the stall and recovery process.

3 Results and Discussion

3.1 Aerodynamic Lift Test Result

The aerodynamic lift test data of the SPICS morphing wing was evaluated against conventional wing simulation data [8] and plain flaps simulation data [8], as shown in Figs. 7 and 8.

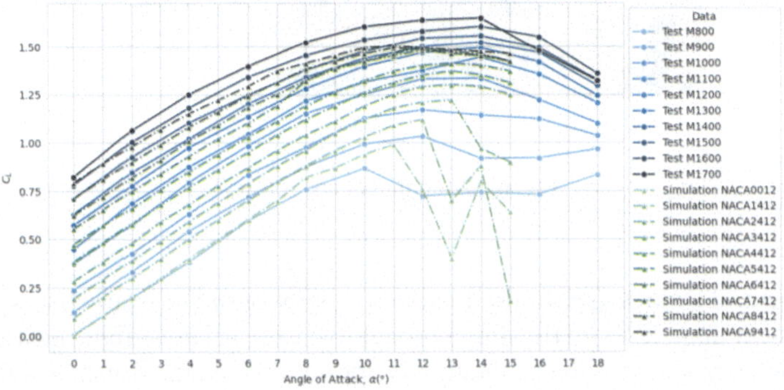

Fig. 7. SPICS C_L data with simulated NACA series data from previous research.

Table 2 summarizes the maximum lift coefficient $C_{L_{max}}$ and stall angles α_{stall} for the SPICS wing, conventional wings, and plain flaps from previous research by T. Majid [8]. At the baseline scenario (M800), the SPICS wing exhibited a slightly lower stall angle

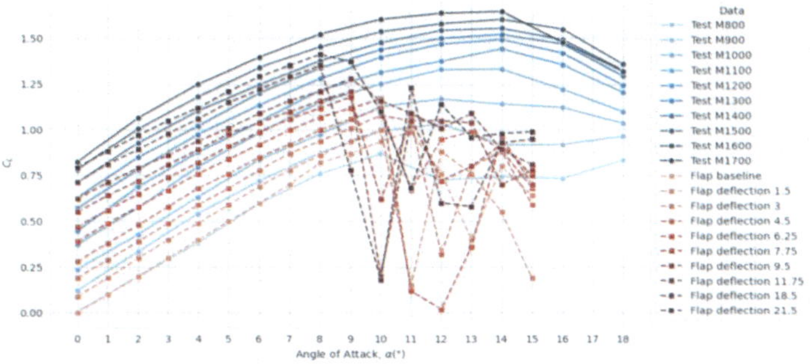

Fig. 8. SPICS C_L data with simulated NACA0012 with plain flaps data from previous research.

(10°) compared to the plain flaps (11°). However, for other configurations, the SPICS consistently achieved higher stall angles, reaching up to 14° at higher deflection angles, whereas the plain flaps stalled between 8° and 11°. This demonstrates the SPICS wing's stall resistance.

Table 2. SPICS $C_{L_{max}}$ and stall angle compared to conventional wing.

Morphing SPICS			NACA0012 with Plain Flaps [8]			Conventional Wing [8]		
Test Series	$C_{L_{max}}$	α_{stall}	Flaps Deflection	$C_{L_{max}}$	α_{stall}	NACA Series	$C_{L_{max}}$	α_{stall}
M800	0.88	10°	0°	0.99	11°	NACA0012	0.99	11°
M900	1.03	12°	1.5°	1.00	11°	NACA1412	1.12	11.8°
M1000	1.17	12°	3°	1.08	10.6°	NACA2412	1.22	12.6°
M1100	1.33	14°	4.5°	1.12	10°	NACA3412	1.30	13°
M1200	1.44	14°	6.25°	1.17	9°	NACA4412	1.37	13°
M1300	1.49	14°	7.75°	1.18	9°	NACA5412	1.41	13°
M1400	1.52	14°	9.5°	1.21	9°	NACA6412	1.48	13°
M1500	1.55	14°	11.75°	1.28	8°	NACA7412	1.49	12°
M1600	1.60	14°	18.5°	1.35	8°	NACA8412	1.49	12°
M1700	1.65	14°	21.5°	1.41	8°	NACA9412	1.50	11°

At higher flap deflections, the SPICS wing significantly outperformed plain flaps in terms of both $C_{L_{max}}$ and α_{stall}, indicating its capability to provide higher lift and better stall resistance across various configurations. This improvement is primarily due to the SPICS design incorporating a 'slats'-like feature in the leading edge, which helps to delay stall by maintaining airflow attachment at higher angles of attack [10]. Additionally, the parabolic nature of the SPICS mechanism ensures there are no discontinuities in the camber line, which reduces the likelihood of premature flow separation. This fulfils the design requirements and objectives stated earlier.

3.2 Stall Mitigation Result

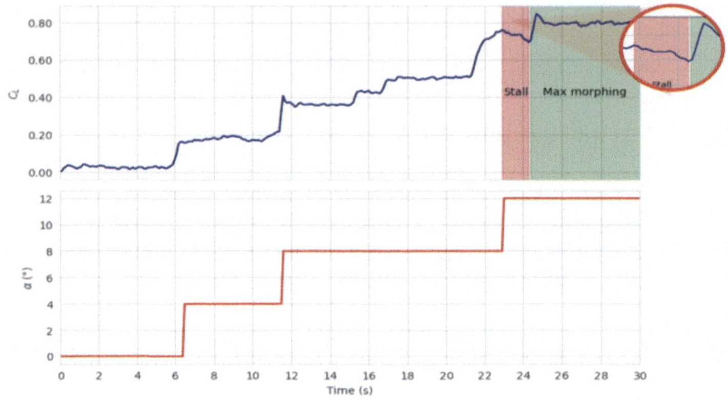

Fig. 9. SPICS C_L response to α input from 0° to stall angle 12°

As shown in Fig. 9, the SPICS wing demonstrated clear recovery in lift after experiencing a stall at an angle of attack α of 12°. Initially, the lift coefficient (C_L) dropped from 0.76 to 0.69, indicating the onset of a stall. During the first 24 s, no PWM input was given to the servo, meaning the wing profile remained unchanged. However, as the angle of attack was manually adjusted from 8° to 12°, slight unintended changes in the angle of attack caused the lift force to fluctuate, with a noticeable increase after the 15th second and a sharp rise at the 21st second. These fluctuations were due to the manual adjustment process. After the stall occurred around after 24 s, a maximum PWM input (set to 1700) was applied, transforming the SPICS into its M1700 configuration, which features increased camber. This adjustment allowed the wing to recover from the stall, with the C_L stabilizing around 0.79. The SPICS model effectively maintained aerodynamic lift, not only recovering from the stall but also enhancing the wing's profile to prevent subsequent stalls under similar conditions. This outcome aligns with the design requirements and objectives outlined earlier.

4 Conclusion

The SPICS morphing wing demonstrated superior performance compared to articulated flaps, achieving stall angles up to 14° compared to 11° for plain flaps. This improvement is due to the SPICS design, which mimics the effect of leading-edge slats by morphing the entire camber of the airfoil.

Innovative 3D-printed morphing wings with parabolic designs were successfully engineered. These wings enhanced aerodynamic lift performance and mitigated stall risks, aligning with the Design Requirements and Objectives (DRO) and proving the feasibility of integrating such mechanisms into small UAVs.

Wind tunnel tests validated the SPICS wing's effectiveness, showing significant improvements in stall resistance and lift performance, with a maximum lift coefficient $C_{L_{max}}$ of 1.65 and the ability to prevent stall at angles up to 14°. These results confirm the practical viability of morphing wings under varied conditions, highlighting their potential to advance UAV aerodynamics.

Further research should focus on improving manufacturing quality and automating data acquisition to enhance accuracy. Comprehensive drag analysis is needed to optimize designs for reduced drag and improved performance. Simplifying the morphing mechanism could facilitate cost-efficient production and improve reliability. Optimizing structural integration, refining control systems, and conducting real-world tests will ensure successful implementation in practical UAV applications.

References

1. Crane, D.: Dictionary of Aeronautical Terms, third edition, p. 486. Aviation Supplies & Academics (1997). ISBN 1-56027-287-2
2. Hunsaker, D.F., Reid, J.T., Joo, J.J.: Geometric definition and ideal aerodynamic performance of parabolic trailing-edge flaps. Int. J. Astronaut. Aeronaut. Eng. **4**(1), 1–17 (2019). https://doi.org/10.35840/2631-5009/7526
3. Hunsaker, D. F., Reid, J.T., Moorthamers, B., Joo, J.J.: Geometry and aerodynamic performance of parabolic trailing-edge flaps. In: 2018 AIAA Aerospace Sciences Meeting (2018)
4. Joo, J.J., Marks, C.R., Zientarski, L., Culler, A.: Variable camber compliant wing – design. In: 23rd AIAA/AHS Adaptive Structures Conference (2015). https://doi.org/10.2514/6.2015-1050
5. Woods, B.K.S., Friswell, M.I.: Structural characterization of the fish bone active camber morphing airfoil. In: 22nd AIAA/ASME/AHS Adaptive Structures Conference (2014). https://doi.org/10.2514/6.2014-1122
6. Moulton, B.C., Hunsaker, D.F.: 3D-printed wings with morphing trailing-edge technology. In: AIAA Scitech 2021 Forum (2021). https://doi.org/10.2514/6.2021-0351
7. Ozbek, E., Ekici, S., Karakoc, T.H.: Unleashing the potential of morphing wings: a novel cost effective morphing method for UAV surfaces, rear spar articulated wing camber. Drones **7**, 379 (2023). https://doi.org/10.3390/drones7060379
8. Majid, T., Jo, B.W.: Comparative aerodynamic performance analysis of camber morphing and conventional airfoils. Appl. Sci. **11**(22), 10663 (2021). https://doi.org/10.3390/app112210663
9. Park, J.K., Jung, H.K., Lee, J.R.: Development and evaluation of fall impact protection pads using additive manufacturing. Materials **12**(20) (2019) https://doi.org/10.3390/ma12203440
10. Abbott, I.H., Von Doenhoff, A.E.: Theory of Wing Sections: Including a Summary of Airfoil data. Press, vol. 11 (1959)

Flutter Analysis of Fighter Aircraft Wing with Installed Armament

Alfrito Mikhael Carlo Hutagaol[1(✉)], Rianto Adhy Sasongko[1], Leonardo Gunawan[1], Muhammad Kusni[1], and Mahesa Akbar[2]

[1] Fakultas Teknik Mesin dan Dirgantara, Institut Teknologi Bandung, Jalan Ganesa 10, Bandung 40132, Indonesia
alfritohutagaol@gmail.com

[2] College of Business Technology and Engineering, Sheffield Hallam University, City Campus, Sheffield S1 1WB, UK

Abstract. In the present work, computational analyses on aeroelastic instability of an aircraft wing are conducted. A jet fighter wing with different armament configurations is evaluated. Aeroelastic instability, i.e., divergence and flutter, occurs as the aerodynamic loads interact with the structure and may lead to a catastrophic phenomenon, i.e., amplified deformation or oscillation. In a case of a jet fighter, the armaments affect the inertial, and further, the structural dynamic characteristics of the aircraft or wing. In the current study, the variation of armament's placement and computational model is investigated. The structure of the wing and its armaments is modelled by means of Finite Element Method (FEM). The aerodynamic model of the wing is constructed using frequency domain – based methods, Doublet Lattice Method (DLM) and Piston theory for subsonic and supersonic regimes, respectively. The shifting of armament center of gravity further from elastic axis causes flutter to occur in a lower speed. In addition, the increase in mass moment of inertia also gives the same effects.

Keywords: Jet Fighter · Wing · Armament · Aeroelastic · FEM · Flutter

1 Introduction

Aeroelasticity is a subfield of aeromechanics that studies the interaction between dynamics, elasticity, and aerodynamics in aircraft design including the phenomena caused in this subfield. Flutter is dynamic instability where a lifting surface abruptly self-exciting flutter with rapidly growing amplitudes due to the interaction between aerodynamics, structural, and inertial forces [1].

Countries have been conducting research to create the best military technology during the time of war and keep the integrity of their territory or even to aggress other countries' territory. Fighter, jet trainer, and many other military aircrafts are the products that created for war purposes. Note that most of the time, the armaments and wing store are stored under the wing structure of the aircraft for the purpose of war. Warheads were also developed for war, such as guns, air-dropped bombs, air-launched missiles,

E. Fitrianingsih et al. (Eds.): ISAST 2024, SPPHY 416, pp. 364–374, 2025.
https://doi.org/10.1007/978-981-96-1344-1_39

air-launched rockets, and air-launched torpedoes. Furthermore, the abilities to fly in high velocity are needed for trainer to evade missiles during air-to-air combat, pursuit targeted aircraft, and intercept bomber aircrafts. Several military aircrafts are designed to fly in high speed, reaching supersonic speed for most fighter aircrafts. Maneuverability of these aircraft is also a major consideration to maneuver during air-to-air combat. In sum, these conditions experienced by military aircraft conducted a need for lightweight structure that can overcome these loads. The condition experienced by fighter might causes aeroelastic problems such as divergence and flutter; static and dynamic instability that affects structure and has the potential of structural failure.

The instalment of armament is proven to affect the aeroelastic characteristics through research conducted by Chesta [2], Baird and Clark [3]. Research in the 1960s in terms of flutter were conducted using wind tunnel test, which results in the conclusion that he instalment of underwing armament affects the flutter characteristics since it has certain effects to the change of inertial forces which is one of the three main forces reviewed in aeroelasticity. Aeroelastic characteristics, specifically flutter, are critical in determining the safe flight envelope for military aircraft [4]. Thus, the study of flutter analysis is crucial. In modern days, due to the complexity of the wing structure, several structural dynamic and aeroelasticity analyses for wing with store were conducted using numerical method, such as Abdullah and Sulaeman [5], Akpinar [6], and Serin [7].

The wing structure model used in this paper are based on Abdullah and Sulaeman. The model used for wing structure is limited only in the wing-box part. The wing-box is modelled using quadrilateral shell elements. For the armaments, only one type of armament used and placed on the tip of the wing. The armament will be modelled as concentrated mass and bar elements. Armament launcher also modelled using bar elements.

The aerodynamic model used in this paper is constructed using frequency domain-based methods, such as Doublet-Lattice Method (DLM) for the subsonic regime and Piston theory for the supersonic regime. For the subsonic regime, the wing lifting surface is divided into chordwise and spanwise directions using a number of trapezoidal panel elements, whereas for the supersonic regime, it is only divided into the spanwise direction. Note that even though the wing is designed for supersonic regime, aeroelastic analyses also needed to be conducted in the subsonic regime to ensure the flutter phenomenon does not occur in the designated flight envelope.

In this paper, a parametric study is conducted on the effect of armament on the aeroelastic instability such as flutter of a jet fighter wing. Subsequently, the flutter analyses are conducted in subsonic and supersonic regimes. The variation of the armament's location and inertia are also investigated in this study.

2 Methods

For this paper, finite elements method is chosen for numerical analysis. Finite element method (FEM) is a numerical method used to address various engineering and mathematical physics problems [8]. The finite element method has the ability solve complex geometric shapes by dividing the structure into small elements with the defined number or also known as mesh. The elements that used in the geometric shapes must be defined, whether its 1-D elements, 2-D elements, 3-D elements, or even a combination of those elements. In this study, FEM is used to review structural dynamics characteristics of the wing structure. The purpose of this analyses is to determine the modal characteristics such as mode shapes and the natural frequencies of the structure.

The supersonic jet fighter wing is based on the configuration given by Abdullah and Sulaeman [5]. The wing is a half span of 3.75 m with aspect ratio of 5 and a double-wedged airfoil which reconstructed based on illustration provided by Abdullah and Sulaeman [5]. The reconstructed wing airfoil model is shown in Fig. 1. The material used for wing structure is aluminum. The elastic axis of wing is located at 0.15C from the leading edge. The leading and trailing edge part of the wing are not included in this paper to avoid the occurrence of local mode shapes. The armament used in this paper is based on AIM-9 Sidewinder, while the launcher used is based on LAU-7.

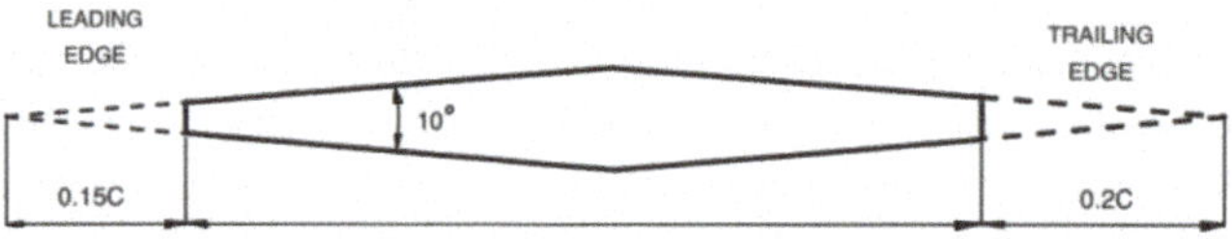

Fig. 1. Double wedge airfoil

The structure of the wing and its armaments is modelled by means of Finite Element Method (FEM). The fighter wing structure modelled using quadrilateral shell elements. The armaments are located at the tip, modelled by two types of elements, i.e., concentrated mass and beam elements, while the launcher are modelled as bar elements. The reconstructed model is shown in Fig. 2.

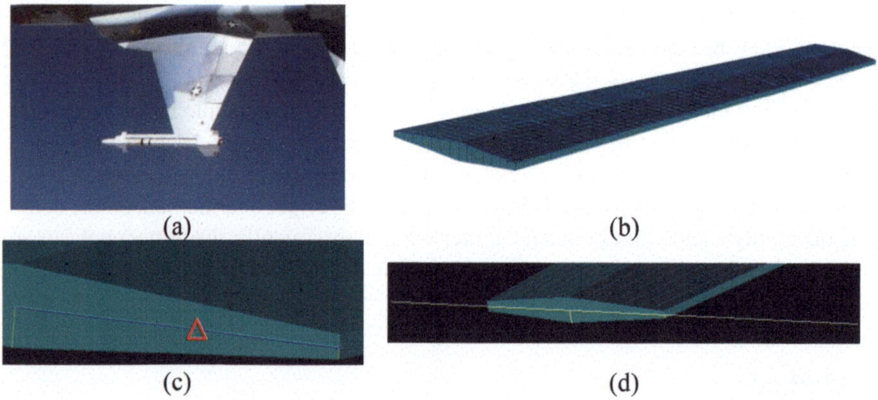

Fig. 2. Reconstructed FEM model: (a) tip-missile in military aircraft (b) wing-box model, (c) concentrated mass, and (d) beam elements.

The aerodynamic model of the wing is constructed using frequency domain – based methods, Doublet Lattice Method (DLM) and Piston theory for subsonic and supersonic regimes, respectively. In this paper, the aerodynamic effect on the armament is neglected. The aerodynamic model used in this paper is shown in Fig. 3.

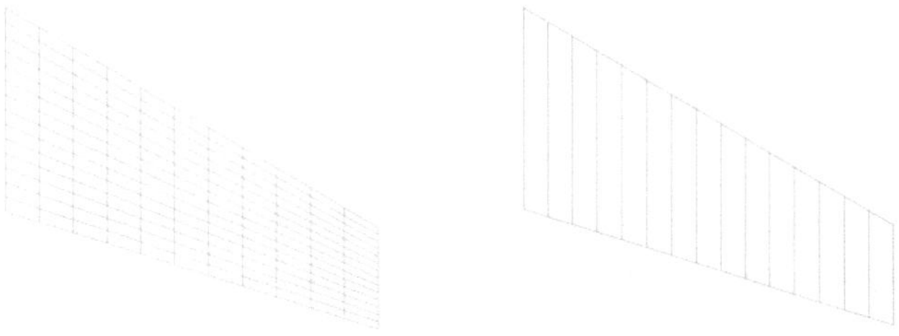

(a) (b)

Fig. 3. Aerodynamic model: (a) DLM and (b) Piston Theory.

Dynamic characteristics such as mode shapes and natural frequency are obtained through formula shown in Eq. 1, where M represent structural mass and K represents structural stiffness. This formula then transformed as eigenproblem where the eigenvalue represents natural frequency while the eigenvector represents mode shapes.

$$[M]\ddot{x}(t) + [K]x(t) = 0 \tag{1}$$

The flutter analyses conducted in FEM are using Eq. 2 to relate between the structural domain and the unsteady aerodynamic loads, where M represent structural mass, C represent the structural damping, K represent the structural stiffness. F_M represents inertial load matrix, F_C represents damping load matrix, and F_K represents stiffness load matrix due to the unsteady aerodynamic from DLM and Piston Theory.

$$[\mathrm{M} - \mathrm{F_M}]\{\ddot{\mathrm{x}}(\mathrm{t})\} + [\mathrm{C} - \mathrm{F_C}]\{\dot{\mathrm{x}}(\mathrm{t})\} + [\mathrm{K} - \mathrm{F_K}]\{\mathrm{x}(\mathrm{t})\} = 0 \tag{2}$$

To obtain the flutter solution of this analysis, p-k method is used. Equation 3 [5] shown the formulation of this method, where Eq. 1 are transformed as an eigenproblem based instability evaluation to get the solution of aeroelasticity complex-roots with M_{hh}, B_{hh}, and k_{hh} are the modal mass, damping, and stiffness matrix each in the form of Mach number and reduced frequency function, respectively. Q_{hh}^R and Q_{hh}^I are the real and imaginary part of the aerodynamic load matrix also in the form of Mach number and reduced frequency function, ρ is the air density at the set altitude, $\bar{c}$ is Mean Aerodynamic Chord (MAC), u_h is modal amplitude vector, and p is the eigenvalue.

$$\left[(M_{hh})p^2 + \left(B_{hh} - \frac{1}{4}\rho\bar{c}VQ_{hh}^I/k\right)p + \left(k_{hh} - \frac{1}{2}\rho V^2 Q_{hh}^R\right)\right]\{u_h\} = 0 \tag{3}$$

Flutter analysis using p-k method is done by calculating the reduced frequency using the natural frequency obtained from the structural dynamics analysis and setting a Mach number value for each airflow regime. Equation 4 shows the formula to calculate reduced frequency and Eq. 5 shows the eigenvalue with complex-roots. The real part of eigenvalue represents damping ratio (g) and the imaginary part represent frequency (f).

The eigenvalue will be calculated for each vibration modes and airspeed to obtain the U-g and U-f graph.

$$k = \frac{\omega c}{2V} \tag{4}$$

$$p = \omega(2g + i) \tag{5}$$

3 Results and Discussion

In this paper, two model of armament are used, which are lumped mass and beam elements. The lumped mass model will be used for the variation of armament location, whereas the beam elements model will be used for the variation of armament mass moment of inertia. To validate this model, a mesh convergence test is conducted and shown in Fig. 4. The result of this test shown that the chosen number of elements converges.

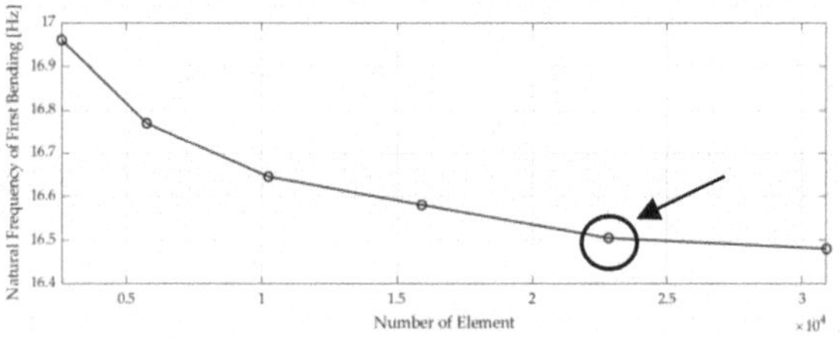

Fig. 4. Mesh convergence test of first bending natural frequency for clean wing configuration.

Table 1. Comparison between simulation results and reference [5].

		Abdullah and Sulaeman (2013)	Simulation	
Natural Frequency [Hz]	Mode 1	3.79	4.52	1B
	Mode 2	6.12	11.62	1T
	Mode 3	11.93	27.14	2B
Critical Flutter Speed [m/s]		570	800	-

From Table 1, it can be seen that the natural frequency obtained in this simulation is higher than the reference. Note that on the reference, three weapons are attached to the wing structure while this simulation only use one weapon in the configuration. The added weapon affects the inertial of the system and thus, decrease the natural frequency and critical flutter speed.

Modal analysis is conducted to obtain the structural dynamic characteristic of the wing, such as natural frequency and mode shape. The mode shapes of each configuration are shown Fig. 5.

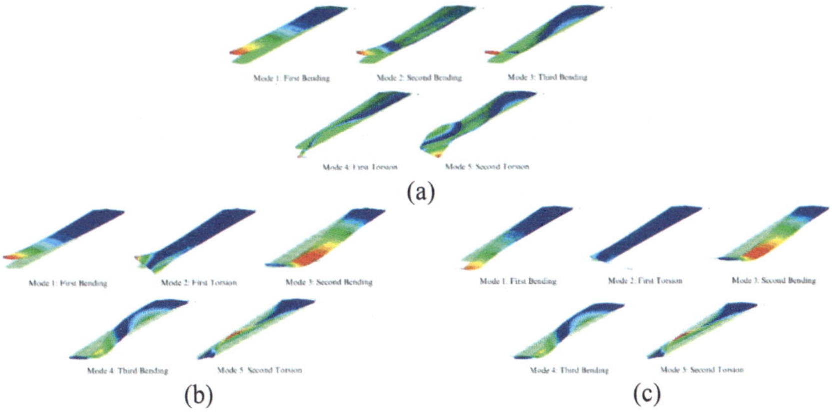

Fig. 5. The first five mode shapes: (a) clean wing configuration, (b) lumped mass configuration, and (c) beam elements configuration.

Table 2. Summary of modal analysis for several armament configuration.

Weapon Config.		Natural Frequency [Hz]														
		Clean		Lumped Mass							Bar Element					
				0.06C	0.18C	0.3C	0.42C	0.54C	0.66C	Mode	0.75I	I	1.25I	1.5I	Mode
Mode	1	16.5	1B	4.3	4.3	4.2	4.2	4.0	4.0	1B	4.5	4.5	4.5	4.5	1B
	2	49.7	2B	10.4	10.8	11.2	11.2	10.4	10.4	1T	11.6	10.0	9.0	8.2	1T
	3	97.5	3B	27.9	27.8	27.8	27.8	27.8	27.8	2B	27.1	27.1	27.1	27.1	2B
	4	110.9	1T	80.5	78.7	79.9	79.9	79.8	79.8	3B	79.3	79.3	77.8	79.6	3B
	5	117.9	2T	116.5	116.5	116.5	116.5	113.7	113.7	2T	116.3	116.5	116.4	116.4	2T

It can be seen from Table 1 that due to the armament instalment, the mode shapes order is changed, whereas the first torsion mode appeared in Mode 2. The natural frequency also shown to be decreased due to the armament instalment, whereas the natural frequency is 16.5 Hz (1st Bending) for clean wing model, 4.2 Hz (1st Bending) for lumped mass model, and 4.5 Hz (1st Bending) for beam model. The decrease in natural frequency proves that armament instalment affected the inertial forces of the system.

The aeroelastic analysis is conducted for the clean wing model. Note that the aerodynamic effect on the armament is neglected in this paper. The results are shown in Figs. 6, 7.

From the result, it can be concluded that there are no instability indications shown for clean wing model in the subsonic regime. Thus, the analysis will further focus on the supersonic regime. However, there are several flutter indications shown through U-g and U-f graphics. The second bending damping ratio is reaching the structural damping value, which is 0, at 1300 m/s in the U-g graph, indicating a structural failure at that velocity. In the U-f graph, second bending, third bending, and first torsion are coupled at 950 m/s. This indicates that the critical flutter speed is 950 m/s.

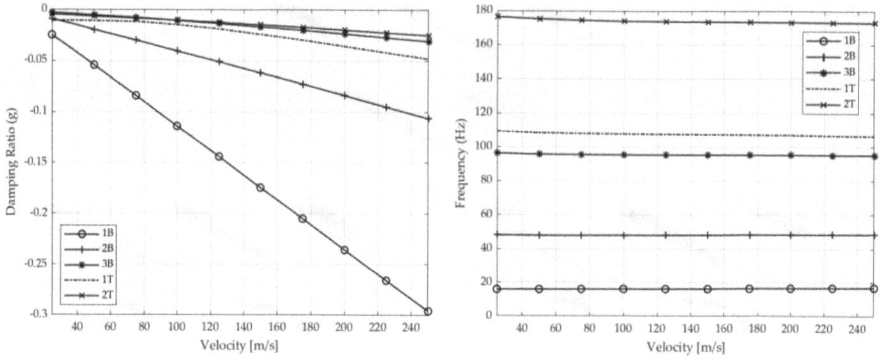

(a) (b)

Fig. 6. Flutter analysis result for clean wing model in subsonic regime: (a) U-g and (b) U-f.

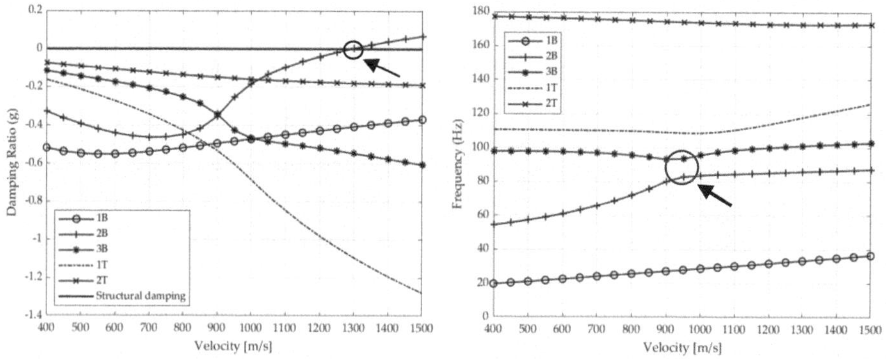

(a) (b)

Fig. 7. Flutter analysis result for clean wing model in supersonic regime: (a) U-g and (b) U-f.

3.1 Variation of Armament Chordwise-Location

Armament location then will be varied into the chordwise direction. The armament will be placed on the tip of the wing and modeled as lumped mass. The location varied from 0.6C, 0.18C, 0.3C, 0.42C, 0.54C, and 0.6C from the leading edge of wing. The flutter results for lumped mass model at 0.3C is shown in Fig. 8.

From the result, there are several flutter indications shown through U-g and U-f graphics. The first torsion damping ratio is reaching 0 at 1180 m/s in the U-g graph, indicating a structural failure at that velocity. In the U-f graph, to determine the flutter speed, the modes that showed instability will be observed, which is the first bending mode. The first bending and first torsion shown indication of coupling since these two modes are approaching each other. The difference between frequencies for first bending and first torsion mode are observed, and smallest difference is located at 750 m/s. This speed then defined as the critical flutter speed. Therefore, this result indicates that the critical flutter speed is 750 m/s. The analysis then conducted for the other armament locations. The critical flutter speed then plotted with the armament location to study

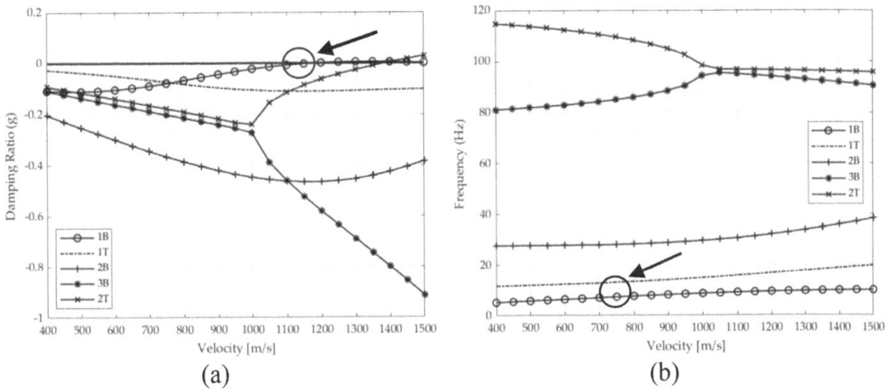

Fig. 8. Flutter analysis result for lumped mass model with armament location at 0.3C in supersonic regime: (a) U-g and (b) U-f.

the effect of armament location on the critical flutter speed and shown in Fig. 9. The summary of critical flutter speed for armament location variation is shown in Table 2.

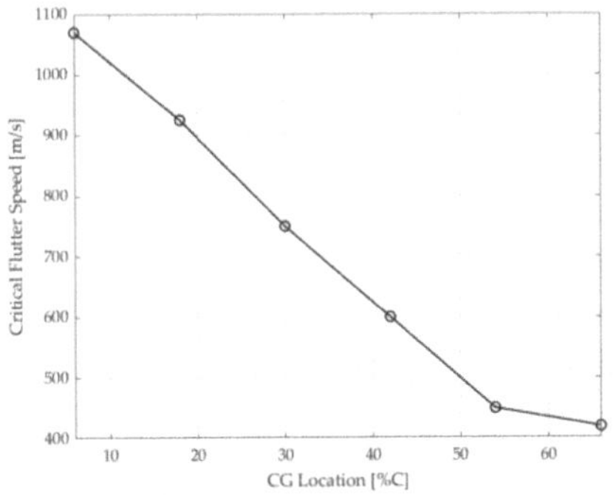

Fig. 9. The effect of armament location on the critical flutter speed.

Table 3. Summary of modal analysis for several armament configuration

Critical Flutter Speed [m/s]						
Clean	Lumped Mass					
	0.06C	0.18C	0.3C	0.42C	0.54C	0.66C
950	1070	925	750	600	450	420

It can be seen from the results that due to the shifting of armament location further from the elastic axis to the trailing edge, the flutter instability is increased. However, it can be seen for the armament location at 0.06C, which is in front of the elastic axis, the critical flutter speed is higher than the clean wing model. Therefore, it can be concluded that the instalment of armament in front of the elastic axis reduces the flutter instability.

3.2 Variation of Armament Mass Moment of Inertia

Armament mass moment of inertia then will be varied. The armament will be placed on the tip of the wing, located at 0.3C and modeled as beam element. The mass moment of inertia varied from 75%, 100%, 125%, and 150% of the initial value. The flutter results for beam model with 100% mass moment of inertia is shown in Fig. 10.

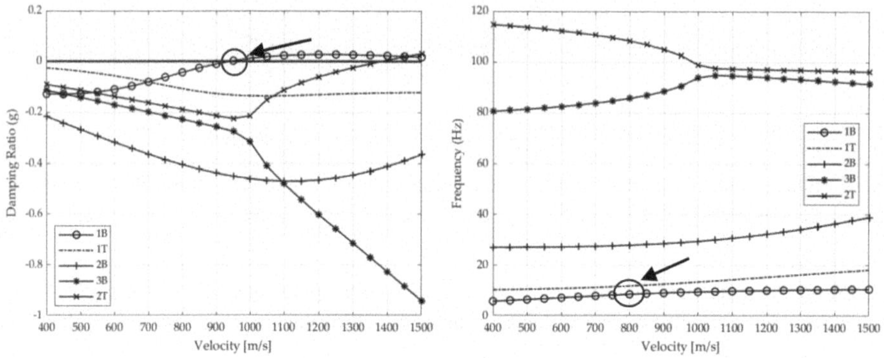

(a) (b)

Fig. 10. Flutter analysis result for beam model with 100% mass moment of inertia in supersonic regime: (a) U-g and (b) U-f.

From the result, there are several flutter indications shown through U-g and U-f graphics. The first bending damping ratio is reaching 0 at 940 m/s in the U-g graph, indicating a structural failure at that velocity. In the U-f graph, first bending and first torsion are coupling at 800 m/s. This indicates that the critical flutter speed is 800 m/s. The analysis then conducted for the other value of mass moment of inertia. The critical flutter speed then plotted with the ratio of mass moment of inertia to study the effect of armament inertia on the critical flutter speed and shown in Fig. 11. The summary of critical flutter speed for armament location variation is shown in Table 3 (Table 4).

It can be seen from the results that due to the increase of armament inertial, the flutter instability is increased. Therefore, it can be concluded that the increase of mass moment of inertia possessed by armament increases the flutter instability.

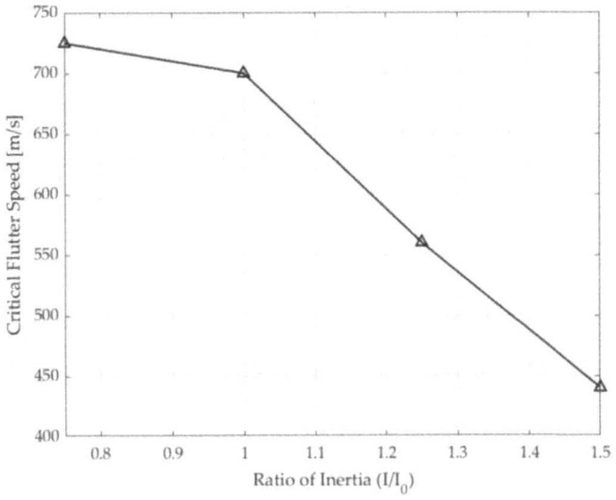

Fig. 11. The effect of armament mass moment of inertia on the critical flutter speed.

Table 4. Summary of modal analysis for several armament configuration

Critical Flutter Speed [m/s]				
Clean	Bar Element			
	0.75I	I	1.25I	1.5I
900	850	800	565	440

4 Conclusion

Based on the current study, the variation of armament's center of gravity location and mass moment of inertia have a significant effect on the flutter characteristics of wing. Although it does not cause great changes of the other natural frequencies, results show that the variation caused a decrease in first torsion. This mode shape tends to couple with the first bending mode and caused earlier occurrence of flutter phenomena. It is also found that the placement of armament in front of the elastic axis reduces the flutter instability of the fighter wing, while the placement behind the elastic axis increases the instability. Moreover, mass moment of inertia also affects the flutter instability, whereas the increased inertia caused the critical flutter speed to decrease.

References

1. Johnson, E.H.: MSC Developments in Aeroelasticity. The MacNeal-Schwender Corporation, Los Angeles (2011)
2. Chesta, L.: A Parametric study of wing store flutter. In: Specialists Meeting on Wing-With-Stores Flutter, Paris (1975)
3. Baird, E.F., Clark, W.B.: Recent observations on external-store flutter. In: Specialists Meeting on Wing-With-Stores Flutter, Paris (1975)
4. Hodghes, D.H., Pierce, G.A.: Introduction to Structural Dynamics and Aeroelasticity, 2nd edn. The Press Syndicate of The University of Cambridge, Cambridge (2002)

5. Abdullah, N.A., Sulaeman, E.: Flutter analysis of supersonic wing with external stores. In: International Conference on Mechanical, Automative, and Aerospace Engineering, Kuala Lumpur (2013)
6. Akpinar, A.: Structural Structural and Aeroelastic Flutter Analysis of Wing Store Systems and Structural Modification Approach in Aeroelasticity. Middle East Technical University, Ankara (2021)
7. Serin, Ö.: Aeroelastic Analyses and Tests of A Wing Like Structure with Payload. Ankara (2019)
8. Logan, D.L.: A First Course in the Finite Element Method. Thompson Canada Limited, Toronto (2007)
9. Wright, J.R., Cooper, J.E.: Introduction to Aircraft Aeroelasticity and Loads. John Wiley & Sons Ltd, Chichester (2007)
10. Yates, E.C.: Modified-strip-analysis method for predicting wing flutter from subsonic to hypersonic speeds. J. Aircr. **3**(1), 25–29 (1966)
11. Ashley, H., Zartarian, G.: Piston theory—a new aerodynamic tool for the aeroelastician. J. Aeronaut. Sci. 1109–1118 (1956)
12. Yurkovich, R.: Status of unsteady aerodynamic prediction for flutter of high-performance aircraft. J. Aircraft, pp. 832–842 (2003)
13. Albano, E., Rodden, W.P.: A doublet-lattice method for calculating the lift distributions on oscillating surfaces in subsonic flows. AIAA J. **7**(2), 279–285 (1969)

A Comparative Analysis of Fluid-Structure Interaction Simulations and Wind-Tunnel Experiments on Cantilever Plate Flutter Phenomena

Firza Fadlan Ekadj(✉) and Idris Eko Putro

Research Organization for Aeronautics and Space, National Research and Innovation Agency, Jakarta, Indonesia
firz001@brin.go.id

Abstract. Fluid-structure interaction (FSI) simulations have been widely conducted on thin plates. However, the conditions in some studies do not adequately replicate those relevant to airplane wings and rocket fins, which are typically attached to the sides of the main bodies and can be subjects for flutter phenomena. The simulations used Cartesian orthogonal meshes and employed the FSI method, transferring force and displacement data between fluid and structure solvers. The results are compared with experimental data obtained from wind tunnel tests, which were documented in two separate studies. The first case involved a plexiglass plate with a dimension of 5 × 131 × 562 mm, while the second case used a polycarbonate model with a dimension of 5 × 125 × 1000 mm. The results demonstrate that the flutter speed obtained from the simulations for the first case was 66.6 m/s, which closely aligns with the experimental result of 72.5 m/s. The simulations for the second case yielded a flutter speed of 29.8 m/s, compared to the experimental result of 24.89 m/s. The study suggests that the FSI method using Cartesian orthogonal meshes has the potential for approximating the flutter speed of a simple rectangular cantilever plate.

Keywords: Flutter · Fluid-Structure Interaction · Plate · Subsonic Flow

1 Introduction

Flutter is a self-excited oscillation and frequently leads to structural failure [1]. This phenomenon can occur in slender aerospace structures, including high aspect ratio wings and the fins of rockets [2, 3]. However, it is perhaps the hardest to predict of all the aeroelastic phenomena. Conducting wind tunnel tests to investigate this structural behavior is both energy-intensive and cost-consuming.

Flutter involves the behavior of both aerodynamics and structure. Consequently, with regard to the problem above, it is essential to have a simulation method that can predict flutter speeds on slender structures in relation to the interaction between fluid and structure. Fluid-structure interaction (FSI) refers to the dynamic interplay between

E. Fitrianingsih et al. (Eds.): ISAST 2024, SPPHY 416, pp. 375–381, 2025.
https://doi.org/10.1007/978-981-96-1344-1_40

a solid structure and the fluid flow that surrounds or permeates it [4]. This interaction encompasses complex phenomena critical in various engineering and scientific applications, as it involves the coupling of fluid dynamics with structural mechanics. In this paper, the FSI method was implemented to predict flutter speeds on two simple slender specimens i.e. rectangular plates. Studies using fluid-structure interaction simulations and thin plates have been conducted [5–8]. However, ref [5, 6], and [7] have fixed sides of the plates facing the upcoming winds while ref [8] has thickness sides parallel to the wind. In the application using aircraft wings and rocket fins, the spanwise directions and the leading edges are perpendicular to the wind. This study conducted simulations using specimens with the spanwise direction parallel to the freestream and one of the thickness sides facing the stream. Chen et al. have conducted a study using a 3-D wing with the FSI method and an O-type grid [9]. Conversely, this study used flat plate specimens and Cartesian orthogonal meshes.

Alizedah et al. have conducted flutter experiments using a cantilever plate [10]. Mevlut and Kemal also did a similar experiment with a cantilever plate made of a different material [11]. This study aims to compare the results of these experiments to those of a fluid-structure interaction.

2 Materials and Method

Two models used in the wind tunnel tests are a plexiglass plate with a dimension of 5 × 31 × 562 mm and a polycarbonate plate with a dimension of 5 × 125 × 1000 mm [10, 11]. The properties of the materials are represented in Table 1.

Table 1. Material properties of the plates used for the simulations.

Material property	Plexiglass (case 1)	Polycarbonate (case 2)
Density (kg/m^3)	1208	1200
Elastic modulus (MPa)	3270	3500
Poisson Ratio	0.35	0.35

The simulations utilize the solid domains of the structural plates and the fluid domains of the surrounding air, represented by Cartesian orthogonal meshes (Fig. 1) [12]. The surfaces of the plates, except for the fixed sides, serve as the interface regions for coupling. The fluid-structure interaction (FSI) method is employed for the simulations, with force and displacement data being exchanged between the solid and fluid regions [13].

To validate the structural solver, a modal analysis simulation for one of the cases is conducted and the result is then compared to the experimental result.

2.1 Boundary Conditions

The outer boundary is categorized into walls, upstream, and downstream segments. For the upstream boundary, the streamwise velocity is uniform, while the transverse velocity

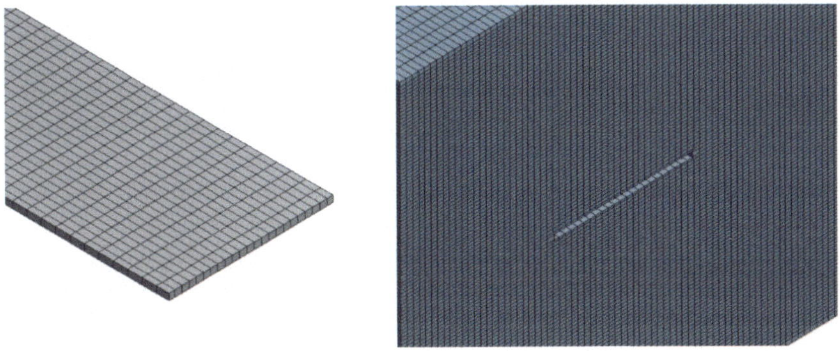

Fig. 1. An example of meshes for a structural plate and a fluid domain.

and spanwise velocity are both zero. For the downstream boundary condition, all flow variables are extrapolated from the interior, except for the pressure, which is maintained at its freestream value.

2.2 Governing Equations

The Reynolds-averaged Navier Stokes models provide the most cost-effective method for calculating complex turbulent flows in industrial applications [14]. The equations can be written as [15]:

$$\frac{\partial \rho}{\partial t} + \frac{\partial}{\partial x_i}(\rho u_i) = 0 \tag{1}$$

$$\frac{\partial}{\partial t}(\rho u_i) + \frac{\partial}{\partial x_j}(\rho u_i u_j) = -\frac{\partial p}{\partial x_i} + \frac{\partial}{\partial x_j}\left[\mu\left(\frac{\partial u_i}{\partial x_j} + \frac{\partial u_j}{\partial x_i} - \frac{2}{3}\delta_{ij}\frac{\partial u_l}{\partial x_l}\right)\right] + \frac{\partial}{\partial x_j}\left(-\rho\overline{u_i' u_j'}\right) \tag{2}$$

where ρ is a density, t is time, x_i are coordinate components, and u_i are velocity components.

Transient dynamic analysis, also known as time-history analysis, is a method used to assess a structure's dynamic response to various time-dependent loads [16]. This technique helps determine the time-dependent displacements, strains, stresses, and forces in a structure subjected to a mix of static, transient, and harmonic loads. The fundamental equation of motion addressed in a transient dynamic analysis can be written as:

$$(M)\{\ddot{u}\} + (C)\{\dot{u}\} + (K)\{u\} = \{F(t)\} \tag{3}$$

where (M) is a mass matrix, (C) is a damping matrix, (K) is a stiffness matrix, $\{\ddot{u}\}$ is a nodal acceleration vector, $\{\dot{u}\}$ is a nodal velocity vector, $\{u\}$ is a nodal displacement vector, and $\{F(t)\}$ is a load vector.

3 Results and Discussion

3.1 First Case Modal Analysis

Table 2 presents the results of the evaluation of the natural frequencies of the first case model. The first column displays the results of the finite element (FE) analysis, while the second column contains the experimental values obtained from vibration testing. A good correlation is observed between the FE analysis and the experimental results for the first three bending modes and the first torsion mode. Some discrepancies may be attributed to differences in the material properties used in the software analysis compared to those in the actual experiments.

Table 2. Comparison of the natural frequency evaluation of the first case model with FE analysis and experimental results.

Mode	FE frequency	Experimental frequency [10]
1st bending	4.19 Hz	4.36 Hz
2nd bending	26.22 Hz	28.84 Hz
1st torsion	34.95 Hz	39.44 Hz
3rd bending	73.62 Hz	79.74 Hz

No experimental results of a modal analysis for the second case are provided.

3.2 Flutter on Two Cantilevered Plates

From the conducted simulations, we can observe the desired displacement data for certain structural mesh elements [17]. These elements are located at the free ends of the plates, where they are subject to oscillations. The directional displacement data, perpendicular to the freestream and the spanwise direction, for two freestream speeds on each specimen, are presented in Figs. 2 and 3 as functions of time. This approach of a freestream speed magnitude variation enables the identification of the specific freestream speed at which the plate begins to exhibit flutter.

The results indicate that the flutter phenomenon occurred on the plexiglass plate at a freestream speed of 66.6 m/s. Employing the following equation

$$f = \frac{N}{t} \tag{4}$$

with f is a frequency, N is the number of cycles, and t is the elapsed time, the oscillation frequency is calculated to be 17.3 Hz. For simulations conducted on the polycarbonate plate, the flutter speed is determined to be 29.8 m/s, with the corresponding oscillation frequency being 13 Hz. A comparison between these results and those of the experiments is represented in Table 3.

It can be seen that in the case of the plexiglass plate, the simulation results closely align with the experiment data. However, there are noticeable differences in the results for the case of the polycarbonate specimen.

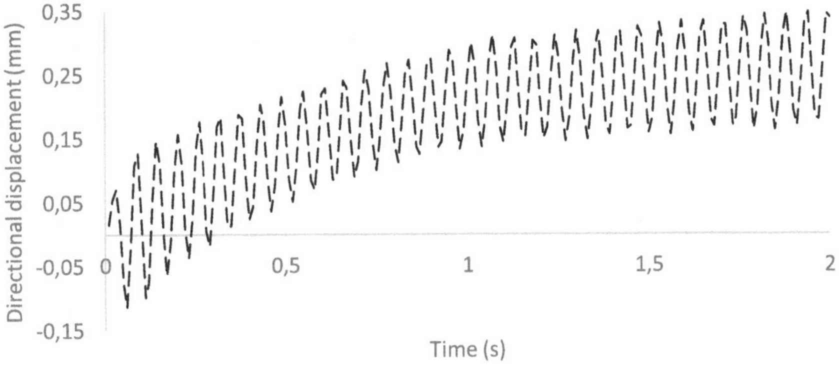

(a)

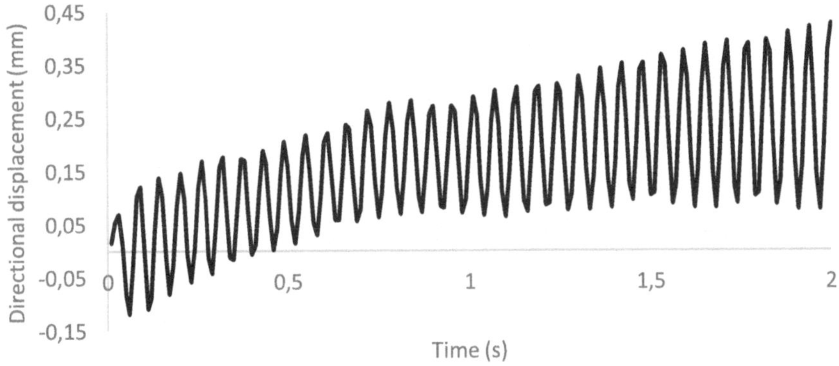

(b)

Fig. 2. Directional displacement data of a plexiglass Plate's mesh element for freestream speeds of 66.5 m/s and 66.6 m/s.

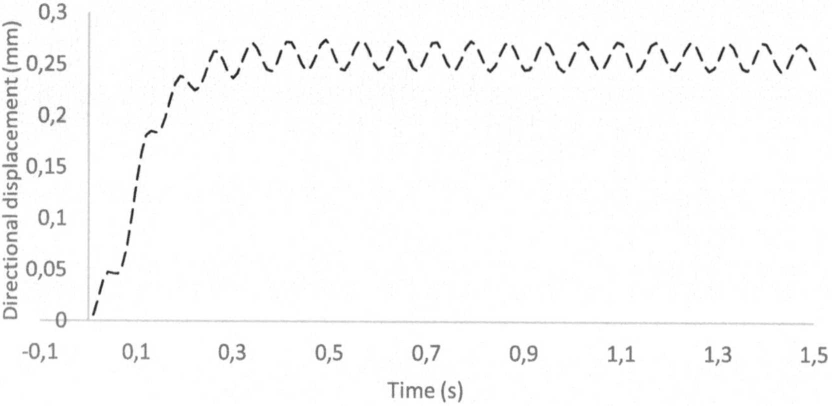

(a)

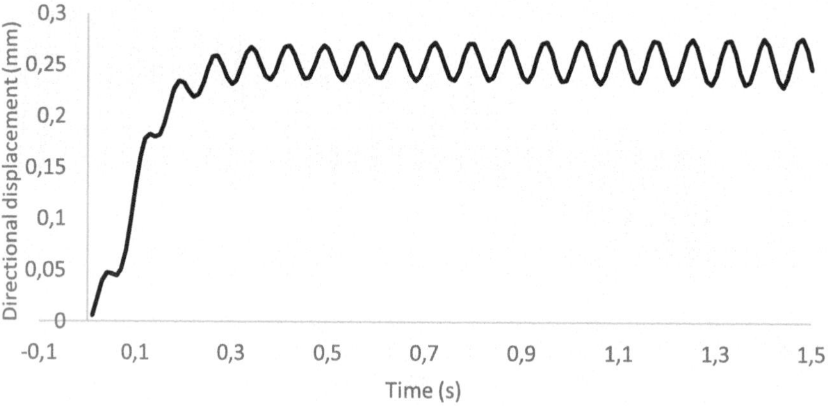

(b)

Fig. 3. Directional displacement data of a polycarbonate plate's mesh element for freestream speeds of (a) 29.7 m/s, and (b) 29.8 m/s.

Table 3. A comparison of flutter speeds and frequencies between FSI simulation results and wind tunnel experiment results on two cantilever plates.

	Plexiglass specimen		Polycarbonate specimen	
	FSI simulation	Wind tunnel experiment [10]	FSI simulation	Wind tunnel experiment [11]
Flutter speed (m/s)	66.6	72.5	29.8	24.89
Flutter frequency (Hz)	17.2	17.08	13.2	8.8

4 Conclusion

Fluid-structure interaction simulations using Cartesian orthogonal meshes have been conducted to investigate flutter phenomena on the plexiglass and polycarbonate plates. The simulations using a cantilever plate with plexiglass material show a flutter speed and a flutter frequency that are very close to the experimental data. However, considerable differences remain in the case with the polycarbonate specimen. These findings suggest that the FSI method using Cartesian orthogonal meshes can be beneficial for approximating the flutter speed of a simple rectangular cantilever plate, suggesting the potential to reduce the necessity for energy-intensive wind tunnel experiments. Further research is necessary to identify potential causes of these differences and to explore whether a more complex geometry, equivalent to an aircraft wing or a rocket fin, can be employed using the same method.

References

1. Wright, J.R., Cooper, J.E.: Introduction to Aircraft Aeroelasticity and Load. John Wiley & Sons Ltd, England (2007)
2. Cestino, E., Frulla, G., Spina, M., Catelani, D., Linari, M.: Numerical simulation and experimental validation of slender wings flutter behaviour. J. Aerosp. Eng. **233**(16), 5913–5928 (2019)
3. da Silva, R.G.A., Damilano, J.G., Azevedo, J.L.F.: A sensitivity investigation on the aeroelastic dynamic stability of slender spinning sounding rockets. J. Aerosp. Technol. Manag. **5**(1), 15–26 (2013)
4. de Vaucorbeil, A., Nguyen, V.P., Sinaie, S., Wu, J.Y.: Chapter two - material point method after 25 years: theory, implementation, and applications. Adv. Appl. Mech. **53**, 185–398 (2020)
5. Bin, J., Oates, W.S., Hussaini, M.Y.: Fluid–structure interactions of photo-responsive polymer cantilevers. J. Fluids Struct. **37**, 34–61 (2013)
6. Rahtika, I.P.G.S., Wardana, I.N.G., Sonief, A.A., Siswanto, E.: Numerical and experimental investigation on the flutter of cantilevered plates with free leading edge in axial flow. Adv. Acoustics Vibr. 1–8 (2017)
7. Jendoubi, A., Deteix, J., Fortin, A.: A simple mesh-update procedure for fluid–structure interaction problems. Comput. Struct. **169**, 13–23 (2016)
8. Nemer, R., Larcher, A., Hachem, E.: Adaptive Immersed Mesh Method (AIMM) for Fluid-Structure Interaction (2022)
9. Chen, X., Zha, G., Yang, M.: Numerical simulation of 3-D wing flutter with fully coupled fluid-structural interaction. Comput. Fluids **36**(5), 856–867 (2007)
10. Alizadeh, A., Ebrahimi, Z., Mazidi, A., Fazelzadeh, S.A.: Experimental nonlinear flutter analysis of a cantilever wing/store. Int. J. Struct. Stab. Dyn. **20**(7) (2020)
11. Dalmış, M.B., Yaman, K.: Numerical and experimental investigations of cantilever rectangular plate structure on subsonic flutter. Int. J. Aerosp. Mech. Eng. **9**(11), 2031–2037 (2015)
12. Girfoglio, M., Quaini, A., Rozza, G.: Fluid-structure interaction simulations with a LES filtering approach in solids4Foam. Commun. Appl. Indust. Math. **13**(1), 13–28 (2022)
13. ANSYS, Inc.: System Coupling User's Guide. ANSYS, Inc. (2018)
14. ANSYS, Inc.: ANSYS FLUENT User's Guide. ANSYS, Inc. (2010)
15. ANSYS, Inc.: Ansys Fluent Theory Guide. ANSYS, Inc. (2021)
16. ANSYS, Inc.: ANSYS Mechanical APDL Structural Analysis Guide. ANSYS, Inc. (2013)
17. Currao, G.M.D., Neely, A.J., Buttsworth, D.R., Gai, S.L.: Hypersonic fluid-structure interaction on a cantilevered plate. In: 7th European Conference for Aeronautics and Space Sciences (EUCASS) (2017)

Analysis of Hardware-in-the-Loop Simulation (HILS) on Thrust Vectoring Control Design for a Ducted Fan UAV

Octavianus Bagaswara Adi(✉) and Yazdi Ibrahim Jenie

Faculty of Mechanical and Aerospace Engineering, Institut Teknologi Bandung, Bandung, Indonesia
octavianus.bagaswara@gmail.com

Abstract. Ducted fan Unmanned Aerial Vehicles (UAV) face unique control challenges, leading to varied parameters between simulations and real-world testing. Hardware-in-the-Loop Simulation (HILS) bridges this gap by validating control algorithms in real-time. This study integrates Pixhawk 1 as the flight controller, managing UAV dynamics with designed algorithms. Host computers run Autopilot Controller and Plant Simulation apps alongside Pixhawk, linking via MAVLink to a Ground Control Station (GCS). Simulations demonstrate quicker responses and lower errors than HILS, affected by sensor noise and synchronization delays. Differences in Integral of Time-Weighted Absolute Error, ITAE (45.69%) and amplitude (50.87%) highlight HILS's longer settling and rise times (averaging 17.1% and 36.99% more than simulation), emphasizing its role in refining UAV control for robust practical performance.

Keywords: Hardware-in-the-loop Simulation · Thrust Vectoring Control · Ducted Fan UAV

1 Introduction

Unmanned Aerial Vehicles (UAVs) have emerged as versatile tools for various applications including surveillance, agriculture, logistics, military operations, and environmental monitoring. Among the various types of UAVs, ducted fan UAVs are particularly notable in military applications due to their vertical take-off and landing capabilities, precise control, and exceptional maneuverability. These UAVs utilize a thrust vectoring control system, which is more complex but offers superior performance and reliability compared to conventional control methods [1]. Ducted fan UAVs encounter inherent instability and control complexities due to their unique design, often necessitating different control parameters between simulations and real-world testing. To address this disparity, Hardware-in-the-Loop Simulation (HILS) serves as a crucial bridge, facilitating real-time validation of control algorithms under realistic conditions. This study focuses on HILS analysis for thrust vectoring control design in ducted fan UAVs, aiming to refine control system design through HILS and thereby enhance the stability and performance of ducted fan UAVs.

E. Fitrianingsih et al. (Eds.): ISAST 2024, SPPHY 416, pp. 382–391, 2025.
https://doi.org/10.1007/978-981-96-1344-1_41

2 State-of-the-Art

Previous research in UAV control systems has utilized Pixhawk and Simulink for multicopter configurations, allowing for the integration of sophisticated control algorithms and real-time hardware testing [2,3]. However, when it comes to Thrust Vectoring Control, prior work has primarily focused on single-degree-of-freedom implementations using 8-bit microcontrollers [4]. These studies have successfully demonstrated that HILS system produces responses that have a good fit to the actual system responses. This research aims to advance the state-of-the-art by integrating Pixhawk and Simulink for a more complex 3-DoF TVC system.

3 HILS Architecture

HILS testing involves applying control algorithms to real hardware while simulating components like power converters, generators, and feedback sensors [5]. This method relies on bidirectional closed-loop interactions between physical and virtual components. Figure 1 illustrates the HILS architecture, where two applications will run on a host computer alongside hardware components, which is Pixhawk as the flight controller. The Autopilot Controller application will be embedded in the Pixhawk, utilizing μORB messaging for data communication. The Pixhawk will then interface with the host computer to execute the Ground Control Station (GCS) and Ducted Fan UAV or Plant Simulation applications. The GCS plays a crucial role in linking the Simulink-based plant simulation with the Pixhawk. Within this setup, the Plant Simulation models the dynamics of the ducted fan UAV and responds to input-output signals received from the Pixhawk.

3.1 Connection Between Pixhawk, Simulink, and GCS

A system diagram illustrating HILS is presented in Fig. 2. This diagram outlines the communication framework involving the PX4 autopilot, Simulink Simulator, and QGroundControl as GCS. In HILS mode, the PX4 autopilot serves as the flight controller with embedded control algorithms managing real-time flight dynamics via Pixhawk 1. Communication between the PX4 autopilot and Simulink Simulator occurs over USB using the MAVLink protocol. The Simulink Simulator simulates UAV dynamics, processes MAVLink messages, and bridges data with QGroundControl via UDP (port 14550), ensuring real-time status updates and command issuance. QGroundControl acts as the ground control station, facilitating MAVLink communication with Simulink and other devices over serial connections. During HILS operation, sensors are inactive, and actuators are disabled to prevent interference with the simulation environment.

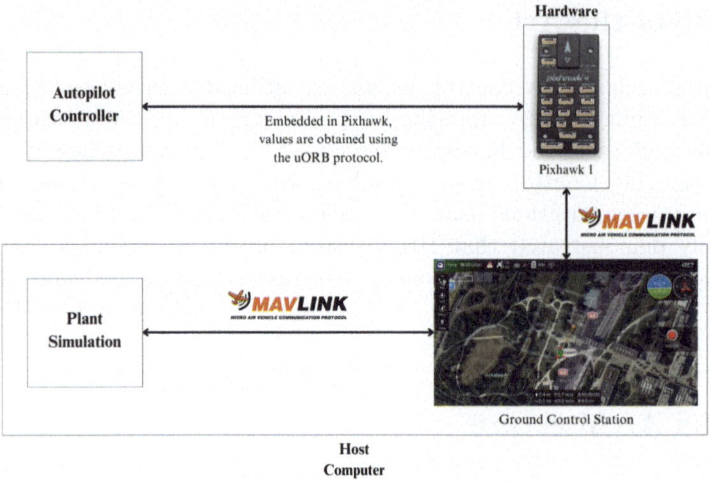

Fig. 1. HILS architecture depicting two applications running on a host computer alongside hardware components, with Pixhawk serving as the flight controller.

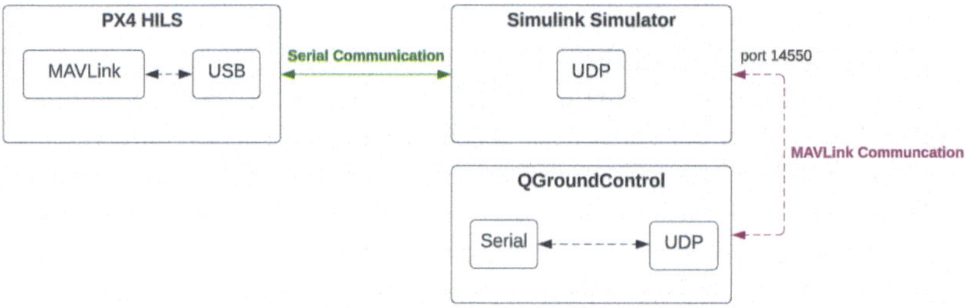

Fig. 2. System diagram illustrating the HILS setup.

3.2 μORB Messaging

PX4's internal communication relies on μORB (Micro Object Request Broker), a publish-subscribe messaging framework that enables efficient data exchange among modules by allowing processes to publish and subscribe to topics. This ensures decoupled and scalable communication, so sensor data can be utilized by multiple modules like flight controllers or logging systems without direct dependencies. In this model, publishers create and publish messages to specific topics, while subscribers receive these messages by subscribing to the topics. Messages are stored in circular buffers associated with their topics, and subscribers access them directly from these buffers upon notification. For example, in Fig. 3, an estimator module can publish state estimates to μORB as a topic, which other modules can then subscribe to for real-time state data. This bidirectional communication allows modules to publish and receive updates as needed, enhancing

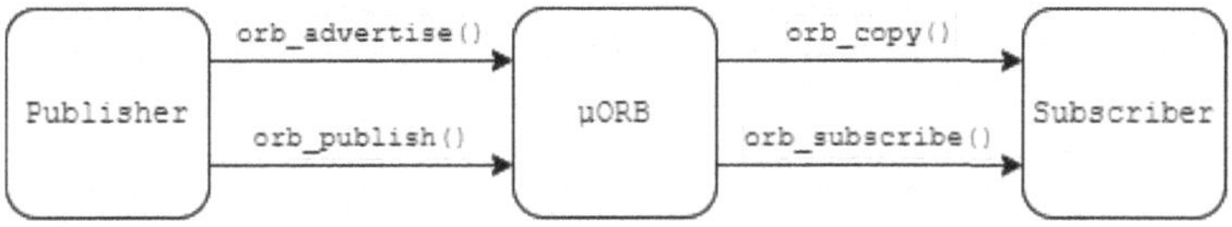

Fig. 3. Example of the Micro Object Request Broker (μORB) architecture, inspired by [6]

system efficiency through asynchronous interactions and reduced interdependencies [6].

3.3 MAVLink Messaging

The Pixhawk communication protocol relies on MAVLink, a lightweight messaging protocol for unmanned vehicles like drones, enabling efficient bidirectional communication for real-time telemetry, command, and control data. MAVLink's robustness and versatility make it ideal for aerospace and robotics applications [7]. It defines message structure and serialization at the application layer, which are transmitted through lower layers. MAVLink 2.0, the latest version, uses 11 fields. Messages range from 12 bytes for simple acknowledgment packets to 280 bytes for signed messages with a full payload. Key components include a start-of-text marker (0xFD), payload length, compatibility and incompatibility flags affecting message structure and features like packet signing, sequence number for message loss detection, system and component IDs for sender details, message type ID, and checksum for ensuring message integrity. MAVLink 2.0 optionally includes a 13-byte Signature field for added security against tampering.

4 Ducted Fan UAV Simulation

Fig. 4. Setup of the Ducted Fan UAV Model Dynamics subsystem within the plant simulation.

The plant simulation integrates the dynamics of the ducted fan UAV model, interconnected via MAVLink Bridge Sink and Source blocks to enable communication with the GCS using the MAVLink protocol. Figure 4 illustrates the setup of the Ducted Fan UAV Model Dynamics subsystem within the plant simulation for this HILS research. This diagram comprises three subsystem blocks:

Data Deserialization, Actuator Model, and Plant Dynamics and Serialization. The simulation runs in real-time using Simulink. Initially, data is fetched from the GCS through the MAVLink Bridge Source block, and subsequently, data is transmitted back to the GCS via the MAVLink Bridge Sink block. Configuring parameters such as Serial Port for the Pixhawk, Data Length set to 1024, and adjusting the default remote address and port type to UDP within the MAVLink Bridge block ensures smooth data exchange between Simulink and the GCS.

The ducted fan UAV design, based on [8], features three degrees of freedom, with two translational and one rotational motion, and a total weight of 5.1 kg. Its lightweight construction, optimized for stability and maneuverability, allows for precise control of thrust magnitude and direction. The UAV is composed of interconnected components without relative rigid body displacements, making it suitable for a multibody modeling approach. This method represents the system as a network of interconnected bodies linked by joints and constraints, accurately depicting motion, forces, and interactions between components.

5 Autopilot Controller

Given that Pixhawk replaces the Flight Control System in this setup, the control algorithms developed in simulation must be implemented into the Pixhawk firmware. This integration is achieved by creating a Simulink program Autopilot Controller and utilizing the μORB messaging protocol to fetch values or states computed by the Pixhawk. Figure 5 provides an overview of the controller diagram, comprising three subsystems: Estimator Output, Controller, and Send to Actuator. The Estimator Output subsystem retrieves ducted fan UAV states and conditions calculated by the Pixhawk using the Extended Kalman Filter (EKF) algorithm, which processes sensor data to estimate quaternion, velocity, and position. The Send to Actuator subsystem then transmits thrust commands and input deflection angles as actuator data. This transmission employs the μORB messaging protocol with the topic actuator_controls.

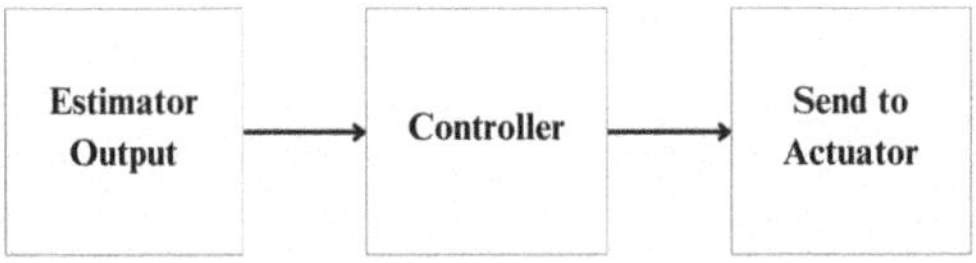

Fig. 5. Overview of the Autopilot Controller diagram for the Ducted Fan UAV.

In this study, PID control was selected to align with Pixhawk 1 using PX4. However, the control diagram must handle three states (x position, z position, and pitch angle) with two inputs (thrust and gimbal deflection angle), making the ducted fan UAV MIMO while PID is inherently SISO. To simplify this complexity, transfer functions were analyzed during linearization to relate thrust and deflection angle inputs to outputs such as x position, z position, pitch angle, x velocity, z velocity, and pitch rate. This necessitates a two-loop control system: thrust controls z position, while gimbal deflection angle controls x position and pitch angle. The complete control diagram is depicted in Fig. 6.

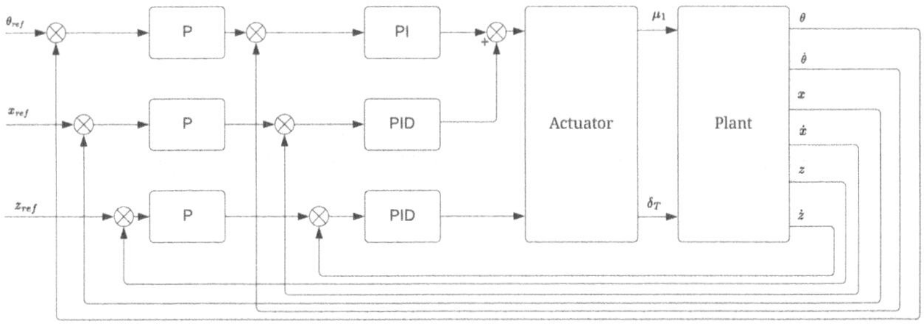

Fig. 6. Complete controller block diagram for the ducted fan UAV.

After several tuning processes, the parameters listed in Table 1 were chosen as the final PID parameters for controlling the ducted fan UAV, ensuring effective tracking and stabilization. In the tracking test, the UAV began from initial conditions of zero position and velocity. At $t = 0$, it was directed towards a setpoint of $z = -1$ m, $x = 1$ m, and a pitch angle of $0°$ to achieve stabilization.

Table 1. Final PID parameters selected for controlling the ducted fan UAV.

Controller	K_p	K_i	K_d	N
Position Z	2	–	–	–
Velocity Z	30	84	0.5	30
Position X	1.5	–	–	–
Velocity X	−0.015	−0.005	−0.9	0.01
Pitch Angle	2	–	–	–
Pitch Rate	−0.06	−0.07	–	–

6 Real-Time Simulation Using Simulink Pacing

In the context of HILS systems, achieving real-time performance is essential for accurately testing and validating control strategies. Traditional approaches often rely on Real-Time Operating Systems (RTOS) or specialized toolboxes like Simulink Real-Time Workshop to maintain synchronization between the simulation and physical hardware. However, such methods can be complex and resource-intensive. In this research, Simulink pacing is utilized as a practical alternative to create a real-time-like environment without the need for dedicated real-time systems. By controlling the simulation's execution speed, Simulink pacing allows the plant model to run in sync with the actual hardware.

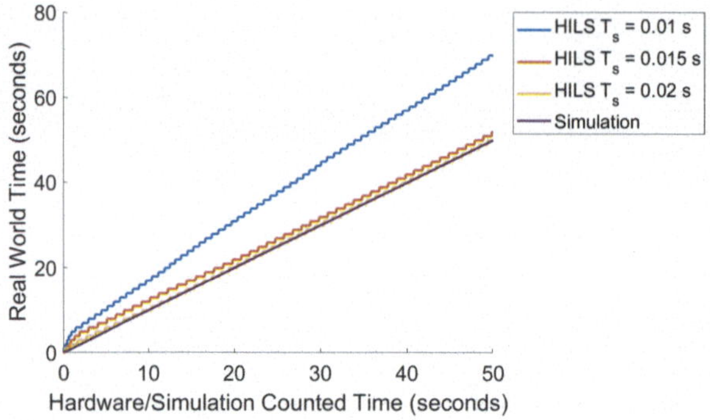

Fig. 7. Comparison of real-world time with time measured by hardware and simulation.

Accurate time synchronization between the Pixhawk flight controller and the simulation environment is essential in HILS setups for reliable real-time data exchange. In Ducted Fan UAV HILS, selecting an appropriate sampling time is critical for accuracy and reliability. Sampling time, the interval at which system states are updated, affects how well the simulation mirrors real-world UAV dynamics. Proper sampling ensures precise performance analysis and system validation. Figure 7 shows that perfect synchronization results in the same slope. However, different sampling times affect this relationship. At a 0.01-s interval, the simulation may lag due to processing demands, while a 0.015-s interval can improve alignment but with increased bias. A 0.02-s sampling time, however, closely matches the Pixhawk's timing and offers minimal bias, making it the optimal choice for accurate synchronization in this research.

Figure 8 shows the UAV's response and input under HILS with three different sampling times: 0.01 s, 0.015 s, and 0.02 s. Shorter sampling times can cause more oscillatory responses, especially when synchronization issues arise. While trends remain consistent, lower sampling times often lead to longer settling times due to synchronization challenges between the Pixhawk 1 hardware and the simulation, resulting in inaccurate control values.

The observation is supported by the significant oscillations in input thrust and deflection angle at lower sampling times. These oscillations stem from the Pixhawk 1 controller updating faster than Simulink's state calculations, leading to mismatches when synchronization is off.

7 Comparative Analysis Between HILS and Simulation

An essential aspect of this research is comparing simulation and HILS to validate control systems' performance and reliability. In both HILS and simulation, the program will execute with setpoints assigned to the ducted fan UAV: $z = -1$ m, $x = 1$ m, and a pitch angle of 0° to achieve stabilization.

Figure 8 illustrates the distinct response and input of the ducted fan UAV between HILS and simulation environments. Performance indices are used to evaluate the controller's efficacy in HILS with a sampling time of 0.02 s when synchronization occurs. Table 2 summarizes these indices for the controller. Simulations generally exhibit quicker responses in transient conditions due to their idealized, noise-free environment, optimizing controller performance. In contrast, HILS integrates real hardware with sensor noise and communication delays, resulting in slower response times.

Table 2. Comparison of controller performance indices between Hardware-in-the-Loop Simulation (HILS) and simulation.

Controller	ITAE		AMP		Settling Time (seconds)		Rise Time (seconds)	
	HILS	Simulation	HILS	Simulation	HILS	Simulation	HILS	Simulation
Position Z	2.9787	0.49	1.0049	0	2.1695	2.49	0.8523	1.33
Velocity Z	1.2197	0.789	–	–	2.4322	2.63	–	–
Position X	10.13	2.164	1.1957	1.175	5.9925	5.89	1.0722	0.88
Velocity X	3.1641	2.274	–	–	6.6195	4.59	–	–
Pitch Angle	0.566	0.421	–	–	6.4909	5.16	–	–
Pitch Rate	1.2647	0.975	–	–	5.9857	4.38	–	–

The Integral of Time-Weighted Absolute Error (ITAE) is a crucial metric for evaluating cumulative error over time, giving higher importance to errors occurring later in the process. Comparing HILS and simulation results reveals that simulation consistently shows lower ITAE values by an average of 45.69% compared to HILS. This disparity suggests that the simulation system effectively reduces errors over time, benefiting from its idealized conditions and absence of real-world noise and disturbances present in HILS. Noise sources include sensors, the GPS model, and the observer or Extended Kalman Filter (EKF) in Pixhawk.

The performance index AMP measures the peak deviation of the UAV response from its desired trajectory in dimensionless units. In simulations, AMP values are typically lower by 50.87%, indicating that the UAV's responses are more controlled and less prone to overshoot. This advantage is due to the idealized and noise-free environment of simulations, allowing the controller to operate optimally, similar to the reason for lower ITAE values.

Rise time and settling time are also critical metrics. Simulations generally show faster rise times averaging at 36.99%, highlighting the controller's ability to achieve rapid responses under ideal conditions. In contrast, HILS exhibits slower rise times averaging at 17.1%, attributable to inherent latency and slower response rates in physical systems like the Pixhawk.

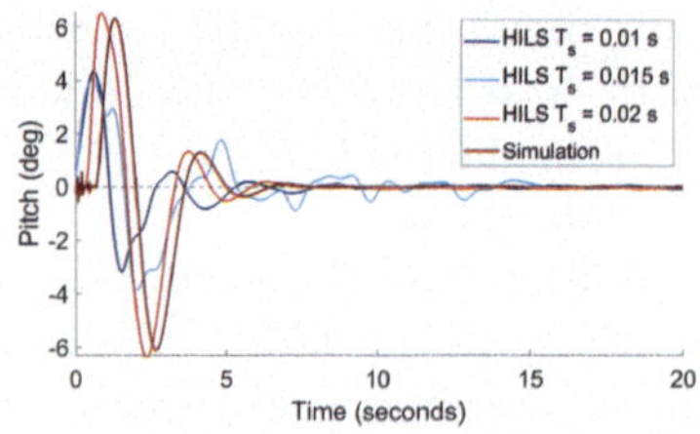

(a) Pitch Angle

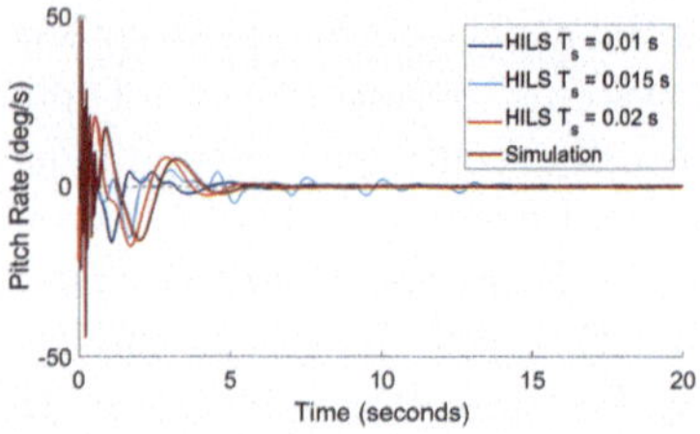

(b) Pitch Rate

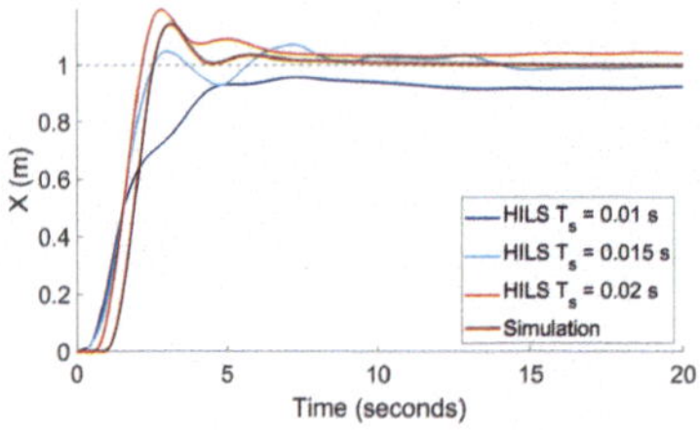

(c) Position X

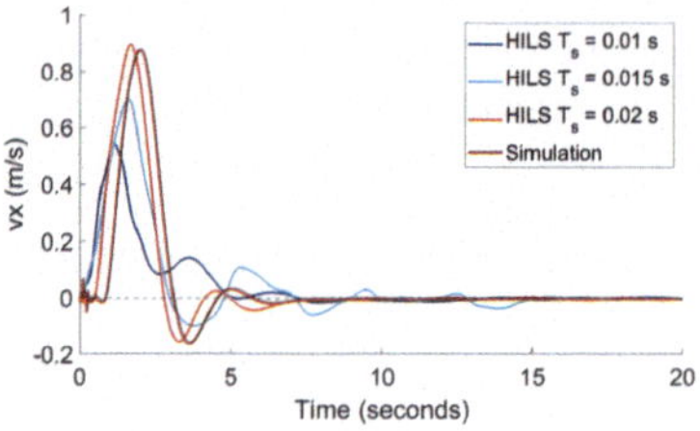

(d) Velocity X

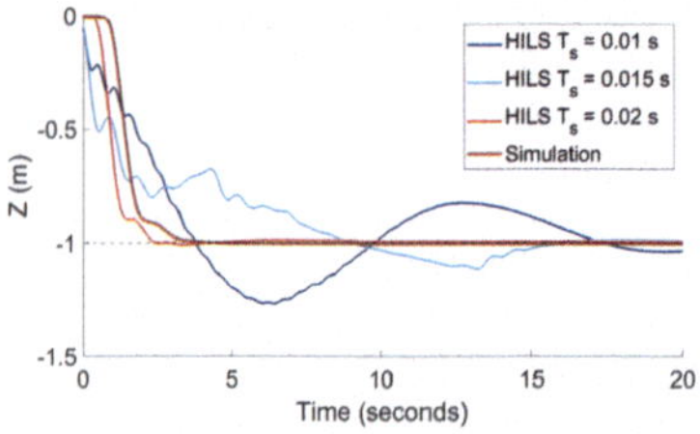

(e) Position Z

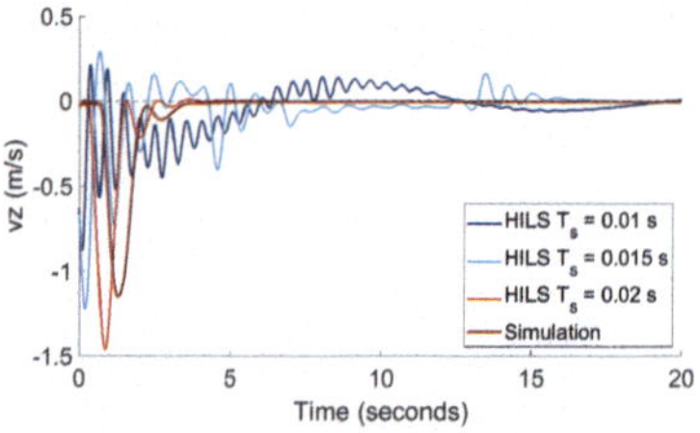

(f) Velocity Z

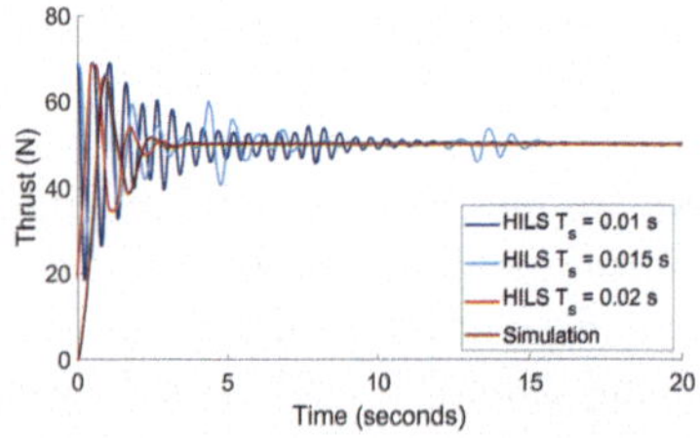

(g) Thrust

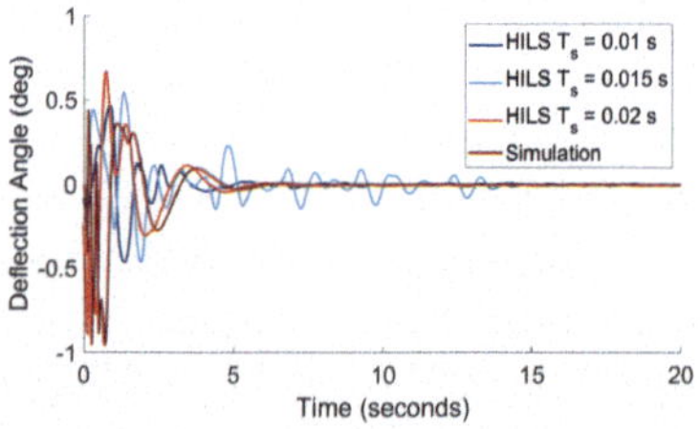

(h) Deflection Angle

Fig. 8. Comparison of the ducted fan UAV response (a–f) and input (g–h) between Hardware-in-the-Loop Simulation (HILS) with different sample times and simulation results.

The data path in HILS, involving multiple steps via MAVLink, GCS, and Pixhawk, introduces significant latency, contributing to slower response times compared to simulations. Simulations also typically show shorter settling times, implying quicker stabilization in an idealized environment. In contrast, HILS has longer settling times due to sensor noise delaying the stabilization process.

8 Conclusions

This study illustrates the effectiveness of HILS for TVC systems with the Pixhawk flight controller, emphasizing real-time data acquisition, software-hardware synchronization, and optimal sampling interval selection. The research identifies a 0.02-s sampling time as ideal for achieving precise synchronization. Furthermore, HILS with this 0.02-s interval closely matches simulation results for the Ducted Fan UAV's response and input data, confirming effective synchronization. These results highlight HILS as a crucial bridge between theoretical simulations and practical experimentation, revealing performance discrepancies. The process of sequential tuning from simulation to HILS, with a focus on optimal sampling intervals, ensures robust controller optimization and improves performance and reliability across various operational scenarios.

References

1. Nguyen, T.H., Nguyen, N.T., Van Huynh, K.: Control system design and model development for a ducted-fan singlecopter. Sci. Technol. Dev. J. **20**(K3), 38–44 (2017). https://doi.org/10.32508/stdj.v20iK3.1089
2. Saif, E., Eminoğlu, I.: Modelling of quad-rotor dynamics and hardware-in-the-loop simulation. J. Eng. 937–950 (2022). https://doi.org/10.1049/tje2.12152
3. Wang, S., Dai, X., Ke, C., Quan, Q.: RflySim: a rapid multicopter development platform for education and research based on pixhawk and MATLAB. In: 2021 International Conference on Unmanned Aircraft Systems (ICUAS), pp. 1587–1594 (2021). https://doi.org/10.1109/ICUAS51884.2021.9476786
4. Yulnandi, R.A., Machbub, C., Prihatmanto, A.S., Hidayat, E.M.I.: Design and implementation of hardware in the loop simulation for electric ducted fan rocket control system using 8-bit microcontroller and real-time open source middleware. J. Mechatron. Electr. Power Veh. Technol. **8**(1), 60–69 (2017). https://doi.org/10.14203/j.mev.2017.v8.60-69
5. Ellis, G.: Control System Design Guide, 4th edn. ButterworthHeinemann, Boston (2012)
6. Amadi, C.A.: Design and implementation of a model predictive control on a pixhawk flight controller. Stellenbosch University, Stellenbosch (2018)
7. Koubâa, A., Allouch, A., Alajlan, M., Javed, Y., Belghith, A., and Khalgui, M.: Micro air vehicle link (MAVlink) in a nutshell: a survey. IEEE Access (7), 87658–87680. IEEE Press (2019). https://doi.org/10.1109/ACCESS.2019.2924410
8. Kurniawan, J.L., Adi, O.B., Baylon, P.B.A., Jenie, Y.I., Arifianto, O.: Robust controller design for planar rocket with thrust vectoring control. In: 2023 IEEE International Conference on Aerospace Electronics and Remote Sensing Technology (ICARES), pp. 1–7. IEEE Press (2023). https://doi.org/10.1109/ICARES60489.2023.10329891

Study of Wing Fairing Effects on the Drag Coefficient of a Low Subsonic UAV

Nathan Valencio Hu[1], Eugene Timothy Clay[2], Solahudin Al Ayubbi[1], Nicholas Patrick[1], Muhamad Novian Akbar[1], and Ema Amalia[1](✉)

[1] Faculty of Mechanical and Aerospace Engineering, Bandung Institute of Technology, Bandung, Indonesia
{13623011,13622008,13122091,13122103}@mahasiswa.itb.ac.id, ema_amalia@itb.ac.id

[2] Faculty of Mathematics and Natural Sciences, Bandung Institute of Technology, Bandung, Indonesia
10323055@mahasiswa.itb.ac.id

Abstract. Wing fairing variations may improve aerodynamic performance by allowing airflow streamlines to meet gradually, reducing interference drag. However, the addition of wing fairings may instead increase skin friction drag as well as form drag. This study investigates the influence of wing fairings towards the total drag of a UAV. Two test models based on Aksantara ITB's LELA and RP Teams' existing UAV designs are analyzed in this research. A total of four variations were made, with one having no wing fairing and the remaining having symmetrical fillets with varying radii. These models were analyzed using Computational Fluid Dynamics, employing the Reynolds Averaged Navier-Stokes method with the SST k-ω turbulence model. The angle of attack varies from −4 to 20° with flow velocity set as 12 m/s for the LELA derived models and 68 m/s for the RP derived models. Results show that the addition of wing fairings may reduce drag generation. However, certain wing fairing sizes increase drag, instead, due to form drag. This suggests that there is an optimal fillet size that reduces interference drag without increasing too much form drag.

Keywords: Wing Fairing · Drag · Unmanned Aerial Vehicle

1 Introduction

UAV technology has developed very rapidly nowadays. Its ease in completing various needs, from military to commercial, makes it an attractive alternative to take because of its cost-effectiveness and speed. Because its use has mushroomed, many aspects of UAVs have been developed further so that they can be more relevant in meeting the needs of their users. One such aspect of UAVs that has been studied extensively is the aerodynamic and flight performance. The aerodynamics of UAVs has always been a major concern to improve the flight performance and efficiency of the UAV. It is influenced by the external geometry of the UAV itself because it interacts directly with air as it travels. A part of the UAV that influences aerodynamic efficiency is the wing-fuselage juncture, or wing fairing, namely the part where the wing connects to the fuselage [1].

E. Fitrianingsih et al. (Eds.): ISAST 2024, SPPHY 416, pp. 392–401, 2025.
https://doi.org/10.1007/978-981-96-1344-1_42

Wing fairings, also known as wing root fairings, are structures found on aircraft, specifically located where the wing meets the fuselage. Its primary function is to produce a smooth outline and reduce drag [2]. These structures cover the gaps and spaces between the wing and the fuselage. The wing root fairing consists of a small, rounded edge on top and below the wing. At the leading and trailing edge, it consists of a much larger taper that smooths out the pressure differences. The use of wing root fairings has been credited with achieving more favorable flight characteristics at high and low speeds.

Wing fairings are commonly used on high-speed aircraft because they reduce the effects of interference drag. Interference drag is one of many types of drag. Interference drag is generated when airflows around one component of an aircraft is forcefully mixed with the airflow across an adjacent or proximal component. The wing fairings on an aircraft reduce interference drag between the wing and fuselage by providing a continuous surface that eases the transition between components. Fairings and fillets use curved surfaces to soften the transition at the junction of two aircraft components. This, in turn, allows the airflow streamlines to meet gradually rather than abruptly, reducing the amount of turbulent airflow, and thus interference drag, generated [3, 4]. The design and composition of wing fairings can vary across different species and aircraft, but their fundamental function remains the same [2]. These fairings are a trade-off in advantages, as they increase the frontal and surface area, but also provide a smooth surface and a faired nose and tail for laminar flow. Though the increase in frontal and surface area contributes to higher form and skin friction drag, the benefits they provide in terms of reduced drag and enhanced lift make them an integral part of aircraft design [5].

There is a significant amount of literature [1, 3, 6, 7] that proves the enormous benefits of using wing fairings in the high subsonic to supersonic regime. However, there are very few literature sources that discuss the role of wing fairings on the scale of UAVs [8]. UAVs generally travel in the low subsonic regime, and many aerodynamic effects are ignored. Thus, the use of wing fairings may not necessarily have a significant impact, as the effect of increased skin friction drag as well as form drag may be greater than the interference drag reduced by using a wing fairing.

Therefore, this research aims to determine the extent to which the use of wing fairings influences the total drag of the UAV in the low subsonic regime. In order to examine the impact of wing fairings on low Reynolds UAVs, the Coefficients of Lift (C_L) and Drag (C_D), will be evaluated, as well as the aerodynamic efficiency, or the lift-to-drag ratio (C_L/C_D). New findings made in this research will be used in future aircraft design considerations.

2 Methodology

This study's methodology is shown on the flowchart in Fig. 1.

Computer Fluid Dynamics (CFD) is a part of fluid mechanics which applies numerical analysis to solve problems regarding fluid flows. CFD computes data such as the boundary condition, the geometric constructions, and other factors that may affect the fluid flow depending on the model of CFD we are using. Such models are commonly derived from a base formula, often a differential equation, and then applied numerically using methods such as finite differences or finite elements.

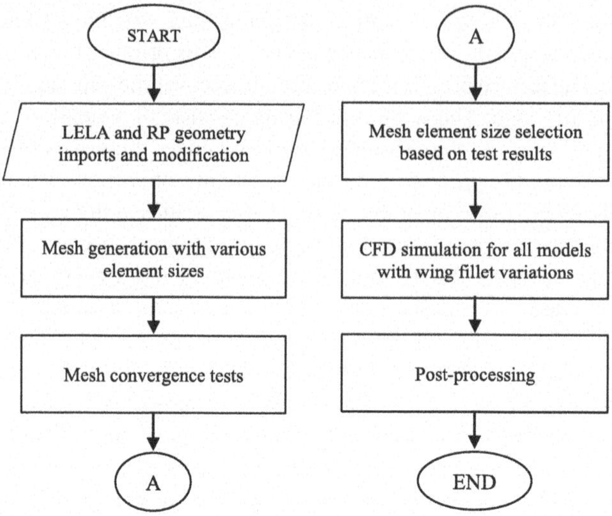

Fig. 1. Flowchart of the study.

To obtain the aerodynamic properties of wing fairings, computational fluid dynamics is done in the commercially available solver ANSYS Fluent using the Reynolds-Averaged Navier-Stokes equations, commonly referred to as the RANS equations. The RANS equations are simplifications of the direct numerical approach of the Navier-Stokes equations where instantaneous time elements are decomposed to a set of time-averaged fluctuating quantities [9]. The RANS equations are commonly used to describe turbulent flow models over a certain period of time and are shown in Eqs. (1) and (2).

$$\frac{\partial \rho}{\partial t} + \frac{\partial}{\partial x_j}(\rho U_j) = 0 \tag{1}$$

$$\frac{\partial}{\partial t}(\rho U_j) + \frac{\partial}{\partial x_j}(\rho U_i U_j) = -\frac{\partial p}{\partial x_i} + \frac{\partial}{\partial x_j}\left[\tau_{ij} - \rho\overline{u_i u_j}\right] + S_M \tag{2}$$

In these equations, ρ refers to the fluid density, U is the velocity, τ represents the molecular stress tensor, $\rho\overline{u_i u_j}$ represents Reynolds stresses, and S_M is the sum of the body forces [10]. The presence of Reynolds stresses shows that the simplified solving process using RANS equations introduces new unknown Reynolds Stresses and Fluxes. These unknown variables can be approximated using turbulence models [9]. The SST k-ω turbulence model is used in this research. The SST k-ω turbulence model is a two-equation eddy-viscosity turbulence model that combines the k-ε and k-ω turbulence models to achieve the better combination of the former's ability to predict flow behavior in regions outside of the viscous sub-layer (off the walls) and the latter's ability to simulate flow within the viscous sub-layer. This results in a better model if the separation layer isn't too extreme [11]. The SST k-ω turbulence model is mostly used for general cases as it falls short for niche cases, such as predicting the eddy currents of a turbulent flow.

The SST k-ω turbulence model was suggested by Menter [11] in Eqs. (3) and (4):

$$\frac{\partial(\rho k)}{\partial t} + \frac{\partial(\rho U_i k)}{\partial x_i} = \frac{\partial}{\partial x_j}\left[\left(\mu + \frac{\mu_t}{\sigma_\epsilon}\right)\frac{\partial k}{\partial x_j}\right] + P_k + P_b - \rho\epsilon + S_k \tag{3}$$

$$\frac{\partial(\rho\epsilon)}{\partial t} + \frac{\partial(\rho U_i \epsilon)}{\partial x_i} = \frac{\partial}{\partial x_j}\left[\left(\mu + \frac{\mu_t}{\sigma_\epsilon}\right)\frac{\partial \epsilon}{\partial x_j}\right] + \frac{\gamma}{\nu_t}P_k - \beta\rho\omega^2 + 2(1 - F_1)\frac{\rho\sigma_{\omega^2}}{\omega}\Delta k : \Delta\omega \tag{4}$$

In the previous equations, k represents the dissipation of turbulent kinetic energy, μ is the molecular viscosity, μ_t is the turbulent viscosity, P_k is the production of k due to mean velocity shear, P_b shows the production of k due to buoyancy, ϵ is the dissipation rate, S_k is a user-defined source, σ_ϵ is the turbulent Prandtl number for ϵ, γ is intermittency, v_t is the ratio of turbulent viscosity to density, β is a function of Menter's blending function, ω represents the specific dissipation rate, $\frac{\rho\sigma_{\omega^2}}{\omega}\Delta k : \Delta\omega$ is an additional term appearing as part of substitution, and F_1 is the blending function [12].

This research utilizes computational fluid dynamics using ANSYS Fluent software. To study the effects of wing fairings on the aerodynamic performance of low subsonic UAVs, analysis will be done on two different UAVs with different regimes. The UAVs to be analyzed are based on Aksantara ITB's Low Endurance Low Altitude Team (LELA) and Racing Plane Team's (RP) UAVs. To minimize meshing error risks, both geometries were simplified by removing the boom and tail for the LELA geometry, the canard for the RP geometry, and the propellers for both geometries. Additionally, the fuselage of the RP geometry was slightly modified to accommodate the addition of a wing fairing. The simplified models can be observed in Fig. 2.

ANSYS Mechanical is used in the mesh generation process. An example of the mesh generated for each UAV model's computational model can be observed in Fig. 3. The mesh only captures the starboard side due to both geometries being symmetrical. The computational domain for each UAV model has four main components, i.e.: inlet, farfield, symmetry, and outlet. The inlet and farfield components of the computational domain are set as velocity inlets with air velocities set as 12 m/s for the LELA model and 68 m/s for the RP model based on each UAV's intended cruising speed. The inlet, the location where fluid enters, is located in the front of each respective UAV. The farfield component is placed perpendicular to the inlet and outlet components, surrounding the UAV model. The symmetry component is situated at the fuselage's midpoint, and the outlet, i.e., where fluid flow shall exit the computational domain, is placed towards the back of the UAV. The mesh element number is varied to perform a Mesh Convergence Test. This test shall be done to ensure that computational fluid dynamics simulation results do not depend on the number of elements of a given mesh. For each geometry, three additional models are evaluated. A fillet is added between the fuselage and wing with varying radii of 6, 10, and 12 mm for each derived model. Each derived model is designed using Dassault Systèmes SolidWorks computer-aided design (CAD) software. An example of a derived model with the fillet added can be observed in Fig. 4.

After completing the Mesh Convergence Test, the aerodynamic coefficients C_L, C_D, and C_L/C_D are evaluated against the angle of attack (α) using ANSYS Fluent. The velocity components in the x and z directions are varied to simulate a change in α, while conforming to the aerodynamic coordinate system (i.e., the positive x-axis pointing

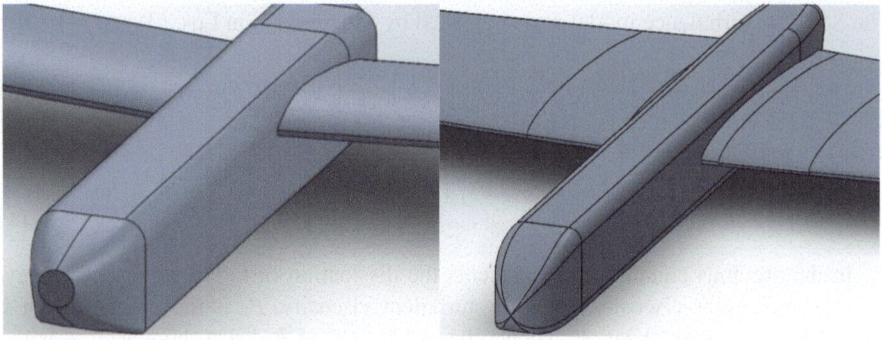

Fig. 2. Close-ups of the simplified LELA (left) and RP (right) geometries to be used.

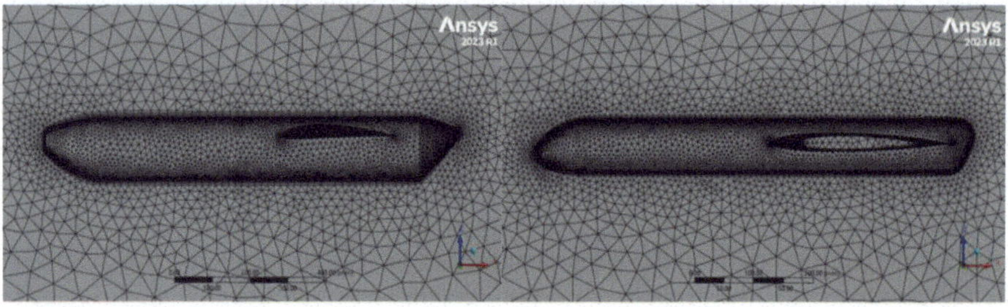

Fig. 3. Computational mesh of the simplified LELA (left) and RP (right) geometries to be used.

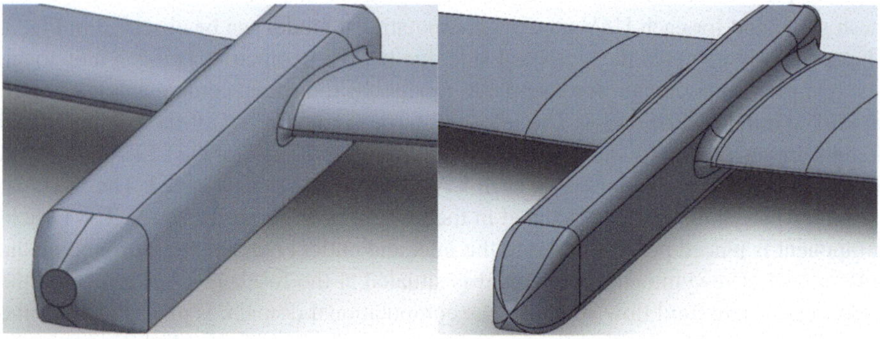

Fig. 4. Close-ups of the derived LELA (left) and RP (right) models with added wing fairings.

towards the tail of the aircraft, the positive y-axis running to the right-side wing, and the positive z-axis pointing to the direction of lift).

3 Results and Discussion

In accordance with the flowchart in the Methodology section, a series of Mesh Convergence Tests were done on the LELA and RP models with no fairings and a fillet with radii of 6, 10, and 12 mm (R6, R10, and R12), as shown in Fig. 5 for LELA and Fig. 6 for RP. Mesh element variations and the amount of mesh elements chosen can be observed in Table 1.

Next, fluid simulations were done to determine the aerodynamic performance of all derived models. The no fairing, R6, R10, and R12 models for each UAV travelling in different regimes are analyzed using ANSYS Fluent. The results can be observed in Figs. 7, 8 and 9.

Figure 7 shows the Coefficient of Lift (C_L) of all geometries for various angles of attack (α). Examining the LELA-based geometry's C_L against α curve reveals interesting results. The R6 model displays a unique property of delaying stall. The no fairing, R10, and R12 LELA models stalls around $\alpha = 12°$, while the R6 model stalls at around $\alpha = 16°$. Another notable result is that the R12 model generates higher lift compared to the rest of the models. It can be observed at α values not less than 4° that the R12 models consistently generate the most lift. Around $\alpha = 4°$, the R12 model shows the highest C_L value, albeit not by much. The R12 model's superior lift generation, however, is more prevalent at greater angles of attack. This occurrence is not present in the RP models, however. Examining the RP-based geometry's C_L against α curve shows that lift generation between the various derived models do not differ significantly. All of the RP-based geometries stall at the same angle of attack, i.e. 12°. At lower angles of attack, the addition of wing fairings does not affect lift generation, and at angles of attack greater than 8°, only minor differences can be observed.

The numerical computations for each wing fairing variation for the Coefficient of Drag (C_D) for both geometries can be observed in Fig. 8. Examining the figure reveals that the addition of fillets at lower angles of attack, i.e. at $\alpha \leq 8°$, does not affect drag generation significantly. At greater angles of attack, however, small changes can be noticed. In the LELA geometry, the R6 model shows the lowest C_D results among the four models at $\alpha = 16°$, and the R12 model shows the lowest C_D value at $\alpha = 12°$. Curiously, the R10 model exhibits slightly higher drag generation compared to other models at $\alpha = 12°$ and $\alpha = 16°$. The C_D against α curve for the RP model shows a result that is more in line with theory [3]. Examining the no fairing, R6, and R10 curves show that the addition of wing fairings does, in fact, affect drag generation. The presence of wing fairings reduces drag generation, with a larger fairing resulting in larger drag generation. This generalization falls apart, however, after observing the R12 curve. The R12 curve showcases the highest drag generation, even surpassing the no fairing curve, at $\alpha = 12°$, though it still generates less drag at $\alpha = 16°$ than the no fairing curve. This increased overall drag occurs due to form drag. The form drag generated from the wing fairing exceeds the interference drag reduced by it.

The Aerodynamic Efficiency (C_L/C_D) for each derived model can be observed in Fig. 9. Examining the C_L/C_D against α curve for the LELA-based models first, it can be observed that the R10 model shows the overall worst aerodynamic performance. This poor performance is attributed to its worse lift generation as shown in Fig. 6 as well as higher drag generation as per Fig. 7. On the other hand, the R12 shows the best

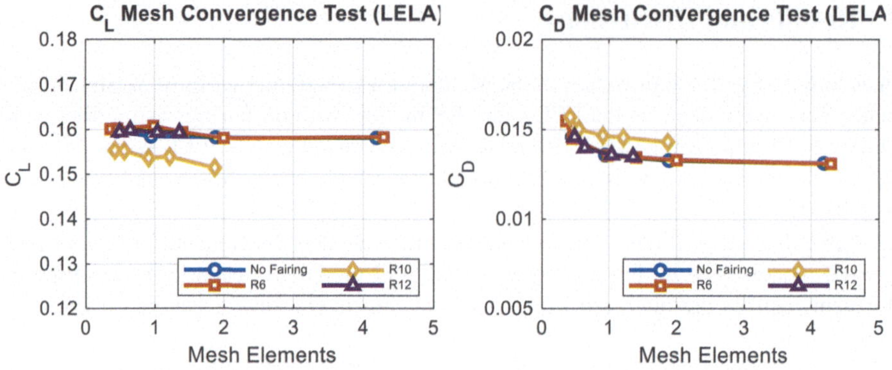

Fig. 5. Mesh Convergence Test results of the Coefficient of Lift (left) and the Coefficient of Drag (right) for the LELA-based geometry.

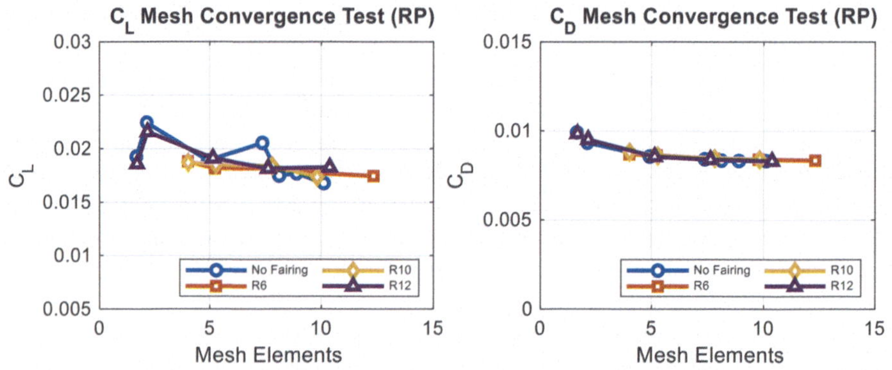

Fig. 6. Mesh Convergence Test results of the Coefficient of Lift (left) and the Coefficient of Drag (right) for the RP-based geometry.

aerodynamic performance due to its uniquely excellent lift generation, even though it does not display the best drag reduction among the rest of the LELA derived models. The C_L/C_D against α curve for the RP-based models show different results. The values for C_L/C_D are nearly identical across all angles of attack, with the exception occurring at $\alpha = 12°$. At that angle of attack, it is observed that the R10 model shows the best aerodynamic efficiency, while the R12 model displays the opposite.

Table 1. Mesh element variations for each geometry.

Geometry	Mesh Elements		
	Lower Bound	Upper Bound	Chosen Amount
LELA No Fairing	449225	4187689	946827
LELA R6	365491	4290742	1409559
LELA R10	424490	1876375	1221344
LELA R12	488676	1362317	1043531
RP No Fairing	164534	1011338	812835
RP R6	402096	1234230	978643
RP R10	403606	982909	782886
RP R12	168209	1038353	763932

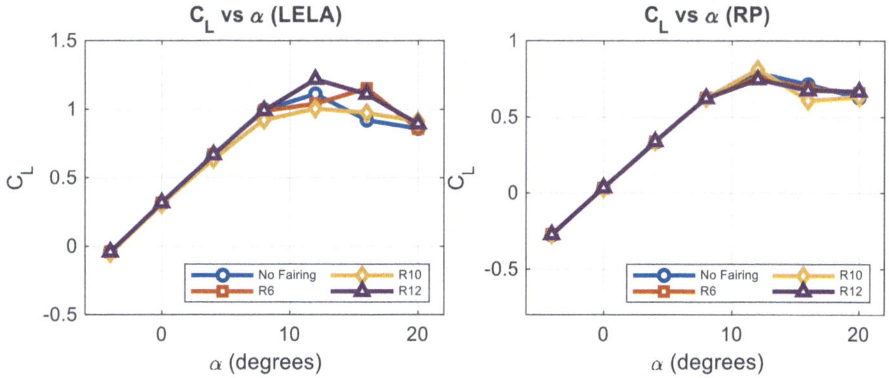

Fig. 7. The Coefficient of Lift (C_L) against the angle of attack (α) curves for the LELA-based geometry (left) and the RP-based geometry (right) for wing fairing variations.

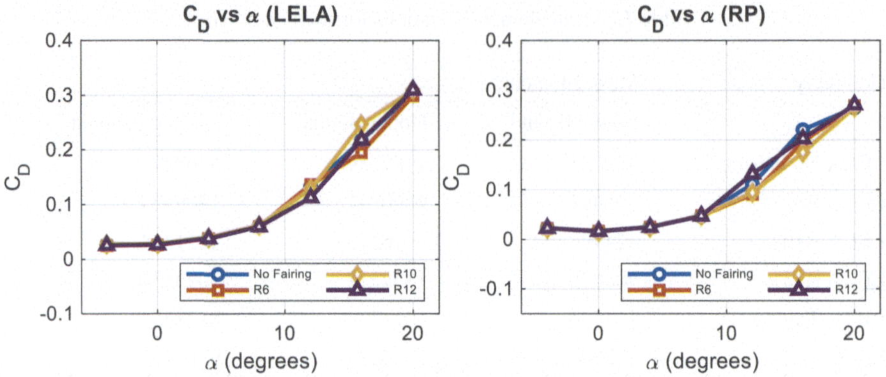

Fig. 8. The Coefficient of Drag (C_D) against the angle of attack (α) curves for the LELA-based geometry (left) and the RP-based geometry (right) for wing fairing variations.

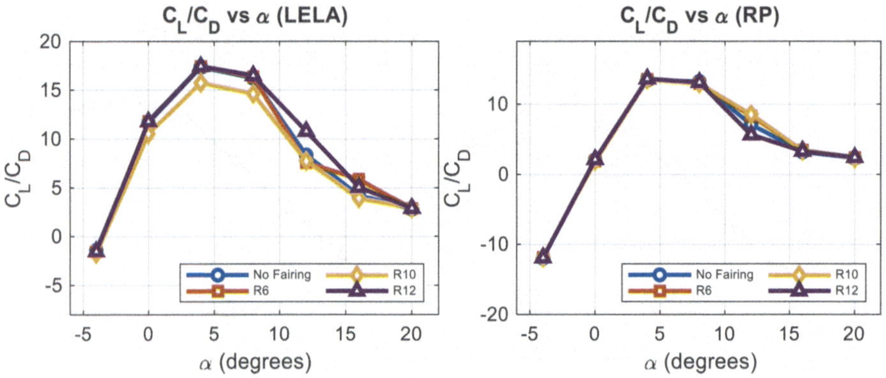

Fig. 9. The Aerodynamic Efficiency (C_L/C_D) against the angle of attack (α) curves for the LELA-based geometry (left) and the RP-based geometry (right) for wing fairing variations.

4 Conclusion

The study of wing fairing variations effects on the aerodynamic performance of low subsonic UAVs has been completed with several interesting outcomes that may be considered for future aircraft design considerations. Numerical methods have been employed to discover that the addition of wing fairings does not significantly impact lift nor drag generation at lower angles of attack. The influence of wing fairings towards lift and drag generation can generally only be observed at higher angles of attack. For a UAV traveling at a velocity of 12 m/s, a sufficiently large wing fillet notably increased lift generation. However, analysis of a UAV traveling at a faster velocity, i.e., 68 m/s, reveals that this difference in lift generation is not shared across different regimes. The addition of a wing fillet may reduce interference drag, leading to lower CD values. However, certain wing fillet sizes may instead exhibit greater CD values. The higher CD value can be attributed to form drag. The form drag generated from the increase in an aircraft's cross-section may exceed the interference drag reduced from the implementation of a wing fillet. This suggests that there is an optimal fillet size that reduces the most interference drag without increasing too much form drag. Further research must be done on different wing fairing shapes while also considering structural constraints.

Acknowledgments. The authors would like to thank Bandung Institute of Technology's Aksantara UAV Kontes Robot Terbang Indonesia (KRTI) Long Endurance Low Altitude Team and Racing Plane Team for lending their respective UAVs to be the basis of this research. The authors would also like to thank others whose names cannot be mentioned individually for their assistance in this study.

References

1. Gudmundsson, S.: General Aviation Aircraft Design: Applied Methods and Procedures. Embry-Riddle Aeronautical University, Elsevier (2014)
2. Belligoli, Z., Koers, A.J., Dwight, R.P., Eitelberg, G.: Using an anti-fairing to reduce drag at wing/body junctions. AIAA J. 1–13 (2019)
3. Sadraey, M.: Aircraft Design: A Systems Engineering Approach. John Wiley & Sons, West Sussex (2013)
4. Zhang, R., Cui, D., Song, G.: Fairing optimization design of wing structure. In: 30th Congress of the International Council of the Aeronautical Sciences. Korea (2016)
5. Pittman, M., Barlow, L.A., Kaye, T.G., Habib, M.B.: Pterosaurs evolved a muscular wing–body junction providing multifaceted flight performance benefits: advanced aerodynamic smoothing, sophisticated wing root control, and wing force generation. Proc. Nat. Acad. Sci. Unit. Stat. Am. **118**(44) (2021)
6. Raymer, D.P.: Aircraft Design: A Conceptual Approach. American Institute of Aeronautics and Astronautics Inc., Virginia (2018)
7. Klein, A.L.: Effects of fillets on wing-fuselage interference. Trans. Am. Soc. Mech. Eng. **56**(1), 1–7 (1934)
8. Park, S., Kwak, E., Lee, S.: Design and analysis of wing-tip and wing-body fairings. J. Korean Soc. Aeronaut. Space Sci. **39**, 289–296 (2011)
9. Ansys, Inc.: Ansys Fluent User's Guide 13.0. United States of America (2010)
10. Ansys, Inc.: ANSYS CFX-Solver Theory Guide 2021 R2. United States of America (2021)
11. Menter, F.R.: Two-equation eddy-viscosity turbulence models for engineering applications. AIAA J. **32**(8), 1598–1605 (1994)
12. Könözsy, L.: The k-ω Shear-Stress Transport (SST) turbulence model. In: A New Hypothesis on the Anisotropic Reynolds Stress Tensor for Turbulent Flows. Fluid Mechanics and its Applications, vol. 120. Springer, Cham. (2019)

3D-Printed Thin-Walled Multi-cell Structures: A Crashworthiness Study

Dony Hidayat[1,2(✉)], Riki Ardiansyah[1], Redha Akbar Ramadhan[1], Jos Istiyanto[2], Danardono Agus Sumarsono[2], Endarto Tri Wibowo[3], Farohaji Kurniawan[1], Abian Nurrohmad[1], and Aryandi Marta[1]

[1] Research Center for Aeronautics Technology, BRIN, Bogor, Indonesia
dony.hidayat@brin.go.id

[2] Department of Mechanical Engineering, Faculty of Engineering, Universitas Indonesia, Jakarta, Indonesia

[3] Research Center for Energy and Manufacturing Technology, BRIN, Bogor, Indonesia

Abstract. This research investigates the crashworthiness characteristics of 3D-printed Thin-Walled Multi-Cell Structures (TWMCS) through experimental testing and numerical modeling. Numerical modeling was developed based on quasi-static experimental testing of TWMCS fabricated from 3D-printed PLA material. The comparative analysis reveals differences in Crushing Force Efficiency (CFE) and Specific Energy Absorption (SEA) values of 6.89% and 7.82%, respectively, between experimental and numerical results. The TWMCS, optimized through numerical simulations, exhibited an increase in energy absorption capacity per unit mass from 6.94 J/g to 10.22 J/g. Additionally, the CFE value improved by 47.32% compared to the initial design. The findings provide valuable insights into the optimization of crashworthiness in 3D-printed TWMCS, contributing to advancements in the design of aeronautics technology.

Keywords: Thin-Walled Multi-Cell Structures (TWMCS) · Crashworthiness · 3D-printed · Polylactic Acid (PLA)

1 Introduction

Thin-walled (TW) structures are extensively utilized across various engineering sectors such as automotive [1], railway [2], aerospace [3], and shipbuilding [4]. This widespread adoption is attributed to their superior energy dissipation capabilities, cost-effectiveness, and ease of installation. Numerous researchers have investigated energy absorbers with diverse structural design configurations [5, 6] to attain optimal crashworthiness while minimizing weight. The development of these structural configurations started with simpler shapes, including circular, square, and triangular tubes, as illustrated by Baroutaji et al. [7]. Subsequently, more complex designs emerged, including multi-corner tubes, filled tubes [8], multi-cell structures [9, 10], and filled multi-cell tubes [11]. The latest developments include bio-inspired structures, where the design configurations are inspired by natural structures. An example of such research is the work conducted by Zhang et al. [12].

E. Fitrianingsih et al. (Eds.): ISAST 2024, SPPHY 416, pp. 402–411, 2025.
https://doi.org/10.1007/978-981-96-1344-1_43

Crashworthiness is a vital aspect in the design and functionality of contemporary public transportation systems, playing a key role in safeguarding passengers during crash incidents. 3D printing has proven particularly advantageous in the development of protective structures and devices for various applications. Research utilizing the 3D-printed method has been extensively conducted and proven to play a role as an energy absorber, enhancing crashworthiness values [13]. This method offers numerous advantages and eases the specimen manufacturing process. It is known for its simplicity, requiring no molds and expensive tooling equipment, making the manufacturing process more straightforward.

Prior research has explored the crashworthiness characteristics of 3D-printed polymeric materials [14]. Tsouknidas et al. [15] explored the influence of printing parameters on SEA, highlighting infill density as a significant factor while finding minimal impact from layer height and negligible influence from filling pattern.

This research aims to explore the crashworthiness performance of 3D-printed TWMCS. This study involves experimental testing and numerical modeling of TWMCS fabricated from 3D-printed PLA material, aiming to optimize their crashworthiness performance by varying the tube wall thickness. The PLA (Polylactic Acid) material was chosen because it is more environmentally friendly and has an affordable material price. This investigation is expected to unveil valuable knowledge regarding the optimization of crashworthiness in 3D-printed TWMCS, fostering advancements in the design of aeronautics technology.

2 Methods

2.1 Experimental

In this research, PLA filaments from eSUN with a 1.75 mm diameter were extruded through a 0.6 mm nozzle using a roller mechanism. The printing parameters were set to a 0.1 mm layer height, printing speed of 60 mm/s, 100% infill density with a line pattern, a build plate temperature of 60 °C, and a nozzle temperature of 210 °C.

The experimental testing involved applying axial quasi-static compressive forces using an RTF 2410 Tensilon universal testing machine (Fig. 1a). The crosshead speed was set to 5 mm/min [16], and the data acquisition system recorded the crushing load and the corresponding displacement.

2.2 Simulation

Numerical simulations were conducted using Abaqus 2022 software to evaluate the crushing force and the corresponding displacement of thin-walled multi-cell structures (Fig. 1b). The simulation process comprises three main stages: pre-processing (preparation), processing (simulation execution), and post-processing (results analysis). The flowchart of the numerical simulation is shown in Fig. 2. The PLA material used has an elastic modulus of 4.1 GPa and a Poisson's ratio of 0.35. The plastic behavior was characterized using true stress-strain data, which were derived from engineering stress-strain data obtained from tensile tests.

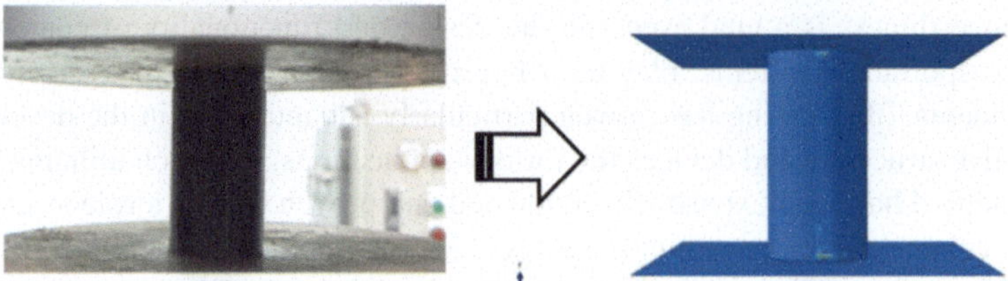

Fig. 1. Set-up of research: (a) experimental testing and (b) numerical simulation.

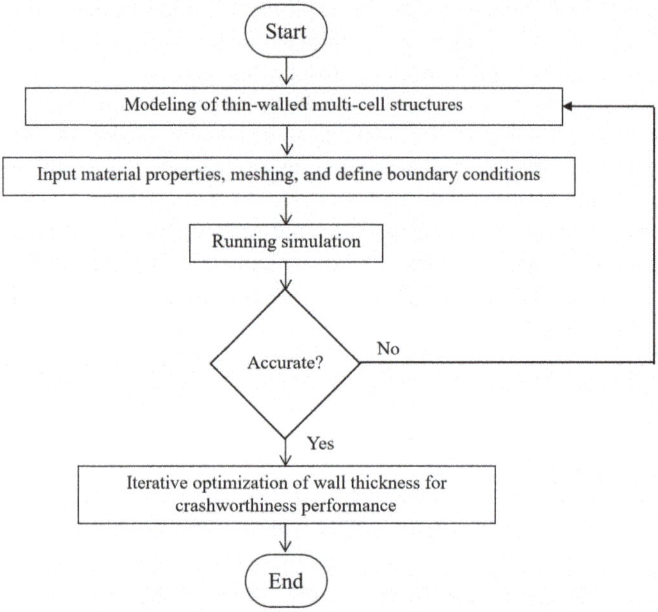

Fig. 2. Flowchart numerical modeling of TWMCS.

3 Result and Discussion

3.1 Experimental

The material properties of PLA in its plastic condition were identified through the processing of tensile test data (engineering stress-strain). This data was subsequently converted into true stress-strain data using Eqs. 1 and 2. Where ε_T represents the true strain and ε_E represents the engineering strain while σ_T is the true stress and σ_E is the engineering stress. The stress-strain curve for the PLA material, which was utilized in the numerical simulation of thin-walled multi-cell tubes, is depicted in Fig. 3. For the hardening conditions, the isotropic type of true stress-strain data was employed to characterize the material properties.

$$\varepsilon_T = \ln(1 + \varepsilon_E) \tag{1}$$

$$\sigma_T = \sigma_E.(1 + \varepsilon_E) \quad (2)$$

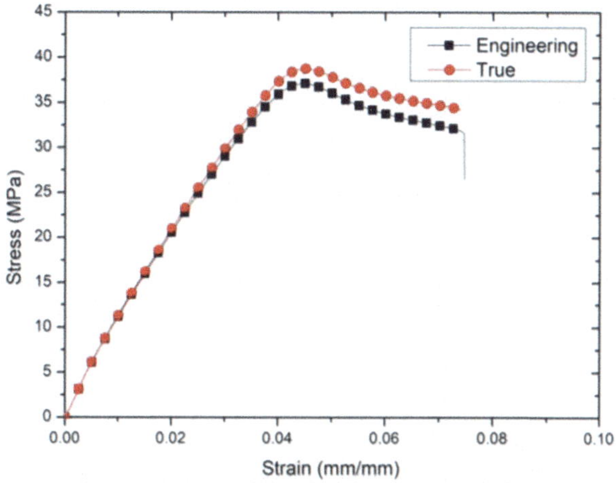

Fig. 3. Stress-strain graph of PLA material.

The force-displacement curves and crushing behavior of PLA thin-walled multi-cell structures under axial quasi-static loading are illustrated in Fig. 4. The peak force is seen at displacements of 2–4 mm. Following the peak force, these structures progressively collapsed during the compression phase.

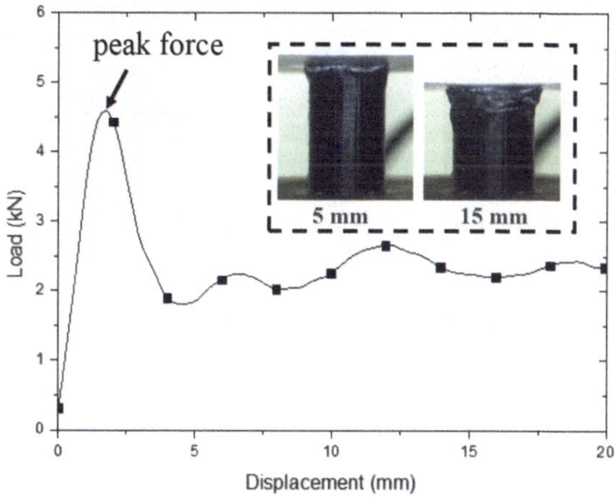

Fig. 4. Force-displacement curves of PLA 3D-printed thin-walled multi-cell structures.

Throughout the compression process, plastic deformation was observed to be continuous, steady, and demonstrating a progressive collapse mode. Previous studies [9] have indicated that the failure mode in PLA is characterized by the beginning of progressive folding at the upper end of the tubes, leading to asymmetric deformation. This behavior subsequently enhances the overall crashworthiness performance.

3.2 Numerical Simulation

The stress distribution resulting from numerical simulations of thin-walled multi-cell structures made from PLA material under axial loading is illustrated in Fig. 5. The regions experiencing maximum stress are highlighted in red (38.65 MPa), indicating that the tensile strength of the PLA material has been exceeded in these areas. As deformation progresses, stress levels increase until they surpass the material's tensile strength, leading to the formation of folds in the tube. The initial formation of these folds occurs at the locations of maximum stress (indicated in red). In thin-walled, diamond-shaped, multi-cell PLA tubes, these folds are characterized by the formation of sequential triangular patterns [17], as depicted in Fig. 5, with a deflection of 15 mm.

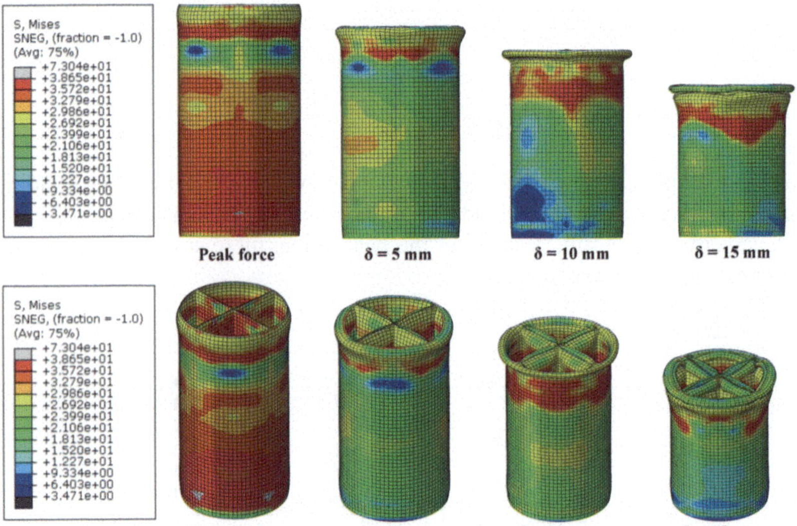

Fig. 5. Stress distribution in thin-walled multi-cell structures in numerical simulations with axial loading.

A comparison of the experimental and numerical crushing processes of thin-walled multi-cell polymer tubes is depicted in Fig. 6. The results indicate that the entire crushing process of the thin-walled multi-cell tube exhibits a ductile fracture failure mode at one end of the tube. The von Mises stress is initially concentrated at the site of the first fold. Subsequently, a new stress concentration appears below this initial fold, leading to the formation of a second fold. This pattern continues iteratively until the entire quasi-static testing process is completed.

A comparison of the load-displacement curves between experimental test results and numerical simulations is illustrated in Fig. 7. The overall trend of the numerical

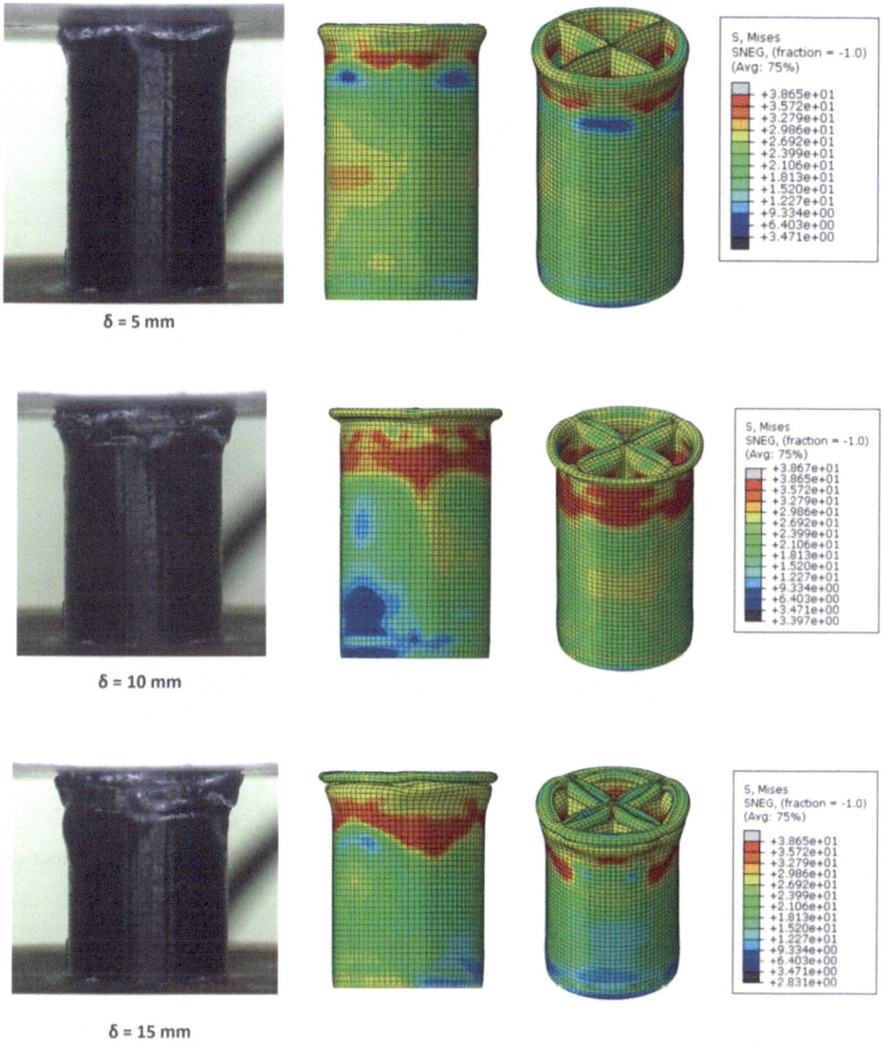

Fig. 6. Comparison of numerical simulation and experimental testing of TWMCS.

simulation curve aligns closely with the experimental results. In the elastic phase, the simulated multi-cell tube exhibits greater stiffness compared to the experimental results, and the peak force appears earlier in the numerical simulation. The peak force value from the simulation is 15.95% higher than that from the experimental results. As the structure becomes unstable and enters the plastic phase, the crushing force begins to fluctuate around the mean crushing force. However, discrepancies remain between the numerical simulations and experimental results in the plastic phase, primarily due to the elimination of failed elements [5].

The crashworthiness performance, as determined from experimental tests and numerical simulations, is presented in Table 1. This comparison considers displacements up

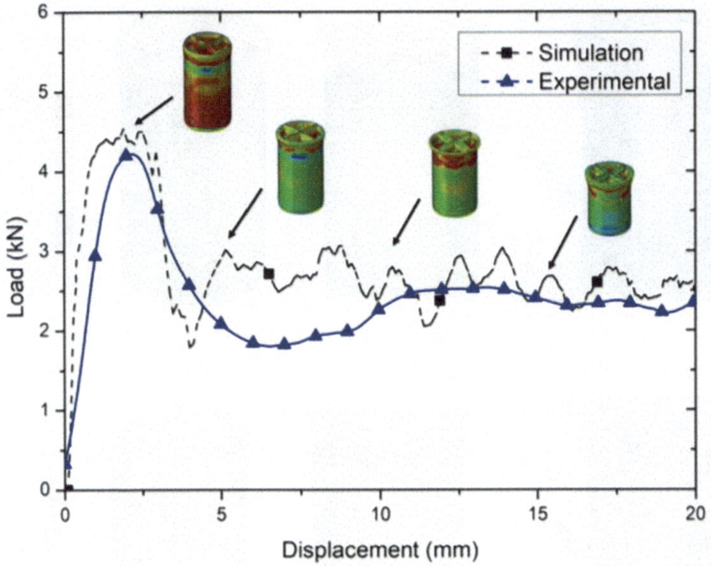

Fig. 7. Load-displacement curve experimental test results and numerical simulation.

to 20 mm and reveals differences in CFE and SEA values of 6.89% and 7.82%, respectively. The close correlation between the experimental and numerical simulation results indicates that the numerical model is robust and reliable, suggesting its suitability for further parametric studies on thin-walled multi-cell structures.

Table 1. Comparison of crashworthiness performance from experimental tests and numerical simulations.

Crashworthiness performance	Experimental	Numerical simulation	Difference (%)
Crushing force efficiency	0.57	0.61	6.89
Specific energy absorbtion (J/g)	6.44	6.94	7.82

3.3 Optimization Iterations of Tube Wall Thickness in TWMCS

To optimize a thin-walled multi-cell tube's CFE, the peak force must be reduced, or the average force increased, thereby enhancing the total absorbed energy. This optimization can be achieved by increasing the tube's thickness by augmenting the average force. In this study, the initial design of a specimen with a wall thickness of 1 mm, then in the first iteration, the tube thickness was increased to 1.5 mm (t_2) at a position of 2.5 mm from the top end of the tube, as illustrated in Fig. 8. In the second iteration, the tube from the first iteration was modified by increasing its thickness to 2 mm at a position 5 mm from the end, as shown in Fig. 8. The mass increased by 37.5% in the first iteration compared to the initial design, whereas in the second iteration, the mass increase was 75% compared to the initial design.

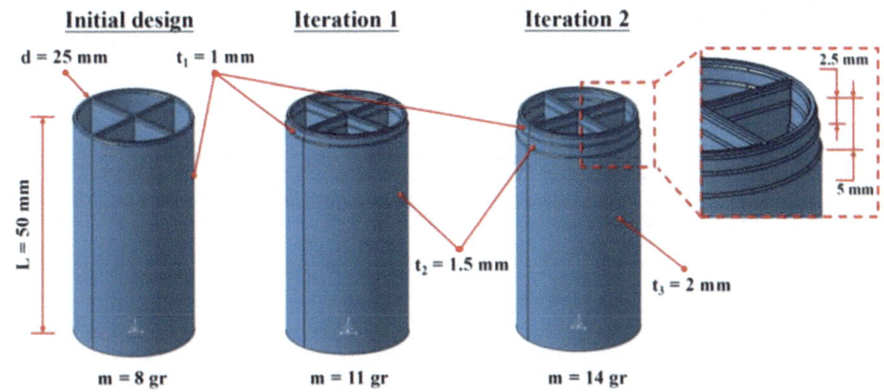

Fig. 8. Iteration stages for optimizing the CFE value.

Figure 9 depicts the relationship between load and displacement for a simulated crashworthiness scenario involving a thin-walled multi-cell structure (TWMCS). The graph presents two curves: one representing the initial design of the TWMCS and another representing the optimized design obtained through iterations. The first iteration of the simulation displays different behavior from the initial design. There is a marked increase in peak load, reaching higher values compared to the initial design, and even after peak load, there is an increase in load due to the effect of increasing tube thickness. However, the load fluctuations suggest that further refinements are needed to achieve a more stable and efficient energy absorption profile.

The final iteration demonstrates the most significant improvements, as shown in Fig. 9. The curve maintains higher load values across a broader range of displacements, suggesting an enhanced energy absorption capability. This iteration appears to have the closest resemblance to the desired crashworthiness performance, indicating successful optimization.

The optimized TWMCS demonstrated an increase in energy absorption capacity per unit mass from 6.94 J/g to 10.22 J/g. Concurrently, the CFE value improved by 47.32% compared to the initial design, as shown in Table 2.

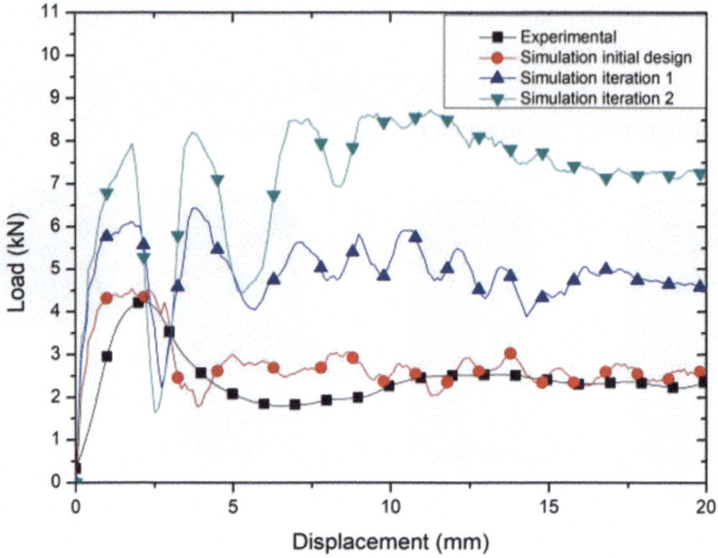

Fig. 9. Comparative analysis of load-displacement curves: initial design versus optimization iterations.

Table 2. Comparison of crashworthiness performance across optimization iterations.

Crashworthiness performance	Initial design	Iteration 1	Iteration 2
Crushing force efficiency	0.61	0.80	0.90
Energy absorbtion (J)	55.51	97.50	143.13
Specific energy absorbtion (J/g)	6.94	8.86	10.22

4 Conclusion

This research investigates the crashworthiness characteristics of 3D-printed TWMCS through experimental testing and numerical modeling. TWMCS tubes progressively collapsed during the compression phase. Throughout the compression process, plastic deformation was observed to be continuous, steady, and demonstrating a progressive collapse mode.

Numerical modeling was developed based on quasi-static experimental testing of TWMCS. The comparative analysis reveals differences in Crushing Force Efficiency (CFE) and Specific Energy Absorption (SEA) values of 6.89% and 7.82%, respectively, between experimental and numerical results. The TWMCS, optimized through numerical simulations, exhibited an increase in energy absorption capacity per unit mass from 6.94 J/g to 10.22 J/g. Additionally, the CFE value improved by 47.32% compared to the initial design.

In the numerical simulation conducted in this study, the PLA material properties were modeled as isotropic despite the 3D printing process inherently producing anisotropic multi-cell tubes due to the layered formation. Therefore, further investigations using micro-scale numerical simulations are recommended to observe the delamination phenomena between the layers of 3D-printed PLA.

References

1. Zhang, Y., et al.: Crashworthiness design of car threshold based on aluminium foam sandwich structure. Int. J. Crashworth. 1–12 (2021)
2. Zhu, T., et al.: Rail vehicle crashworthiness based on collision energy management: an overview. Int. J. Rail Transp. **9**(2), 101–131 (2021)
3. Mou, H., et al.: Impact test and numerical simulation of typical sub-cargo fuselage section of civil aircraft. Aerosp. Sci. Technol. **107**, 106305 (2020)
4. Ko, Y.G., Kim, S.J., Paik, J.K.: Effects of a deformable striking ship's bow on the structural crashworthiness in ship–ship collisions. Ships Offshore Struct. **13**(sup1), 228–250 (2018)
5. Zhang, Z., et al.: Parametric study on the crashworthiness of the Al/CFRP/GFRP hybrid tubes under quasi-static crushing. Thin-Walled Struct. **192**, 111156 (2023)
6. Hidayat, D., et al.: Experimental study on crashworthiness characteristics of composite hybrid tube utilise axial quasi-static crushing test. In: AIP Conference Proceedings. AIP Publishing (2023)
7. Baroutaji, A., Sajjia, M., Olabi, A.-G.: On the crashworthiness performance of thin-walled energy absorbers: recent advances and future developments. Thin-Walled Struct. **118**, 137–163 (2017)
8. Sun, G., et al.: Experimental study on crashworthiness of empty/aluminum foam/honeycomb-filled CFRP tubes. Compos. Struct. **152**, 969–993 (2016)
9. Hidayat, D., et al.: Investigation on the crashworthiness performance of thin-walled multi-cell PLA 3D-printed tubes: a multi-parameter analysis. Designs **7**(5), 108 (2023)
10. Liu, Y., et al.: Effects of loading rate and temperature on crushing behaviors of 3D printed multi-cell composite tubes. Thin-Walled Struct. **182**, 110311 (2023)
11. Sun, G., et al.: Topological configuration analysis and design for foam filled multi-cell tubes. Eng. Struct. **155**, 235–250 (2018)
12. Zhang, L., Bai, Z., Bai, F.: Crashworthiness design for bio-inspired multi-cell tubes with quadrilateral, hexagonal and octagonal sections. Thin-Walled Struct. **122**, 42–51 (2018)
13. Fu, X., Zhang, X., Huang, Z.: Axial crushing of Nylon and Al/Nylon hybrid tubes by FDM 3D printing. Compos. Struct. **256**, 113055 (2021)
14. Hidayat, D., et al.: Experimental investigation on axial quasi-static crushing of Al/PLA hybrid tubes. In: Journal of Physics: Conference Series. IOP Publishing (2023)
15. Tsouknidas, A., et al.: Impact absorption capacity of 3D-printed components fabricated by fused deposition modelling. Mater. Des. **102**, 41–44 (2016)
16. Hidayat, D., et al.: Experimental investigation on crashworthiness characteristics of e-glass–lycal composite tube and local product aluminum tube using quasi-static crushing test. In: AIP Conference Proceedings. AIP Publishing LLC (2020)
17. Bhutada, S., Goel, M.: Crashworthiness parameters and their improvement using tubes as an energy absorbing structure: an overview. Int. J. Crashworth. 1–32 (2021)

Sensitivity Analysis of CFRP Monocoque Structural Response Subjected to Loitering Munition Explosion

Bevan Bintang Setiawarman(✉), Djoko Bayu Murtie, Ditho Ardiansyah Pulungan, and Tatacipta Dirgantara

Institut Teknologi Bandung, Bandung, Jawa Barat 40116, Indonesia
{13621054,13621083}@mahasiswa.itb.ac.id, {pulungan.ditho, tdirgantara}@itb.ac.id

Abstract. Loitering munitions can be considered a disruptive technology and a non-traditional ranged weapon. Filling the role between cruise missiles and unmanned combat aerial vehicles (UCAVs), they possess the low operational and procurement costs of UCAVs while obtaining the capabilities of cruise missiles. As with other aerial vehicles, loitering munitions were designed with weight optimization in mind. Weight optimization can be achieved by using lightweight composite materials, such as carbon-fiber reinforced polymers (CFRP). Due to its high strength-to-weight ratio, however, CFRP can potentially dampen the explosive power of a loitering munition's warhead. Thus, it lowers its damage performance significantly. This paper aims to investigate the effect of composite parameters, namely the type of fibers, fiber directions, and overall composite thickness on the damage performance of loitering munitions. The HESA Shahed 136 model, equipped with a high-explosive warhead, is used for analysis in this paper. The parameters used to assess damage performance are the damage dissipation energy absorbed by the composite shell and the maximum velocity generated by the explosion. We performed sensitivity analysis by varying various material parameters. The simulation results show that the CFRP shell indeed dampens the explosion and therefore, it is very critical to design the monocoque with suitable material properties and dimensions.

Keywords: Explosion · Loitering Munition · Suicide Drone · Composite · Unmanned Aerial Vehicle

1 Introduction

Loitering munitions—also commonly known as suicide drones, kamikaze drones, or exploding drones—are a type of unmanned aerial vehicle (UAV) with a built-in warhead [1]. These UAVs are typically designed to loiter around an area until a target is identified, then attack the target by crashing into it. Loitering munitions allow for quicker responses to briefly appearing hidden targets without needing to position expensive assets close to the target area [1].

E. Fitrianingsih et al. (Eds.): ISAST 2024, SPPHY 416, pp. 412–421, 2025.
https://doi.org/10.1007/978-981-96-1344-1_44

As with commercial drones, or other UAVs, loitering munitions are built with lightweight materials to achieve an optimal payload-to-gross take-off weight (GTOW) ratio. By optimizing the weight of their structure, they can carry more payload, which in turn improves their damage output performance. This weight optimization can be achieved by using lightweight composite materials, such as fiberglass reinforced polymers (FRP) and carbon-fiber reinforced polymers (CFRP) [2].

Composite materials, widely used in aerospace and defense industries, offer numerous benefits due to their high strength-to-weight ratio, corrosion resistance, and design flexibility. Fiberglass reinforced polymers (FRP) are known for their excellent impact resistance and low cost, making them suitable for various applications. Carbon-fiber reinforced polymers (CFRP), on the other hand, provide superior stiffness and strength, albeit at a higher cost. These materials can be tailored to meet specific performance requirements by varying the type and orientation of the fibers, the resin matrix [3], and the manufacturing process [4].

However, despite the advantages of composite materials, which are known for their great strength-to-weight ratio, this property can backfire on a loitering munition design that desires an optimal payload-to-GTOW ratio. The strength of composite materials can dampen the explosive power of the warhead of a loitering munition, thus reducing its damage performance. Therefore, this research aims to determine the effect of composite structure thickness variations on loitering munition damage performance. This study compares the energy output magnitude and maximum velocity generated by the warheads based on FEA using Abaqus CAE. Therefore, this research aims to contribute to the consideration of optimal monocoque composite structure thickness when designing a loitering munition.

1.1 UAV Model

A model chosen to represent loitering munitions for this analysis is the HESA Shahed 136, designed and manufactured by the Iranian state-owned corporation HESA in association with Shahed Aviation Industries. The Shahed 136 has been used in multiple conflicts around the world, such as the Yemeni Civil War, the Russo-Ukrainian War of 2022 [5], and most recently, the 2024 Iranian Attack on Israel [6] (Fig. 1).

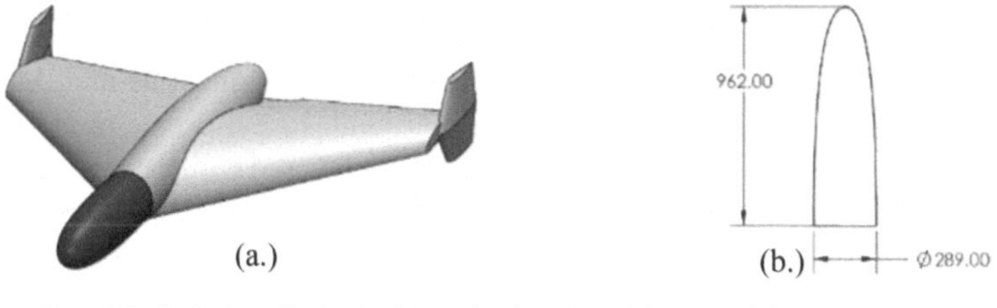

Fig. 1. (a) Simplified HESA Shahed 136 Loitering Munition Model and (b) Shahed warhead dimension.

The Shahed 136 is an autonomous pusher-propeller UAV with a delta wing planform and blended wing roots, lacking a vertical tail plane (VTP). Instead of a VTP for vertical axis stabilization, it has large wing tip devices that protrude from the upper and lower surfaces of the wing. Its warhead is located at the nose of the airframe, separated by a bulkhead from the avionics and other systems mounted in the middle of the fuselage. To simplify the simulation, we will only concern ourselves with the nose of the HESA Shahed 136 which contains an explosive load. The warhead will have a diameter of 289 mm with a length of 962 mm.

1.2 Composite Model

Finite element modeling of a composite presents itself with a challenge. In general, modeling composites can be categorized into three length-scales [7]. In the micro scale, composites are modeled from its constituent parts. In the case of fiber reinforced plastics, these parts include the matrix and the fiber itself. For this, we model the interactions between the fiber and the matrix, when subjected to different types of loads (Fig. 2).

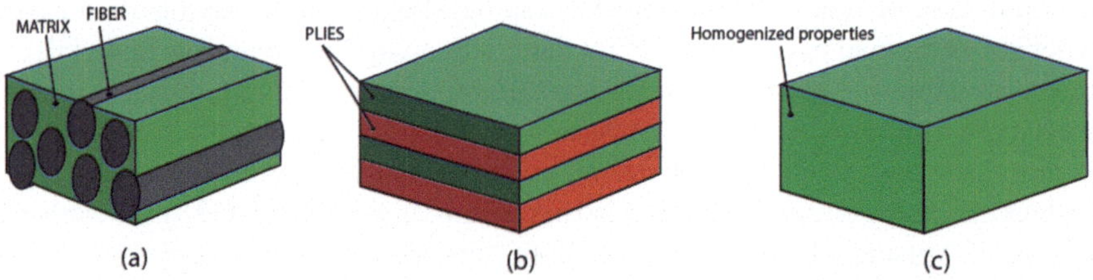

Fig. 2. Types of composite modelling in FEA: (a) microscale; (b) mesoscale; (c) macroscale.

Micro modeling is good at simulating microscopic failures and damage, especially in the beginning stages of damage initiation. However, depending on the objective of the research, microscale modeling can be unnecessarily complex with how much information it provides. Most importantly, it also results in a higher computational time and complexity.

Meanwhile, macroscale modeling provides the simplest representation of a composite where only the homogenized properties of the material is taken into consideration. Sometimes, this can lead to failure modes not predicted correctly. Especially in the case of carbon fibers with multiple constituent plies, cases of delamination simply will not occur.

Hence, a middle ground between simplicity and accuracy must be met. Mesoscale modeling, as the name suggests, is this middle ground. In mesoscale modeling, each composite plies is modeled macro. These plies are then layered with varying orientations [8]. This method provides a good enough representation of the anisotropic nature of fiber composites and its damage modes while keeping complexity relatively low compared to microscale.

More commonly, composite orientation has been done quasi-isotropically, where plies are angled 0°, ±45°, 90°. Nevertheless, regular methods of strengthening quasi-isotropic can be bulky and relatively heavy. Therefore, a new way of orientating composites named the Double-Double (DD) is introduced by Tsai where similar properties can be achieved with less layers [9].

Hence this research will investigate the effects of these two configurations (Quasi-isotropic and DD) while varying the thickness and types of fiber on the explosive qualities of a loitering munition system (Table 1).

Table 1. Properties of composite materials used in the simulation.

Elastic Constants (Mechanical Properties)	CFRP [10]	GFRP [11]	Units
Density	1800	2000	kg/m3
Modulus of elasticity in the X direction	40	36	GPa
Modulus of elasticity in the Y direction	10	5.1	GPa
Modulus of elasticity in the Z direction	10	5.1	GPa
Poisson's ratio XY	0.3	0.28	-
Poisson's ratio YZ	0.21	0.28	-
Poisson's ratio XZ	0.3	0.28	-
XY Shear module	3150	3000	MPa
YZ Shear module	4320	190	MPa
XZ Shear module	3150	3000	MPa
Hashin Damage			
Longitudinal Tensile Strength	988	402	MPa
Longitudinal Compressive Strength	1432	389	MPa
Transverse Tensile Strength	44	39	MPa
Transverse Compressive Strength	285	101	MPa
Longitudinal Shear Strength	60.6	26	MPa
Transverse Shear Strength	22	26	MPa
Damage Evolution			
Longitudinal Tensile Fracture Energy	70	12.5	N/mm
Longitudinal Compressive Fracture Energy	70	12.5	N/mm
Transverse Tensile Fracture Energy	70	1	N/mm
Transverse Compressive Fracture Energy	70	1	N/mm

1.3 High Explosive Warhead

A high explosive (HE) warhead is a type of explosive warhead that uses a chemical compound to produce a rapid and violent release of energy. When detonated, the high explosive undergoes a chemical reaction that produces a high-pressure shock wave and generates intense heat. This explosive force can cause significant damage to structures, vehicles, and personnel. High explosive warheads are commonly used in various types of munitions, including missiles, bombs, artillery shells, and loitering munitions, to achieve destructive effects against a wide range of targets (Fig. 3).

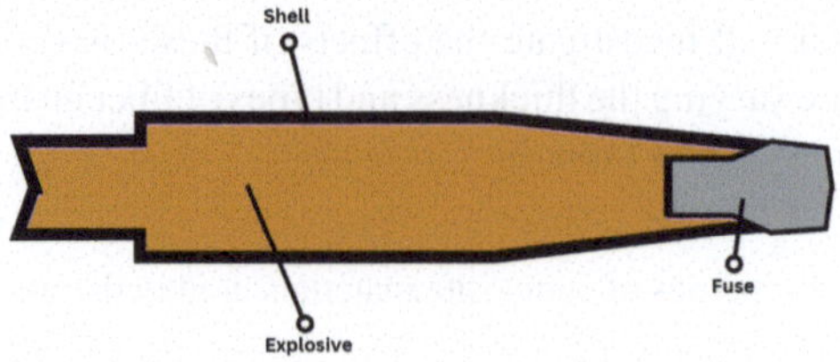

Fig. 3. Typical high explosive warhead () adapted from Ref. [12]

With Abaqus FEA, the explosive will be modeled using the JWL equations of state [10]. The explosive used is Composition-B and modelled by using the TNT-equivalent method [10] (Table 2).

Table 2. Composition-B explosive properties used for JWL equations of state.

Parameter	Value	Units
Density	1630	kg/m3
JWL Parameter A	373.8	GPa
JWL Parameter B	3.7	GPa
JWL Parameter R1	4.15	-
JWL Parameter R2	0.9	-
JWL Parameter w	0.35	-
Detonation Speed	6930	m/s
Energy Density	4.5	MJ/kg
CJ Pressure	21	GPa

1.4 Finite Element Method

Before heading into the Abaqus software, a 3D CAD model of the shell and explosive was created in SolidWorks. The model is then imported into Abaqus as a *.stp* file. Partitions are made to achieve hex and quad elements during meshing. Hex elements are preferred since they offer quicker running time due to lower element count with more accurate results for linear simulations [13]. Figure 4 shows one of the resulting mesh.

The explosive has solid elements, while the composite shell is modeled as shell elements. Shell elements are chosen to simplify the simulation process since a readily available failure model exists in the software, namely the Hashin damage criterion. Finally, fiber orientation is defined accordingly.

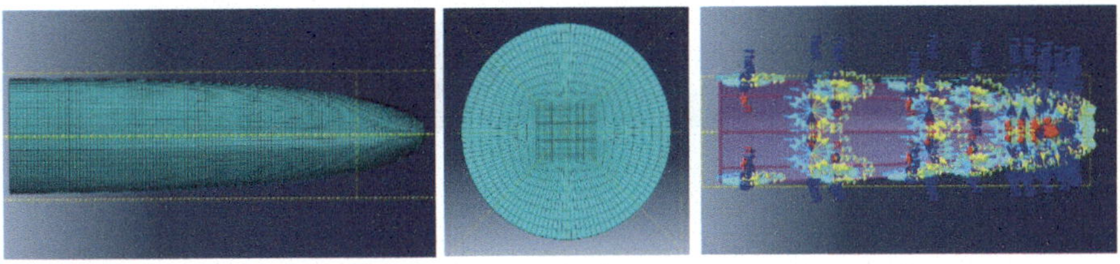

(a) (b) (c)

Fig. 4. Shell mesh from a) side view; b) bottom view; c) isotropic fiber orientation in Abaqus software.

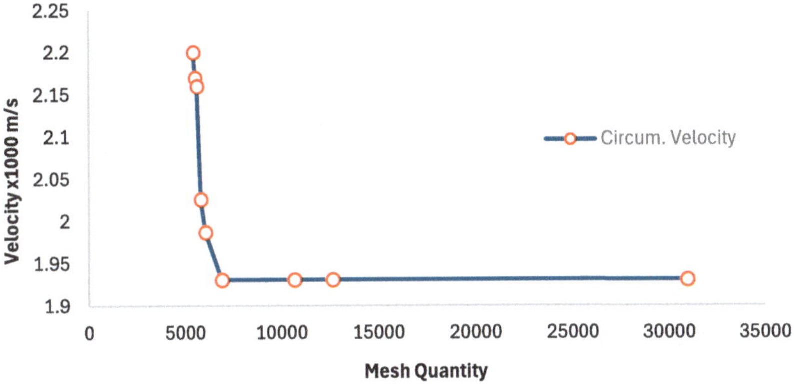

Fig. 5. Mesh convergence test, Radial velocity vs number of elements.

2 Results and Analysis

2.1 Mesh Convergence Test

To make sure that the obtained results are consistent, we start the research by conducting a mesh convergence test. Variations of mesh size is simulated which resulted in various element quantities. As can be seen in Fig. 5, the results converge as the number of elements increases. With 6914 elements, the velocity has converged enough, that it is decided to go with 6914 elements for the whole simulation.

2.2 Thickness Variation

In the thickness variation test, composites of CFRP and GFRP with quasi-isotropic orientation is varied in thickness ranging from 1 to 25 mm (Fig. 6 and Table 3).

To better understand the trend, we can plot the data in a plot of velocity vs thickness as follows. From Fig. 7 we can see that as the composite thickness is increased, the velocity of the composite explosion will decrease, which shows that the composite will dampen the energy of the explosion. The most drastic drop is shown from the data between 1- and 5-mm. The base layer velocity of a 1 mm CFRP shell reaches 2610 m/s, while for a 5 mm CFRP shell, that velocity drops by 45% to only 1430 m/s.

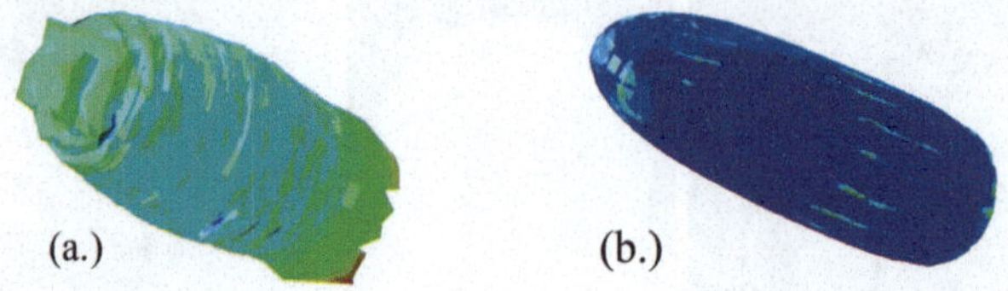

Fig. 6. (a.) 1 mm-thick shell explosion and (b.) 25 mm-thick shell explosion at the same time frame of 2×10^{-4} s.

Table 3. Results of velocity due to thickness variations.

Composite Type	Thickness (mm)	Velocity (m/s)		
		Base Layer	Circumference	Tip Layer
CFRP	1	2610	2180	2070
	5	1430	1360	1280
	10	975	1050	896
	15	638	858	749
	20	501	743	664
	25	457	594	578
GFRP	1	3075	1931	2242
	5	1933	1516	1666
	10	1466	1229	1237
	15	1174	995	888
	20	782	829	767
	25	555	699	585

An interesting point can be observed on the maximum velocities of the different composites. For CFRP, apart from the base layer where the explosion started, similar velocities can be seen between the tip and circumference. Meanwhile for GFRP, the tip layer experiences a significantly higher velocity of around 300 m/s than the circumference. This shows a more directed blast occurring using the GFRP composite.

Overall, GFRP scores higher velocities across the board. This may be attributed to its lower mechanical strength. When designing a loitering munition, it is important to consider the loads that may be experienced by the munition during flight, but also maximize the blast produced by the explosive. We need to also remember that loitering munitions are flying vehicles which suffer major penalty from heavier components as the GFRP is heaviest between the two composites.

2.3 Fiber Orientation

From Table 4, we see that DD (config. 1) results in the fastest shell velocity, while the QI configuration is observed to dampen the explosion the most. Interestingly, using DD

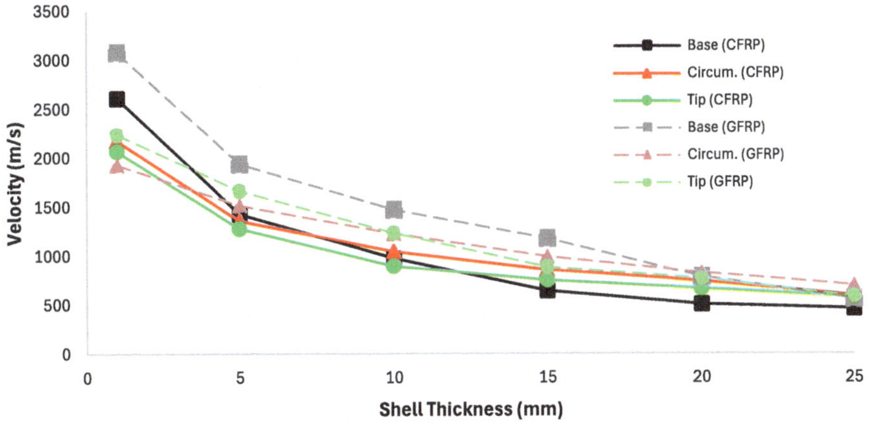

Fig. 7. Plot of velocity vs thickness.

(config. 1) configuration, CFRP can direct more of its explosive energy radially, resulting in higher circumferential velocity than the GFRP. This may be attributed to the base layer of the CFRP being stronger than GFRP, thus limiting explosive energy at the base, and directing it more to the front of the munition. Meanwhile, both config. 2 and 3 of DD have resulted in the worst explosive performance.

Table 4. Velocity data with fiber orientation variation.

Composite Type	Orientation	Velocity (m/s)		
		Base Layer	Circumference	Tip Layer
CFRP	Isotropic[0/±45/90/±22.5/±67.5]	2676	1931	1949
	QI [0/±45/90]	2809	2027	1966
	DD [±22.5/±67.5] (config. 1)	3026	2927	2112
	DD [±0/±52] (config. 2)	2804	2006	1960
	DD [±0/±37] (config. 3)	2872	2090	2070
GFRP	Isotropic[0/±45/90/±22.5/±67.5]	3399	2590	2236
	QI [0/±45/90]	3747	2598	2354
	DD [±22.5/±67.5] (config. 1)	4278	2615	2365
	DD [±0/±52] (config. 2)	3720	2596	2349
	DD [±0/±37] (config. 3)	3372	2527	2236

More comprehensively, the following table provides the explosive velocity drop of the different configurations with respect to a baseline. As DD (config. 1) results in the fastest explosion, it will be set as the baseline.

The colors in Table 5 shows the amount of velocity dampening. More red means the velocity is dampened more compared to the reference configuration (GFRP-DD config. 1), while blue means less dampening. It is understood that fiber orientations give negligible effect toward the circumference and tip velocities, except for config. 1 of CFRP for the reasons mentioned in the previous paragraph. Base velocities of CFRP also does not seem to be affected by orientation, while for GFRP, it varies. This may mean that orientation is not as dominant in determining the overall structural integrity when compared to the actual material strength in the case of CFRP; meanwhile for GFRP, orientations play a more crucial role.

Table 5. Velocity dampening data with respect to the fiber orientation.

Composite Type	Orientation	Velocity dampening		
		Base	Circumference	Tip
CFRP	Isotropic	37.45%	26.16%	17.59%
	QI	34.34%	22.49%	16.87%
	DD (config. 1)	29.27%	-11.93%	10.70%
	DD (config. 2)	34.46%	23.29%	17.12%
	DD (config. 3)	32.87%	20.08%	12.47%
GFRP	Isotropic	20.55%	0.96%	5.45%
	QI	12.41%	0.65%	0.47%
	DD (config. 1)	0.00%	0.00%	0.00%
	DD (config. 2)	13.04%	0.73%	0.68%
	DD (config. 3)	21.18%	3.37%	5.45%

3 Conclusion

In conclusion, a simulation of a high explosive warhead explosion has been done using the Abaqus software. The effects of thickness, fiber orientation, and fiber types have also been analyzed. The results show that as thickness increases, the explosion velocity decreases rapidly until around 10 mm, after which the velocity somewhat decreases asymptotically. The types of fiber also affected the explosive velocity. Fibers with higher mechanical strength such as CFRP dampen the explosion more severely compared to GFRP. This is to be considered when designing the structural shell of a warhead, making sure that the structure is strong enough to withstand loads during launch and cruise conditions, while not dampening the explosion too much as it can reduce the potential damage caused by the explosion. Otherwise, the Double-Double fiber orientation shows that it resulted in the highest shell velocity, implying a higher blast energy.

References

1. Loitering Munitions - In Focus. In Drone Center. Center for the Study of Drone (2017)

2. Rekatsinas, C.S., Dimitriou, D.K., Chrysochoidis, N.A.: A road map to the structural optimization of a Type-C Composite UAV. Aerospace **11**(3), 211 (2024). https://doi.org/10.3390/aerospace11030211
3. Pulungan, D., Lubineau, G., Yudhanto, A., Yaldiz, R., Schijve, W.: Identifying design parameters controlling damage behaviors of continuous fiber-reinforced thermoplastic composites using micromechanics as a virtual testing tool. Int. J. Solids Struct. **117**, 177–190 (2017). https://doi.org/10.1016/j.ijsolstr.2017.03.026
4. Pulungan, D., Andika, S., Dirgantara, T., Wirawan, R., Judawisastra, H., Wicaksono, S.: Promoting bridging in adhesively bonded composites with polymer inserts: a computational perspective. Compos. A Appl. Sci. Manuf. **180**, 108107 (2024). https://doi.org/10.1016/j.compositesa.2024.108107
5. How are Iran's shahed drones used on Ukraine's frontlines? euronews. (n.d.). https://www.euronews.com/next/2024/04/22/how-irans-kamikaze-shahed-drones-are-being-used-in-ukraine
6. Bezhan, F.: Explainer: The "kamikaze" drones Iran used to attack Israel. RadioFreeEurope/RadioLiberty (2024). https://www.rferl.org/a/iran-shahed-drones-israel-attack/32904882.html
7. Lam, D., et al.: 5th International LS-DYNA ® Users Conference Meso-scale Modeling of Carbon Fiber Composites for Crash Simulation. 15th International LS-DYNA Users Conference, pp. 1–9 (2018). https://www.dynalook.com/conferences/15th-international-ls-dyna-conference/composites/meso-scale-modeling-of-carbon-fiber-composites-for-crash-simulation
8. Wriggers, P., Hain, M.: Micro-meso-macro modelling of composite materials. In: Oñate, E., Owen, R. (eds.) Computational Plasticity. Computational Methods in Applied Sciences, vol. 7, pp. 105–122. Springer, Dordrecht (2007). https://doi.org/10.1007/978-1-4020-6577-4_7
9. Tsai, S.W.: Double–double: new family of composite laminates. AIAA J. **59**(11), 4293–4305 (2021). https://doi.org/10.2514/1.j060659
10. Gargano, A., Das, R., Mouritz, A.P.: Finite element modelling of the explosive blast response of carbon fibre-polymer laminates. Compos. B Eng. **177**, 107412 (2019). https://doi.org/10.1016/j.compositesb.2019.107412
11. Navaratnam, S., Small, D.W., Corradi, M., Gatheeshgar, P., Poologanathan, K., Higgins, C.: Numerical modelling of timber beams with GFRP pultruded reinforcement. Buildings **12**, 1992 (2022)
12. Thales Belgium (BL VTS) – Rocket System 70 mm (2.75"): Portfolio of warheads: FZ71 HE, FZ319 HEAP, FZ181 Flash Signature, FZ120 Inert Practice, FZ122 Flechette, FZ149 MULTIDART. (n.d.). https://fz.be/products.php?p=11
13. Schneider, T., Hu, Y., Gao, X., Dumas, J., Zorin, D., Panozzo, D.: A large scale comparison of tetrahedral and hexahedral elements for finite element analysis. J. Comput. Methods Appl. Mech. Eng. (2019)

Respiratory Symptoms in Aerotoxic Syndrome Due to Long-Term Exposure of Contaminated Bleed Air – A Literature Review

A. A. S. A. Santhi Sueningrum[1(✉)], Amilya Agustina[1], and Triya Damayanti[2]

[1] Aerospace Medicine Residency Program, Department of Community Medicine, Faculty of Medicine, Universitas Indonesia, Jakarta, Indonesia
santhisueningrum.aasa@gmail.com

[2] Department Pulmonology and Respiratory Medicine, Faculty of Medicine, Universitas Indonesia, Persahabatan Hospital, Jakarta, Indonesia

Abstract. Aerotoxic syndrome is associated to the aircraft's bleed air system which provides air supply in the cockpit and cabin. This system is susceptible to contamination due to the leak of engine oil and hydraulic fluid, resulting in fume events. The contaminated air often contains organophosphates, thus, posing potential hazards to aircrews. Although exposure levels of these substances are considered low, prolonged exposure is believed to correlate with the onset of symptoms in aircrews. Aerotoxic syndrome encompasses a broad spectrum of symptoms that are frequently reported by aircrews, including respiratory symptoms. Prolonged exposure to low-doses of organophosphates has been linked to hyperreactivity and inflammation of the respiratory tract, thereby precipitating both acute and chronic respiratory symptoms. These symptoms could potentially cause sudden incapacitation, which might compromise flight safety, or result in permanent disability for aircrews. Despite the significant impact it can have, there is still ongoing debates within the aviation community regarding the effects of organophosphate exposure during fume events on symptoms associated with aerotoxic syndrome. Furthermore, there is no universally acknowledged medical protocol to guide clinicians in diagnosing aerotoxic syndrome. Consequently, many clinicians might not be aware of the risks associated with aerotoxic syndrome, leading to delays in diagnosis and inappropriate treatment. This literature review aims to raise awareness among the aviation community and clinicians about aerotoxic syndrome. It is hoped that this could spur collaborative efforts across multiple disciplines to better understand the extent of health problems related to contaminated cabin air, especially regarding respiratory symptoms in aerotoxic syndrome.

Keywords: Aerotoxic syndrome · Bleed air · Contaminated air · Respiratory symptoms

1 Introduction

The aviation environment has distinctive characteristics that differ from those of the terrestrial environment. The differences notably affect the air quality within aircraft due to the impact of extreme conditions encountered in the aviation environment [1]. For

E. Fitrianingsih et al. (Eds.): ISAST 2024, SPPHY 416, pp. 422–430, 2025.
https://doi.org/10.1007/978-981-96-1344-1_45

over the past two decades, there has been an increasing number of aircrews reporting experiencing symptoms related to air contamination exposure in the aircraft, including respiratory symptoms. These symptoms are thought to result from the leaks of engine oil or hydraulic fluid into the cabin air. Therefore, air-conditioning systems of an aircraft plays an important role in ensuring air quality inside the cabin and cockpit [2, 3].

Most commercial and military aircraft nowadays utilize bleed air and recirculated air to supply air circulation inside the cabin and cockpit. The bleed air is unfiltered air coming from the gas turbine engine or the auxiliary power unit (APU) [2, 4]. Engine oil and hydraulic fluid might contaminate the bleed air as it could leak even in normal flight operations. One of the contributing factors to the leak is the oil seal design; it is not an absolute design to prevent leakages, thus enabling small leakages of oil [5]. The contaminated bleed air could enter the aircraft's air circulation and trigger the fume events. Notably, the reported fume events range from visible smoke to the smell of old socks [4].

During fume events, there is potential exposure of organophosphates that present in additives used in engine oil and hydraulic fluid. Tricresyl phosphate (TCP) is detected among those additives, which comprises three organophosphate isomeric forms. One of the isomers of TCP is tri-ortho-cresyl phosphate (TOCP) and is known as the most neurotoxic isomer. Long-term and low-dose exposure to TCP is believed to be associated with the etiology of aerotoxic syndrome. Several studies conducted on the incidence of fume events show varying results, such as 1 in every 66 flights or as low as 1 in every 2000 flights. This suggests that many incidents might go unreported. In addition, the doses of TCP exposure during fume events are also debatable as no sensors have been installed to detect air contamination inside the aircraft [6–9].

Studies related to respiratory problems among aircrews are relatively sparse, including the respiratory symptoms related to aerotoxic syndrome. Specifically, due to the uncertainty of exact pathophysiology, the wide range of symptoms experienced by the individuals, and the lack of current universally acknowledged medical protocols, it is difficult to recognize aerotoxic syndrome. Moreover, airlines and aircraft manufacturers debated that the exposure of contaminated aircraft bleed air do not cause symptoms related to aerotoxic syndrome [6, 9, 10]. This literature review aims to provide a better understanding of aerotoxic syndrome, particularly the respiratory symptoms. Hence, it is hoped could raise awareness of aerotoxic syndrome among the aviation community and healthcare professionals.

2 Methods

In this literature review, we elaborate the existing scientific references to further understand about aerotoxic syndrome, especially respiratory problem in aerotoxic syndrome. PubMed, Science Direct, and Google Scholar was used to conduct the literature searching. We include original researches, literature reviews and case reports without year limitation. We also only use literature that written in English. The keyword we use are combination of aerotoxic syndrome, respiratory symptoms, trycresyl phosphate, triorthocresyl phosphate, asthma irritant and organophosphate exposure.

3 Results and Discussion

3.1 Aerotoxic Syndrome

Aerotoxic syndrome is a constellation of symptoms experienced by individuals following exposure to fume events in the aircraft. The symptoms might be acute, appearing in 48 h post-exposure, or chronic which persisting for more than one month. The acute symptoms could jeopardize the flight safety as they can lead to sudden incapacitation of the pilot. Other than that, the symptoms might be severe enough to cause permanent disability of the aircrews. The symptoms of aerotoxic syndrome usually manifest gradually and might begin months or years after initial exposure [4, 9–11]. The aircrews with aerotoxic syndrome would experience worsening of symptoms during flight duty and symptoms would improve or even diminish when off duty. The nervous system, respiratory tract, gastrointestinal tract, skin, eyes, or cardiovascular could involve in aerotoxic syndrome. Neurological symptoms are the most commonly reported symptoms and followed by respiratory symptoms [2, 11].

A case report by Creeden et al. (2023) presented a commercial pilot who developed aerotoxic syndrome with some neurologic symptoms after exposed to the odor of dirty socks while on duty. He then underwent multiple therapies which were collaboration of some specialty services, such as occupational medicine, optometry, and physical and occupational therapy. His signs and symptoms were completely resolved nine months after exposure, at which point he felt his health returned to what it was before exposure. His flight duty resumed at twenty months after exposure. However, after several flights, he experienced another episode of symptoms which rendered him unfit to fly as a commercial pilot and possibly permanently disqualified from performing his duties. He did not notice any unusual odor or smoke before he re-experienced the symptoms. Thus, in addition to possible of re-exposure to the vapors, the significant load while performing duties as pilot is also feared as one of the causes of the recurrence of the symptoms [9].

As mentioned above, during fume events there is a risk of TCP exposure as this is one of additives used in engine oil and hydraulic fluid. One of isomers in TCP is TOCP which known for its neurotoxicity as it could disrupt axonal transport leads to irreversible degradation of the neuron and neuronal functional loss [6, 12, 13]. The safe limit of TOCP exposure by inhalation is 0.1 mg/m^3 for 30 min in individual weight 70 kg. The TOCP exposure doses detected in the aircraft during fume events vary, but most show very low-doses and below detection limit of 0.5 ng/m^3. The lack of accurate data for measuring TOCP exposure makes it difficult to determine actual TOCP exposure dose during fume events. Exposure doses that are very low and relatively below the safe limit are one of the debatable reasons for specifying aerotoxic syndrome as a diagnosis of occupational diseases [6, 14, 15].

Even though the exposure doses of TOCP appears to be within the range of safe exposure limits, it is necessary to consider the other types of chemical substances that might also enter the aircraft air circulation during fume events. As it is thought that interactions between chemicals might increase toxicity [10]. The possibility of symptoms occurs due to long term low-dose exposure of TOCP or other chemical subtances in aerotoxic syndrome also to be considered. This could be seen on differences of symptoms experience by passengers and aircrews. Oftentimes the health of aircrew become systematically deteriorating and requires further medical supervision, while only few passengers that likely experience symptoms more severe than irritation after fume events [10, 16].

In addition, an in vitro study by He et al. (2021) used a mini bleed-air contaminant simulator (Mini-BACS) to simulate an inhalation exposure of fume events in the aircraft. Fume samples were obtained from four types of engine oil and two types of hydraulic fluid. This study exposed human lung cell models to fumes from each source. Chemical profiling study of the respective fume found that hydraulic fluid fumes had relatively high organophosphate level compared to engine oil fumes. The level of cytotoxicity and inflammation response within 24 h after exposure were studied using benchmark dose analysis (BMD). From the analysis, it was indicated that high doses of engine oil and hydraulic fluid fumes might induce cytotoxicity by reduce transepithelial electrical resistance (TEER) and viabilities of the cells. This then induced cytotoxicity and increased the release of proinflammatory cytokines. Hydraulic fluid fumes appear to be more toxic and were suggested due to higher level of organophosphate that might induced higher toxicity [17]. However, the cytotoxicity and inflammation response in this study were only measured for 24 h, therefore, further studies are needed to look at the long-term effect of fume events to human lungs.

3.2 Respiratory Symptoms in Aerotoxic Syndrome

Respiratory symptoms in aerotoxic syndrome experienced by the aircrews range from mild to more severe symptoms. The symptoms duration might last for days, weeks or even longer to months or years. Cough, dyspnea, tightness of chest, and wheezing are the common respiratory symptoms in aerotoxic syndrome. Some cases of aerotoxic syndrome has been linked to irritant-induced asthma or reactive airway distress syndrome (RADS) [7]. A case report by Roig et al. (2011) reported three cases of aircrews who experience respiratory symptoms after fume events, such as dry cough and shortness of breath. The symptoms tend to be persistent and progressive in nature, also would worsening when on duty and improve during off duty. All of them had history of chronic intermittent exposure to fume events during flight duties and had no history of respiratory disease, such as asthma [11].

From that case series, the spirometry testing revealed an increase in forced expiratory volume in 1 s (FEV1) post bronchodilator compared to baseline spirometry values. However, only one case showed an increase greater than 20% post-bronchodilator test. The other two cases showed increase only 10% and 11% post-bronchodilator test with normal spirometry baseline [11]. Based on American Thoracic Society (ATS) and European Respiratory Society (ERS) guidelines, positive criteria of bronchodilator response (BDR) require both 200 ml and 12% increase FEV1 [18]. Thus, the cases present above did not meet the criteria to be diagnose as asthma.

Nevertheless, the normal result of baseline spirometry in the cases reported above should be considered. Hansen et al. (2019) proposed grading categories of bronchodilator response (BDR) by using FEV1 percentage or volume change, which are negative ($\leqq$0.00% or ($\leqq$0.00 L), minimal (> 0.00% to $\leqq$9.00% or >0.00 L to $\leqq$9.00 L), mild (> 9.00% to $\leqq$16.00% or >9.00 L to $\leqq$16.00 L), moderate (> 16.00% to $\leqq$26.00% or >16.00 L to $\leqq$26.00 L), and marked (>26.00% or >26.00 L). This grading system take into consideration the need to separate volume and percentage BDR change in patient with small or large FEV1 [18].

The three cases presented by Roig et al. (2021) [11] showed that the aircrews might experience aerotoxic syndrome with predominant respiratory symptoms. It is due to history of several fume events exposure experience by the aircrews prior their symptoms onset. During fume events there is a possibility of exposure to irritant which is the chemical substances contain in engine oil or hydraulic fluid, such as TCP. The baseline spirometry test result in two patients were normal, but the bronchodilator test showing increase of FEV1 although less than 12%. Considering altogether the illness history, the exposure history to irritant and the normal baseline FEV1, the BDR grading by Hansen et al. (2019) [18] could be used as basis for confirming irritant-induced asthma related to aerotoxic syndrome in that case report. Moreover, aerotoxic syndrome could still be considered as special cases that yet need further study regarding the diagnostic criteria.

3.3 Possible Mechanism of Organophosphate Induce Respiratory Symptoms

The TCP is an organophosphate substance which use in additive substances for engine oil and hydraulic fluid. This substance was one of chemical substances that found in the cabin air after fume events. Organophosphate is an acetylcholinesterase (AChE) inhibitor leading to accumulation of acetylcholine at nicotinic and muscarinic receptors of the target tissue [19, 20]. Low-dose organophosphate poisoning that does not lead to cholinergic crisis is known to potentially trigger asthma or asthma-like symptoms in adulthood [19]. Respiratory tract hyperreactivity due to low-dose organophosphate exposure is associated with neural mechanism, inflammation, oxidative stress and neuronal plasticity [19, 21].

In the neural mechanism, it is stated that organophosphate exposure causes bronchoconstriction by strengthening the activity of M3 muscarinic receptors in airway smooth muscle or inhibiting the autoinhibitory work of M2 receptors of airway parasympathetic nerves. Exposure of low-dose organophosphate also triggers airway inflammation due to increased activity of inflammatory mediator such as histamine and prostaglandin; inflammation cells, such as eosinophils and macrophages; and mRNA expression of proinflammatory cytokine such as interleukin-1β, interleukin-6, and tumor necrosis factor-α [19, 21].

Tissue damage that occurs in low-dose organophosphate exposure results in increased oxidative stress. The production of reactive oxygen species (ROS) and reacting nitrogen species (RNS) exceeds the antioxidant system capacity resulting in imbalance that triggers an inflammatory response. Airway tissue remodeling also occurs due to activation of M3 muscarinic receptors characterized by proliferation of airway smooth muscle resulting in accumulation of smooth muscle in the airway. Hyperinnervation in the lung is also believed to be associated with low-dose organophosphate exposure resulting in airway hyperreactivity. The cause of hyperinnervation is related to neuronal plasticity, both morphologically and neurochemically [19, 21].

3.4 Hierarchy of Control to the Risk of Aerotoxic Syndrome

Mawdsley (2022) had mentioned in his paper several events that related to cabin air contamination and how it affects the aircrews. The issue of reporting fume events also being highlighted as some aircrews might hesitate to report fume events as they concern for their jobs or lack of knowledge regarding the issues. Besides, it is said that some cases of fume events did not always forward to regulators by the airlines. There is also disagreement between aircrews and airlines industry physician regarding consideration of aerotoxic syndrome as an occupational disease. At the end of his research, he emphasized that the science of aerotoxic syndrome is still debated, although of growing number of peer-reviewed studies support the recognition of aerotoxic syndrome [6]. In study by He et al. (2021), it was emphasized that fume events are difficult to capture in real time, partly due to unclear understanding under which conditions they occur. As a result, knowledge of the composition and extents of fume inhaled during a fume event in an aircraft cabin is limited [17].

Mitigating the risk of aerotoxic syndrome could use the hierarchy of control as a guide. Multi-discipline collaboration is required to ensure the five components of the hierarchy of control work effectively. Airlines, aircraft manufacturers, engineers, aircrews, aviation medicine, pulmonology, toxicology, environmental health, and other field in medicine are some of the areas that have important role in developing strategies to mitigate the risk aerotoxic syndrome. The purpose of applying the hierarchy of control is to understand the effectiveness of various strategies implemented in order to reduce the occurrence of risks. In addition, it is expected that the hierarchy of control could help determine the effective and most feasible solutions. Elimination, substitution, engineering controls, administrative control, and personal protective equipment (PPE) are the five components of the hierarchy of control framework and depicted in an inverted pyramid, with elimination at the top and PPE at the bottom [22, 23].

Multi-disciplines collaboration is needed to further understand the exact causes and pathophysiologic mechanisms of the symptoms related to aerotoxic syndrome. Since the symptoms are said to be caused by exposure of chemical substances related to aircraft technology, the expertises from aeronautical engineers, environmental engineers, toxicologists are deemed important to prove that certain chemical substances at certain doses and durations might be harmful to human health, in this context the health of aircrews. Understanding how the symptoms occur will help clinicians to immediately diagnose aerotoxic syndrome in patient who exposed to fume events and provide appropriate treatment. Prompt diagnosis and appropriate management are essential to minimize disability and achieve better outcomes. Participation of aircrews and airlines is also very important as they are the first to be exposed and aware of the occurrence of fume events. Moreover, the aircraft manufacturers involvement is also significant as they have role in developing the technology used in the aircraft.

Based on five components in hierarchy of controls, some strategies might be considered to mitigate the risk of aerotoxic syndrome. Eliminate the bleed air in the aircraft might be the key for elimination component, for examples manufacture more aircraft that does not apply bleed air technology. Substitution to safer chemical substances that do not contain TCP to use in engine oil or hydraulic fluid might also be a way to reduce risk of

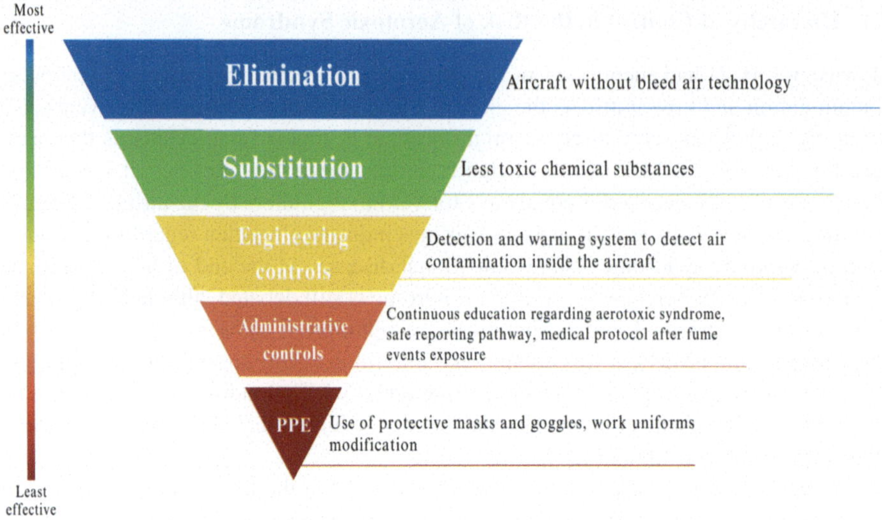

Fig. 1. Proposed hierarchy of control to mitigate the risk of aerotoxic syndrome.

aerotoxic syndrome. Develop detection devices or warnings systems for cabin air contamination events, particularly fume events could also be implemented as engineering control to reduce the risk of aerotoxic syndrome. Aircraft-installed detection devices and warning systems for air contamination allow real-time monitoring of exposure doses to hazardous chemicals during fume events. Such information is crucial not only for determining preventive measures and mitigating exposure impacts but also contributes to the advancement of research related to aerotoxic syndrome (Fig. 1).

Administrative controls strategy, in general, aim to raise the awareness of aerotoxic syndrome within the aviation and healthcare communities. This could be done by providing continuous education regarding cabin air quality and aerotoxic syndrome. A safe reporting pathway should also be ensured so that aircrew would not worry about their jobs when reporting the fume events. Also, medical department in airline involvement is required to provide medical protocol for management of aircrews and passengers who exposed to fume events, either with or without symptoms. The use of PPE as the last step in hierarchy of control in mitigating risk of aerotoxic syndrome should aim to minimize the exposure of chemical substances which mostly by inhalation or exposure to skin. The use of protective masks and goggles could be advised during fume events. Modification of work uniforms could also consider to minimize skin exposure, for example by wearing long-sleeved shirt and trousers.

4 Conclusion

Long-term low-dose exposure to TCP which containing organophosphates is believed to be the cause of symptoms associated with aerotoxic syndrome, especially in this context of respiratory symptoms. Often aircrew who exposed to fume events and experience symptoms related to aerotoxic syndrome would experience health impairment over period of time. The negative impact that aerotoxic syndrome might have on aircrews' health is quite concerning as it could cause aircrew to lose their jobs due to permanent disability. In addition, the acute symptoms of aerotoxic syndrome are also worrying as it might cause pilots become incapacitate and consequently endanger the flight safety. Therefore, multi-discipline collaboration is needed to address issues related to aerotoxic syndrome to find effective strategies to mitigate the risk.

References

1. Shehadi, M., Jones, B., Hosni, M.: Characterization of the frequency and nature of bleed air contamination events in commercial aircraft. Indoor Air **26**, 478–488 (2016). https://doi.org/10.1111/ina.12211
2. Hageman, G., van Broekhuizen, P., Nihom, J.: The role of carbon monoxide in aerotoxic syndrome. Neurotoxicology **100**, 107–116 (2024). https://doi.org/10.1016/j.neuro.2023.12.008
3. da Silva, M.G., Broday, E.E., Ruivo, C.R.: Indoor climate quality assessment in civil aircraft cabins: a field study. Thermal Sci. Eng. Progress. **37**, 101581 (2023). https://doi.org/10.1016/j.tsep.2022.101581
4. Anderson, J.: Sources of onboard fumes and smoke reported by U.S. Airlines. Aerospace **8**, 122 (2021). https://doi.org/10.3390/aerospace8050122
5. Michaelis, S., Loraine, T., Howard, C.V.: Ultrafine particle levels measured on board short-haul commercial passenger jet aircraft. Environ. Health **20**, 89 (2021). https://doi.org/10.1186/s12940-021-00770-7
6. Mawdsley, S.E.: Burden of proof: the debate surrounding aerotoxic syndrome. J. Contemp. Hist. **57**, 959–974 (2022). https://doi.org/10.1177/00220094221074819
7. Roig, J., Michaelis, S., Burdon, J.: The lung in aerotoxic syndrome. Arch. Bronconeumol. **58**, 787–789 (2022). https://doi.org/10.1016/j.arbres.2022.03.021
8. De Boer, J., Antelo, A., van der Veen, I., Brandsma, S., Lammertse, N.: Tricresyl phosphate and the aerotoxic syndrome of flight crew members – current gaps in knowledge. Chemosphere **119**, S58–S61 (2015). https://doi.org/10.1016/j.chemosphere.2014.05.015
9. Creeden, R., Blonien, N., Schultz, J.K., Wheeler, J., Haltson, E.L., McKinney, Z.J.: Prolonged disability following re-exposure after complete recovery from aerotoxic syndrome: a case report. IJERPH **20**, 7156 (2023). https://doi.org/10.3390/ijerph20247156
10. Hageman, G., Pal, T.M., Nihom, J., Ross, S.J.M., van den Berg, M.: Aerotoxic syndrome, discussion of possible diagnostic criteria. Clin. Toxicol. **58**, 414–416 (2020). https://doi.org/10.1080/15563650.2019.1649419
11. Roig, J., Domingo, C., Burdon, J., Michaelis, S.: Irritant-induced asthma caused by aerotoxic syndrome. Lung **199**, 165–170 (2021). https://doi.org/10.1007/s00408-021-00431-z
12. Naughton, S.X., Terry, A.V., Jr.: Neurotoxicity in acute and repeated organophosphate exposure. Toxicology **408**, 101–112 (2018). https://doi.org/10.1016/j.tox.2018.08.011
13. Liyasova, M., et al.: Exposure to tri-o-cresyl phosphate detected in jet airplane passengers. Toxicol. Appl. Pharmacol. **256**, 337–347 (2011). https://doi.org/10.1016/j.taap.2011.06.016
14. Schopfer, L.M., Furlong, C.E., Lockridge, O.: Development of diagnostics in the search for an explanation of aerotoxic syndrome. Anal. Biochem. **404**, 64–74 (2010). https://doi.org/10.1016/j.ab.2010.04.032
15. de Ree, H., et al.: Health risk assessment of exposure to TriCresyl Phosphates (TCPs) in aircraft: a commentary. Neurotoxicology **45**, 209–215 (2014). https://doi.org/10.1016/j.neuro.2014.08.011

16. Burdon, J., et al.: Health consequences of exposure to aircraft contaminated air and fume events: a narrative review and medical protocol for the investigation of exposed aircrew and passengers. Environ. Health **22**, 43 (2023). https://doi.org/10.1186/s12940-023-00987-8
17. He, R.W., Houtzager, M.M.G., Jongeneel, W.P., Westerink, R.H.S., Cassee, F.R.: In vitro hazard characterization of simulated aircraft cabin bleed-air contamination in lung models using an air-liquid interface (ALI) exposure system. Environ. Int. **156**, 106718 (2021). https://doi.org/10.1016/j.envint.2021.106718
18. Hansen, J.E., et al.: A new bronchodilator response grading strategy identifies distinct patient populations. Ann. ATS **16**, 1504–1517 (2019). https://doi.org/10.1513/AnnalsATS.201901-030OC
19. Shaffo, F.C., Grodzki, A.C., Fryer, A.D., Lein, P.J.: Mechanisms of organophosphorus pesticide toxicity in the context of airway hyperreactivity and asthma. Am. J. Physiol.-Lung Cellular Mol. Physiol. **315**, L485–L501 (2018). https://doi.org/10.1152/ajplung.00211.2018
20. Heutelbeck, A.R.R., Bornemann, C., Lange, M., Seeckts, A., Müller, M.M.: Acetylcholinesterase and neuropathy target esterase activities in 11 cases of symptomatic flight crew members after fume events. J. Toxicol. Environ. Health A **79**, 1050–1056 (2016). https://doi.org/10.1080/15287394.2016.1219561
21. Gosens, R., Gross, N.: The mode of action of anticholinergics in asthma. Eur. Respir. J. **52**, 1701247 (2018). https://doi.org/10.1183/13993003.01247-2017
22. Sehgal, N.J., Milton, D.K.: Applying the hierarchy of controls: what occupational safety can teach us about safely navigating the next phase of the global COVID-19 pandemic. Front. Public Health **9**, 747894 (2021). https://doi.org/10.3389/fpubh.2021.747894
23. Ajslev, J.Z.N., Møller, J.L., Andersen, M.F., Pirzadeh, P., Lingard, H.: The hierarchy of controls as an approach to visualize the impact of occupational safety and health coordination. IJERPH **19**, 2731 (2022). https://doi.org/10.3390/ijerph19052731

Advanced Aerospace Technology in Combating Climate Change

Canard Fin Failure Trajectory Prediction of Ballistic Missile with Cruciform Configuration

Oka Sudiana(✉), Azizul Hanif, Idris Eko Putro, Ahmad Riyadl, Arif Nur Hakim, and Rika Andriarti

Research Center for Rocket Technology, National Research and Innovation Agency (BRIN), Bogor, Indonesia
Oka.sudiana@brin.go.id

Abstract. The precise prediction of ballistic missile trajectories is critical to the operational efficacy. Nevertheless, a ballistic missile's trajectory remains vulnerable to various external disturbances or internal malfunctions that could lead to deviations from the intended flight. Comprehensive analyses of missile performance in failure scenarios are necessary to solve these issues. This study employs a 200 mm caliber missile, fitted with canards specifically engineered to manage body dynamics and perform trajectory correction tasks. The canards are arranged in a cruciform configuration. This research presents a comprehensive analysis of a 6-degree-of-freedom (6 DOF) simulation model designed to predict the dynamic flight behavior of a rocket in the event of canard failures. The research investigates how canard failures affect trajectory deviations, taking into account situations in which canard fins 1, 2, and 3 malfunction. The research investigates how these failures affect critical flight parameters. Notably, results show that the trajectory exhibits a large divergence, with characteristics like the maximum impact point deviation reaching up to 1.63 km, in the configuration where all three canard fins fail. On the other hand, when two canard fins fail, the highest side deviation that is seen is much smaller, about 15 m.

Keywords: Ballistic Missile · Trajectory Prediction · Canard Fin/s Failure

1 Introduction

The prediction of ballistic missile trajectories is critical to air defense systems. Typically, the trajectory of a ballistic missile is divided into three primary phases: boost, mid-course, and re-entry (terminal) [1]. Most existing trajectory prediction methods concentrate on mid-course and reentry phases, where the trajectories are often modeled as ellipses, or the state components are propagated using dynamic integral equations over specific time scales [2–5]. This predictive capability is essential during the preliminary design phase to estimate the missile's flight trajectory and impact point, considering the launcher's elevation angle and azimuth.

E. Fitrianingsih et al. (Eds.): ISAST 2024, SPPHY 416, pp. 433–442, 2025.
https://doi.org/10.1007/978-981-96-1344-1_46

Throughout the ballistic missile flight, various environmental disturbances and internal malfunctions may arise, leading to deviations from its intended trajectory. These disturbances and failures can significantly affect the missile flight path and potentially jeopardize mission objectives and efficiency.

This study utilizes a 200 mm caliber missile, equipped with canards specifically designed to control body dynamics and execute trajectory correction missions. Canards exhibit asymmetric deflection relative to the angle of attack plane, resulting in asymmetric downwash/upwash effects on the fins [6]. This induces substantial reverse moments, which in certain scenarios may surpass the rolling, pitching, and/or yawing moments generated by the canards, potentially leading to control failures [6, 7]. Hence, research by Daniel J. Lesieutre et al. on aerodynamic coupling effects and methods for decoupling between canards and fins holds significant importance [8, 9]. Silton et al. [10–12] conducted advanced computational aerodynamic analyses and parameter sensitivity studies to investigate the impact of canard interactions on the aerodynamic performance of fin-stabilized projectiles. Their work included analyses of canard trailing-edge vortex interactions with tail fins.

This research is used the 6-degree-of-freedom rocket dynamic simulation to assess the missile performance in case of canard failure. This study aims to examine the missile flight performance when deviation occurs from the original path due to canard-fin failures. This study focuses on trajectory parameters under various scenarios of fin failure, including situations where one, two, and three fins fail. These simulations offer valuable insights into the prediction of missile behavior in challenging conditions. This paper does not address the design of a control system to mitigate fin failure and restore the missile's flight to its original trajectory. Any specialized mitigation strategies or fault-tolerant schemes are not covered in this study and will be explored in future research.

2 Methodology

2.1 Characteristics of Missile in Ballistic Flight

According to [2], the flight performance of a rocket in six degrees of freedom (6 DOF) is determined from the fundamental equations of force and moment as described by Newton's second law [3, 13–17]. This study flight performance simulation utilizes the nonlinear flight dynamics equations of a 6 DOF rocket, which have been implemented in Matlab/Simulink software, as detailed in [2, 4, 18, 19]. Canard control movement with Cruciform configuration.

The missile control canard uses an X-configuration (cruciform) to enhance maneuverability and stability during flight. This configuration involves four fins arranged in a cruciform pattern, allowing for precise control over the missile pitch, yaw, and roll.

Figure 1 illustrates the movement of the canard fins in an X-configuration. Each fin can move independently to adjust the missile's flight path and maintain stability. This illustration can be seen below (see Fig. 1).

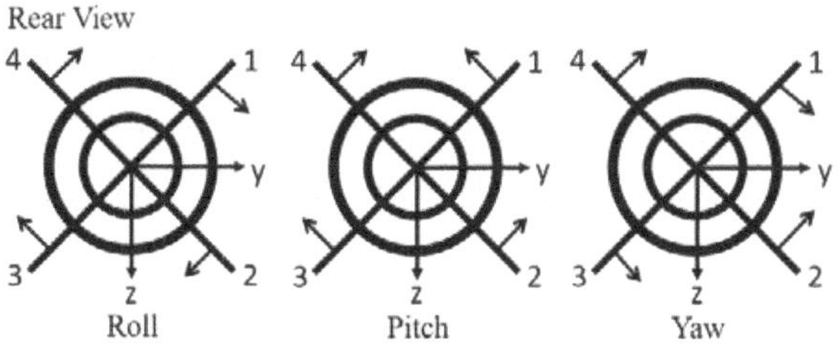

Fig. 1. Illustration of canard fin movement in an x-configuration [5].

From Fig. 1, it is needed to convert control surface deflection ($\delta_1, \delta_2, \delta_3, \delta_4$) into roll, pitch, and yaw control surface $\left(\delta_r, \delta_p, \delta_y\right)$ by using conversion matrix [5, 19] as:

$$\begin{bmatrix} \delta_r \\ \delta_p \\ \delta_y \end{bmatrix} = \frac{1}{4}\begin{bmatrix} 1 & 1 & 1 & 1 \\ -1 & -1 & 1 & 1 \\ 1 & -1 & -1 & 1 \end{bmatrix}\begin{bmatrix} \delta_1 \\ \delta_2 \\ \delta_3 \\ \delta_4 \end{bmatrix} \tag{1}$$

2.2 Trajectory Prediction When Canard Fin Failure

This stage of the research presents a detailed trajectory analysis of a ballistic missile in the event of canard fin failures. The study involves simulations of eight different failure scenarios. In this case study, the canard fins are uncontrolled or fixed, simulating conditions where the fins fail after the propellant burns out at the 10th second of the flight. The sequence failure of canard fin/s can be seen bellow (see Table 1).

Table 1. The sequence failure of canard fin/s.

Case 1	Fin 1 fail	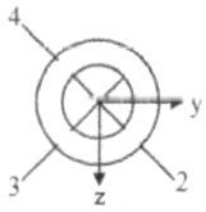	Case 5	Fin 1,4 fail	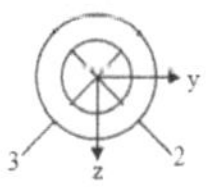
Case 2	Fin 2 fail	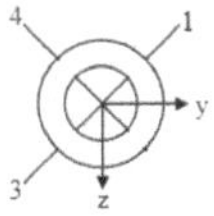	Case 6	Fin 2,3 fail	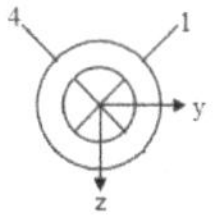
Case 3	Fin 1,2 fail	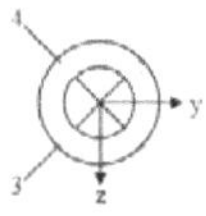	Case 7	Fin 1,3,4 fail	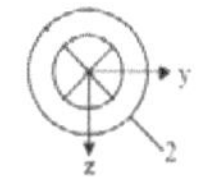
Case 4	Fin 1,3 fail	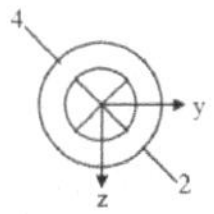	Case 8	Fin 2,3,4 fail	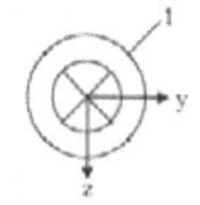

3 Simulation

The trajectories of ballistic missiles are simulated using MATLAB/SIMULINK, a simulation tool widely utilized for modeling dynamic systems.

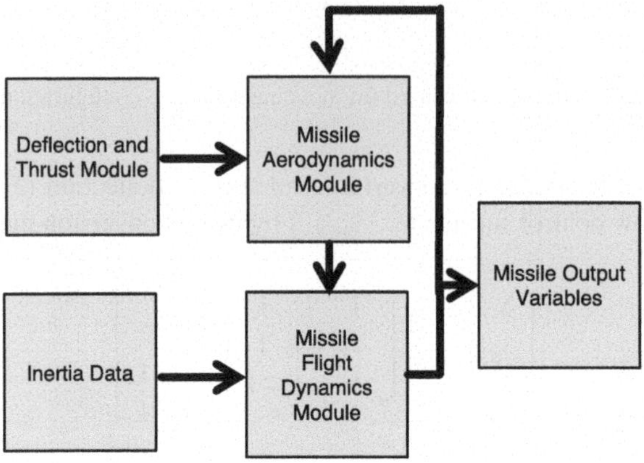

Fig. 2. Module and workflow of the mathematical model of a ballistic missile.

Figure 2 illustrates the modular structure and workflow of the mathematical model employed to simulate the behavior of a ballistic missile. This simulation presents a structured module layout and workflow for the mathematical calculations involved in simulating the trajectory of a ballistic missile. Utilizing MATLAB/Simulink, this computational framework integrates various modules and data inputs to the missile flight dynamics model. This simulation is centered on modeling and analyzing missile deflection solely as a result of canard failure. Wind disturbances are excluded from the analysis to minimize the complexity of parameters in the modeling that can affect changes in missile deflection.

The model and its parameters followed the definition from references [2, 4]. The ballistic missile specifications used in the research can be seen in Table 2. The simulation are configured with the ODE45 numerical solver, which is based on the Dormand-Prince method; and used a variable step size configuration.

4 Results and Discussion

The simulation results of the ballistic missile model under various fin failure scenarios are presented in this section.

Figure 3 illustrates the trajectories of the ballistic missile when experiencing different fin failures. Each trajectory shown in Fig. 3 corresponds to a specific configuration of fin failure, including cases where one, two, or three fins fail. For instance, the trajectory labeled "1 fin fail" shows the path of the missile when a single fin is no operated. Similarly, the "2 fin fail" and "3 fin fail" trajectories depict the missile flight when two or three fins are compromised, respectively.

Table 2. Ballistic missile data specification [5].

Parameter	Value	Unit	Parameter	Value	Unit
Length	4.15	m	Inertia		
Max Diameter	0.2	m	I_{xx}	1.3	kg.m^2
Lift-off Mass	205	kg	I_{yy}	201	kg.m^2
Propellant Mass	76	kg	I_{zz}	201	kg.m^2
Burning Time	9.64	s	I_{xz}	1.3	kg.m^2
I_{SP}	227	s			

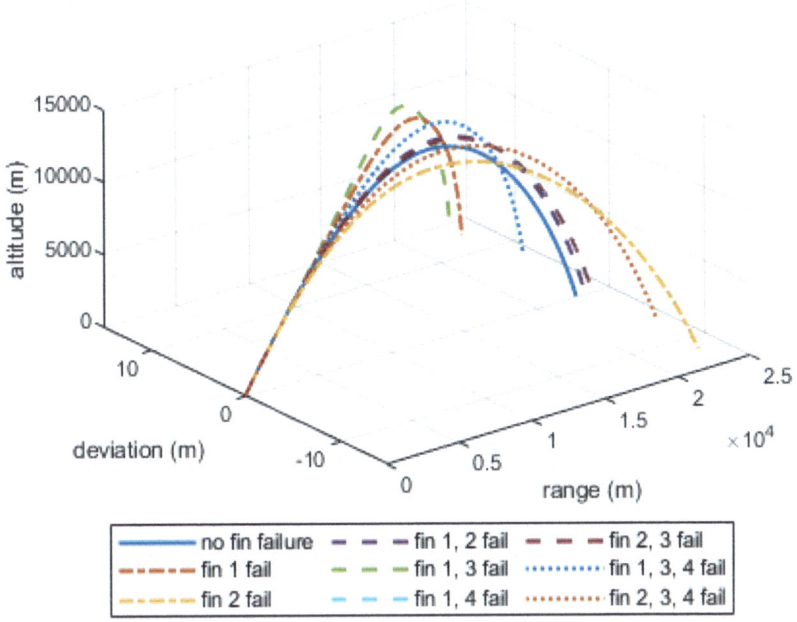

Fig. 3. Trajectory of ballistic missile with cruciform canard fin failure cases variation.

A failure occurred in the canard fin at the 10th second of flight, coinciding with the burnout of the missile fuel. At this critical juncture, the missile reaches its maximum speed just before it transitions to free flight without the aid of thrust. The canard fin failure induces a deflection in the flight path, resulting in a change of direction. This alteration in trajectory impacts not only the direction but also the missile apogee (the peak height reached during flight) and its eventual impact point.

In Fig. 4, the data reveals that when a failure occurs in the canard fin during ballistic flight, both the peak height and the distance from the drop point to the launch point increase. In scenarios where three canard fins fail, the peak height experiences a significant increase of 588 m, reaching a maximum altitude of 13.73 km above the launcher

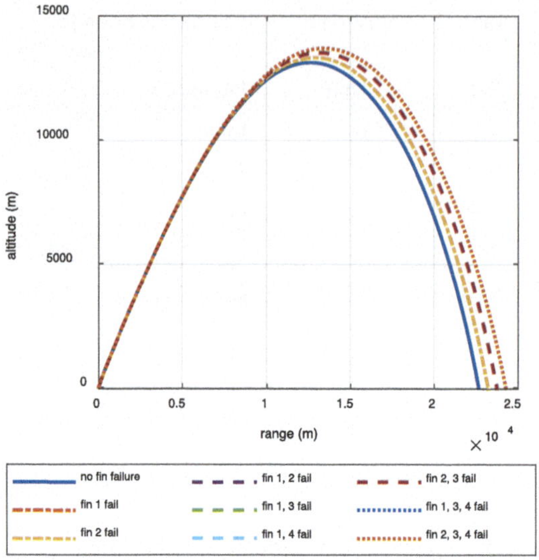

Fig. 4. A comparative of the height and flight distance of a ballistic missile under conditions of canard fin failure.

level. Concurrently, the impact point of the missile shifts considerably, moving 1.63 km further than the original impact point, resulting in a total distance of 24.32 km from the launch point. In contrast, when only one canard fin fails, the deviations in peak height and the distance from the impact point is minimal. The missile maintains a more stable and predictable trajectory, with the smallest recorded deviations in both peak height and impact distance.

Figures 5 and 6 depict a different trend regarding the maximum side deviation resulting from canard fin failures. Notably, Fig. 6 showcases that the most significant side deviation occurs when both canard fins (referred to as fin 1, 3 fail) move 15 m from their intended position. However, an intriguing divergence in behavior is observed in cases where individual canard fins experience failure, such as fin 1, 2 fail, fin 1, 4 fail, and fin 2, 3 fail. Despite these failures, the impact point only marginally deviates from the reference, these individual failure cases are relatively close, the extent of deviation from the reference varies.

Similarly, Figs. 5 and 6 illustrate comparable trends in side deviation for other fin failure configurations. For instance, the deviations observed in cases of fin 1 fail and fin 2 fail hover around 13 m, indicating a moderate impact on the missile flight path. Furthermore, configurations where multiple canard fins fail, such as fin 1, 3, 4 fail and fin 2, 3, 4 fail, exhibit deviations of approximately 8 m, reflecting a relatively minor influence on the missile trajectory.

As previously discussed, Fig. 6 illustrates that an increase in the number of damaged or failed fins results in a greater distance from the missile intended impact point. However, the directional deviation resulting from fin failure or damage is influenced by both the number and the position of the failing fins. This phenomenon is particularly evident in cases where both missile fins fail. For instance, when fins 1 and 3 fail symmetrically,

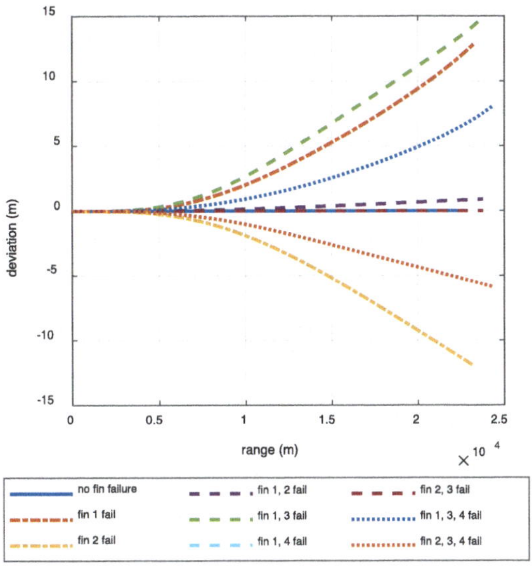

Fig. 5. A comparative of the side deviation and flight distance of a ballistic missile under conditions of canard fin failure.

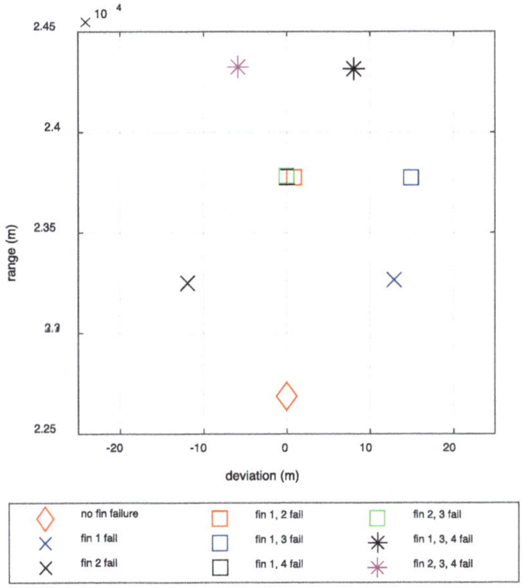

Fig. 6. Deviation distribution of fin failure case on ballistic missile with cruciform canard.

there is a significant directional deviation. Conversely, when the lower fin (fin 2), the upper fin (fin 4), and the side fin (fin 2) fail (as in the configurations where fins 2, 3 fail, fin 1, 4 fail, or fin 1, 2 fail), there are minimal directional deviation.

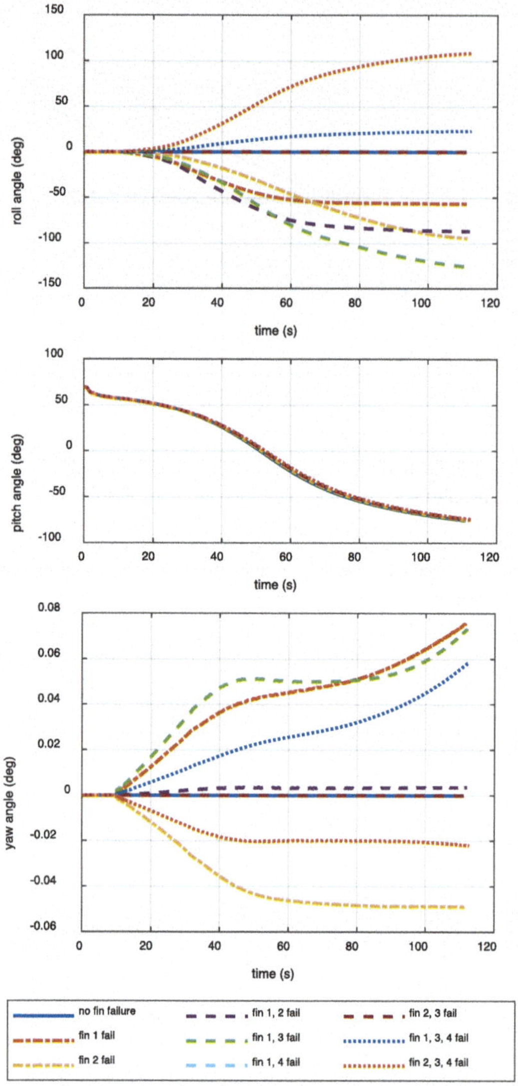

Fig. 7. Angular position of ballistic missile against time for fin failure case.

Figure 7 provides a comprehensive depiction of the attitude dynamics exhibited by a ballistic missile in scenarios characterized by canard fin failure. The figure meticulously plots the angular displacement of the missile over time, specifically highlighting the significant alterations in the roll angle attributable to fin failure. Notably, the most pronounced change in roll angle is observed when fin 3 fails, resulting in a peak value of 109°. Conversely, the minimum roll angle is recorded when fins 2, 3, and 4 fail, yielding a roll angle of $-125°$. These observations suggest that the isolated failure of 3 fins exert substantial influences on the missile roll dynamics, while the concurrent failure of multiple fins appears to attenuate this effect. Furthermore, there is a discernible shift in the pitch angle relative to the reference pitch angle, with the most significant deviation occurring when three fins fail, resulting in a deviation of 1.25° from the reference pitch angle.

5 Conclusion

The analysis provides the views of how the missile flight path is affected by fin failures, demonstrating deviations from the intended trajectory under various conditions. The results demonstrate that the trajectory has a significant divergence, with maximum impact point deviations of up to 1.63 km and peak height deviations of up to 588 m in the configuration where all three canard fins fail. The most significant side deviation occurs when both canard fins (1, 3 fail) and it is seen around 15 m. The most significant impact of fin failure is observed in the roll angle, which is more pronounced compared to other attitude determinations. This outcome leads to spinning rockets, with the most significant variation in roll angle occurring when fin 3 experiences failure, yielding a maximum value of 109°. Furthermore, the most significant shift in the pitch angle deviation occurring when three fins fail, resulting in a deviation of 1.25° from the reference pitch angle.

Acknowledgments. The authors are immensely grateful to the National Research and Innovation Agency of Indonesia (BRIN) and the Research Centre for Rocket Technology for their support in this research activity.

References

1. Ji, R., Liang, Y., Xu, L., Wei, Z.: Trajectory prediction of ballistic missiles using Gaussian process error model. Chin. J. Aeronaut. **35**(1), 458–469 (2022). https://doi.org/10.1016/j.cja.2021.05.011. ISSN 1000-9361
2. Sudarto, D.A., Dzaki Fayyadh, M.S., Sembiring, J., Putro, I.E.: Dynamic characteristics analysis of missile in ballistic flight. AIP Conf. Proc. **2941**(1), 020051 (2023). https://doi.org/10.1063/5.0182689
3. Etkin, B.: Dynamics of Atmospheric Flight, pp. 143–151. Dover Publications, Mineola (2012)
4. Cook, M.V.: Flight Dynamics Principles: A Linear Approach to Aircraft Stability and Control, 2nd edn., pp. 85–90. Butterworth-Heinemann, London (2007)
5. Hanif, A.: Modeling and simulation of a canard-based sounding rocket dynamics at climb and descent phases. AIP Conf. Proc. **2941**(1), 020033 (2023). https://doi.org/10.1063/5.0181512
6. Zhang, J., Lei, J., Yin, J., Niu, J.: The numerical investigation on the rolling decoupling of a canard-controlled missile using the jet control system. Eng. Appl. Comput. Fluid Mech. **14**(1), 1062–1077 (2020). https://doi.org/10.1080/19942060.2020.1799866
7. Lesieutre, D.J.: Prediction of sparrow missile aerodynamic characteristics with a nonlinear engineering level missile prediction method. In: AIAA 2017-3399. 35th AIAA Applied Aerodynamics Conference, Denver, Colorado, USA (2017). https://doi.org/10.2514/6.2017-3399
8. Lesieutre, D., Mendenhall, M., Dillenius, M.: Prediction of induced roll-on conventional missiles with cruciform fin sections. In: AIAA 1988-529. 26th Aerospace Sciences Meeting, Reno, NV, U.S.A (1988). https://doi.org/10.2514/6.1988-529

9. Mirzaei, M.: Roll reversal phenomenon control in flight vehicles. Aerosp. Sci. Technol. **79**, 413–425 (2018). https://doi.org/10.1016/j.ast.2018.05.059. ISSN 1270-9638
10. Silton, S.I., Coyle, C.J.: Effect of canard deflection for roll control on fin performance of a fin-stabilized projectile. In: AIAA 2016-0309. 54th AIAA Aerospace Sciences Meeting, San Diego, California, USA (2016). https://doi.org/10.2514/6.2016-0309
11. Silton, S.I., Coyle, C.J.: Effect of canard deflection on fin performance of a fin-stabilized projectile. In: AIAA 2015-2586. 33rd AIAA Applied Aerodynamics Conference, Dallas, TX, USA (2015). https://doi.org/10.2514/6.2015-2586
12. Silton, S.I., Fresconi, F.: Effect of canard interactions on aerodynamic performance of a fin-stabilized projectile. J. Spacecr. Rockets **52**(5), 1430–1442 (2015). https://doi.org/10.2514/1.A33219
13. de Oliveira Junior, W.L., et al.: 6 degrees of freedom simulation of an unguided sounding rocket using Matlab/Simulink. No. 2023-36-0095. SAE Technical Paper (2024). https://doi.org/10.4271/2023-36-0095
14. Srour, K., Abdulkerim, S.: Simulation of a 6DOF nonlinear missile. In: Karakoc, T.H., Rohács, J., Rohács, D., Ekici, S., Dalkiran, A., Kale, U. (eds.) ISATECH 2021, pp. 415–423. Springer, Cham (2024). https://doi.org/10.1007/978-3-031-38446-2_46
15. Shalaby, A.M.A.: Evaluation of the Performance of a Rocket under Oscillatory Thrust. California State University, Long Beach (2019)
16. Bashir, M., Khan, S.A., Udayagiri, L., Noor, A.: Dynamic stability of unguided projectile with 6-DOF trajectory modeling. In: 2nd International Conference for Convergence in Technology (I2CT), Mumbai, India, pp. 1002–1009 (2017). https://doi.org/10.1109/I2CT.2017.8226280
17. Siouris, G.M.: Missile Guidance and Control Systems. Springer Science & Business Media, New York (2006). https://doi.org/10.1007/b97614
18. Spirito, J.D., Vaughn, M.E., Washington, W.D.: Numerical investigation of canard-controlled missile with planar and grid fins. J. Spacecr. Rocket. **40**(3), 363–370 (2003). https://doi.org/10.2514/2.3971
19. Song, C., Lee, Y.I., Kim, S.H., Kim, P.S.: A new approach to structure of aerodynamic fin control system for STT missiles. In: The 3rd International Conference on Control, Automation, and Systems, Gyeongju, Korea, pp. 537–541 (2003)

Study on Aqueous Phenolic Matrix Composites: Physical, Mechanical Properties, and Morphological Characteristics

Haryadi Abrizal[1,2], Ariadne Laksmidevi Juwono[1](✉), Bagus Hayatul Jihad[2], Yudi Nugraha Thaha[3], and M. Dito Saputra[2]

[1] Department of Physics, Faculty of Mathematics and Natural Sciences, Universitas Indonesia, Depok, Indonesia
ariadne.laksmidevi@ui.ac.id

[2] Aviation and Space Research Organizations, National Research and Innovation Agency, Jakarta, Indonesia

[3] Research Centre for Metallurgy, National Research and Innovation Agency, Jakarta, Indonesia

Abstract. This study investigates aqueous phenolic matrix composites, focusing on their physical, mechanical properties, and morphology. Aqueous phenolic resins are the most widely available type on the market, primarily used in the wood industry, while solvent-based phenolics are mainly used in specific markets like aerospace. Despite extensive research on solvent-based phenolic matrices, aqueous phenolic resins remain relatively underexplored. This research evaluates four curing cycles using Dover Chemical Factory's LMPF-505™ resin and Jushi's ESC558™ e-glass fibers. Samples (PFG-1 to PFG-5) underwent hot compression molding with varied curing cycles (CC1 to CC4) and resin pre-conditioning to optimize water content. Analysis reveals that resin viscosity significantly influences distribution; higher water content leads to uneven distribution, while preconditioned resin shows improved uniformity. Despite employing four curing cycles, all samples exhibited macro voids, indicating suboptimal consolidation due to insufficient molding pressure and the absence of degassing.

Keywords: Aqueous Phenolic Resin · Composite Fabrication · Curing Cycles · Mechanical Properties · Morphological Characteristics

1 Introduction

Phenolic resins have emerged as highly sought-after materials for composite applications due to their exceptional performance at elevated temperatures. They exhibit superior thermal stability, chemical resistance, and dimensional integrity compared to other resin types such as polyester and epoxy, making them particularly well-suited for thermal protection systems across various industries [1, 2]. Phenolic resins are typically classified into two main categories: resole (liquid) and novolac (solid) [3]. Resoles are predominantly used in the wood adhesive sector, while novolacs used in the application in automotive and aerospace sectors for thermal protection systems, abrasive materials, and molding compound.

E. Fitrianingsih et al. (Eds.): ISAST 2024, SPPHY 416, pp. 443–454, 2025.
https://doi.org/10.1007/978-981-96-1344-1_47

Resoles are available in both solvent-based and aqueous formulations. Solvent-based resoles, exemplified by products like Durite™ SC-1008 with isopropanol solvent [4], are commonly utilized in composite laminate manufacturing. Conversely, aqueous resoles are typically employed in refractory materials and wood adhesives [5]. Notably, resoles are considered single-stage resins, requiring only heat for curing without the need for additional catalysts. This curing process, known as polycondensation, results in the release of water as a byproduct during crosslinking, posing challenges in composite fabrication due to potential issues with resin distribution uniformity [5].

Beyond these general applications, phenolic resins play a crucial role in ablative composite materials designed for rocket nozzles. These materials protect rocket nozzle surfaces against extreme temperatures and pressures during launch and operational phases [6, 7]. The outstanding heat resistance and structural integrity of phenolic resins render them exceptionally suitable for such demanding applications [7, 8]. However, the use of aqueous phenolic resins in composite fabrication remains relatively uncommon, primarily due to challenges associated with viscosity control and ensuring consistent resin distribution—a critical factor influencing the quality and performance of the final composite material [9].

Extensive research has explored how curing parameters and component ratios influence the mechanical properties of phenolic resin-based composites. Guney et al. [10] investigated the impact of curing conditions on silica-SC1008 phenolic composites using hot-pressing with varying temperatures (125 °C–205 °C), durations (0.5–24 h), and pressures (8–400 bar). They determined that an optimal curing temperature of 165 °C reduced void content and enhanced mechanical properties. The ideal curing duration was found to be 1 h, with longer durations offering negligible additional benefits. Notably, the highest curing pressure (400 bar) yielded the greatest mechanical strength. Kim et al. [11] developed compaction techniques for carbon/phenolic composites using an autoclave and compressive jig, varying jig pressures between 0.6 MPa, 1.2 MPa, and 1.8 MPa. Their findings highlighted that increasing jig pressure decreased void content, correlating directly with improved mechanical strength. This method proved effective in mitigating interlaminar failures in composites utilized for robust applications such as rocket nozzles. Fu et al. [12] investigated the effects of curing conditions on the mechanical and thermal properties of 2.5D quartz glass/boron-modified phenolic composites, examining variations in curing temperature (130 °C–180 °C) and curing duration (30–180 min). Their research underscored that elevating both curing temperature and duration up to 180 °C and 180 min, respectively, heightened the degree of curing, thereby augmenting the flexural strength of the composite material. Houdart et al. [13] conducted experiments to determine the optimal fabrication method for high-performance glass-phenolic composites, comparing continuous and non-continuous approaches and analyzing process variables such as drying, molding, after-cure, and solvent properties. Their study utilized phenolic resin with ethanol solvent. Elwan et al. [14] explored the influence of phenolic resin-to-ethanol solution ratios and short e-glass fiber on the mechanical properties of composites. They achieved this by mixing resin and fiber and subsequently subjecting them to compression molding at 10 MPa and 175 °C for 8–12 min. Similar to Houdart et al., their study employed phenolic resin with an ethanol solvent. Their findings revealed that a fiber weight fraction of 55 wt.% yielded optimal mechanical strength, while higher fractions led to decreased mechanical properties due to inadequate resin for effective adhesion.

While significant research has been devoted to the curing process of solvent-based phenolic matrix composites, studies focusing on aqueous phenolic resin as a composite matrix remain limited. This study aims to evaluate several fabrication processes by varying four types of curing cycles in glass-phenolic composites using aqueous phenolic resin and short e-glass fiber. The primary objective is to explore four distinct curing cycles to identify methods that yield optimal morphology and mechanical properties. This research is anticipated to significantly advance the understanding and application of water-based phenolic resins in the composite industry, potentially addressing challenges related to resin distribution encountered during the fabrication process.

2 Experimental Details

2.1 Materials

In this study, two raw materials used to create composite samples are phenolic resin and short e-glass fiber. The aqueous resole phenolic resin was obtained from Dover Chemical Factory with the code LMPF-505™, having a solid content ranging from 50–55%. Further details regarding the resin properties can be found in Table 1. The glass fiber used is chopped e-glass fiber from the Jushi brand, type ESC558™, with a length of 6 mm commonly used in bulk molding compound (BMC) manufacturing.

Table 1. Characteristics of Phenolic Resin LMPF-505.

Characteristic	Value	Unit
Density	1.185	g/cm^3
Solid Content	46	%
Viscosity	32	cPs
PH	7–8	-

2.2 Fabrication Process

The fabrication involved four distinct curing cycles, labeled CC1 to CC4. Additionally, the effect of pre-conditioning phenolic resin by reducing its water content before proceeding to the next stage was investigated. Five composite samples were fabricated, coded as PFG-1 (Phenolic Fibre Glass) to PFG-5. PFG-5 represented a sample processed using the CC4 method but with resin pre-conditioned to reduce its water content to 20%, resulting in an increase in solid content up to 70% through a vacuum dehydration process. Details regarding the four types of curing cycles and the five types of samples are presented in Fig. 1 and Table 2.

Table 2. Details of five samples and the types of curing cycles used.

Sample	Curing cycle	Matrix	Reinforcement
PFG-1	CC1	LMPF-505	Chopped Glass
PFG-2	CC2	LMPF-505	Chopped Glass
PFG-3	CC3	LMPF-505	Chopped Glass
PFG-4	CC4	LMPF-505	Chopped Glass
PFG-5	CC4	Pre-conditioning LMPF-505	Chopped Glass

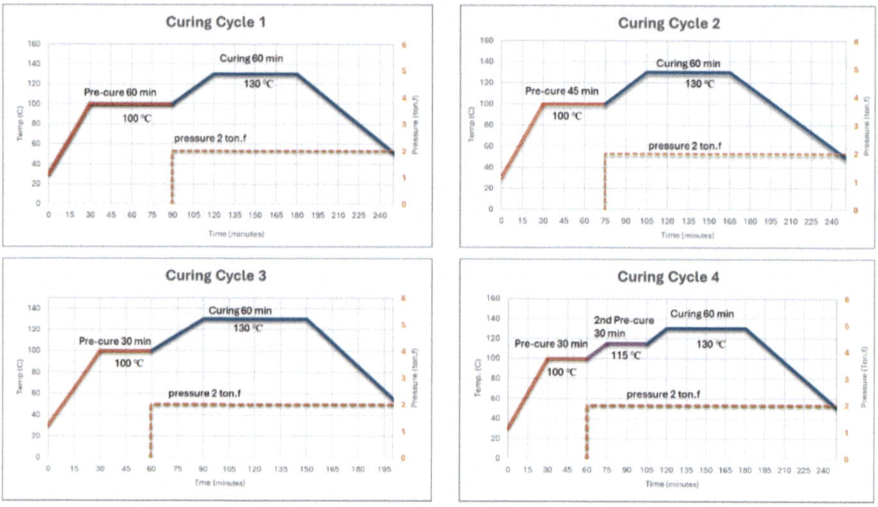

Fig. 1. Four curing cycles were chosen for the fabrication process.

In this study, the fabrication process employed hot compression molding using a compression jig with a screw mechanism, as illustrated in Fig. 2. The process involved molding a mixture of fibers and resin inside a metal mold, subjected to heat and mechanical pressure. The initial step in sample preparation involved weighing the fibers and resin according to predetermined ratios. Both materials were then combined in a beaker and stirred using a mechanical stirrer for 5 min. The resulting fiber-resin mixture was poured into an aluminum mold cavity treated with a mold release agent. The open mold was placed in an oven for the pre-curing stage, during which water dissolved in the phenolic resin evaporated, and pre-crosslinking of the resin molecules occurred [6]. Following the pre-curing phase, the mold was closed with a mold punch and subjected to mechanical pressure using the compression jig with an M16x2 screw mechanism. The required clamping force was determined by applying a torque of 100 N·m. Using the formula [15]:

$$T = Fv \cdot [0.16 \cdot p + 0.58 \cdot \mu_t \cdot d_2 + 0.5 \cdot \mu_u \cdot d_u] \tag{1}$$

where p is the pitch of 2 mm, μ_t and μ_u are both 0.25, d_2 is 14.701 mm, and d_u 16 mm, the clamping force F_V calculated to be approximately 22,473 N, or about 2-ton force. The closed mold was then heated to 130 °C with specific durations and heating rates according to the curing cycle's conditions. Finally, the mold was allowed to cool by natural convection to room temperature, and the composite samples were extracted from the mold.

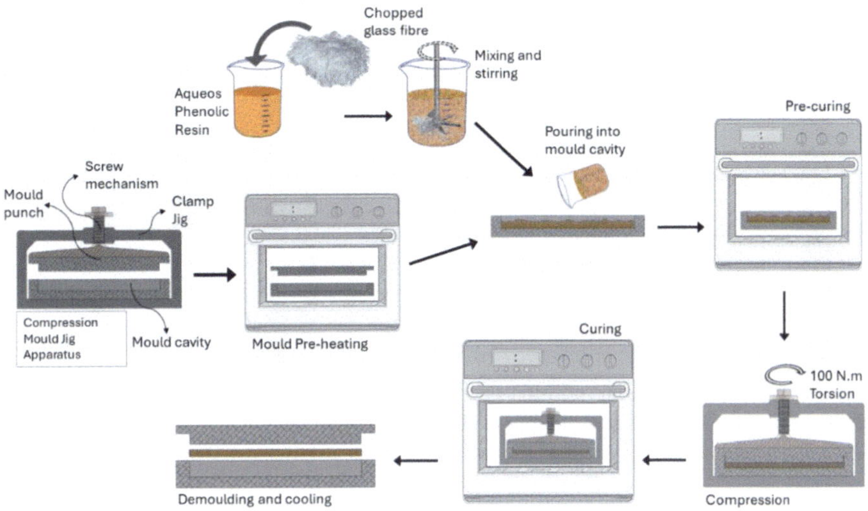

Fig. 2. Fabrication process of aqueous phenolic-glass fibre composites.

3 Characterizations

3.1 Physical Properties Tests

Physical property testing included density testing and a burn-off test. Density testing was conducted using a gravimetric buoyancy densitometer on each sample. Meanwhile, the burn-off test was conducted to assess the homogeneity of the composite samples by calculating the matrix content in specific segments. Specifically, the test was performed on the edges and center of the sample plate, with five specimens taken from each edge and the center area. The standard method employed was ASTM D3171 [7], involving heating the composite specimen to 500 °C for 30 min to remove the polymer matrix and leave only the fibers. Subsequently, the initial and final masses of the specimens were compared to determine the percentage reduction.

3.2 Mechanical Properties Tests

Tensile Test. Tensile testing was conducted using a Universal Testing Machine Shimadzu AG-X Plus 50 kN in accordance with ASTM D638 [8] standard. The prepared sample size was type 1 in dog bone shape, with a total length of 165 mm, a middle width of 13 mm, and an end width of 19 mm. The testing speed was 5 mm/minute with a grip-to-grip distance of 115 mm. From this testing, the average tensile strength values were obtained for 5 specimens of each composite sample (Fig. 3). Out of the five samples, only three samples coded as PFG-3, PFG-4, and PFG-5 were suitable for tensile testing. The remaining two samples, PFG-1 and PFG-2, were deemed unsuitable for tensile testing due to inadequate consolidation of the samples.

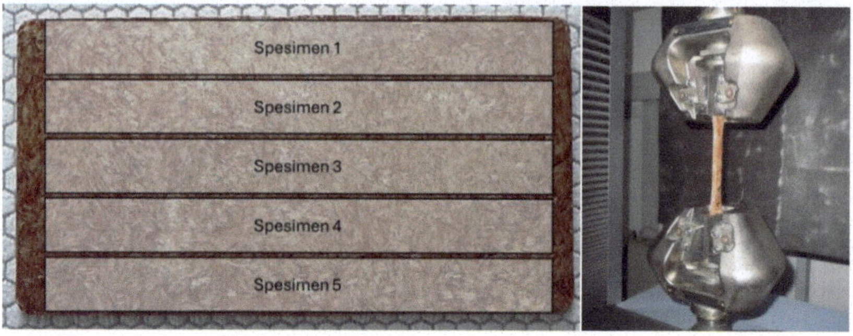

Fig. 3. Left: specimen cutting position on the composite plate sample, Right: tensile testing conditions of the composite according to ASTM D638 standard.

Compression Test. Compression tests were conducted using the Shimadzu AG-X Plus 50 kN Universal Testing Machine according to ASTM D695 standards. The prepared samples were thin plates measuring 7.94 cm in total length, 1.9 cm in total width, and 1.27 cm in middle width. The compression rate was 1.3 mm/min. The average compressive strength values were obtained from 5 specimens of each composite sample. Among the five samples, only four labeled as PFG-2, PFG-3, PFG-4, and PFG-5 were suitable for compression testing. Sample PFG-1 was excluded from compression testing due to inadequate sample consolidation.

3.3 Morphology Tests

Morphological observations were conducted using a Phenom ProX Desktop SEM machine on the cross-sectional area of the glass-phenolic composite samples. The observations focused on the presence of voids within the middle region per composite thickness.

4 Result and Discussion

4.1 Physical Characterization

As commonly understood in the mechanics of composite structures, the void content in fiber-reinforced polymer matrix composites is inversely proportional to their mechanical strength. Higher void content leads to reduced mechanical strength. Table 3 provides detailed physical properties of the five samples tested. Among the four curing cycles employed, samples processed with curing cycles 1 and 2 exhibited the poorest physical condition, with fibers not properly compacted or consolidated. In contrast, samples processed with curing cycles 3, 4, and 5 showed significantly better physical conditions and appeared satisfactory upon surface inspection, as illustrated in Fig. 4.

Table 3. Physical properties and density test results.

Sample	Density (g/cc)	Fibre: matrix wt. Ratio (%)	Description
PFG-1	1.61 ± 0.02	65:35	The fiber and matrix did not consolidate well and contained significant voids
PFG-2	1.64 ± 0.04	65:35	The fiber and matrix did not consolidate well and contained significant voids
PFG-3	1.5 ± 0.03	70:30	The sample consolidated well, but some resin extruded from the mold when pressure was applied
PFG-4	1.82 ± 0.02	63:37	The sample consolidated well
PFG-5	1.8 ± 0.01	56:44	The sample consolidated well

As shown in Table 3, the heaviest sample, PFG-5, resulted from the pre-conditioning process that produced the highest matrix ratio, with a density of 1.8 g/cc. Despite this, PFG-4 had a density 0.05 greater than that of PFG-5. Examination indicated that the thickness of PFG-5 was greater than that of PFG-4. Thus, under identical curing conditions, PFG-4 exhibited the highest density. This suggests that while PFG-5 had a larger overall volume, the material composition of PFG-4 was more compact, leading to its higher density.

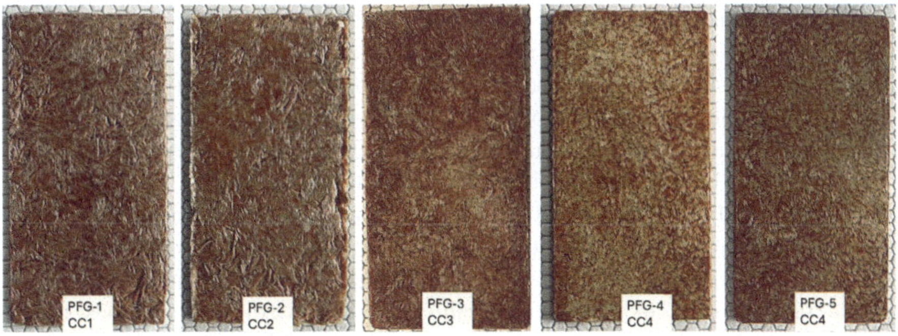

Fig. 4. Comparison of glass-phenolic composite samples with different curing cycles.

The burn-off test data analysis depicted in Fig. 5 for the composite samples revealed notable variations in resin matrix distribution across different sample areas. The test, conducted following the ASTM D3171 method, involved heating composite specimens to 500 °C for 30 min to eliminate the polymer matrix, leaving only the fibers. By comparing the initial and final masses of the specimens, the percentage of matrix reduction was determined.

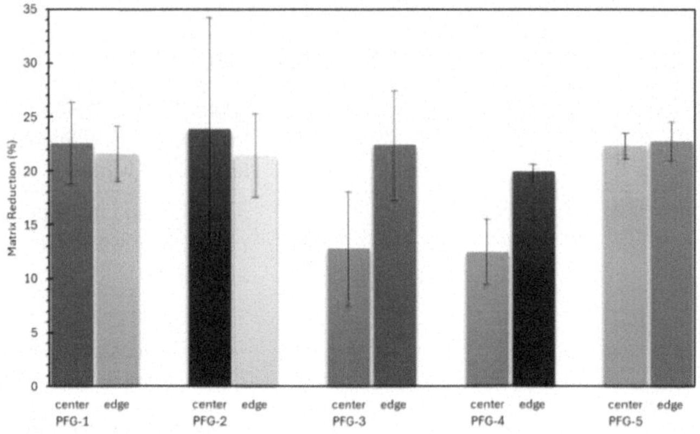

Fig. 5. Matrix reduction comparison on edge and center of sample.

The results showed that specimens taken from the edges of the samples generally exhibited a lower percentage reduction compared to specimens from the center areas. This suggests that the edge areas had a higher resin content, indicating an uneven distribution during fabrication. Specimens from samples PFG-3 and PFG-4 displayed a significant difference between their edge and center areas, suggesting that the resin tended to migrate towards the edges during fabrication. This phenomenon was likely influenced by the low viscosity of the resin used. In contrast, sample PFG-5, which utilized a resin with higher viscosity, demonstrated a more uniform resin distribution. This was evident from the consistent percentage reduction values observed across all specimens. The increased viscosity of the resin in PFG-5 effectively minimized resin migration towards the edges, resulting in a more even distribution and enhancing the consistency of tensile strength. These findings are consistent with previous studies, such as those by Groom [15] and Ujianto [16] which reported that low viscosity resins tend to migrate and cause uneven distribution during the fabrication process.

As shown on Fig. 6, compressive and tensile strength analysis of the composite samples PFG-2, PFG-3, PFG-4, and PFG-5 reveals significant variations in mechanical properties and measurement deviations. PFG-5 shows the highest average compressive strength at approximately 40 MPa, with relatively short error bars indicating consistent material properties and fabrication quality. PFG-5 also exhibits the highest tensile strength, reinforcing the effectiveness of using higher viscosity resin in achieving uniform composite properties. In contrast, PFG-2 and PFG-3 exhibit substantial variability, with compressive strength values ranging from 20 MPa to 50 MPa and tensile strengths

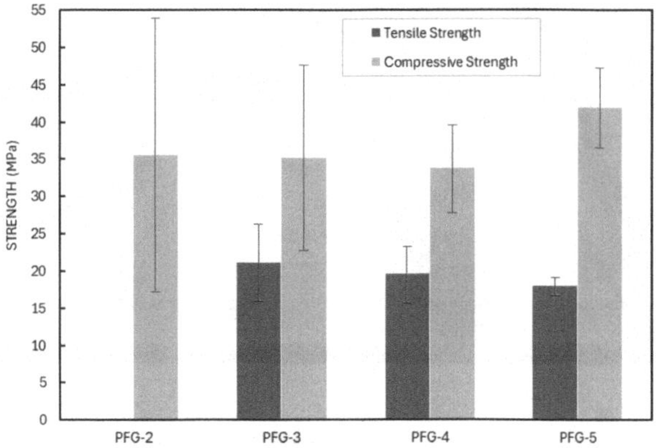

Fig. 6. Tensile and compressive strength of chopped glass-phenolic composite.

showing similar trends. This reflects significant inconsistencies due to uneven resin distribution during fabrication, attributed to the low viscosity resin used, which tended to migrate towards the edges. PFG-4 demonstrates intermediate strength and variability, suggesting moderate inconsistency in resin distribution. Notably, the use of higher viscosity resin in PFG-5 resulted in a more uniform distribution, minimizing edge migration and enhancing consistency, as evidenced by smaller deviation errors. These findings align with previous studies that emphasize the importance of resin viscosity in achieving uniform composite properties. For instance, research by Hassan et al. [17] and Ujianto et al. [16] similarly concluded that higher viscosity resins lead to more uniform mechanical properties and reduced measurement deviations in composite materials. However, it is important to note that the compressive strength values obtained in this study are still lower compared to other studies in the field. Additionally, based on references [10, 18, 19] and commercial products like [20, 21], the tensile strength of phenolic composites with chopped-fibre reinforcement ranges from about 30–57 MPa, while the compressive strength ranges from about 180–250 MPa. This further indicates the need for improved fabrication techniques in this study.

The results of the Scanning Electron Microscope (SEM) analysis of the five composite samples PFG-1 to PFG-5 are presented in Fig. 7. Further detailed observations revealed clusters of poorly consolidated fibers, with intra-fiber voids ranging from 150 to 400 μm. At 320× magnification, the matrix areas did not show significant voids, indicating adequate control over the curing conditions. However, smaller voids were still visible at 500× magnification. These observations suggest that the presence of voids between fiber clusters is due to insufficient pressure applied during the fabrication process. The pressure of 2 tons-force is still too low compared to pressures used in other studies, typically ranging from 10 to 20 tons-force [10, 13, 22, 23]. This suboptimal pressure results in incomplete consolidation of fibers and resin, leading to void formation that diminishes the mechanical properties of the composite. However, it is important to note that excessively high pressure can cause resin to extrude from the fibers through molding gaps. Therefore, precise and optimal control of resin viscosity is essential to ensure uniform distribution.

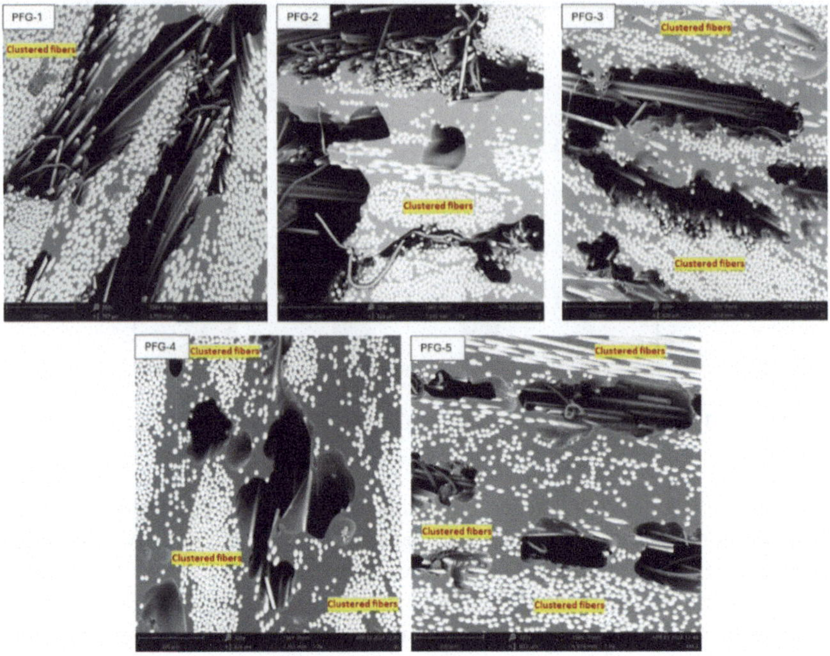

Fig. 7. Morphology of interlaminar voids in cross sections for all specimens.

In another SEM observations using a magnification of 1000x, it can be observed in Fig. 8 that in the resin-rich area of the cross-section of the composite sample, there are micro voids approximately 3–4 µm in size. Additionally, initial intrafiber cracking is also apparent, which has the potential to reduce the mechanical properties of the composite. However, these micro voids are not prominently present throughout the entire resin-rich area.

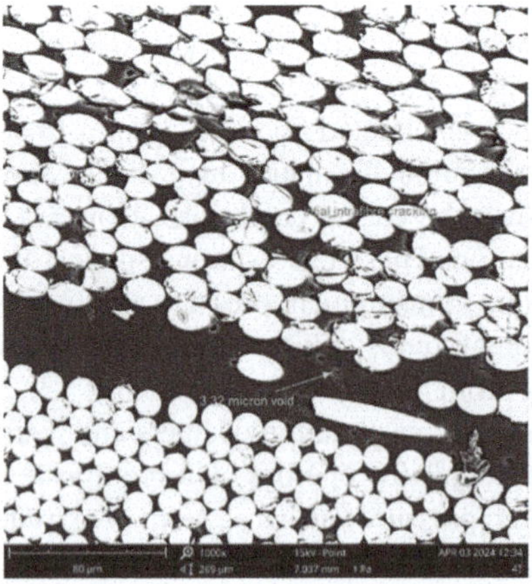

Fig. 8. SEM imaging of micro voids in resin-rich area of PFG-4.

5 Conclusion

The use of aqueous resin in the fabrication of composite materials with hot compression molding is highly dependent on the viscosity properties of the resin to achieve a good result. Resin with high water content on initial process tends to cause uneven distribution of the resin throughout the sample, whereas resin that has been pre-conditioned by reducing its water content and increasing its viscosity shows more uniform distribution. The four curing cycles used in this research have not yet produced optimal composites, as indicated by the presence of macro voids clearly visible in the cross-sectional observations of all samples. Therefore, excessively thin resin tends to cause uneven distribution throughout the composite.

Moreover, although the effect of molding pressure was not discussed in detail in this study, other research indicates that the molding pressure used here is insufficient to adequately compress the chopped fibers and matrix to create good quality of composite. This issue can be addressed in future research by using more adequate equipment to achieve higher pressures and by implementing a degassing process during the curing cycle. Consequently, future research should focus on optimizing resin viscosity, increasing molding pressure, and applying degassing techniques to produce composites with better mechanical properties.

References

1. Srebrenkoska, V., Bogoeva-Gaceva, G., Dimeski, D.: Composite material based on an ablative phenolic resin and carbon fibers. J. Serb. Chem. Soc. **74**(4), 441–453 (2009). https://doi.org/10.2298/JSC0904441S
2. Kumar, C.V., Kandasubramanian, B.: Advances in ablative composites of carbon-based materials: a review. Ind. Eng. Chem. Res. **58**(51), 22663–22701 (2019). https://doi.org/10.1021/acs.iecr.9b04625
3. Umoren, S.A., Solomon, M.M., Saji, V.S.: Resin based polymers. In: Umoren, S.A., Solomon, M.M., Saji, V.S. (eds.) Polymeric Materials in Corrosion Inhibition, chap. 18, pp. 419–441. Elsevier (2022). https://doi.org/10.1016/B978-0-12-823854-7.00008-4
4. Phenolic—Resin—Durite-SC-1008-_Hexion_
5. Ibeh, C.C.: 2 - Phenol-Formaldehyde Resins. In: Goodman, S.H. (ed.) Handbook of Thermoset Plastics, 2nd edn., pp. 23–71. William Andrew Publishing, Westwood (1998). https://doi.org/10.1016/B978-081551421-3.50005-9
6. Pulci, G., Tirillò, J., Marra, F., Fossati, F., Bartuli, C., Valente, T.: Carbon–phenolic ablative materials for re-entry space vehicles: manufacturing and properties. Compos. Part A Appl. Sci. Manuf. **41**(10), 1483–1490 (2010). https://doi.org/10.1016/j.compositesa.2010.06.010
7. Dixit, D., Pal, R., Kapoor, G., Stabenau, M.: 6 - Lightweight composite materials processing. In: Bhatnagar, A., (ed.) Lightweight Ballistic Composites. Series in Composites Science and Engineering, 2nd edn., pp. 157–216. Woodhead Publishing (2016). https://doi.org/10.1016/B978-0-08-100406-7.00006-4
8. Wu, H., Koo, J.H.: Characterization of high-temperature polymers for extreme environments. In: Vahabi, H., Saeb, M.R., Malucelli, G. (eds.) Analysis of Flame Retardancy in Polymer Science, chap. 8, pp. 299–331. Elsevier (2022). https://doi.org/10.1016/B978-0-12-824045-8.00008-3
9. Knop, A., Pilato, L.A.: Phenolic Resins. Springer, Heidelberg (1985). https://doi.org/10.1007/978-3-662-02429-4

10. Güney, Ö., Demirkesen, E.: Influence of curing conditions on the mechanical properties of silica-phenolic composites: a comprehensive study. IARJSET **3**, 1–6 (2015). https://doi.org/10.17148/IARJSET.2015.21101
11. Kim, J.W., Kim, H.G., Lee, D.G.: Compaction of thick carbon/phenolic fabric composites with autoclave method. Compos. Struct. **66**(1–4), 467–477 (2004). https://doi.org/10.1016/j.compstruct.2004.04.070
12. Fu, H., Qin, Y., He, X., Meng, X., Zhong, Y., Zou, Z.: Effect of curing degree on mechanical and thermal properties of 2.5D quartz fiber reinforced boron phenolic composites. E-Polymers **19**(1), 462–469 (2019). https://doi.org/10.1515/epoly-2019-0048
13. Houdart, P.-R., Fujii, T., Okubo, K.: Optimization of Manufacturing Process of Glass Fibers/Phenol Composites. Effects of Solidification Conditions, Fiber Length and Additional Materials on their Mechanical Properties (2003). https://doi.org/10.4271/2003-01-1128
14. Elwan, I., Jabra, R.: Preparation and Evaluation of Mechanical and Physical Properties of Random Silica Fiber/Modified Resole Resin Composites (2017). https://api.semanticscholar.org/CorpusID:88501596
15. Groom, L., So, C.-L., Elder, T., Pesacreta, T., Rials, T.: Effect of refining pressure and resin viscosity on resin flow, distribution, and penetration of MDF Fibers. In: The 7th Pacific Rim Bio-Based Composites Symposium, vol. 1, no. 2, pp. 227–239 (2004)
16. Ujianto, O., Nugroho, A., Setijotomo, H.S., Bintoro, A.: Effect of fabrication techniques, resin types and fiber combinations on mechanical properties and morphology of glass fiber composites. Int. J. Mater. Mech. Manuf. **8**(1), 17–21 (2020). https://doi.org/10.18178/ijmmm.2020.8.1.477
17. Hassan, S.S., Sh, H.S.: An Evaluation of Mechanical Properties of Different Types of Composite Resins (An in vitro study) (2014). www.rafidaindentj.net
18. Ku, H., Jacobson, W., Trada, M., Cardona, F., Rogers, D.: Tensile tests of phenol formaldehyde SLG reinforced composites: pilot study. J. Compos. Mater. **42**(26), 2783–2793 (2008). https://doi.org/10.1177/0021998308096334
19. Asim, M., Saba, N., Jawaid, M., Nasir, M., Pervaiz, M., Alothman, O.Y.: A review on phenolic resin and its composites. Curr. Anal. Chem. **14**(3), 185–197 (2018). https://doi.org/10.2174/1573411013666171003154410
20. Matweb. Overview of materials for Phenolic; Unreinforced; Molded
21. Resinoid: Resinoid 1550 Phenolic Molding, Glass Reinforced. https://www.matweb.com/search/datasheet.aspx?matguid=43c57e2ed79d455a8dbefc11e65e7d4d&ckck=1
22. Tanaka, K., Yamada, T., Moriito, K., Katayama, T.: The effect of molding pressure on the mechanical properties of CFRTP using paper-type intermediate material. In: High Performance and Optimum Design of Structures and Materials II, pp. 307–315. WIT Press (2016). https://doi.org/10.2495/hpsm160281
23. Xie, J., Wang, S., Cui, Z., Wu, J.: Process optimization for compression molding of carbon fiber-reinforced thermosetting polymer. Materials **12**, 2430 (2019). https://doi.org/10.3390/ma12152430

The Impact of Hydrostatic Weighing Immersion Time on the Estimation of an AP-Based Composite Propellant Density

Luthfia Hajar Abdillah[1](✉), Retno Ardianingsih[1], Anita Pinalia[1], Afni Restasari[2], Prawita Dhewi[1], and Hamonangan Sitompul[1]

[1] Research Center for Rocket Technology BRIN, Bogor 16350, Indonesia
{luth002,retn014,anita.pinalia,praw004,hamo001}@brin.go.id
[2] Research Center for Biomass and Bioproducts BRIN, Bogor 16915, Indonesia
afni.restasari@brin.go.id

Abstract. Density is one of the essential characteristics of composite propellants. A common method for estimating the density of materials is hydrostatic weighing, which is based on Archimedes' principle. The technique, often known as underwater weighing, may also be used to determine the density of composite propellants, which are categorized as porous solid materials. Although the method is simple, the stability issue of water weighing needs to be considered since the propellant's component may dissolve in water. This work investigates the impact of hydrostatic weighing immersion time on the density of an AP-based composite propellant. It was done by adjusting the immersion time of samples to assess propellant density and comparing the result with the theoretical value. The study found that the loss of density value and accuracy increased with immersion time. Setting the immersion times below five minutes is recommended to achieve the density value with more than 99% accuracy.

Keywords: Propellant Density · Hydrostatic Weighing · Immersion Time

1 Introduction

1.1 Composite Propellant

A composite propellant is composed of liquid and solid components. AP-based composite propellant's solid material is ammonium perchlorate (AP), an oxidizing agent that makes up around 70–80% of the weight (w). With various particle sizes (20 - 400 microns), AP is typically used as multimodal with varying coarse-to-fine ratio [1]. Another solid material is aluminum powder, which is a fuel component and is comprised of 15–20%w. The liquid materials are hydroxyl-terminated polybutadiene (HTPB) polymer, which is a superior binder that can withstand high solid loading (86–90%w) [3]; isocyanate compounds, i.e., toluene diisocyanate (TDI), isophorone diisocyanate (IPDI) as curing agents; dioctyl adipate (DOA) as a plasticizer. Other additives include chain extenders, curing catalysts, burning rate catalysts, bonding agents, and so on [4].

E. Fitrianingsih et al. (Eds.): ISAST 2024, SPPHY 416, pp. 455–465, 2025.
https://doi.org/10.1007/978-981-96-1344-1_48

In composite solid propellants, metal particles and oxidizing agents are attached to the polymer matrix. One of the superior polymers as a propellant matrix is polyurethane. Polyurethanes result from polymerization between HTPB and diisocyanate compounds [5]. The polymer matrix occurs from the crosslinks that bind polymer chains to form networks [6]. Process variables (mixing time, speed, sequence, temperature, type of materials, and composition) must be considered to create propellant samples as required. Previous studies investigated these parameters through rheological analysis to prevent the occurrence of damage in the propellant samples [2, 7–10].

1.2 Composite Propellant Density

The propellant density is an important characteristic used in the simulation process and design of a rocket motor to obtain the desired performance of the rocket [11]. The determination of solid material density has been traditionally performed by the hydrostatic weighing technique [12]. This technique is known as the displacement method and is based on Archimedes' principles, also known as underwater weighing. When a solid object replaces a volume in a fluid, the buoyancy force exerted on the solid object in the fluid is equal to the weight of the fluid [13].

Various propellant density measurements using the hydrostatic weighing technique were reported. In one study, 95% ethanol was applied as a suspending medium for propellant density measurement based on Archimedes' principle [14]. Using a thin plastic layer to coat the propellant sample was reported in density measurement by hydrostatic weighing to avoid dissolving AP in water [15]. However, this method could increase the density measurement deviation value due to the presence of plastic weight and possibly trapped air in the plastic.

Theoretically, the propellant density can be calculated from each component, which is known as true density, as in Eq. (1) [16]:

$$\rho_{true} = \frac{\sum_N m_j}{\sum_N \frac{m_j}{\rho_j}} \tag{1}$$

ρ_{true}: true density
m_j: mass of the j-th component
ρ_j: density of the j-th component
N: number of components

The true density is defined as the mass of material per unit real volume (without considering the volume of all pores). In contrast, the apparent density is defined as the mass of material per unit external volume, which considers the volume of all pores [17]. The propellant density value measured in this study was categorized as apparent density.

B. Prianto, et al. proposed a new modification for the theoretical method calculation of propellant density by considering the volume fraction of the liquid material, which is applied to liquid volume fractions equal to or more than 0.26 [18].

$$\rho_{theo} = \frac{0.74}{\left(\left(\sum_{j=1}^{N_S}\left(\frac{Y_j}{\rho_j}\right)\right) + \left(\sum_{j=1}^{N_L}\left(\frac{Z_j}{\rho_j}\right)\right) - \left(\frac{0.26}{0.74} * \sum_{j=1}^{N_S}\left(\frac{Y_j}{\rho_j}\right)\right)\right)} \tag{2}$$

ρ_{theo}: theoretical density
Y_j: mass of the j-th solid component in propellant
Z_j: mass of the j-th liquid component in propellant
ρ_j: density of the j-th component
N_S: number of solid components in propellant
N_L: number of liquid components in propellant

In the solid material's density measurement by displacement method, the sample and the solid density standard are placed in the liquid with low surface tension, and the buoyancy forces acting on them are measured using a balance. It provides much better stability for the buoyancy force measurement [13]. However, this stability issue means increasing the immersion time, so it should be considered when applied to propellants as it affects the density measurement value. This work investigates the impact of hydrostatic weighing on the density of an AP-based composite propellant by adjusting the immersion time of the samples to obtain an appropriate immersion time limit that gives a non-deviating value.

2 Material and Method

2.1 Material

The propellant used in this study is an AP-based composite solid propellant with a liquid volume fraction of 0.28. The composite propellant consisted of materials as listed in Table 1. Total solid materials were 85%w with 8%w of Al powder. AP particles are used in bimodal size distribution with a coarse-to-fine ratio 1/1. The liquid materials used were HTPB/TDI with a weight ratio of 15/1 and 3%w of DOA.

Table 1. Properties of AP-Based composite materials.

Materials	Properties
Ammonium Perchlorate (AP)	• avg. particle size = 200 μm and 50 μm • density value = 1.95 g/cm^3
Hydroxy-terminated polybutadiene (HTPB)	• molecular weight = 2300 – 2800 g/mol • viscosity at 30°C = 5000 - 6000 centipoise • density = 0.9 g/cm^3
Aluminum powder (Al)	• avg. diameter = 10 μm • density = 2.7 g/cm^3
Dioctyl adipate (DOA)	• specific gravity = 0.926
Toluene diisocyanate (TDI)	• density = 1.21 g/cm^3

2.2 Method

Sample Preparation. Each raw material was mixed for fifteen minutes sequentially in a horizontal-blade mixer at 55 °C. Liquid materials, except TDI, were mixed and followed by solid materials. TDI is the last liquid component to be incorporated because of its reaction with the polymer to form a network that will cure the propellant.

Propellant samples were prepared from six mixing batches to produce six propellant cylinders. The propellants were made with the same composition but differed in the mixing time before pouring TDI. The mixing time variables before mixing TDI were 50 and 20 min. The 50-min time was used to obtain homogeneous mixing results [19], while the 20-min time was used as a comparison. Each of these times, the mixing was carried out in three batches to obtain three propellant cylinders. The mixtures were cured in the tube at 60 °C for 72 h to be prepared as samples. Each mixing batch produces one cylinder of cured propellant, which is then cut into fifteen pieces of a cube with a volume of 1 cm^3 for density measurement.

Density Measurement. Utilizing military standards [20] for assessing propellant characteristics here was impractical. Due to the availability of the equipment, the density measurement was approached using the density test method of solid plastic by displacement [21]. The instrument was a densitometer AND HR-200, equipped with a thermometer and beaker glass. Distillate water of about 300 ml was used as the immersion liquid. The samples' weight in water was recorded at various immersion times (t): 0, 2, 5, and 10 s. In addition, the data was collected for a 60-s immersion time to observe the sample's weight stability.

Data Analysis. The density value was calculated and analyzed statistically by the *t*-test method [22] to determine any differences in the density value at each immersion time. This analysis establishes the maximum immersion time sufficient for a non-deviating density value.

First, setting the null hypothesis (H_0), as the density value is not significantly different at the specified immersion time, then calculating the t_{calc} using Eq. (3). The value of $t_{calc.}$ Was then compared with the value of t_{table} at a degree of freedom (dk) equal to 28, and a level of significance (α) of 5%. If t_{calc} is smaller than t_{table}, , then the null hypothesis (H_0) can be accepted. At that dk value and significance level, the value of t_{table} is 1.701 [22].

$$t_{calc} = \frac{\overline{y_1} - \overline{y_2}}{\sqrt{\frac{s_1^2}{n_1} + \frac{s_2^2}{n_2}}} \tag{3}$$

t_{calc}: calculated value of t
$\overline{y_1}$: mean density at time t_0
$\overline{y_2}$: mean density at time $t \neq t_0$
s_1^2: variance for time t_0
s_2^2: variance for time $t \neq t_0$
n_1: number of samples of time t_0
n_2: number of samples of time $t \neq t_0$

3 Result and Discussion

The actual density of propellants calculated by Eq. (1) is 1.6998 g/ml, assuming that all materials blend into one another without considering the existence of the pores as a result of the reaction between these materials. This value is higher than the measured and theoretical value calculated using Eq. (2). The measured density was apparent because pores were formed in the structure of the propellant. Therefore, Eq. (2) provides a better mathematical model concerning the estimation of the propellant density. For liquid volume fractions of 0.28, the propellant's theoretical density is 1.6850 g/ml.

The experiment used six propellants of the same composition. The difference lies in the times before TDI mixing to study its effect on density. The first three cylinders were mixed using 50 min as a variable time before TDI mixing, and the last three were 20 min. The propellant density was then measured by the displacement method [21]. The propellant density values at various immersion times for each cylinder are shown in Table 2.

Table 2. Propellant density at various immersion times.

Number of cylinders	Density (g/ml)					
	t_0	t_2	t_5	t_{10}	t_{60}	t_{120}
1	1.6703	1.6703	1.6700	1.6682	1,6538	1,6428
2	1.6687	1.6686	1.6680	1.6656	1,6347	1,6238
3	1.6647	1.6646	1.6638	1.6616	1,6413	1,6300
4	1.6726	1.6724	1.6717	1.6692	1,6518	1,6401
5	1.6729	1.6729	1.6720	1.6696	1,6422	1,6289
6	1.6676	1.6676	1.6668	1.6641	1,6390	1,6274

The most prominent weight loss occurs within the first several seconds of immersion, gradually decreasing through the first minute and beyond. The significance of the difference up to t_{10} was calculated first because it is the time when the sample starts to lose its weight. It was determined through statistical analysis to find the significant differences in density value, as shown in Table 3 [22].

A statistical analysis using the *t*-test was performed on three immersion time variables (t_2, t_5, and t_{10}), to discover the time at which significant change occurs. It is observed that there were no significant differences in the density at t_2 or t_5, compared with the values at t_0, while a significant difference occurred at t_{10}. . At later times (t_{60} and t_{120}), without being statistically calculated, the differences will certainly be significant. It is known that the sample weight did not stabilize until t_{120}. So, when measuring propellant density using hydrostatic weighing, the stability of the sample weight does not need to be considered while measuring buoyancy force because taking so increases the immersion time.

The significant differences at t_{10} occur in four propellant samples out of the six. It indicates at t_{10}, the AP particles were already being dissolved by immersing liquid at a previous time. Immersing the samples for no more than five seconds will be preferable. The results from the experiments suggest that AP particle dissolution only takes place at the sample surface and not within the sample due to the limited immersion time. This is indicated by the propellant density values that were not significantly deviated. The result "not significantly different" in Table 3 means the density value can be assumed as non-deviating value, even if the number not exactly the same. Since t_{10} is the edge for considering the significant differences in density values; a discussion will be provided to gain insight into this significant result of density values.

Table 3. Statistical results of the significance of differences in density values at specified immersion times.

Number of cylinders	t_0 and t_2		t_0 and t_5		t_0 and t_{10}	
	t_{calc}	Result	t_{calc}	Result	t_{calc}	Result
1	0.014	*	0.174	*	1.201	*
2	0.055	*	0.562	*	2.635	**
3	0.058	*	0.314	*	1.043	*
4	0.244	*	1.144	*	4.180	**
5	0.028	*	0.611	*	2.149	**
6	0.008	*	0.460	*	2.058	**

*not significantly different
**significantly different

Density changes before t_{10} can be divided into three regions, as shown in Fig. 1. Regions I, II, and III represent the immersion times elapsed from t_0 to t_2, t_2 to t_5, and t_5 to t_{10}, respectively. Figure 1 shows a decreasing shift in all the curves. The curves initially display an essentially linear line in Region I, start to decrease in Region II, and then exhibit a steep line in Region III.

The pores structure in the propellants allows the water molecules to penetrate deep into the propellant sample and then start to dissolve AP particles, which results in density loss, as shown in Table 4. The lowest rate of density loss occurs in Region I where there is about considerably small. In this very short time, water just started to cover the sample surface; therefore, only a very small amount of AP particles dissolved.

In Region II, more AP particles were dissolved in the water. The samples started to lose a small fraction of weight of 0.5633 mg with a solubility rate of 0.1878 mg/s. The particles attached to the sample surfaces, as seen in Fig. 2, have the biggest potential to dissolve in water first compared to the particles that are not on the surface. Water molecules infiltrated the inner part of the sample and triggered the steep line in Region III, meaning more AP particles were dissolved.

The surface of the propellant sample in Fig. 2 (a) was porous and rough. The network generated by the polymer's reaction with the curing agent was responsible for these pores, whereas the particles attached to the network were the reason for the rough texture [5, 6]. Most particles with larger sizes, as seen in Fig. 2 (b), are identified as AP, while

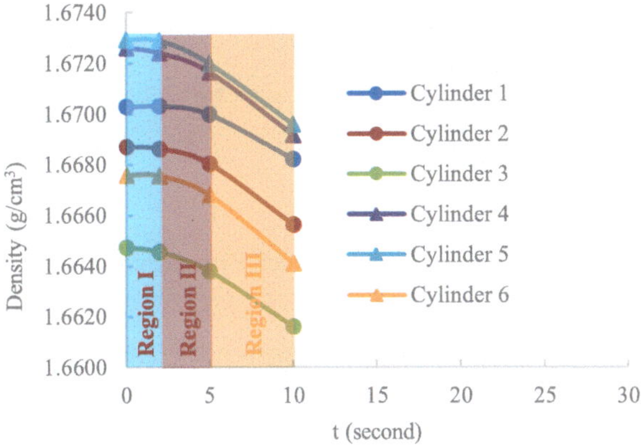

Fig. 1. Grouped regions according to immersion time.

Table 4. Parameters according to the immersion times regions.

Parameters	Region		
	I	II	III
Δt (s)	2	3	5
Density loss rate (g/ml.s)	0.004%	0.023%	0.047%
Weight loss (mg)	0,1333	0,5633	1,8533
Solubility rate (mg/s)	0,0667	0,1878	0,3707

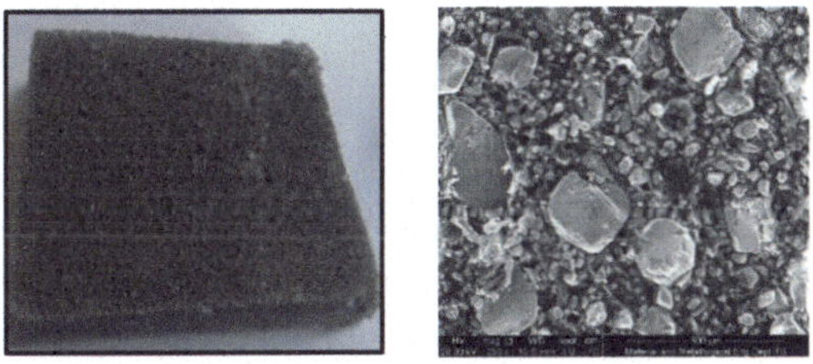

(a) (b)

Fig. 2. (a) Propellant sample's surface structure (b) SEM of the propellant sample surface.

the remaining particles are aluminum. Water can easily separate or dissolve the loosely attached particles on the sample's surface.

Most of the AP particles are relatively spherical/round. However, some are slightly tapered, as shown in Fig. 3. In addition to impurities, the cooling system configuration during the crystallization process affects the shape of the AP particles [23]. The size and shape of the particles are related to the dissolving process in the water. The smaller the particle size, the larger the surface area, the more interaction of particles with the water, so the solubility will increase [24].

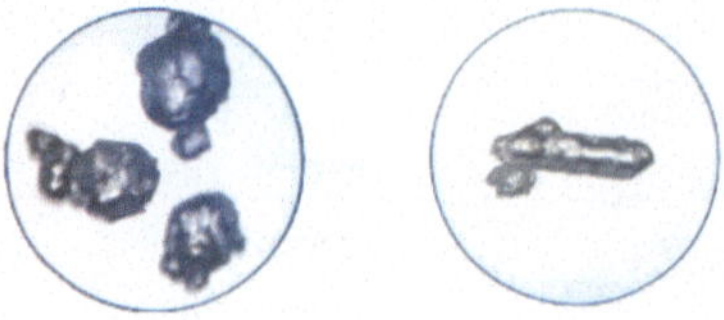

(a) (b)

Fig. 3. AP particle shape: (a) relatively round. (b) slightly tapered [23].

Figure 1 shows density variations for the six batches. This could result from variations in the size and shape of AP particles attached to the propellant matrix in each sample. Due to the various sizes and non-uniform shapes attached to the surface of each sample, the loss of AP weight will vary when the samples are immersed in the water. The more AP particles dissolve, the more significant the impact will be on density.

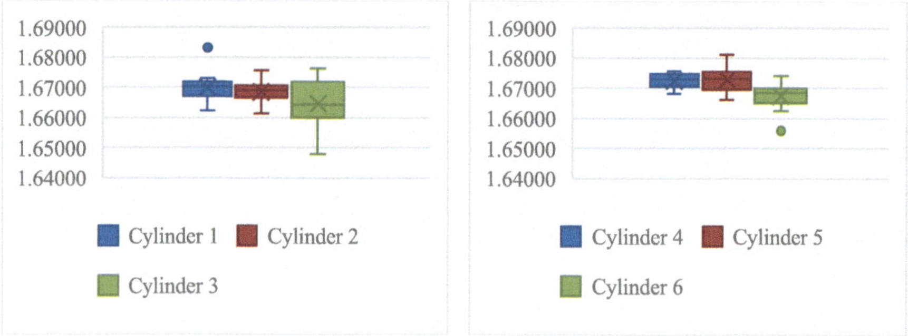

(a) (b)

Fig. 4. (a) Density at mixing time before TDI 50 min. (b) Density at mixing time before TDI 20 min

Cylinders 1, 2, and 3 were produced using mixing time less than TDI 50 min, while mixing time before TDI 20 min produced cylinders 4, 5, and 6. Previously, it is shown in Fig. 1 that the density values in each cylinder differ. According to statistical test results presented in Table 3, the density of cylinders 1 and 2 are reported to be not significantly different. In contrast, cylinder 3's value is stated to be significantly different from each. Cylinders 4 and 5 are stated to not differ significantly from each other; however, cylinder 6 is reported to differ significantly from cylinders 4 and 5. Based on this, one out of three batches for each mixing time variable before TDI has a different density value. This is also observed in Fig. 4; cylinder 3 and cylinder 6 are depicted as having different average density values compared to the other two.

As previously mentioned, three batches are used to repeat the procedure, which is common when one of the three batches brings about different products. The potential factor is the air humidity during the mixing process. TDI and polyurethane prepolymers containing NCO compounds at the end of the chain/structure can react with water vapor (H_2O). This water vapor can be generated from the surrounding humidity and then absorbed into the prepolymer. This reaction produces primary amines and CO_2. The formation of CO_2 should be avoided because it leaves cavities in the polyurethane, affecting the hardness and density [5].

Table 5. The accuracy of the propellant density measurement against the theoretical value.

Number of cylinders	Accuracy (%)			
	t_0	t_2	t_5	t_{10}
1	99.12	99.13	99.11	99.00
2	99.03	99.03	98.99	98.85
3	98.80	98.79	98.74	98.61
4	99.26	99.25	99.21	99.06
5	99.28	99.28	99.23	99.08
6	98.96	98.96	98.92	98.76
Average	99.08	99.07	99.03	98.89

The density values at all immersion times were compared to the theoretical value to obtain accuracy. The accuracy decreased as more AP particles dissolved in the water as a result of increasing the immersion time. Since there are considerably small dissolved AP particles were present when the measurements were made, the density measurement at t_0 had the highest theoretical accuracy value, as listed in Table 5.

4 Conclusion

The research that we conducted confirms that the hydrostatic weighing method is still applicable to measure propellant density; however, the result is affected by the immersion times. The immersion time on the hydrostatic weighing procedure has an impact; the longer the immersion time, the more significant loss will be experienced, and the less propellant density and accuracy will be achieved. It was discovered that there were no significant differences in propellant density value at a maximum immersion time of five seconds (t_5) with an accuracy level greater than 99% and a density loss rate of 0.023%. However, immersion times longer than five seconds will considerably decrease the density value.

References

1. Isert, S., Hedman, T.D., Lucht, R.P., Son, S.F.: Oxidizer coarse-to-fine ratio effect on microscale flame structure in a bimodal composite propellant. Combust. Flame **163**, 406–413 (2016). https://doi.org/10.1016/j.combustflame.2015.10.015
2. Restasari, A., et al.: Effect of aluminium content on pseudoplasticity involving wall slip of high density composite solid propellant (HD-CSP) slurries. Iran. Polym. J. (English Ed.) **32**(10), 1347–1357 (2023). https://doi.org/10.1007/s13726-023-01213-5

3. Gupta, B.L., Kumar, V., Shivhare, N.: Rheological Studies On Virgin, Plasticized And Solid Filled HTPB Binder System. Glob. J. Adv. Eng. Technol. Sci. **1**(2), 41–48 (2014). http://www.gjaets.com
4. Abdillah, L.H., Restasari, A., Hartaya, K., Budiman, Y.: The selection of composite solid propellant process condition based on flow characteristics of propellant slurry. In: AIP Conference Proceedings. AIP Publishing (2020). https://doi.org/10.1063/5.0002342
5. Restasari, A., Abdillah, L.H.: Pengaruh clearance impeller dan viskositas HTPB terhadap polimerisasi binder propelan. In: Astuti, R.A., Marta, A. (eds.) Prosiding SIPTEKGAN XXI, Bogor: Pusat Teknologi Penerbangan Deputi Bidang Teknologi Penerbangan dan Antariksa Lembaga Penerbangan dan Antariksa Nasional (LAPAN), pp. 331–338 (2017)
6. Sperling, L.H.: Introduction to Physical Polymer Science, 4th edn. Wiley, Hoboken (2006)
7. Abdillah, L.H., Winardi, S., Sumarno, S., Nurtono, T.: Effect of mixing time to homogeneity of propellant slurry. In: IPTEK Journal of Proceedings Series, p. 94 (2018). https://doi.org/10.12962/j23546026.y2018i1.3515
8. Muthiah, R., Manjari, R., Krishnamurthy, V., Gupta, B.: Rheology of HTPB propellant: effect of mixing speed and mixing time. Def. Sci. J. **43**(2), 167–172 (1993)
9. Muthiah, R., Krishnamurthy, V.N., Gupta, B.R.: Rheology of HTPB propellant. I. Effect of solid loading, oxidizer particle size, and aluminum content. J. Appl. Polym. Sci. **44**(11), 2043–2052 (1992). https://doi.org/10.1002/app.1992.070441120
10. Nair, C.P.R., Devi, C.H., Prasad, V., Ninan, K.N.: Effect of process parameters on the viscosity of AP / Al / HTPB based solid propellant slurry. J. Energy Chem. Eng. **1**(1), 1–9 (2013)
11. Platzek, H.: Preliminary solid rocket motor design techniques, California (1979). https://doi.org/10.1017/CBO9781107415324.004
12. Davidson, S., Perkin, M.: An investigation of density determination methods for porous materials small samples and particulates. In: IMEKO 2010 TC3, TC5 and TC22 Conferences Metrology in Modern Context, pp. 69–72 (2010)
13. Fujii, K.: Precision density measurements of solid materials by hydrostatic weighing. Meas. Sci. Technol. **17**(10), 2551–2559 (2006). https://doi.org/10.1088/0957-0233/17/10/004
14. Trache, D., et al.: Effect of amide-based compounds on the combustion characteristics of composite solid rocket propellants. Arab. J. Chem. **12**, 3639–3651 (2015). https://doi.org/10.1016/j.arabjc.2015.11.016
15. Verma, S., Ramakrishna, P.A.: Dependence of density and burning rate of composite solid propellant on mixer size. Acta Astronaut. **93**, 130–137 (2014). https://doi.org/10.1016/j.actaastro.2013.07.016
16. Toft, H.O.: Voids in solid propellants, no. June. DANSK AMATOR Raket Klub (2002)
17. Van Keulen, J.: Density of porous solids. Matériaux Constr. **6**(33), 181–183 (1973). https://doi.org/10.1007/BF02479031
18. Prianto, B., Setyaningsih, H., Pinalia, A.: Preliminary results of the particle packing density theory for improving the theoretical density of composite propellant. In: AIP Conference Proceedings, vol. 2366, no. September (2021). https://doi.org/10.1063/5.0060448
19. Hartaya, K., Budi, R.S., Budiman, Y., Restasari, A., Ardianingsih, R., Abdillah, L.H.: Metode pembuatan propelan padat komposit dengan pengaduk bilah horisontal (horizontal blade mixer) Rpm 60, IDS0000029 (2020)
20. D. S. A Military Standard Propellants, Solid : Sampling, Examination and Testing, no. October (2010)
21. ASTM D792-08: Standard test methods for density and specific gravity (Relative density) of plastics by displacement. Am. Soc. Test. Mater. 6 (2008)
22. Sugiyono, Statistika Untuk Penelitian, vol. 19, no. September. Bandung: Alfabeta (2011)

23. Pinalia, A.: Kristalisasi Ammonium Perklorat (Ap) dengan sistem pendinginan terkontrol untuk menghasilkan kristal berbentuk bulat. J. Teknol. Dirgant. **9**(2), 124–131 (2011). https://doi.org/10.30536/j.jtd.2011.v9.a1680
24. Savjani, K.T., Gajjar, A.K., Savjani, J.K.: Drug solubility: importance and enhancement techniques. ISRN Pharm. **2012**(100 mL), 1–10 (2012). https://doi.org/10.5402/2012/195727

Curve Analysis of Motor Rocket Static Test Results of K-Round Scale Composite Solid Propellant

Kendra Hartaya[1], Retno Ardianingsih[1(✉)], Luthfia Hajar Abdillah[1], Yudha Budiman[1], Afni Restasari[2], Bagus Wicaksono[1], and Hamonangan R. Sitompul[1]

[1] Research Center for Rocket Technology BRIN, Bogor 16350, Indonesia
{kend001,retn014,luth002,yudha.budiman,bagu005, hamo001}@brin.go.id

[2] Research Center for Biomass and Bioproducts BRIN, Bogor 16915, Indonesia
afni.restasari@brin.go.id

Abstract. An analysis of the thrust versus combustion time curve has been carried out from the static test results of a rocket motor with a small-size solid propellant (K-round). It was conducted by observing the similarity of the shape of the test result's curve, the area under the curve, the slope of the line of sudden pressure increase, and the appearance at the beginning of the sudden pressure increase. As a comparison, the curve from the results of the static test of a rocket motor with a homogeneous propellant mixture is presented using a standardized process. From the analysis, we obtained that the rocket motor with the homogeneous propellant mixture of the same propellant composition did not give similar static test curve shapes, thus yielding a different area under the curve due to the Isp value difference. We also found that the slopes of the sudden pressure rise line are not the same, although the appearance at the beginning of the sudden rise is similar. The value (occurrence) of the start of the sudden rise is suspected to be related to the initial igniter, not to the propellant.

Keywords: Curve Analysis · Static Test · Composite Solid Propellant

1 Introduction

Propellant is a type of rocket fuel known for its high energy output. It exists in two primary forms: solid and liquid. Solid propellants can be classified into homogeneous and heterogeneous (or composite) types. Composite propellants are favored for long-range and large rockets, whereas homogeneous propellants are typically employed in short-range, high-speed rockets like missiles and defense projectiles. Composite propellants are usually made of several components that vary depending on the intended use. Based on the composition, it consists of major and minor components. Based on the composition, it consists of major and minor components. Major components include an oxidizer, binder, and metal, while minor components include a curing catalyst, plasticizer, bonding agent, and burn rate modifier. Based on its phase, it consists of liquid and

E. Fitrianingsih et al. (Eds.): ISAST 2024, SPPHY 416, pp. 466–477, 2025.
https://doi.org/10.1007/978-981-96-1344-1_49

solid content. Solid contents include particles of oxidation and metal. Meanwhile, liquid content or binder systems include a polymer component that builds a matrix network [1]. Variations of components are carried out to achieve the desired properties, such as the value of the burn rate, the value of Isp, the viscosity of the mixture, homogeneity, etc. The variations involve the shape and size of the constituent components, purity, and ratio between components. The variations in the period (length of time) of the overall process or process stage periods are also carried out [2].

The composite propellant is made of Hydroxyl-terminated polybutadiene (HTPB) as a binder and ammonium perchlorate (AP) as an oxidizing agent. Aluminum powder is added to increase the flame temperature and increase the specific impulse (Isp) [3]. The mixture is then poured into a propellant cast. A mandrel is inserted to shape the propellant grain as desired. The propellant is then cured in an oven at a specific temperature and duration until it hardens and maintains its form. Finally, the propellant is extracted from the mandrel and assembled into a rocket motor for performance evaluation [4]. The use of metallic iron nanoparticle additives has also been reported to increase the combustion rate of this composite propellant [5].

The primary energy characteristic of solid propellants is the specific impulse of the reactive force, which is equal to the increase in thrust that occurs during combustion per unit mass of the propellant. This Isp measurement can be carried out on a lab scale using a device that measures the reactive force of the gasification of the propellant sample with a mass of 1–5 g [6]. In addition, it can also be determined from the performance of the rocket motor through static testing. This performance value is obtained from the propellant that has been integrated with the rocket motor. The rocket motor consists of several main parts, including the rocket motor tube, propellant grain, nozzle and cap, heat shield (liner/inhibitor), and igniter [7]. These parts have varying types and conditions depending on the type of rocket to be manufactured, for example, the shape and size of the tube, nozzle, or cap, the shape of the propellant grain and the composition of the material, liners that are used can also vary in composition. For small rocket motor types, the K-Round scale propellant usually uses the dimensions of the inner diameter, outer diameter, and length of 26 mm, 59 mm, and 200 mm, respectively [8].

In the design of solid rocket motors, thrust stability is a complex phenomenon that is closely linked to the underlying combustion dynamics within the motor. The supply of combustion material, which dictates the thrust output, heavily depends on the conditions in the combustion chamber. Specifically, the pressure and temperature at the propellant surface play a crucial role in determining the rate of propellant supply, which in turn affects the thrust. Furthermore, [9] and [10] reveal the utility of propellant performance testing in measuring key parameters such as thrust, torque, and efficiency curves. Similarly, the shape of the pressure-time curve for the solid rocket motor can provide valuable insights into combustion stability and pressure regulation [11].

Solid rockets are manufactured to achieve a high combustion temperature along with the lowest molecular weight in the exhaust like other rockets. The difficulty with solid oxidizing agents is that most are inorganic and contain metal atoms, which cause high molecular weight in the exhaust. Another problem is that some combustion products can form solid particles at exhaust temperatures. This affects the nozzle's performance in converting heat energy into gas flow. All these things affect the performance of solid rocket motors [12].

Figure 1 shows various internal burning sections, their partially burned shapes, and their resulting thrust profiles. High-thrust rockets, which need a large area of burning surface, are typically of the internal-burning type. For lower thrust, solid cylindrical grains that burn only on the end surface can be used.

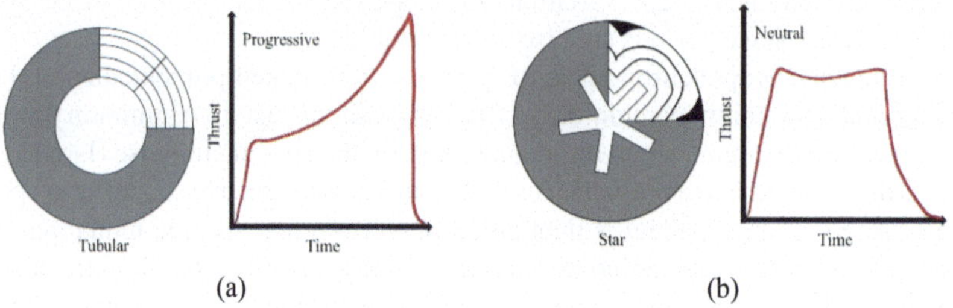

Fig. 1. Illustration of internal-burning charge designs along with their thrust-time profiles for (a) tubular grain; (b) star grain. Modified from [13]

An evaluation of the effect of chamber pressure on the ignition delay has also been carried out. The throat diameter varies according to the igniter's pressure rate and peak pressure. The result is that the ignition delay increases exponentially with the addition of the nozzle diameter [14].

This paper conducts a curved analysis of the results obtained from the static testing of the K-Round rocket motor, which utilizes a composite solid propellant. This analysis is crucial for several reasons, such as serving as a foundation for increasing the Isp, underscoring the need to standardize the technology in manufacturing propellants and igniter fillings, and assessing whether similar curves exist for propellants of identical composition. Through this analysis, one can determine the delay time, evaluate the performance consistency of propellants with the same composition, assess the quality of the igniter performance, and examine the homogeneity of both the propellant and the initial igniter.

2 Methodology

The methodology used in this study includes a careful approach to propellant manufacture and analysis of static test results. The propellant manufacturing process was structured by mixing HTPB and Al in a mixer, followed by timed stirring to achieve optimal consistency. All materials were mixed in a batch system using a planetary-type mixer. They included using a horizontal sigma blade, vacuum equipment, applying temperatures of 50–55 °C, and a high rotational speed of 60 rpm [15]. Subsequently, 200 μm AP was incrementally added into the mixture, each stage requiring precise stirring durations to form propellant slurry [16]. Subsequently, to achieve optimal consistency and performance, TDI was added to the mixture as a curing agent that would react with HTPB to form polyurethane. A polyol reacts with extra diisocyanate to make polyurethane (PU) chains with isocyanate groups at the ends. The PU chains are joined by reactions between isocyanate and amine groups. Polyurethane can be adjusted to achieve the desired mechanical characteristics [17]. Once the mixing phase was completed, the propellant slurry was cast into the mold tube, a critical step in ensuring uniform combustion characteristics. After the propellant slurry is poured into the mold tube, a mandrel with a specific shape is inserted through the center of the slurry column. In the present study, a cylindrical mandrel configuration was used. The maturation of the propellant was then completed carefully through curing process in oven with temperature of 60–70 °C for 24 h. Upon completion of the curing process, the propellant was carefully removed from the mold (decorating process) and the result of propellant form after cured can be seen in Fig. 3 part (a). The propellant underwent rigorous testing, including X-ray analysis and static testing, to assess its structural integrity and performance characteristics. The steps of the process are illustrated in Fig. 2.

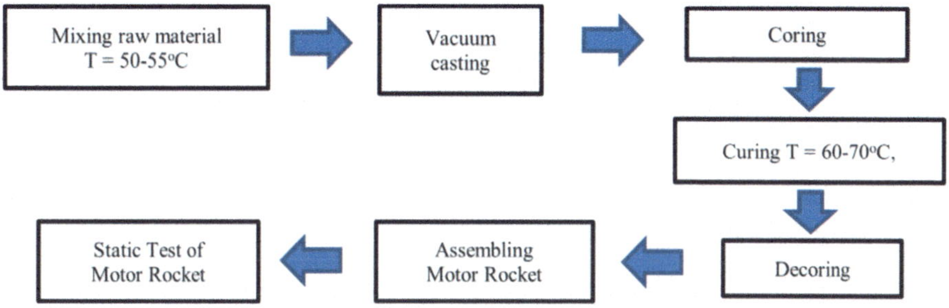

Fig. 2. Flowchart for the K-round propellant production process.

Following the manufacturing phase, the study analyzed the results obtained from the static tests. This analysis involved a comprehensive examination of the curve data generated during static firing tests, aiming to derive crucial insights into the propellant's performance characteristics under various conditions. The analysis encompassed detailed parameters such as thrust development over time, pressure variations, and other critical performance indicators. Through this approach, the study aimed to validate the manufacturing propellant process and deepen understanding of the propellant's performance dynamics. This study's methodological approach to curve analysis involved a systematic evaluation of various parameters inherent in the pressure-time curves obtained during testing. First, the analysis scrutinizes the similarities among multiple curves and investigates what differentiates each, thus revealing the crucial patterns and trends to understanding propellant performance. Additionally, the examination delved into the Isp exhibited by each curve, indicative of propulsion efficiency and effectiveness.

Figure 3 part (b) provides a top view of the motor rocket using the K-round propellant, offering insight into its structure and configuration. Finally, part (c) depicts the motor rocket with K-round propellant positioned on the test bed, fully prepared for the upcoming static test. This analysis provided invaluable data regarding the propellant's capacity to convert chemical energy into thrust under varying operational conditions. Furthermore, the study carefully investigated instances of sudden pressure rises within the curves and identified transient spikes that could signify operational anomalies or characteristics relevant to performance optimization. Moreover, the analysis extends to assessing ignition delay times across the curves, offering detailed insights into the propellant's responsiveness to ignition stimulus. This aspect of the examination lights on crucial time intervals governing the onset and sustainability of combustion processes, thereby influencing overall rocket performance.

(a) (b) (c)

Fig. 3. (a) Casted K-round scale propellant with hollow grain, (b) Top view of motor rocket K-round propellant, (c) Motor rocket of K-round propellant in the test bed, ready for static test.

3 Result and Discussion

Manufacturing composite solid propellants in one batch process can produce three cast propellants. For small-sized propellant samples (K-round), the shape is cylindrical with a hollow grain (the center of the cylinder is hollow). The grain shape of the propellant greatly affects the performance of static test results. Grains with a specific design can influence the burning rate, generated pressure, and combustion stability. For example, tubular grains usually have a larger surface area, which can increase the burning rate compared to hollow grains. By analyzing static test results based on grain shape, engineers can optimize propellant design to achieve desired performance characteristics, such as high efficiency, stable thrust, and appropriate burning duration. Static testing is carried out by assembling a cylindrical propellant into a rocket motor and connecting it to a test device. Data that can be obtained from the results of static tests usually include Isp, thrust, and burning time.

The following discussion analyzes the static test result curves representing the relationship between thrust and burning time. The curves of K-round scale rocket motors with composite solid propellants at the same composition in different batches are depicted in Fig. 4. The analysis of the static test results for a K-round scale composite solid propellant motor rocket is a crucial step in understanding the performance characteristics of the propulsion system. This paper presents a comprehensive study of the curve analysis, focusing on the similarity of the shape of the curve, the area under the curve, the slope of the line of sudden pressure increase, and the appearance at the beginning of the sudden pressure increase.

Theoretically, the increase in the aluminum content in the propellant will increase the Isp value. Increasing the aluminum content will boost the combustion energy released in the reaction.

$$2\,NH4C1O4 + 4\,A1 \rightarrow 2\,A12O3 + 2HC1 + N2 + 2H2O + H2 + 2{,}5\,\text{kcal/gr} \quad (1)$$

However, A rise in Isp might also result from other factors, such as a reduction in the size of the ammonium perchlorate oxidizer or the addition of a combustion catalyst that enhances the burning rate. Therefore, the Isp improvement is not solely attributable to the increased aluminum content in the propellant. The result indicates that changes in aluminum levels do not consistently lead to corresponding increases or decreases in the Isp, suggesting no clear relationship exists between the two. The inconsistent quality of the raw materials, which are not domestically produced, results in a slight specification variation. This also contributes to the behavior where consistent Isp results can not be achieved. As a result, when the product is re-manufactured, the Isp results differ significantly from the previous Isp, even with the same aluminum content.

Aluminum particles can react not only with free oxygen resulting from oxidizer decomposition but also with inert decomposition gaseous products and add much more heat to the combustion process [18, 19]. This series of exothermic combustion reactions could increase the combustion temperature and the average operating pressure inside the rocket motor. Operating pressure and flame temperature are the most critical factors that affect the combustion mechanism [20].

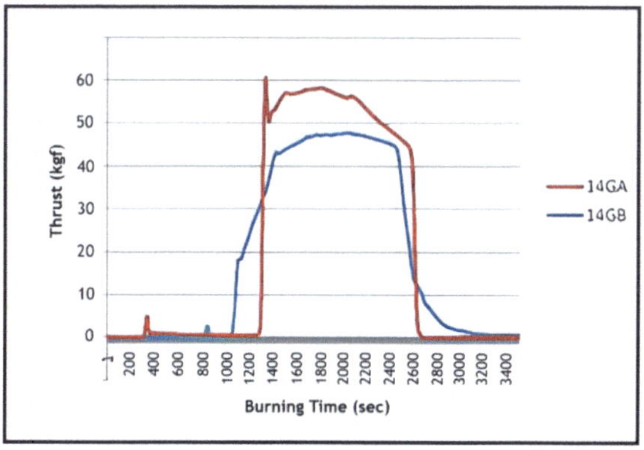

Fig. 4. Curves of K-round scale rocket motors with composite solid propellants with similar compositions (14 GA and 14 GB).

From Fig. 4, the sample with the code 14 GA and 14 GB results from the two-batch propellant process of the same composition. The shape of this curve is not the same, even very different. Regarding burning time between 1366 and 2536, 14 GA has a greater thrust than 14 GB. Table 1 states that the 14 GA propellant sample has an Isp of 206,2441 s compared to 14 GB, which is 172,8744 s. Both samples have a difference in Isp values of more than 5 s even though they come from the same composition and the different batch.

Considering that the samples from different batches had similar compositions, it can be concluded that the propellant mixture was not homogenous. So, the Isp values of 14 GA and 14 GB do not represent the Isp values in the composition because the cylinders of the 14 GA have different Isp values from 14 GB. The inhomogeneity is also shown more clearly in Fig. 5. These two curves are also from the same production batch.

14L MR1 and 14L MR2 have similar shapes; however, both have different areas so the Isp value is different.

Table 1. Static test characteristic value of K-round scale rocket motors with composite solid propellants at the same composition in different batches.

Characteristic	14 GA	14 GB
Average Thrust (kgf)	53.7437	41.5751
Burning time (second)	2.540	2.758
Total Impulse (kgf.s)	136.5089	114.6641
Propellant Weight (kg)	0.66188	0.66328
Specific Impulse (second)	206.2441	172.8744
Maximum Thrust (kgf)	60.662	47.838

The homogeneity of propellant mixtures has directly impacted their performance and safety. One factor that could contribute to the lack of homogeneity in propellant mixtures is the design and implementation of the content uniformity test. Proper formulation and processing considerations are crucial to ensure the homogeneity of the final interactive powder mixtures, as the presence of high-dose units can pose significant risks. Even if propellant mixtures have the same composition, variations across different batches can arise due to factors such as manufacturing variability, ingredient quality, mixing method, environmental conditions, storage and handling, scale of production, and human factors. A systematic evaluation of the synthesis and processing parameters that contribute to the variability of different batches is crucial to address these challenges.

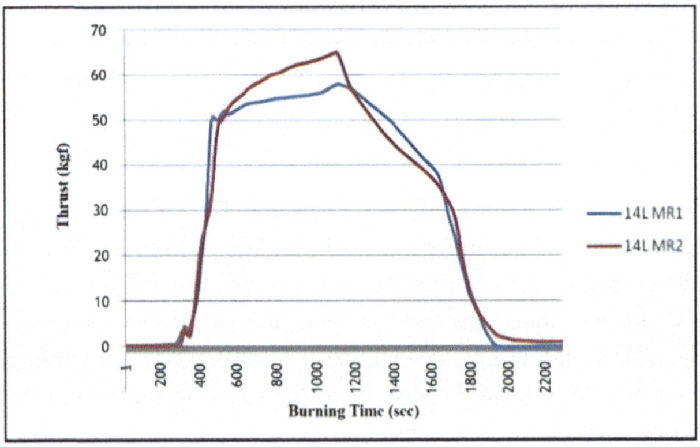

Fig. 5. Curves of K-round scale rocket motors with composite solid propellants with the same composition and batch (14L MR1 and 14L MR2).

Figure 5 illustrates the static test result curve, showing the relationship between thrust and burning time. The curves of K-round scale rocket motors with composite solid propellants at the same composition in the same batch are depicted in Figs. 5, 6 and 7. The Isp is the area inside the curve or the area above the x-axis (burn time) upwards bounded by the thrust curve. It can be estimated that the area under the red curve is proportional to the area under the blue curve in Fig. 4.

The combination of the area on the right and the left between the red and blue curves will cover the area at the top between the red and blue. Therefore, the Isp of these two curves should be close. The difference in Isp values indicates whether the propellant is homogeneous or not. If the difference in Isp values is not more than 5 s, this is considered a homogeneous propellant mixture. The Isp value is 189,3115 s for 14L MR1 propellant, while 14L MR2 is 200,5456 s as mention in Table 2. There has been a difference in the Isp value and also total impulse as in Table 2, which is more than 5 s even though the propellant comes from the same composition and the same production batch. Thus, the propellant with the curve in Fig. 5 is also not a homogeneous mixture because it has an Isp range of more than 5 s.

Table 2. Static test characteristic value of K-round scale rocket motors with composite solid propellants with the same composition from the same batch (14L MR1 and 14L MR2).

Characteristic	14L MR1	14L MR2
Average Thrust (kgf)	49.7074	50.3882
Burning Time (second)	2.588	2.692
Total Impulse (kgf.s)	128.6428	135.645
Propellant Weight (kg)	0.67953	0.67638
Specific Impulse (second)	189.3115	200.5456
Maximum Thrust (kgf)	57.804	64.861

In Fig. 6, the Isp values of the three rocket motors differ by less than 5 s, so it is concluded that the propellant is a homogeneous mixture. One of the three rocket motors has a lower sudden pressure rise slope than the other two. Therefore, it is necessary to study the homogeneity of the igniter filling mixture. The homogeneous propellant is also shown in Fig. 7. The three rocket motors have a similar curve shape, and the difference in Isp value is 5 s.

A sudden increase in pressure is a curve that rises suddenly at the start of combustion. The difference in slope is due to changes in pressure in the combustion chamber. Since most of the propellant's energy is generated from aluminum combustion, this slope change might result from the non-homogeneous propellant mixture. Since the propellant mixture is not homogeneous, the Isp value of propellant sample that depict on Fig. 4 and Fig. 5 differs more than 5 s. In contrast to the propellant samples labeled 16F1 MR1, 16F1 MR2, and 16F1 MR3 as illustrated on Fig. 6, which are all derived from the same composition and production batch, the static test results indicate that their Isp values differ by no more than 5 s, as demonstrated in Table 3, approximately in 204–206 s.

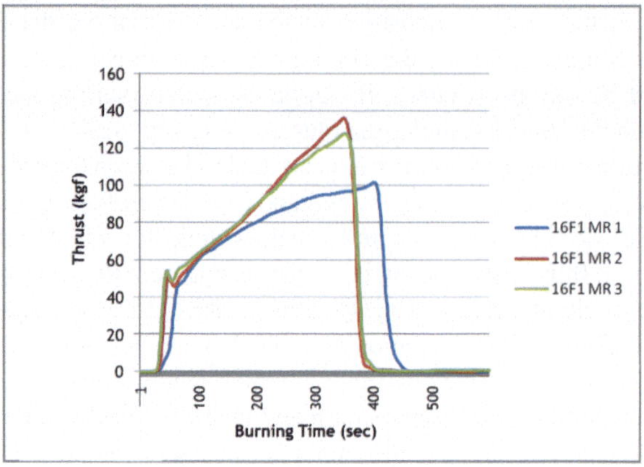

Fig. 6. Curves of K-round scale rocket motors for 16F1 MR1, 16F1 MR2, 16F1 MR3.

The sudden increase in pressure can also measure the performance of the initial combustion igniter if the propellant used in the rocket motor is a homogeneous mixture of propellant with a well-distributed component.

Table 3. Static test characteristic value of K-round scale rocket motors for 16F1 MR1, 16F1 MR2, 16F1 MR3.

Characteristic	16F1 MR1	16F1 MR2	16F1 MR3
Average Thrust (kgf)	79.8496	89.182	88.2883
Burning Time (second)	1.78	1.575	1.605
Total Impulse (kgf.s)	142.1323	140.4616	141.7028
Propellant Weight (kg)	0.68674	0.68808	0.68632
Specific Impulse (second)	206.9666	204.1356	206.4675
Maximum Thrust (kgf)	101.67	136.23	127.97

Homogeneous and heterogeneous propellant mixtures can exhibit distinct performance characteristics in rocket motors. The composition and distribution of the propellant components can impact factors such as pressure changes during combustion, which may serve as a metric for evaluating the performance of the initial ignition source. In a homogeneous propellant blend, the components are evenly distributed, allowing the initial combustion igniter to efficiently ignite the mixture and cause a sudden increase in pressure. This pressure change can then be used to assess the performance of the ignition system. The complex interactions between various factors, including the supply of combustion material, pressure, and temperature, can lead to a rise and spread of instability within the combustion chamber. Accurate control and measurement of propellant flow is essential for the stable operation of solid rocket motors, as it directly impacts the thrust output.

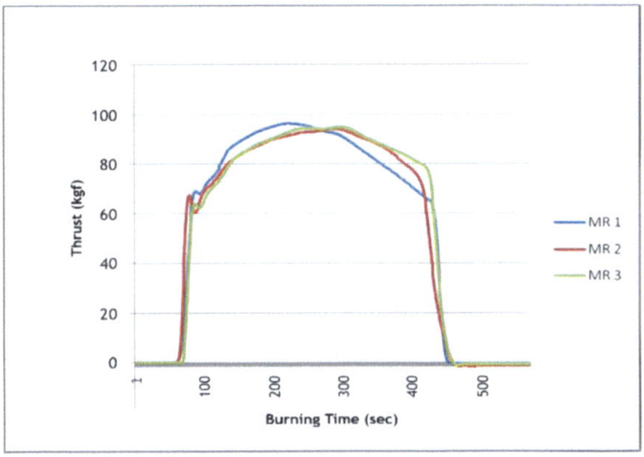

Fig. 7. Curves of K-round scale rocket motors for MR1, MR2, MR3

Table 4. Static test characteristic value of K-round scale rocket motors for MR1, MR2, MR3.

Characteristic	MR1	MR2	MR3
Average Thrust (kgf)	83.0419	84.0419	84.7646
Burning time (second)	1.8	1.73	1.755
Total Impulse (kgf.s)	149.4755	145.4049	148.7619
Propellant Weight (kg)	0.68448	0.6834	0.68588
Specific Impulse (second)	218.3782	213.7669	216.8920
Maximum Thrust (kgf)	96.368	94.191	94.996

Determination of ignition delay time starts from when there is an indication of combustion (the appearance of a curved line) until the appearance of a "sudden increase in pressure" (with a steep line slope approaching 90 degrees). The difference in the time of sudden pressure emergence indicates a pressure difference that produces a fairly large thrust. This can be caused by an inhomogeneous propellant mixture and/or an uneven igniter flame which results in uneven spraying. Therefore, the performance of the igniter must also be homogeneous to obtain the appropriate delay time for the propellant mixture used for testing. It can be seen in Fig. 7 that the startup delay time on sample code MR1, MR2 and MR3 are quite similar. This correlates with the difference in total impulse and also the Isp values of no more than 5 s, as seen in Table 4. All three propellant cylinder have the sudden pressure increase of the three curves starts at the same time (as the end point of the delay time measurement). There is one slope of the curve's starting line, which is less likely due to differences in igniter homogeneity. This is different with Figs. 4 and 5 where the start up delay time is not the same.

The complex interactions between various factors, including the supply of combustion material, pressure, and temperature, can lead to a rise and spread of instability within the combustion chamber. Accurate control and measurement of propellant flow is also essential for the stable operation of solid rocket motors, as it directly impacts the thrust output.

4 Conclusion

From the analysis of the curve of the results of the rocket motor static test with a small-size propellant (K-round), it can be concluded that:

- The composite solid propellant from a homogeneous mixture propellant produces three curves with similar shapes, and the area under the curve is the same as the Isp value, with an Isp value range of 5 s. Also, the slope and "starting time" of the sudden pressure rise line are the same and are independent of the propellant.
- An inhomogeneous mixture of propellants results in an unequal shape and or curve area of the rocket motor static test outcome.
- The appearance of a "starting time" line of sudden pressure rise is thought to be a factor from the initial ignition with an inhomogeneous filling, not from the propellant. It is reflected in the uneven igniter flame, resulting in the slope of the line of sudden pressure rise not being the same at the same propellant composition.

Acknowledgment. The authors would like to thank the Research Center for Rocket Technology, the Research Organization for Aeronautics and Space, National Research and Innovation Agency (BRIN) for facilitating the research.

References

1. Restasari, A., Ardianingsih, R.: Effect of NCO/OH-Aluminum on hardness for mixing evaluation from view of composition design of solid propellant. IP Conf. Proc. **2366**, 040010 (2021). https://doi.org/10.1063/5.0061228.(2021)
2. Hartaya, K.: Penyempurnaan Proses Pembuatan Propelan Komposit Lapan Berdasar Hasil Penelitian Dari India (Process Improvement of Lapan's Composite Propellant Preparation Based on Research Result From India). Jurnal Teknologi Dirgantara **13**(2), 121–130 (2015)
3. Hartaya, K.: Analisis Kandungan Aluminium Powder Propelan Berdasar Energi Pembakaran dari Bomb Kalorimeter. Jurnal Teknologi Dirgantara **14**(1), 73–80 (2016)
4. Susanto, A., Abdillah, L.H.: Propelan Dan Teknologi Pembuatannya. *Berita Dirgantara*, **15**(2), 50–57 (2014) http://103.16.223.15/index.php/berita_dirgantara/article/view/2105/1913
5. Styborski, J.A., Scorza, M.J., Smith, M.N., Oehlschlaeger, M.A.: Iron nanoparticle additives as burning rate enhancers in AP/HTPB composite propellants. Propellants Explos. Pyrotech. **35**, 1–8 (2010). https://doi.org/10.1002/prep.201400078
6. Arkhipov, V.A., Kiskin, A.B., Zarko, V.E., et al.: Laboratory method for measurement of the specific impulse of solid propellants. Combust Explos Shock Waves **50**, 622–624 (2014). https://doi.org/10.1134/S0010508214050177
7. Sutrisno, -. Uji Coba Peluncuran Motor Roket Tahun 2000. Berita Dirgantara, **1**(1), 10–14 (2000). http://kliping.lapan.go.id/index.php/berita_dirgantara/article/view/672
8. Abdillah, L.H., Ardianingsih, R.: Peninjauan Homogenitas Propelan Skala K-Round Berdasarkan Varians Nilai Kalor (Homogeneity Observation of Propellant K-Round Scale Based on Calorific Value Variance). Jurnal Teknologi Dirgantara **12**(1), 11–18 (2014)

9. Snyder, J.S., et al.: Recommended practice for flow control and measurement in electric propulsion testing. J. Propuls. Power (2016). https://doi.org/10.2514/1.B35644
10. Dantsker, O.D., Selig, M.S., Mancuso, R.: A rolling rig for propeller performance testing. In: 35th AIAA Applied Aerodynamics Conference. AIAA, pp. 2017–3745 (2017). https://doi.org/10.2514/6.2017-3745
11. Fahd, A., Mostafa, H.E., Elbasuney, S.: Certain ballistic performance and thermal properties evaluation for extruded modified double-base propellants. Central Europ. J. Energ. Mater. **14**(3), 621–635 (2017). https://doi.org/10.22211/cejem/70206
12. Turner, M.J.L.: Rocket and spacecraft propulsion: principles, practice and new developments. (J. Mason, Ed.) (3rd ed.). Leichester, UK: Springer in association with Praxis Publishing (2009). https://doi.org/10.1017/CBO9781107415324.004
13. Hill, P.G., Peterson, Carl R.: Mechanics And Thermodynamics Of Propulsion. Addison-Wesley Publishing Company (2nd ed.) (2017)
14. Risha, G.A., et al.: Aluminum- Ice (ALICE) Propellants for Hydrogen Generation and Propulsion. American Institute of Aeronautics and Astronautics, (August 2009), 4877–4877. (2009)
15. Restasari, A., et al.: hixotropic Behavior in Defining Particle Packing Density of Highly Filled AP/HTPB-Based Propellant. Symmetry **13**, 1767 (2021). https://doi.org/10.3390/sym13101767.(2021)
16. Isert, S., Hedman, T.D., Lucht, R.P., Son, S.F.: Oxidizer coarse-to-fine ratio effect on microscale flame structure in a bimodal composite propellant. Combust. Flame **163**, 406–413 (2016). https://doi.org/10.1016/j.combustflame.2015.10.015
17. Restasari, A., Naderizadeh, S., Kaur, A., Akutagawa, K., Busfield, J., Tan, Wei., Characterisation of Polyurethane-urea as LeadingEdge Erosion Resistant Materials of Wind Turbine Blades. IOP Conf. Ser: Earth Environ. Sci. **1344**, 012015 (2024). https://doi.org/10.1088/1755-1315/1344/1/012015
18. Yen, N., Wang, L.Y.: Reactive Metals in Explosives. Propellants Explos. Pyrotech **37**, 143–155 (2012)
19. Zarko, V., Gromov, A.: Energetic Nanomaterials. Synthesis, Characterization and Application. Elsevier, Amsterdam 2016 (2016). ISBN 9780128027103
20. Hayri Yaman, V., Degirmenci, E.: Experimental Investigation of the Factors Affecting the Burning Rate of Solid Rocket Propellants. Fuel **2014**(115), 794–803 (2014)

Preliminary Research Optimization of Vacuum Vibration Casting System on Case-Bonded Propellant

Prawita Dhewi(✉), Henny Setyaningsih, and Anita Pinalia

Research Center for Rocket Technology BRIN, Bogor 16350, Indonesia
{praw004,henn002,anita.pinalia}@brin.go.id

Abstract. Composite solid propellants consist of two types of materials: solid and liquid. Solid ingredients include oxidizers, metal fuels, and additives, whereas liquid materials include fuel binders, curing agents, and plasticizers. To produce a composite solid propellant, all these ingredients are mixed homogeneously to make a propellant slurry, which is then poured into a molded tube. The process is known as the casting process. The casting process consisted of two stages: putting the slurry into the mold tube under vacuum conditions and then putting the mandrel into the tube before the propellant slurry solidifies. This conventional vacuum casting technique has limitations when applied to viscous propellant slurries due to the rheological behavior of the slurry. This research aims to enhance vacuum casting by incorporating vibration into the process. This technology allows the composite propellant slurry to fill the gap between the tube and the mandrel, making the molding process easier and more efficient and developing a propellant with extremely low porosity to achieve better mechanical properties.

Keywords: Propellant · Casting · Vacuum · Vibration

1 Introduction

Composite solid propellants are comprised of fuel and an oxidizer, bounded by a synthetic or plastic binder matrix. This specific type of propellant offers a significant amount of propulsive energy, which has made it a well-liked option for usage in rockets, satellites, and cruise missiles [1]. A solid composite propellant has a composite structure composed of fundamental elements from both the liquid and solid phases [2]. Ammonium perchlorate (AP) and HTPB (hydroxyl-terminated polybutadiene) are the primary components of composite solid propellants. Composite solid propellants have a high energy density, are easily produced, and have good mechanical properties. Consequently, they are commonly utilized as propellants for composite solid rockets and missiles [3].

Rocket engines with solid propellants are frequently selected for space launches and weapons systems because of their cost-efficiency, competitiveness, and reliability. Ammonium perchlorate is commonly employed as an oxidizer in propellant formulations because of its substantial energy content. The characteristics of the substance

E. Fitrianingsih et al. (Eds.): ISAST 2024, SPPHY 416, pp. 478–487, 2025.
https://doi.org/10.1007/978-981-96-1344-1_50

include good thermal stability, specific impulse, density, low shock sensitivity, excellent compatibility, and long shelf life [4–7].

The performance of the propellant is determined by various characteristics, precisely the burning speed and specific impulses (Isp). To enhance the combustion temperature, density, and specific impulses, the addition of metal fuel to the propellant is required [8]. The precise impulses of each propellant differ based on the kind, composition, and grain utilized. Two methods for pouring the propellant slurry into the mold are free-standing and case-bonded. However, between the two methods, the case-bonded is more commonly used.

The production method involves casting the propellant slurry in a case-bonded way, carried out in a vacuum environment, a widely used traditional technique. However, there are also additional drawbacks to this method, including the formation of porous propellants. The porous section of the propellant does not fulfill the necessary criteria for a propellant suitable for use in a rocket engine, and the porous component must be disposed of as a byproduct of a composite propellant. Vibration vacuum casting is utilized as a novel manufacturing technology to reduce the size and quantity of pores in a compound propellant. This method employs vacuum and vibration techniques to improve the mechanical strength of the composite propellants. Furthermore, the propellant generated is nearly impermeable, so it effectively minimizes propellant waste since there is no requirement to dispose of a substantial amount of porous propellant. This aligns with prior studies that suggested a combined approach of vacuum and vibratory techniques for molding composite propellants with a seven-star grain structure [9].

This research aims to develop innovative technology in the casting process for producing propellant with little or no porosity. This will decrease propellant waste and production costs, as a significant amount of porous propellant that fails to meet the requirements for rocket engines will be eliminated. Porosity in propellants can induce undesired pressure accumulation due to heightened surface burning during combustion, leading to problems in both testing and operation.

2 Methodology

2.1 Production of a Composite Solid Propellant

The production of a composite solid propellant involves multiple phases, including mixing propellant slurry, casting propellant slurry using vacuum vibrating technology, and curing process. Mixing slurry propellants involves the successive stages of eliminating the moisture from the raw materials for ammonium perchlorate by submitting them to an oven-drying process at a temperature of 100 °C for 2 h. Subsequently, produce a propellant slurry by meticulously combining all the constituent elements, which encompass ammonium perchlorate (AP), hydroxyl-terminated polybutadiene (HTPB), aluminium (AL), isophorone diisocyanate (IPDI), dioctyl adipate (DOA), and triphenyl bismuth (TPB), until they amalgamate into a uniform blend. Before initiating the casting procedure, assessing the slurry propellant's viscosity is crucial as the mixing procedure's concluding step.

In the casting procedure, a mold tube and mandrel are used to mold the composite solid propellant. Mandrel is a molding tool that forms the grain configuration of solid propellant. In traditional casting methods that only involve a vacuum system, rearranging the propellant slurry in the mold relies solely on the forces of gravity and the viscosity of the material. However, by introducing vibration, the propellant slurry can be redistributed more effectively, ensuring that the composite propellant fills the space between the mandrel and the mold tube. Once the mixing phase has been finished, the propellant casting is started. The propellant casting process entails injecting a propellant slurry into a molded tube equipped with a mandrel while maintaining a vacuum environment in the chamber room, requiring special equipment. The equipment inserts the molding tube into a vacuum-sealed and oscillating casting chamber (see Fig. 1). The slurry is poured into a mold tube at a rate of 2–4 kg/h until it is filled. The vibration procedure was conducted for 15–30 min, ensuring the chamber's atmospheric pressure conditions were maintained. The utilized motor module has a specified vibration speed of 1500 rpm, a power output of 0.3 kW, and a vibrating force of 4 kN. The final stage of propellant production is the curing process. The molded tubes filled with propellant slurry are removed from the casting chamber and then heated to 60–120 °C for 72 to 100 h.

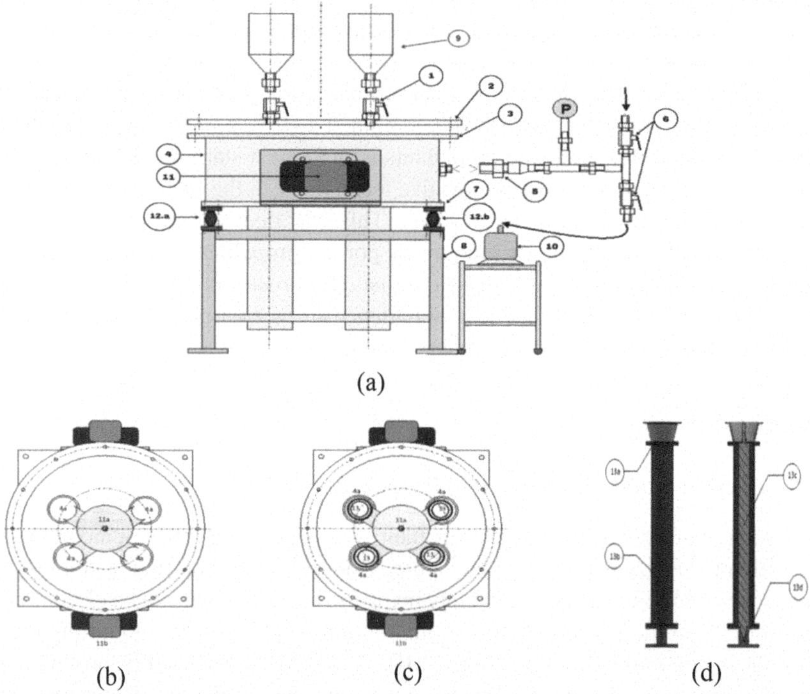

Fig. 1. Casting chamber set: (a) casting chamber, (b) inner chamber (top view), (c) tube position in the chamber (top view), (d) mold tube and mandrel set.

2.2 Composite Solid Propellant Sample Characterization

Tests were conducted to determine the disparities in propellant characteristics resulting from two distinct casting techniques: the conventional approach employing vacuum and the vibrating vacuum method. Four tests were conducted: radiography (x-ray), density test, hardness test, and mechanical properties test.

The radiography test is performed using an X-ray machine employing computed digital radiography (CR) utilizing film sizes 7 × 17 and 14 × 17 in. This tool applies a film image plate that can be used up to 1,000 times for exposure. The scanning process is carried out in approximately 8 min (depending on the desired resolution). The resulting images can be directly evaluated and analyzed using D-Tect X software.

Each composite propellant, manufactured and cured, went through density testing. Ten cubic propellant samples were subjected to density testing and averaged. Each sample of the propellant formulation was cut into cubes with a length of approximately 1 mm. An analytical balance with an additional accessory for weighing the sample in liquid and free air media is employed as the density propellant test instrument. The principle of Archimedes' law is employed to measure propellant density experimentally [10]. The hardness of the propellant samples that had been produced and cured was determined using the Bareiss HPE II shore-A hardness digital tester.

The mechanical property test was performed utilizing a Tensilon RTG-1250 5 kN universal testing machine according to the American Society for Testing and Materials (ASTM) criteria. We utilized uniaxial samples from the Join Army Navy NASA Air Force (JANNAF) for mechanical characteristics testing. As per the JANNAF standard, the maximum tensile stress is determined directly from the uniaxial stress-strain curve. The term refers to the total value of the uniaxial tensile strength of the solid propellant.

3 Result and Discussion

It is essential to comprehend the propellant design process of missiles to understand the impact of propellant design on system performance, such as the thrust profile, maximum velocity, and specific impulse of the rocket motor engine. [11]. Various casting procedures are available for propellant casting [12, 13]. Several variables can contribute to propellant porosity. The propellant slurry's rheological behavior significantly impacts the casting technique selection. A propellant slurry's flow ability is increased by plasticizers, one of the propellant raw components that elongate the polymer chain structure. The gap between the tube and the mandrel can be filled with the composite propellant slurry using the vacuum and vibratory composite propellant molding technique. An oxidizer (such as AP and ammonium nitrate), a combustion agent (such as aluminum powder (Al)), and a polymer binder make up a typical composite solid propellant. However, under slow heating settings, significant levels of AP allow the propellant's porous structure to thermally degrade and break down, resulting in uneven, porous structures and boosting the propellant's combustion rate [14, 15]. The standard method of casting composite solid propellant, with or without vacuum conditions but without vibration, has drawbacks, such as porosity in the final propellant. The molding procedure is more straightforward and effective. By incorporating the vibration technique into the casting process, it is anticipated that a uniform and impermeable propellant can be achieved in every section, thereby enhancing the structural integrity of the composite propellant and reducing the production of waste during the manufacturing process. Two types of propellant samples were tested, one of which was cast entirely in vacuum and the other cast in vacuum and vibration. Both samples' physical and mechanical characteristics were examined using tests for density, hardness, and radiography X-rays.

3.1 Radiography (X-Ray)

The propellant was examined using radiography (X-ray) to assess its porosity level. Propellant-A is manufactured using traditional molding techniques, specifically the vacuum process. Several pores are present in the propellant, both at the topside and bottom sections, denoted by and numbered in green, as depicted in Fig. 2. Propellant-B utilizes the vibration vacuum casting technique. As expected, employing the vibration vacuum casting technique reduces the presence of porous sections, as observed in the radiograph (see Fig. 3).

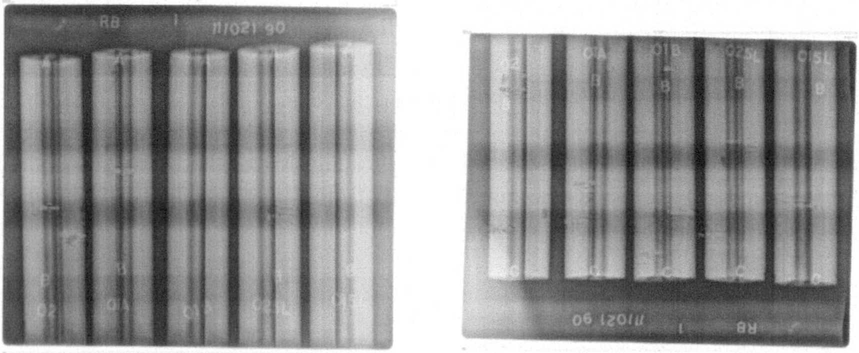

Fig. 2. Propellant-A: (left) topside section, (right) bottom section.

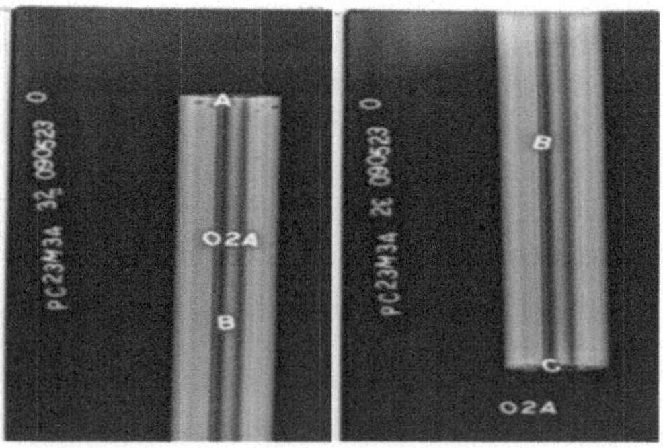

Fig. 3. Propellant-B topside section (left), bottom section (right).

3.2 Density Test

The instrument for the density propellant test is an analytical balance with an extra attachment for weighing the sample in liquid and free air. After manufacture and curing, every composite propellant is ready for experimental or real-density testing. Twelve samples of cubic propellant were chopped into cubes, each measuring about 1 mm long. These samples were then examined for density and averaged. The Archimedes' law principle is used in the experimental propellant density measurement process [16, 17].

Density refers to the extent to which a solid or liquid could fill space. The density value of a propellant with a void will be lower than the value of a propellant without it. As seen in Table 1, propellant A has a 1.4741 gr/cm^3 density. This value is lower than the density value of propellant B, which is 1.707 gr/cm^3 (see Table 2). It can be inferred that the density falls when both the total porosity and effective porosity rise [18]. The density of a material is determined by its porosity, which can be categorized as either open or closed. Pore types are classified based on their interconnectedness, surface connectivity, and shape. Sedimentary rocks, for instance, exhibit three types of compressibility, namely bulk compressibility, matrix compressibility, and pore volume compressibility, which are a result of the decrease in pore volume with increasing depth [19]. For booster stage applications, solid propellants with high densities and energies are preferred; however, they require significant levels of solid loading within a specific polymer matrix [20].

Table 1. The density of propellant-A.

Propellant Sample	W_{air}	W_{water}	W_{air}- W_{water}	ρ (gr/cm^3)
1	0,8467	0,2809	0,5658	1,496316
2	0,9471	0,3023	0,6448	1,468681
3	0,9052	0,3014	0,6038	1,499022
4	0,8414	0,2715	0,5699	1,476252
5	1,1039	0,3307	0,7732	1,42756
6	0,8708	0,2796	0,5912	1,472789
7	1,0781	0,3464	0,7317	1,473271
8	0,7677	0,2456	0,5221	1,470261
9	0,8181	0,2637	0,5544	1,475502
10	1,0992	0,3579	0,7413	1,482652
11	0,8031	0,2522	0,5509	1,457651
12	1,0723	0,3526	0,7197	1,489777
			ρ	**1,474144**

Table 2. The density of propellant-B.

Propellant Sample	W_{air}	W_{water}	W_{air}- W_{water}	ρ (gr/cm^3)
1	1,577	0,665	0,912	1,722
2	1,459	0,611	0,848	1,712
3	1,414	0,591	0,822	1,712
4	1,460	0,612	0,848	1,714
5	1,499	0,626	0,873	1,711
6	1,523	0,630	0,893	1,698
7	1,498	0,622	0,877	1,702
8	1,452	0,607	0,845	1,712
9	1,503	0,626	0,878	1,706
10	1,423	0,591	0,832	1,702
11	1,407	0,575	0,832	1,684
12	1,558	0,649	0,909	1,707
			ρ	**1,707**

3.3 Hardness Test

The propellants must be well-balanced, with good hardness and flexibility, to prevent cracks in the composite matrix. The crack will cause the combustion to become misdirected and explode [21]. The results of the propellant hardness test using the HPE Shore II tool indicate that propellant A has a hardness value of 68.12 Shore-A, while propellant B has a hardness value of 84.53 Shore-A. The result demonstrates that propellant B meets the international standard for hardness, which requires a value of more than 70 Shore-A. The difference in hardness values is directly linked to the material's porosity, where an increase in porosity results in a corresponding decrease in material hardness [22–24].

3.4 Mechanical Properties

To estimate the structural strength of a propellant against the deformation of the forces it experiences, it is necessary to test the mechanical properties of the propellant. These mechanical properties are influenced by various factors, including the composition of the propellant, the additives used, and also the physical form of the propellant [25]. Solid propellants must exhibit significant flexibility across a broad range of operating temperatures. Furthermore, to reduce the frequency of thermal cycling during storage or combustion, efficient material handling is required. Grain distortion may arise from thermal tensions, fractures, or cracks in the material. This can be avoided by attending to the mechanical properties. Typically, the mechanical characteristics are shown by tensile strength, elongation, and hardness values. The propellant must possess a good balance of hardness and flexibility to prevent the formation of cracks in the composite matrix.

Cracks can lead to an unfocused direction of combustion, resulting in an explosion. Hence, it is imperative to uphold the mechanical qualities of the rocket at its designated working temperature. Its components' collective properties determine the propellant's mechanical properties according to the rule of mixture for composites [21].

To achieve valuable mechanical test results, propellant samples have to comply with test standards, which include non-porous propellants. In this study, the propellant A sample had a significant porosity (see Fig. 2), therefore its mechanical properties could not be investigated. The mechanical characteristics test results for propellant B samples are shown in Table 3.

Table 3. Mechanical properties of propellant-B.

Test No	Maximum point Stress (kgf/cm^2)	Maximum point Strain (%GL)	Breakpoint Strain (%GL)	Modulus Young (MPa)
1	5.7858	8.45	12.744	15.016
2	6.6081	8.8688	11.706	17.132
3	6.9962	9.3875	13.894	17.35
4	6.6231	10.025	11.394	18.907
5	6.5114	9.025	10.813	17.027

4 Conclusion

Based on radiography, density test, hardness test, and mechanical properties test findings, the composite solid propellant casting process, which utilizes vacuum and vibration methods, effectively reduces the presence of pores in the propellant. Propellant-A, cast solely using the vacuum technique, exhibits significant porosity. In contrast, propellant-B, cast using vacuum and vibration processes, shows no visible porosity. It has a density value of 1.707 gr/cm^3, as determined by the density test. Propellant-B exhibits a greater hardness of 84.53 Shore-A. Meanwhile, propellant-A fails to meet the requirements for the mechanical properties test due to its number of shafts, indicating that propellant B possesses a relatively high value for mechanical characteristics. It can be concluded that adding vibration in the vacuum casting process can minimize the appearance of pores in composite solid propellant resulting in a better propellant quality in terms of the mechanical properties improvement and a decrease in propellant waste.

Acknowledgment. The authors would like to gratefully acknowledge the Research Center for Rocket Technology, National Research and Innovation Agency (BRIN-Indonesia) for their generous provision of research funding and resources, which enabled the successful execution of this work. We would like to extend our gratitude to all individuals who have contributed to the execution of this research.

References

1. Touidjine, S., et al.: Preparation of ammonium nitrate-based solid composite propellants supplemented with polyurethane/nitrocellulose blends binder and their thermal decomposition behavior. Def Technol. **18**(11), 2023–2033 (2022). https://doi.org/10.1016/j.dt.2021.12.013
2. Prianto, B., Setyaningsih, H., Puspitasari, R.R.: Analysis of composite propellant energy and correlation with specific impulse at vacuum level. AIP Conf. Proc. **2226**(April) (2020). https://doi.org/10.1063/5.0002731
3. Chen, X., Lai, J., Long Chang, X., Zhang, Y., Zhang, L., Wang, C.: Compressive mechanical properties of HTPB propellant at low temperatures and high strain rates. Results Phys. **7**, 4079–4084 (2017). https://doi.org/10.1016/j.rinp.2017.10.034
4. Ashish, J., Swaroop, G., Balasubramanian, K.: Effect of ammonium perchlorate particle size on flow, ballistic, and mechanical properties of composite propellant. Nanomater. Rocket Propuls. Syst. 299–362 (2019). https://doi.org/10.1016/B978-0-12-813908-0.00008-3
5. Banerjee, S., Chakravarthy, S.R.: Ammonium perchlorate-based composite solid propellant formulations with plateau burning rate trends. Combust. Explos. Shock Waves **43**(4), 435–441 (2007). https://doi.org/10.1007/s10573-007-0059-5
6. Wang, X., et al.: Experimental and simulative thermal stability study of HTPE solid propellants in different scales. Case Stud. Therm. Eng. **28**(October) (2021). https://doi.org/10.1016/j.csite.2021.101566
7. Kohga, M.: Mechanical properties and burning characteristics of AP / HTPB composite propellant using polytetrahydrofuran as a plasticizer. Sci. Technol. Energ. Mater. **71**(5–6), 145–150 (2010)
8. Sutton, G.P., Biblarz, O.: Rocket Propulsion Elements, pp. 2–5 (2016)
9. Pinalia, A., Dhewi, P., Setyaningsih, H.: Alat dan Metode Pencetakan Propelan Komposit Tervakum dan Bervibrasi, P00202306936
10. Prianto,B., Setyaningsih, H., Pinalia, A.: Preliminary results of the particle packing density theory for improving the theoretical density of composite propellant. AIP Conf. Proc. **2366**(September) (2021). https://doi.org/10.1063/5.0060448
11. Mahjub, A., Mazlan, N.M., Abdullah, M.Z., Azam, Q.: Design optimization of solid rocket propulsion : a survey of recent design optimization of solid rocket propulsion : a survey of recent advancements, no. November (2019). https://doi.org/10.2514/1.A34594
12. Dombe, G., Jain, M., Singh, P.P., Radhakrishnan, K.K., Bhattacharya, B.: Pressure casting of composite propellant. Indian J. Chem. Technol. **15**(4), 420–423 (2008)
13. Bi, Y., et al.: Safety and reliability analysis of the solid propellant casting molding process based on FFTA and PSO-BPNN. Process. Saf. Environ. Prot. **164**, 528–538 (2022). https://doi.org/10.1016/J.PSEP.2022.06.032
14. Kohga, M.: Burning rate characteristics of ammonium perchlorate-based composite propellant using bimodal ammonium perchlorate J. Propul. Power (2008).https://doi.org/10.2514/1.27107
15. Liu, L., et al.: Combustion promotion and agglomeration reduction of the composite propellant using graphene. Aerosp. Sci. Technol. **118** (2021). https://doi.org/10.1016/j.ast.2021.106988
16. Trache, D., et al.: Effect of amide-based compounds on the combustion characteristics of composite solid rocket propellants. Arab. J. Chem. **12**(8), 3639–3651 (2019). https://doi.org/10.1016/j.arabjc.2015.11.016
17. Sangtyani, R., Kumar, A., Kumar, A., Gupta, M.: Optimization of network forming agents for different types of composite propellant grain. Cent. Eur. J. Energ. Mater. **10**(3) (2013)
18. Liu, H., Luo, G., Wei, H., Yu, H.: Strength, permeability, and freeze-thaw durability of pervious concrete with different aggregate sizes, porosities, and water-binder ratios. Appl. Sci. **8**(8) (2018). https://doi.org/10.3390/app8081217

19. Bagdassarov, N.: Density and Porosity. In: Fundamentals of Rock Physics, Cambridge University Press, pp. 28–65. https://doi.org/10.1017/9781108380713.003
20. Suresh Babu, K.V., Kanaka Raju, P., Thomas, C.R., Syed Hamed, A., Ninan, K.N.: Studies on composite solid propellant with tri-modal ammonium perchlorate containing an ultrafine fraction. Def. Technol. **13**(4), 239–245 (2017). https://doi.org/10.1016/j.dt.2017.06.001
21. Wibowo H.B., Wibowo, R.S.M.: Non energetic binder application for RDBP propellant based large caliber munition (MKB). Int. Semin. Aerosp. Sci. Technol. **IV**, 82–91 (2016)
22. Cherry, J.A., Davies, H.M., Mehmood, S., Lavery, N.P., Brown, S.G.R., Sienz, J.: Investigation into the effect of process parameters on microstructural and physical properties of 316L stainless steel parts by selective laser melting. Int. J. Adv. Manuf. Technol. **76**(5–8), 869–879 (2015). https://doi.org/10.1007/s00170-014-6297-2
23. Jang, B.K., Matsubara, H.: Influence of porosity on hardness and Young's modulus of nanoporous EB-PVD TBCs by nanoindentation. Mater. Lett. **59**(27), 3462–3466 (2005). https://doi.org/10.1016/j.matlet.2005.06.014
24. Kamal Jayaraj, R., Malarvizhi, S., Balasubramanian, V.: Optimizing the micro-arc oxidation (MAO) parameters to attain coatings with minimum porosity and maximum hardness on the friction stir welded AA6061 aluminium alloy welds. Def. Technol. **13**(2), 111–117 (2017). https://doi.org/10.1016/j.dt.2017.03.003
25. Pinalia, A., Prianto, B., Setyaningsih, H.: Preliminary results of DOA plasticizer effect on AP / HTPB composite propellant characteristics Preliminary Results of DOA Plasticizer Effect on AP / HTPB Composite Propellant Characteristics, vol. 040009, no. September (2021)

Environmental Impacts of Rocket Emissions: Case Study in Indonesia and Potential Solutions

Soleh Fajar Junjunan(✉), Bagus Hayatul Jihad, Bayu Prianto, Starida Sofia Moranova, and Saeri

Research Center for Rocket Technology, BRIN, Bogor, Indonesia
jun.aero05@gmail.com, {bagu001,bayu006,star001, saer001}@brin.go.id

Abstract. This study assesses the environmental impacts of various rocket propellants developed in Indonesia, including composite, double base, and liquid types. It focuses on their contributions to atmospheric pollution, climate change, ozone depletion, and surface environmental health. Our analysis reveals that current propellants developed in Indonesia, particularly composite types, generate considerably significant harmful emissions. Composite propellants produce high levels of carbon monoxide (CO) and aluminum oxide (Al_2O_3), which are detrimental to air quality and can contribute to ozone depletion. While less harmful, liquid propellants still contribute significantly to greenhouse gas emissions and atmospheric pollution. LNG-based propellants show considerable promise as a more environmentally friendly alternative. Our findings indicate that LNG/LOX propellants produce significantly lower emissions of harmful substances. This reduction in emissions highlights LNG's potential to mitigate the environmental footprint of rocket launches in Indonesia. By adopting LNG as a green propellant, Indonesia could leverage its natural gas resources to reduce its rocket programs' adverse environmental impacts while fostering economic growth and technological advancement.

Keywords: Rocket Emissions · Environmental Impacts · LNG propellant · Atmospheric Pollution · Climate Change

1 Introduction

Rocket technology is pivotal, impacting daily life and our understanding of the universe. It is fundamental for space exploration, enabling missions that deploy telescopes like Hubble, probes to Mars and Venus, and study cosmic phenomena. Rockets launch satellites crucial for telecommunications, GPS, weather prediction, and environmental monitoring. They are vital for national security, economic growth, and technological advancements, driving jobs and innovation in various fields. It inspires future STEM careers and extends human capabilities beyond Earth, as seen with the ISS and plans for lunar [1–3].

E. Fitrianingsih et al. (Eds.): ISAST 2024, SPPHY 416, pp. 488–497, 2025.
https://doi.org/10.1007/978-981-96-1344-1_51

However, there are challenges in rocket development. The challenges in rocket development encompass various realms, including technical, economic, regulatory, and environmental aspects. Significant technical hurdles include reliable and efficient propulsion, miniaturization of components, and ensuring astronaut safety on long missions. Financial challenges include high costs of development and launches, with companies like SpaceX working on reusable rockets to reduce expenses. Regulatory challenges involve navigating complex international and national regulations to ensure safe launches and space operations. Nowadays, environmental impact is a growing concern. Emissions from chemical rockets contribute to atmospheric pollution, climate change, and ozone depletion, posing risks to air quality and the Earth's surface environment [4–7].

Although numerous studies are available on this topic, there is a lack of references discussing the specific study case in Indonesia and the discussion on possible solutions that can generate multiple benefits, not only about improving the quality of the environment. Therefore, this study aims to evaluate and compare the environmental impacts of various rocket propellants that have developed in Indonesia, including composite, double base, liquid, and LNG/LOX types, by analyzing combustion products and assessing their effects on atmospheric pollution, climate change, ozone depletion, and surface environmental health. Additionally, the study aims to demonstrate the benefits of LNG-based propellants in reducing harmful emissions compared to existing options.

2 Method

The methodology involved systematically analyzing peer-reviewed articles, technical reports, and conference proceedings published but not limited to in the last two decades. The selected studies were evaluated based on their relevance, scientific rigor, and contribution to understanding the environmental impacts of rocket emissions. As we also propose a solution for the ecological issue in rocketry activities in Indonesia, we also select publications related to our solution. Combustion products produced by existing rockets developed in Indonesia are also computed. We use a rocket caliber 450 with Indonesian-sounding rocket geometry as a case study for comparison purposes. We calculate the combustion product in the exit nozzle using CEA software at 40 bar chamber pressure. Details about the rocket caliber 450 geometry are available in the literature [8, 9], with the Ae/At ratio known to be about 5.54.

We selected different types of propellants for comparison. First, we chose the composite propellant used for the RX450 rocket. Then, we included a double-base propellant developed in Indonesia using NC/NG-based formulations. For liquid rockets, we selected kerosene and HNO_3 as the fuel and oxidizer, respectively. Additionally, we considered LNG with LOX as the oxidizer since we see this as a potential propellant for future development in Indonesia. The composition of each propellant is presented in Table 1. As we have several compositions, we chose 60% nitroglycerin for the double base propellant for this computation. There are differences between the design and experimental oxidizer-to-fuel ratio for the liquid rocket due to the pressure drop; we chose the OF ratio presented in the experiment. For LNG/LOX propellant, we chose the OF ratio of 3 as there is literature [10] available for that, so it will be useful to compare the results, although we use different geometry.

Table 1. Composition of propellant [11–14].

Reactant	Composite [11] (%)	Double base [12] (%)	Liquid [13] (%)	LNG + LOX [14] (%)
Ammonium perchlorate	75	–	–	–
HTPB	16.41	–	–	–
Al	7.5	–	–	–
TDI	1.09	–	–	–
Nitroglycerin	–	60	–	–
Nitrocellulose	–	35	–	–
Dibutyl phthalate	–	3.4	–	–
Ethyl Centralite	–	1.6	–	–
HNO_3	–	–	66	–
Kerosene	–	–	34	–
LOX	–	–	–	75
CH_4	–	–	–	22.535
C_2H_6	–	–	–	1.365
C_3H_8	–	–	–	0.745
C_4H_{10}	–	–	–	0.3525
N_2	–	–	–	0.0025

There are many variations of LNG composition percentages, and perhaps they change due to several factors, such as storage and aging [15]. As our context is to develop LNG base propellant in Indonesia, we chose the LNG with a composition that produces in Indonesia, which has approximately about 90.14% of CH_4, 5.46% of C_2H_6, 2.98% of C_3H_8, 1.41% of C_4H_{10} and 0.01% of N_2 [14]. In this study, we have focused on propellants developed in Indonesia. Future research may explore other propellants for a broader environmental impact assessment.

We compared our results with the existing literature. Unfortunately, very few references presented the calculation of the mole fraction of combustion products at the nozzle exit for cases similar to our work. Therefore, we only compare the specific impulse (ISP) presented in Table 2. A significant difference was found in the liquid rocket case, and it may be because the literature shows the Isp with the ideal OF ratio as designed without recalculating using the actual OF ratio due to the pressure drop.

Table 2. ISP comparison with literature.

Case	ISP in this study	ISP in Literature
Composite	231	255 theoretical, 209.91 experimental [16]
Double base	231.2	234.5 [12]
Liquid [13]	192.5	237 [13]
LNG + LOX	280	289 for 92.596% CH_4 [10]

3 Results and Discussion

3.1 Comparison of Combustion Product

The environmental impact of rocket propulsion systems varies depending on the type of propellant used. Different types of propellant will generate different combustion products and then generate different environmental implications. Table 3 presents the mole fraction of combustion products by several types of propellant that developed in Indonesia with additional LNG-based propellant. Based on the table, there are some potentially dangerous products such as CO, HCl, and Al_2O_3 that we still consider significant (more than 0.00001 in mole fraction) at the rocket exit nozzle.

Table 3. Mole fraction of combustion product.

Product	Composite	Double base	Liquid	LNG + LOX
CO_2	0.04995	0.21640	0.10844	0.14372
H_2O	0.16159	0.25929	0.08140	0.52726
CO	0.24500	0.27903	0.28934	0.19564
H_2	0.28454	0.10014	0.35121	0.12755
H	0.00002	0.00004	–	0.00326
OH	–	0.00001	–	0.00244
O	–	–	–	0.00004
O_2	–	–	–	0.00007
N_2	0.07640	0.14510	0.09791	0.00002
HCl	0.14986	–	–	–
Cl	0.00001	–	–	–
$Al_2O_{3(a)}$	0.03263	–	–	–
CH_4	–	–	0.02039	–
NH_3	–	–	0.00006	–
$C_{(gr)}$	–	–	0.05125	–

3.2 Atmospheric Pollution and Air Quality Degradation

Rocket activities can cause atmospheric pollution, primarily through the emissions produced during launches and tests. Aluminum is often used as a composition in solid propellant to increase performance and combustion stability, but it can produce particulate material composed of aluminum oxide (Al_2O_3). Particulate matter can harm health as it can penetrate deep into the lungs and even enter the bloodstream. CO is almost always significantly emitted by solid, hybrid, or liquid propellant rockets. A colorless, odorless gas produced by the incomplete combustion of fuels at high levels can interfere with the blood's ability to carry oxygen, posing serious health risks [4, 17–19].

Although in Table 3, NO_x is not mentioned (due to less than 0.00001 in mole fraction), it does not mean that it does not exist. In the other type of propellant, NO_x may be significantly emitted; moreover, some propellants use NOx as an oxidizer. Nitrogen oxides contribute to the formation of ground-level ozone and acid rain, and they can cause respiratory problems. Rocket fuel vapors release VOCs into the atmosphere, particularly from liquid propellants like kerosene. VOCs can react with NOx to form ground-level ozone, a harmful air pollutant. As a secondary pollutant, ground-level ozone can cause respiratory problems and damage crops and other vegetation [20, 21].

3.3 Climate Change

Rocket activities can contribute to climate change primarily through the emissions they produce during launches. These emissions include greenhouse gases and other pollutants that can alter atmospheric conditions and contribute to global warming. CO_2 and H_2O emitted by any chemical rocket as a combustion product or greenhouse gas will significantly contribute to climate change. Rocket combustion also produces soot, which, if released into the stratosphere, absorbs the radiation of sunlight, leading to local heating of the atmosphere and warming the planet. Ground-level ozone, besides its atmospheric pollution, ozone in the lower atmosphere can act as a greenhouse gas [4, 5].

3.4 Ozone Depletion

Rocket activities can contribute to ozone depletion. Solid rocket motors often use propellants like ammonium perchlorate, which release chlorine upon combustion, as confirmed in our composite propellant combustion product presented in Table 3. The chlorine then participates in ozone-depleting reactions in the stratosphere, as rockets can release these harmful chemicals directly into the stratosphere, where they have a more immediate and pronounced effect on ozone depletion than surface-level emissions. Besides that, aluminum oxide (Al_2O_3) emitted in an Al-contained propellant can contribute to the catalytic destruction of ozone [22, 23]. The high-energy radiation emitted by rocket plumes enhances the catalytic destruction of ozone molecules, contributing to ozone layer thinning and potential long-term implications for ultraviolet radiation exposure at the Earth's surface [6].

Researchers globally have made efforts to mitigate the environmental impact of solid rocket motor emissions, including developing novel plume treaters to reduce temperature and eliminate gaseous pollutants [24]. Research has also explored using filled elastomers as thermal insulators to improve performance and safety [25]. Numerical simulations and optical diagnostics are essential and have been done to understand solid rocket motors' internal flow dynamics, ignition processes, and combustion characteristics. These studies help characterize solid rocket plumes and assess the particles' impact on ozone depletion post-combustion [26–29].

3.5 Effects on Earth's Surface Environment

Rocket emissions impact the atmosphere and the Earth's surface environment. For example, hydrazine and its derivatives are highly toxic and can cause severe health issues such as liver and blood system damage. These compounds are absorbed by soil through cation-exchange mechanisms. While moderate doses of hydrazine are not toxic to microorganisms and plants, high concentrations can be harmful and toxic to plant life. Implementing sodium selenate can improve contaminated soils' enzymatic and nitrification activity, but they do not eliminate the negative effects [30]. Combustion instabilities and pressure oscillations are common challenges associated with solid rocket motors, which can generate high noise levels near rockets and potentially affect the surface earth environment [31, 32].

3.6 Rocket Conversion Program: Utilizing LNG as a Green Propellant

Liquefied Natural Gas (LNG) is considered a potentially green propellant due to its cleaner combustion properties than traditional rocket fuels. Composed primarily of methane (CH_4) with small amounts of ethane, propane, and other hydrocarbons, LNG's combustion produces fewer pollutants such as sulfur oxides and particulate matter. This makes it an attractive option for reducing the ecological footprint of rocket launches. There are also some other potential green propellants, such as LOX/H_2 propellants, but LNG seems more economically competitive than LOX/H_2, especially in Indonesia [33–35]. Given Indonesia's rich LNG reserves, developing an LNG-based rocket technology program presents a unique opportunity. By leveraging LNG, Indonesia could reduce the environmental impact of its rocket launches while utilizing a readily available resource. This initiative supports Indonesia's energy sector by creating new demand for natural gas, fostering economic growth, and driving technological advancement [36, 37]. Therefore, we propose using LNG as a green propellant for the rocket conversion program.

As shown in Table 3, using LNG as fuel and LOX as an oxidizer, there is no or at least insignificant (less than 0.00001 in mole fraction) of HCl and Al_2O_3, which are produced significantly by a composite propellant developed in Indonesia. Besides that, the table also shows that using LNG can significantly reduce CO emissions compared to existing rocket programs in the country. At the rocket nozzle, compared with the existing composite, double base, and HNO_3/Kerosene rocket program, the mole fraction reduces by 20.15%, 29.88%, and 32.38%, respectively. Developing LNG-based rockets can also help Indonesia reduce its dependency on imported raw materials for rocket development. Utilizing domestically sourced LNG enhances national security and technological sovereignty, stimulates the local economy, and fosters innovation.

Despite the potential, there are significant challenges in developing LNG-based rocket technology in Indonesia, requiring considerable time and investment. In the context of technical challenges, the miscibility of LOX with LNG poses a risk of condensed-phase detonation, resulting in higher overpressures than LOX/ LH_2 or LOX/RP-1. Small-scale, unconfined LOX/LNG mixtures have shown a broad detonable range and explosive yields surpassing TNT [38]. Additionally, LNG's composition can change due to evaporation during storage and transportation, complicating the consistency of rocket engine performance [15].

However, immediate engagement in environmentally friendly rocket activities can still be conducted. Several near-term actions can mitigate environmental impacts, such as installing water mist systems near static test facilities, which can help reduce particulate emissions. At the same time, noise barriers can minimize sound pollution. Mechanisms to control the direction of exhaust gases can prevent them from affecting nearby populated areas. Establishing a ground-to-ground launch site is another crucial step. Currently, rockets launched from the launch site head to the sea, posing risks to marine life as they drown without retrieval. A land-based launch site would allow for the recovery of small sounding rockets, reducing the impact on sea biota. Indonesia's vast land area and numerous islands offer suitable locations for such sites, particularly outside the densely populated Java Island. Perhaps it may also need to mitigate population density near rocketry facilities; while challenging due to human and population dynamics complexities, it is achievable with government commitment. Relocating residents or implementing stringent zoning regulations can create a safer and more sustainable environment for rocket development.

4 Conclusion

Our study highlights that the rocket launches contribute to atmospheric pollution through emissions like CO, CO_2, and particulate matter. The existing rocket developed in Indonesia considerably poses a relatively high environmental impact, particularly for solid propellant rockets. LNG/LOX propellant produces significantly lower emissions of harmful substances such as CO and HCl than existing propellants. For the CO emissions, compared to the existing composite, double base, and HNO_3/Kerosene rocket program, the mole fraction at the exit nozzle reduces by 20.15%, 29.88%, and 32.38%, respectively. LNG emerges as a promising green alternative due to its cleaner combustion properties. Utilizing LNG-based rockets could significantly reduce emissions of harmful substances, enhance Indonesia's energy security, and leverage local natural gas resources.

Acknowledgment. This research was supported by the Research Center for Rocket Technology, National Research and Innovation Agency (BRIN).

Contributorship Statement. All authors contributed equally to this research. All authors participated in the conceptualization, methodology development, data collection, analysis, interpretation of results, and manuscript writing. All authors have read and approved the final manuscript.

References

1. Pillai, A.S.: Introduction to Rocket Science and Space Exploration. CRC Press (2022)
2. Turner, M.J.L.: Rocket and Spacecraft Propulsion: Principles, Practice and New Developments. Springer Science & Business Media (2008)
3. Erickson, A.S.: China's space development history: a comparison of the rocket and satellite sectors. Acta Astronaut. **103**, 142–167 (2014)
4. Kokkinakis, I.W., Drikakis, D.: Atmospheric pollution from rockets. Phys. Fluids **34** (2022). https://doi.org/10.1063/5.0090017
5. Ross, M., James A.V.: The policy and science of rocket emissions (2018)
6. Ross, M.N., et al.: Observation of stratospheric ozone depletion associated with Delta II rocket emissions. Geophys. Res. Lett. **27**, 2209–2212 (2000). https://doi.org/10.1029/1999GL011159
7. Nebylov, A., Nebylov, V.: Reusable space planes challenges and control problems. IFAC-PapersOnLine **49**, 480–485 (2016)
8. Al Farizi, M.F., Bura, R.O., Junjunan, S.F., Jihad, H.B.: Grain propellant optimization using Real Code Genetic Algorithm (RCGA). J. Phys. Conf. Ser. **1005**, 012033 (2018). https://doi.org/10.1088/1742-6596/1005/1/012033
9. Abyan, M.I., Nuryaman, A., Jihad, B.H., Junjunan, S.F., Asmiati, A.: Design optimization of propellant grain and nozzle contour to improve performance of solid rocket propulsion. J. Eng. Technol. Sci. **54**, 220508 (2022). https://doi.org/10.5614/j.eng.technol.sci.2022.54.5.8
10. Han, P.-G., Lee, S.-W., Kim, K.-H., Yoon, Y.: Performance analysis of the thrust chamber in liquid rocket engine using liquefied natural gas as a fuel. In: 40th AIAA/ASME/SAE/ASEE Joint Propulsion Conference and Exhibit. American Institute of Aeronautics and Astronautics, Reston, Virigina (2004). https://doi.org/10.2514/6.2004-3860
11. Pinalia, A., Prianto, B., Setyaningsih, H., Dhewi, P.: HTPB/IPDI-based polyurethane: effect of hydroxyl value on the physical and mechanical properties. In: AIP Conference Proceedings (2023)
12. Rosita, G.: Perhitungan Kinerja Propelan Double Base. In: Prosiding SIPTEKGAN XV-2011 Seminar Nasional IPTEK Dirgantara XV Tahun 2011, pp. 119–127 (2011)
13. Hakim, A.N.: RANCANG BANGUN ENJIN ROKET CAIR DENGAN GAYA DORONG 1000 KGF MENGGUNAKAN PROPELAN ASAM NITRAT â€"KEROSEN (DESIGN OF LIQUID ROCKET ENGINE WITH 1000 KGF THRUST USING NITRIC ACID â€"KEROSENE PROPELLANT). Jurnal Teknologi Dirgantara **13** (2015)
14. Kuczyński, S., Łaciak, M., Szurlej, A., Włodek, T.: Impact of liquefied natural gas composition changes on methane number as a fuel quality requirement. Energies (Basel). **13** 5060 (2020). https://doi.org/10.3390/en13195060
15. Arrhenius, K., Karlsson, A., Hakonen, A., Ohlson, L., Yaghooby, H., Büker, O.: Variations of fuel composition during storage at Liquefied Natural Gas refuelling stations. J. Nat. Gas Sci. Eng. **49**, 317–323 (2018). https://doi.org/10.1016/j.jngse.2017.11.014
16. Prianto, B., Setyaningsih, H., Puspitasari, R.R.: Analysis of composite propellant energy and correlation with specific impulse at vacuum level. Presented at the (2020).https://doi.org/10.1063/5.0002731
17. Adeniyi, G.O., Fakinle, B.S., Sholiyi, O.S., Sonibare, J.A., Akeredolu, F.A.: Gaseous emissions characterization from potassium nitrate-based rocket composite propellant combustion. Int. J. Energy Clean Environ. **23**, 75–93 (2022). https://doi.org/10.1615/InterJEnerCleanEnv.2022038877
18. Schuch, D., Fisch, G.: Rocket emissions representation in atmospheric air-quality models: the short-range atmospheric transport and reaction of gases released by solid propellant engines. Meteorol. Appl. **26**, 171–181 (2019). https://doi.org/10.1002/met.1750

19. Brennan, W.D., Hovland, D.L., Netzer, D.W.: Measured particulate behavior in a subscale solid propellant rocket motor. J. Propuls. Power. **8**, 954–960 (1992). https://doi.org/10.2514/3.23578
20. McCaleb, R.: Atmospheric impact of liquid oxygen and kerosene engines. In: Space Programs and Technologies Conference. American Institute of Aeronautics and Astronautics, Reston, Virigina (1995). https://doi.org/10.2514/6.1995-3573
21. Zakirov, V., Sweeting, M., Lawrence, T., Sellers, J.: Nitrous oxide as a rocket propellant. Acta Astronaut. **48**, 353–362 (2001). https://doi.org/10.1016/S0094-5765(01)00047-9
22. Schmid, O., et al.: Size-resolved particle emission indices in the stratospheric plume of an Athena II rocket. J. Geophys. Res.: Atmosph. **108** (2003). https://doi.org/10.1029/2002JD002486
23. Saile, D., et al.: Characterization of SRM plumes with alumina particulate in subscale testing. CEAS Space J. **13**, 247–268 (2021). https://doi.org/10.1007/s12567-020-00338-0
24. Ma, Y., Liu, T., Zhang, Y., Huo, D., Zhang, P., Xie, F.: A novel plumes treater for solid rocket motors during hot-firing tests using ice tunnel structure. Asia-Pacific J. Chem. Eng. **18** (2023). https://doi.org/10.1002/apj.2925
25. Bhuvaneswari, C.M., Kakade, S.D., Deuskar, V.D., Dange, A.B., Gupta, M.: Filled Ethylene-propylene Diene Terpolymer Elastomer as Thermal Insulator for Case-bonded Solid Rocket Motors. Def Sci J. 58, 94–102 (2008). https://doi.org/10.14429/dsj.58.1628
26. Weiser, V., Eisenreich, N., Poller, S., Kelzenberg, S., Lo, R., Adirim, H.: Investigation of the burning behavior of cryogenic solid propellants. Propellants Explos. Pyrotech. **27**, 150–155 (2002)
27. Numerical simulation of internal flow field in solid rocket motor with complex grain. In: 2018 9th International Conference on Civil Engineering, Materials and Machinery (ICCEMM 2018). Francis Academic Press (2018). https://doi.org/10.25236/iccemm.2018.019
28. Li, Z., Liu, J., Zhang, W., Ye, Z., Yang, B.: Numerical simulation of the effect of multiple factors on the ignition process of a solid rocket motor. Propellants, Explosives, Pyrotechnics. 49, (2024). https://doi.org/10.1002/prep.202300198
29. Toscano, A.M., Lato, M.R., Fontanarosa, D., De Giorgi, M.G.: Optical diagnostics for solid rocket plumes characterization: a review. Energies (Basel). **15**, 1470 (2022). https://doi.org/10.3390/en15041470
30. RABINSKIY, L.N., TUSHAVINA, O. V.: Problems of Land Reclamation and Heat Protection of Biological Objects against Contamination by the Aviation and Rocket Launch Site. Journal of Environmental Management and Tourism. 10, 967 (2019). https://doi.org/10.14505//jemt.v10.5(37).03
31. French, A.D., Panelli, M., Di Lorenzo, G., Schettino, A., Paglia, F.: Combustion instability and pressure oscillation numerical simulation in a solid rocket motor. In: 53rd AIAA/SAE/ASEE Joint Propulsion Conference. American Institute of Aeronautics and Astronautics, Reston, Virginia (2017). https://doi.org/10.2514/6.2017-4952
32. Wang, Z., Zhang, W., Liu, Y.: A phenomenological model for the unsteady combustion of solid propellants from a Zel'dovich-Novzhilov approach. Aerospace **10**, 767 (2023). https://doi.org/10.3390/aerospace10090767
33. Haeseler, D., Goetz, A., Froehlich, A.: Non-toxic propellants for future advanced launcher propulsion systems. In: 36th AIAA/ASME/SAE/ASEE Joint Propulsion Conference and Exhibit. American Institute of Aeronautics and Astronautics, Reston, Virigina (2000). https://doi.org/10.2514/6.2000-3687
34. Urbano, A., Nasuti, F.: Numerical study of liquefied natural gas as a coolant in liquid rocket engines. Proc. Inst. Mec. Eng. G J. Aerosp. Eng. **227**, 1130–1143 (2013). https://doi.org/10.1177/0954410012451376

35. Martin, J., Börner, M., Hardi, J., Suslov, D., Oschwald, M.: Experimental investigation of flame anchoring behavior in a LOX/LNG rocket combustor. Aerospace **10**, 542 (2023). https://doi.org/10.3390/aerospace10060542
36. Purwanto, W.W., Muharam, Y., Pratama, Y.W., Hartono, D., Soedirman, H., Anindhito, R.: Status and outlook of natural gas industry development in Indonesia. J Nat Gas Sci Eng. **29**, 55–65 (2016). https://doi.org/10.1016/j.jngse.2015.12.053
37. Dodge, A.: From resource peripheries to emerging markets: reconfiguring positionalities in global production networks. Presented at the (2021).https://doi.org/10.1007/978-3-030-84606-0_4
38. Meyer, M., Haas, J., Eppig, B.: Investigating the explosive Hazard of Liquid Oxygen-Liquefied natural gas rocket propellant. In: 30th Space Cryogenics Workshop (2023)

Constant and Variable Timesteps Evaluation of CFD Simulation for a 200-mm Caliber Sounding Rocket Nozzle

Starida Moranova(✉), Idris Eko Putro, Bayu Prianto, and Soleh Fajar Junjunan

BRIN, Jakarta, Indonesia
staridamaronova@gmail.com

Abstract. A series of CFD simulations with gradually increasing timesteps is considered in this paper to improve efficiency in analyzing a 200-mm caliber sounding rocket nozzle. The accuracy of this method is compared to a simulation using a single timestep size for the entire burning duration. The simulations are conducted using two different calculation approaches. The first set used a constant timestep of 0.01 s throughout the burning period of 10 s, whereas in the second set, the timestep started at 10^{-4} s and increased tenfold up to 0.1 s. The results demonstrated consistent trends in crucial properties, including static pressure, wall total heat flux, and static temperature contour for both methods. Both simulations indicate that the highest heat flux of 3.4 MW/m^2 occurs in the nozzle throat area within the first second.

Keywords: timestep size · numerical timestep · rocket nozzle simulation · rocket nozzle CFD · CFD simulation

1 Introduction

Rocket technology can be used for climate change observation, with one of the most common applications being the launch vehicle for observation payloads. The Solar Orbiter mission was launched by Atlas V launch vehicle to explore the sun and heliosphere, studying the change of solar activities with time [1]. Another solar observation involves studying the energy distribution of solar wind particles [2]. A two-stage observation rocket has also been used to carry a probe to detect plasma irregularities [3].

Sounding rockets are a class of rockets utilized for observation and exploration purposes. A heliophysics study was also conducted using sounding rockets, such as observing the polar night [4] and auroral ionosphere [5]. Sounding rockets can be used for various climate studies, including astrophysical and atmospheric observations [6]. In this paper, a 200-mm caliber rocket nozzle used as a research rocket, was studied. A Computational Fluid Dynamics (CFD) simulation was utilized to analyze the internal flow within the rocket nozzle and its interaction with the structure. This approach allowed for the modeling and analysis of the internal flow dynamics within the nozzle and predicted the complex interactions between the fluid flow and the structural components.

E. Fitrianingsih et al. (Eds.): ISAST 2024, SPPHY 416, pp. 498–507, 2025.
https://doi.org/10.1007/978-981-96-1344-1_52

The timestep size in transient analysis significantly impacts simulation accuracy[7]. Several studies have examined the influence of timestep size on the accuracy of numerical simulations [8, 9]. Studies show that advancing predictive capability relies on accurate simulations, where smaller timesteps yield higher accuracy. However, the challenge lies in the time-consuming aspect of these operations. This paper introduces a series of simulations with varying timestep sizes as the suggested solution for a shorter analysis duration.

The simulations compared in this study focus on a 2D transient analysis of a rocket nozzle with the assumption of a 10-s burning time. The simulations aim to evaluate the influence of fluid on the structure, emphasizing the heat flux and temperature experienced by the nozzle structure. Furthermore, this paper compares a simulation with a constant timestep throughout the entire burning time and a simulation with varying timesteps, which is being introduced. The first simulation begins with a smaller timestep that increases gradually, resulting in a final time of 10 s, while the second simulation uses a constant timestep of 0.01 s, with a total of 1000 timesteps and 1000 iterations per timestep.

2 Material and Methods

2.1 Nozzle Structure Properties

The rocket nozzle used in this study is a 200 mm diameter constructed from two materials: S45C aluminum alloy and molded graphite. Figure 1 shows the nozzle dimensions in the crosssection, side view. The properties of these materials for the nozzle are as in Table 1.

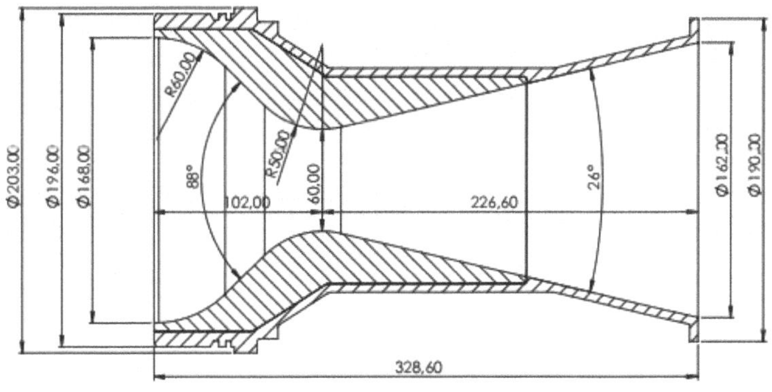

Fig. 1. Technical drawing of the simulated nozzle

In this research, the composite propellant formulation used consists of 75% Ammonium Perchlorate as an oxidizer, 7.5% Aluminum as a fuel, and 17.5% HTPB + TDI as a fuel binder. Based on this formulation data, thermodynamic calculations of the combustion reaction were performed, resulting in data as shown in Table 2. The thermodynamic calculations were carried out using the RPA (Rocket Propulsion Analysis) software.

Table 1. Properties of Materials.

Properties	S-45C	Graphite	Unit
Youngs Modulus	2,05E + 11	1,20E + 10	Pa
Shear Modulus	7,98E + 10	–	Pa
Poisson's Ratio	0,3	0,27	
Density	7850,0	1780,0	Kg/m^3
Thermal Expansion Coefficient	1,3E-05	5,60E-06	m/m.K
Conductivity	49,8	70,19	W/m.K
Specific Heat	486,0	709,97	J/Kg.K
Reference Temperature	20	20	^{0}C

Table 2. Thermodynamics Properties.

Parameter	Chamber	Throat	Exit	Unit
Pressure	4,53	2,55	0,09	MPa
Temperature	2758,15	2514,29	1400,72	K
Molecular Weight	23,52	23,56	23,60	–
Gamma	1,1925	1,2031	1,2369	–

2.2 Numerical Method

A Computational Fluid Dynamics (CFD) simulation was conducted to study the internal flow and predict the interaction between the flow and the rocket structure using ANSYS Fluent [10, 11]. The output parameter of the interaction used in this study is the heat flux transferred from the fluid region to the structure.

i Governing Equation.

The energy equation used in the numerical simulation is given in Eq. [1].

$$\frac{\partial}{\partial t}(\rho E) + \nabla \bullet \left(\vec{v}(\rho E + p)\right) = \nabla \bullet \left(\mathrm{k_{eff}}\nabla T - \sum_j h_j \vec{J}_j + \left(\tau_{eff} \bullet v\right)\right) + S_h \quad (1)$$

where $\mathrm{k_{eff}}$ is the effective conductivity, $\vec{J}_j$ is the diffusion flux of species j.

The heat transport equation is used as Eq [2] for gaining temperature solution.

$$\rho c_p \frac{\partial T}{\partial t} - \nabla \bullet (\mathrm{k}\ \nabla \mathrm{T}) = Q \quad (2)$$

where c_p is the specific heat, k is the thermal conductivity, and Q is the applied heat energy transfer rate.

ii Geometry.

The simulation used 2D computational models constructed based on the nozzle technical drawing (refer to Fig. 1). The domains in the simulations are divided into three zones, depicted in Fig. 2: two solid zones, case and graphite, and the fluid zone. The material settings for case and graphite zones use the properties in Table 1.

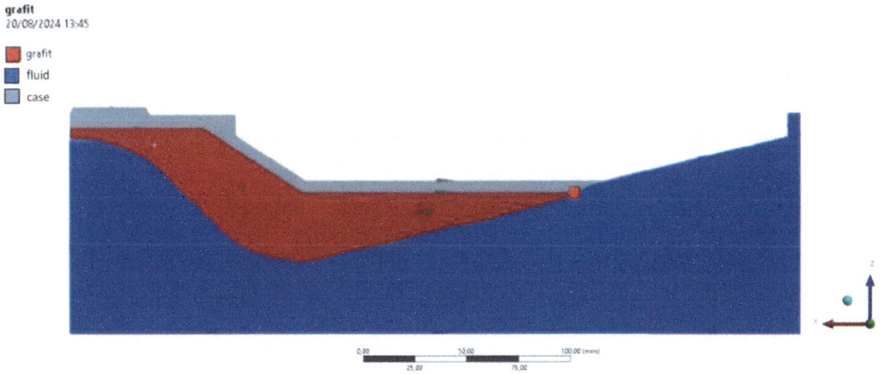

Fig. 2. Simulation domains: case (top), graphite (middle), fluid (bottom)

iii Simulation Settings.

Both simulations are set with the same boundary conditions as listed in Table 3. For the walls interacting with other zones, the thermal condition is set to couple to capture the heat transfer between materials.

In Simulation 1, several simulations were conducted with varying timesteps to compare the differences in results. The first set of simulations locked the number of iterations per timestep, but the number of timesteps was increased tenfold at each step. The second simulation maintained a constant timestep of 0.01 s and was conducted as a single simulation.

In both simulations, initial conditions were set by patching the temperature at 300 K for the structure and 2758 K for the fluid domain (see Fig. 1). This setup allowed for a detailed evaluation of the heat flux and temperature distribution within the nozzle structure (Fig. 3).

Table 3. Boundary conditions.

Boundary conditions	Settings
Inlet	type: pressure-inlet
	Gauge total pressure: 4.5×10^6 Pa
	Initial gauge pressure: 101325 Pa
	Total temperature: 2758 K
Outlet	type: pressure-outlet
	Gauge total pressure: 1 atm
	Total temperature: 1400.72 K
Wall fluid-case	Type: wall
	Material: S45C
	Thermal: Coupling
Wall Fluid-Graphite	Type: wall
	Material: graphite
	Thermal: Coupling
Wall Casing	Type: wall
	Material: S45C
Wall Graphite	Type: wall
	Material: graphite
Wall Case-graphite	Type: wall
	Material: S45C
	Thermal: Coupling
Axis	Axis

3 Simulation Results and Discussion

3.1 Simulation Duration

Table 3 shows the settings for Simulation 1 with the gradually increasing timestep size series, whereas Table 4 provides information on settings in Simulation 2. The duration differed by more than 3 h between the two simulations.

The first two stages of simulation 1, which include timestep sizes smaller than 0.01, were implemented because a larger timestep size, such as 0.1, reduces the chance of convergence in the first second. This is due to a pressure surge occurring between 0 and 1 s, necessitating analysis with a timestep smaller than 0.1 for closer examination (Table 5).

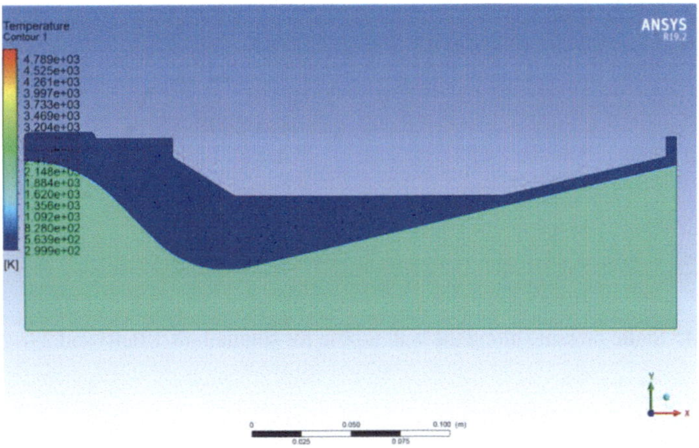

Fig. 3. Patch-initialized simulation domains by temperatures

Table 4. Set of settings and durations in simulation 1.

Stage	Timestep Size	Number of timesteps	Number of iterations per timestep	Final iteration time (s)	Total time spent
1	0.0001	10	100	0.001	5 min
2	0.001	10	1000	0.01	14 min
3	0.01	10	1000	0.1	22 min
4	0.1	100	1000	10.1	38 min
Total duration					79 min

Table 5. Set of settings and durations in simulation 2.

Stage	Timestep Size	Number of timesteps	Number of iterations per timestep	Final iteration time (s)	Total time spent
1	0.01	1000	1000	10	4 h 34 min

3.2 Simulation Results

The simulation results indicate that the static temperature does not change significantly from one second to the next. Differences of less than 10000 Pascals are not visibly noticeable on the graph. Therefore, the average static temperatures between the two simulations are compared.

Figure 4 shows no significant difference by number in static pressure between different transient seconds or between the two simulations and further shows similar trends.

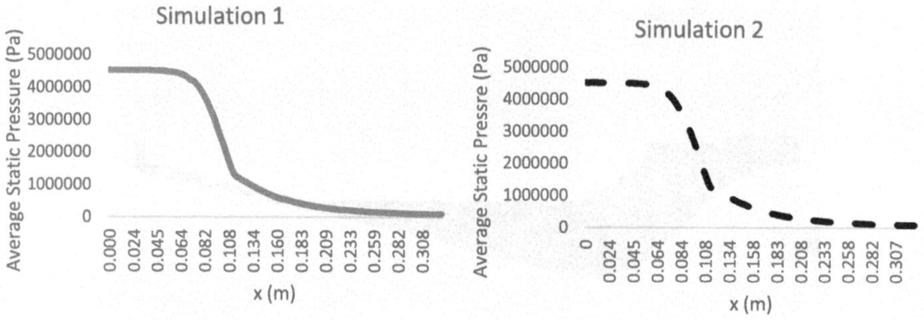

Fig. 4. Static pressure along the wall nozzle for simulations 1 (left) and 2 (right).

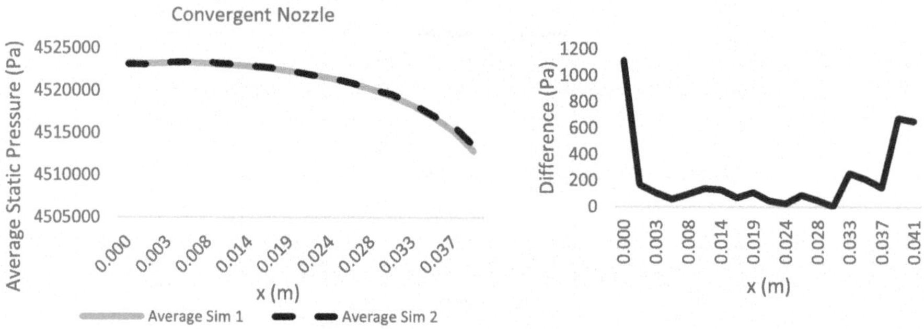

Fig. 5. Static pressure along the wall nozzle in the convergent section for Simulation 1 and Simulation 2 (left) and the difference between both (right).

Regarding this, closer examinations of each nozzle section are conducted to study the differences: convergent, throat, and divergent. The highest difference between simulations 1 and 2 occurs in the nozzle inlet by 1118 Pa, located between points x = 0 and x = 0.014. Figure 5 provides a discrepancy between Fig. 4 in the convergent section (left) and the difference between the two simulation results (right).

Similarly, as shown in **Fig. 6**, the heat flux results do not differ significantly between the two simulations. In both simulations, the highest heat flux occurs in the first second at the nozzle throat area, located 0.11 m from the inlet, with a value of 3,4 MW/m2.

Figure 7 shows the temperature contour in the early (t = 1s), middle (t = 5s), and end of the burning time (t = 10s), with the settings of Simulation 1 (left) and Simulation 2 (right). The results reveal no difference between the temperature range in the nozzle. The structure's temperature in both simulations gradually increased over time until the heat was transferred to the outermost surface, reaching approximately 2100 K by the end of the burning time.

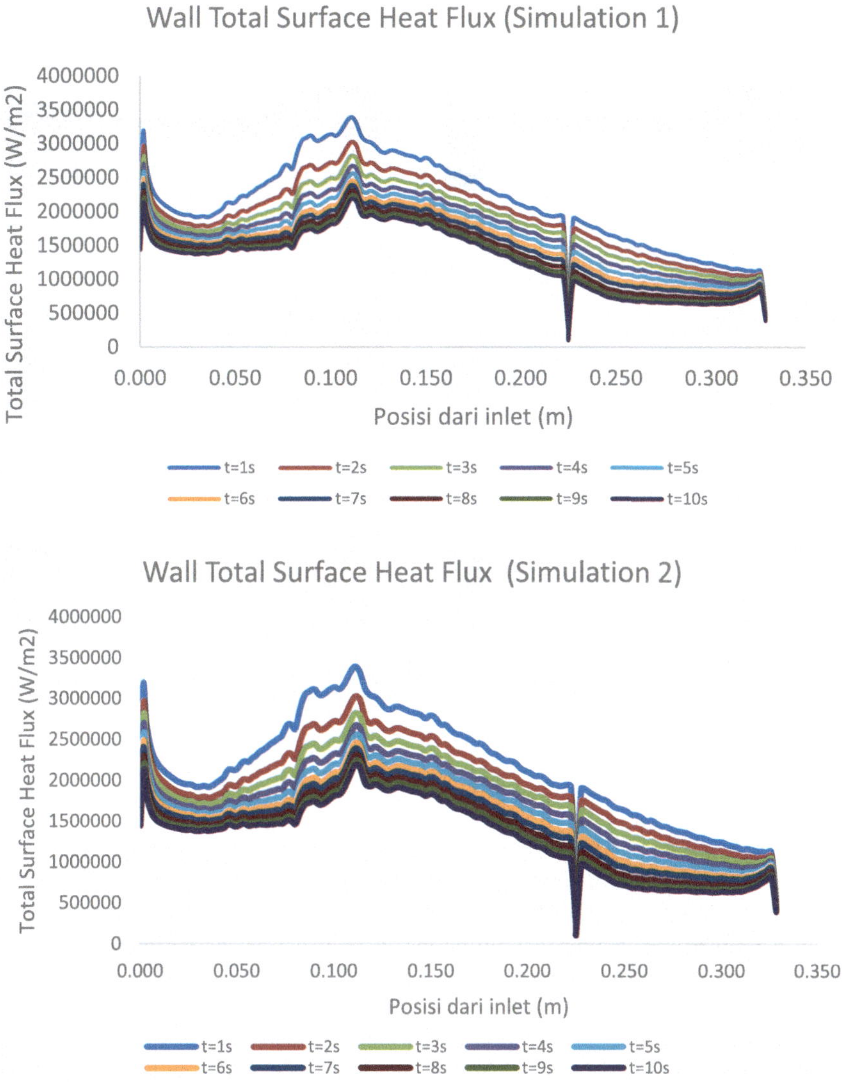

Fig. 6. Total surface heat flux along the nozzle wall in simulations 1 (top) and 2 (bottom).

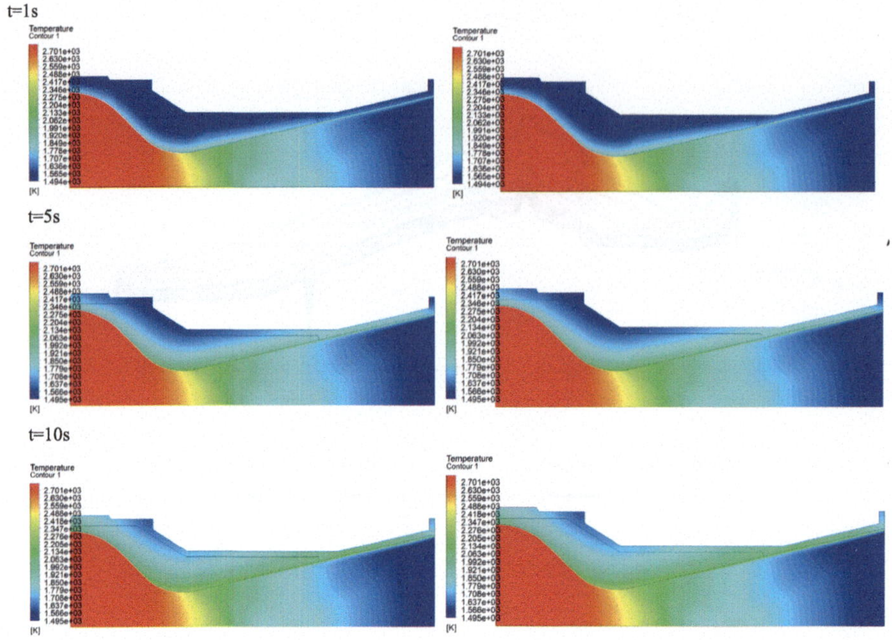

Fig. 7. Temperature contour from t = 1s, t = 5s, and t = 10s in Simulation 1 (left) and Simulation 2 (right).

4 Conclusion

This study has investigated the thermal behavior of the 200-mm caliber rocket nozzle using two different series of simulations. The result shows there are no significant differences. This phenomenon is likely attributable to the iteration used up to 1000 times, which has reached convergence, resulting in minor errors. It can be inferred that with an appropriate number of iterations, the timestep size can be increased up to 0.1 s in internal nozzle simulations for time efficiency while maintaining consistently accurate results. With this result, the simulation series with gradually increasing timestep sizes had a shorter duration of 195 min than the simulation with a constant timestep size of 0.01 s.

References

1. Müller, D., et al.: The solar orbiter mission: science overview. astronomy astrophys. **642**(Special Issue) A1, 1–31 (2020)
2. Goyal, S.K., et al.: Aditya solarwind particle experiment (ASPEX) onboard the Aditya-L1 mission. Planet Space Sci. **163**, 42–55 (2018)
3. Odriozola, S., De Meneses, F.C., Muralikrishna, P., Pimenta, A.P., Kherani, E.A.: Rocket in situ observation of equatorial plasma irregularities in the region between e and F layers over Brazil. Ann. Geophys. **35**(3), 413–422 (2017)

4. Bailey, S.M. et al.: Sounding rocket observation of nitric oxide in the polar night. J. Geophys. Res.: Space Phys. **127**(6) (2022)
5. Namekawa, T., et al.: Rocket observation of sub-relativistic electrons in the quiet dayside auroral ionosphere. J. Geophys. Res.: Space Phys. **126**(7) (2021)
6. Inatani, Y., Ishii, N., Nonaka, S., Abe, T.: Recent activities of sounding rocket and ballooning in Japan. In: 4th International Seminar on Aerospace Science and Technology, pp. 144–152 (2016)
7. Petter, E.: Important Factors for Accurate Scale-Resolving Simulations of Automotive Aerodynamics. Linköping University Electronic Press (2020)
8. Aultman, M., Wang, Z., Duan, L.: Effect of time-step size on flow around generic car models. J. Wind Eng. Indust. Aerodyn. **219** (2021)
9. Satrio, D., Utama, I.K.: Mukhtasor: the influence of time step setting on the CFD simulation result of vertical axis tidal current turbine. J. Mech. Eng. Sci. **12**(1), 3399–3409 (2018)
10. Dîrloman, F.M., et al.: Computational fluid dynamics simulations for composite rocket propellant optimization. In: 9th Modern Technologies in Industrial Engineering. IOP Publishing (2021)
11. Narayana, K.P., Reddy, K.S.: Simulation of convergent divergent rocket nozzle using cfd analysis. IOSR J. Mech. Civ. Eng. **13**(4), 58–65 (2016)

Signal Fidelity Enhancement in Radar Signal Using APSO-MODWPT Framework

Indrawata Wardhana[1(✉)] and Soleh Fajar Junjunan[2]

[1] System Information, UIN STS Jambi, Jambi, Indonesia
indrawataw@uinjambi.ac.id
[2] Rocket Technology Research Center, BRIN, Bogor, Indonesia
soleh.fajar.junjunan@brin.go.id

Abstract. Radar Emitter Signal Classification (RESC) plays a critical role in electronic warfare and communication systems, particularly for accurately identifying signals under challenging conditions, including low Signal-to-Noise Ratio (SNR). This study addresses these challenges by proposing an Advanced Particle Swarm Optimization - Maximum Overlap Discrete Wavelet Packet Transform (APSO-MODWPT) framework. MODWPT offers enhanced frequency resolution and alignment suitable for radar signals; however, parameter optimization remains complex. This issue is addressed using Particle Swarm Optimization (PSO) guided by Kullback-Leibler divergence. Experimental results across BPSK, QPSK, 8-PSK, and 16-QAM signals demonstrate significant SNR improvements. Visual analysis via Continuous Wavelet Transform further validates the framework's effectiveness, significantly reducing noise while preserving echo signal fidelity. This highlights the approach's effectiveness in accurately extracting vital signal features and filtering out unwanted noise, enhancing signal analysis quality and reliability in radar research. The proposed method promises improved signal fidelity and more accurate radar insights.

Keywords: RESC · MODWPT · PSO · SNR

1 Introduction

Radar Emitter Signal Classification (RESC) is crucial in electronic warfare and communication systems. Accurately identifying and distinguishing between different radar signals is essential for effective operation. However, existing algorithms often struggle with low recognition accuracy, especially in environments with low Signal-to-Noise Ratio (SNR). This difficulty in extracting useful features and accurately identifying signal types hampers the performance and reliability of radar systems under challenging conditions. One common strategy to address this issue involves preprocessing steps like feature extraction. Various methods are used to enhance the SNR, including Empirical Mode Decomposition [1], Wavelet Transform [2], Discrete Wavelet Transform (DWT) [3], and Wavelet Packet Transform (WPT) [4]. The DWT and WPT both require the length of the analyzed series to be an integer power of 2 [5].

E. Fitrianingsih et al. (Eds.): ISAST 2024, SPPHY 416, pp. 508–516, 2025.
https://doi.org/10.1007/978-981-96-1344-1_53

In contrast, the Maximum Overlap Discrete Wavelet Packet Transform (MODWPT) does not have this limitation and can analyze sequences of arbitrary length. Additionally, while both MODWPT and DT-CWT are shift-invariant, MODWPT offers better time-frequency resolution than DT-CWT [6]. The MODWPT demonstrates that this decomposition method enhances the SNR by 19.4257 dB when using Symlets12 (sym12) [7]. A limitation of MODWPT is the difficulty in defining optimal parameters. Particle Swarm Optimization (PSO) can be used to optimize the parameters to solve this problem [8]. Furthermore, Kullback–Leibler (KL) - divergence is utilized as an evaluation criterion for parameter optimization between MODWPT and PSO. This paper proposes an APSO-MODWPT method to address the challenge.

2 Method

This section outlines the methodology for developing the Advanced Particle Swarm Optimization - Maximum Overlap Discrete Wavelet Packet Transform (APSO-MODWPT) framework for Radar Emitter Signal Classification (RESC). The process includes signal preprocessing, feature extraction, and parameter optimization.

2.1 Signal Model

The signals used in the following experiments are composed of modulated signals and additive White Gaussian noise (WGN). This paper focuses on common low probability of intercept (LPI) radar signals, including Binary Phase Shift Keying (BPSK), Quadrature Phase Shift Keying (QPSK), 8-Phase Shift Keying (8-PSK), and 16-Quadrature Amplitude Modulation (16-QAM). These signals are processed in the presence of GWN. The signals are generated using the parameter presented in Table 1.

Generally, the signal model can be expressed as: $\vartheta(\theta) = \upsilon(\theta) + \varepsilon(\theta)$, where $\vartheta(\theta)$ is the received signal. $\upsilon(\theta)$ represents the modulated signal, and $\varepsilon(\theta)$ denotes AWGN with power σ_n^2. The modulated signal is described in:

$$\upsilon(\theta) = [A_m \sum\nolimits_n a_n h(t - nT_s)] cos(2\pi(f_c + f_m)t + \xi_0 + \xi_m) \tag{1}$$

where A_m, a_n, T_s, f_c, f_m, ξ_0, ξ_m denote the modulation amplitude, symbol sequence, symbol period, carrier frequency, modulation frequency, initial phase, and modulation phase, respectively, and $h(t)$ is a function expressed as $h(t) = 1$ if $1 \leq t \leq T_s$; 0 other. . The signal-to-noise ratio (SNR) is used to add recognition complexity and is defined as [9]:

$$\text{SNR(dB)} = 10 * log_{10} \frac{E_{signal}}{E_{noise}} \tag{2}$$

where E_{signal} and E_{noise} represent the energies of the signal and noise, respectively. This setup provides a realistic simulation environment for evaluating the performance of radar signal classification algorithms under various SNR conditions.

Table 1. Signal Parameter.

SNR	Center Frequency	Sample Rate	Path Delays	Average Path Gains
10 dB	900 MHz	200 kHz	[0 9.0 × 10^{-6} – 1.7 × 10^{-5}]	[0 –2 –10]

2.2 Modwpt

The DWT [10] is a mathematical tool commonly used in data analysis, especially for handling real-valued time series data, which is typically organized as a column vector (x) containing observations ($x_0, x_1, \ldots, x_{n-1}$). Often, ($N$) is chosen as a power of 2 to simplify the process. The DWT technique employs two key filters: a low-pass filter (g_i) and a high-pass filter (h_i). The low-pass filter is designed to capture the approximate or smooth components of the signal, while the high-pass filter focuses on the detailed aspects [11]. The length of the filter (S) is selected based on the specific requirements of the analysis or the type of wavelet applied.

$$A_{j,t} = \sum\nolimits_{l=0}^{S-1} g_l A_{j-1,2t-l} \tag{3}$$

$$B_{j,t} = \sum\nolimits_{l=0}^{S-1} h_l A_{j-1,2t-l} \tag{4}$$

In this context, ($A_{j,t}$) and ($B_{j,t}$) represent the scaling and wavelet coefficients at stage (j) and time (t), respectively. In the DWT, these filters are applied successively to the time series data, with each level followed by a down-sampling step. Downsampling involves keeping every alternate data point, effectively halving the data size at each level. The low-pass filter satisfies this condition:

$$\sum\nolimits_{l=0}^{S-1} g_l g_{l-2n} = \delta[n]; \sum\nolimits_{l=0}^{S-1} h_l h_{l-2n} = \delta[n]; \sum\nolimits_{l=0}^{S-1} g_l h_{l-2n} = 0 \tag{5}$$

and $\delta[n]$ is the kronecker delta function. The MODWPT can be considered an enhanced version of the DWT [12]. Unlike the DWT, which restricts the sample size to an integer multiple of 2^j at level j, the MODWPT at level j can be applied to sample sizes of any value N. The scaling of the fundamental filters must be adjusted to preserve energy conservation. The MODWPT adapts the DWT to accommodate any sample size by modifying the scaling of the filter coefficients: $\tilde{g}_l = 2^{-j/2} g_l\ \tilde{h}_l = 2^{-j/2} h_l$. The MODWPT coefficients are computed similarly to the DWT coefficients but with the adjusted filters and without downsampling:

$$\tilde{A}_{j,t} = \sum_{l=0}^{S-1} \tilde{g}_l A_{j-1,t-l} \tag{6}$$

$$\tilde{B}_{j,t} = \sum_{l=0}^{S-1} \tilde{h}_l A_{j-1,t-l} \tag{7}$$

In the context of MODWPT, the input signal is broken down into multiple coefficients by employing low-pass and high-pass filters at different decomposition levels, as shown in Fig. 1.

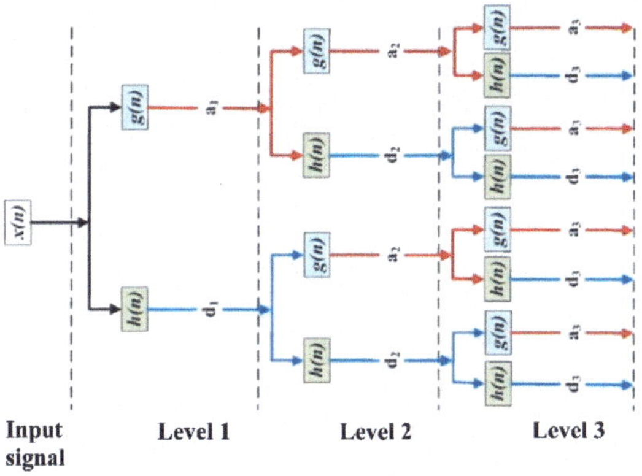

Fig. 1. Three-level examples of signal decomposition using MODWPT [13].

2.3 Particle Swarm Optimizer (PSO)

The PSO arithmetic, developed by Kennedy and Eberhart [14], is a highly efficient search algorithm inspired by the foraging behavior of birds. This technique falls under the category of evolutionary algorithms, akin to methods like simulated annealing, which focuses on iteratively seeking the optimal solution. The key steps involved in PSO arithmetic include [15]:

- Create a population of particles, each with a size N. Assign each particle an initial position pop_i and an initial velocity.
- Calculate the fitness value, *fitness(i)*, for each particle. This fitness function determines how well a particle's position solves the problem.
- Compare each particle's current fitness value, *fitness(i)*, with its personal best fitness value, $p_{best}(i)$. If *fitness(i)* is better, update $p_{best}(i)$ to *fitness(i)*.
- Compare each particle's fitness value, *fitness(i)*, with the global best fitness value, $g_{best}(i)$. If *fitness(i)* exceeds $g_{best}(i)$, update $g_{best}(i)$ to *fitness(i)*.
- Adjust each particle's velocity and particle position pop_i in Eq. 8 9. The term ω represents the inertia weight, which influences how much of the previous velocity is retained. Parameters c_1 and c_2 are learning factors, where c_1 denotes the individual cognitive coefficient and c_2 represents the social acceleration constant for particle i. Additionally, r_1 and r_2 are independent random numbers drawn from a uniform distribution in the range [0,1].

$$v_{id}(t+1) = \omega v_{id}(t) + c_1 r_1 [p_{id} - pop_{id}(t)] + c_2 r_{21} [p_{gd} - pop_{id}(t)], \tag{8}$$

$$pop_{id}(t+1) = pop_{id}(t+1) + v_{id}(t+1) \tag{9}$$

- Determine whether the termination condition has been satisfied. If the solution meets the desired accuracy or if the maximum number of iterations has been reached, the process ends. If not, return to Step 2 to continue the iteration.

2.4 Kullback-Leibler (KL)

KL divergence is a statistical metric used to assess the difference between two probability distributions: the true distribution $p = \{p_i\}$ and a model or candidate distribution $q = \{q_i\}$. The KL divergence is expressed mathematically as [16]:

$$D_{KL}(p||q) = \sum_i p_i log_2 \frac{p_i}{q_i} \tag{10}$$

The main properties of KL divergence include:

- It is always greater than or equal to zero ($D_{KL} \geq 0$).
- It is asymmetric, meaning D_{KL} $(p|| q)$ is not necessarily equal to D_{KL} $(q|| p)$.
- The divergence is zero only when the distributions p and q are identical.
- It can reach infinity if q assigns zero probability to an event that p considers possible.

2.5 Framework

The framework of APSO-MODWPT is described in Fig. 2. This framework is divided into three steps: radar signal generation, noise addition, and the main system. The first step involves generating radar emitter signals, including BPSK, QPSK, 8-PSK, and 16-QAM, using the parameters described in Table 1. The next step adds noise using GWN with initial noise variances. The final step combines three methods: MODWPT, KL-Divergence, and PSO. The signal is first decomposed by MODWPT into a series of wavelet packet coefficients across different levels and frequency bands, using initial wavelet functions. KL-Divergence and PSO are then used to analyze the coefficients obtained from the MODWPT and select the coefficients corresponding to the desired signal characteristics. This process is iterated until the optimal SNR results are achieved. For more details, the framework concepts are defined in the APSO-MODWPT algorithm. Wavelet families used in this research are described in Table 2.

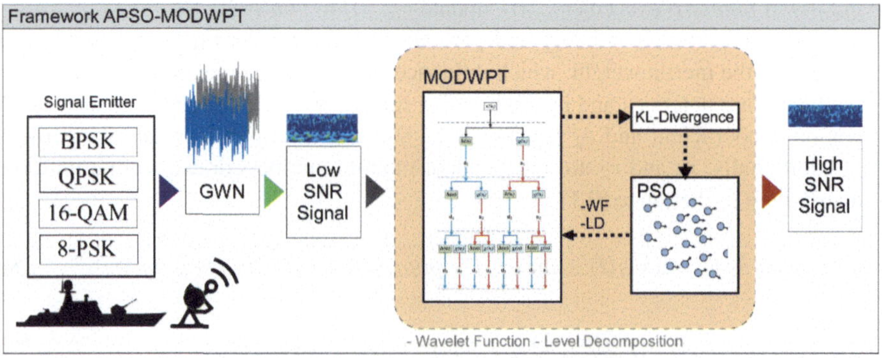

Fig. 2. Framework of APSO-MODWPT.

Algorithm. APSO-MODWPT:
Input: Original Signal, Wavelet Function, Decomposition Levels, Threshold: $x(t)$,wv, level , $\Im$
Steps:

1. Perform the MODWT method on $x(t)$ to get wavelet coefficients for each level.
2. Define the energy of each node, then rank.
3. Use threshold to define the selected node
4. Reconstruct the signal by using inverse MODWT (iMODWT).
5. Calculate the SNR of the reconstructed signal using Eq. 2
6. Calculate the probability distributions of SNR values after decomposition.
7. Compute the KL-divergence to measure the difference between the two distributions using Eq. 10
8. Optimize Thresholds Using PSO using Eq. 8-9. Use PSO to optimize the threshold values to maximize SNR.
9. Initialize Particles: Randomly set initial positions and velocities.
10. Evaluate Fitness: Compute SNR for each particle's threshold set.
11. Update Positions and Velocities: Adjust based on individual and global best solutions.
12. Iterate: Repeat until the convergence or stopping criterion is satisfied.
13. 5. **Output**: Optimal Thresholds, Best SNR Value, Signal Optimize: $\Im_{opt}$, SNR_{best}, $x(t)_{opt}$

Table 2. Wavelet families and filters.

Wavelet Family	Wavelet Filter	Wavelet Filter Length
Daubechies	db2, db3, db4, db5, db6	4,6,8,10,12
Coiflets	coif1, coif2	6,12
Symlets	sym4, sym6	8,12

3 Results and Discussion

In Fig. 3(a), the QPSK signal is zoomed in to show the signal with noise. To reduce the noise and enhance the SNR, we use the APSO-MODWPT framework. The result can be seen in Fig. 3(b), where the noise in the signal is significantly reduced. For more detail, we transformed the original and refined signals using the continuous wavelet transform (CWT), as shown in Fig. 3(c) and Fig. 3(d). It can be clearly seen in Fig. 3(d) that the refined signal, highlighted by the red rectangle, shows stable amplitude in the same frequency range. This CWT image is easier to interpret as a known signal than Fig. 3(c), making it more suitable as a pattern input for machine learning classification.

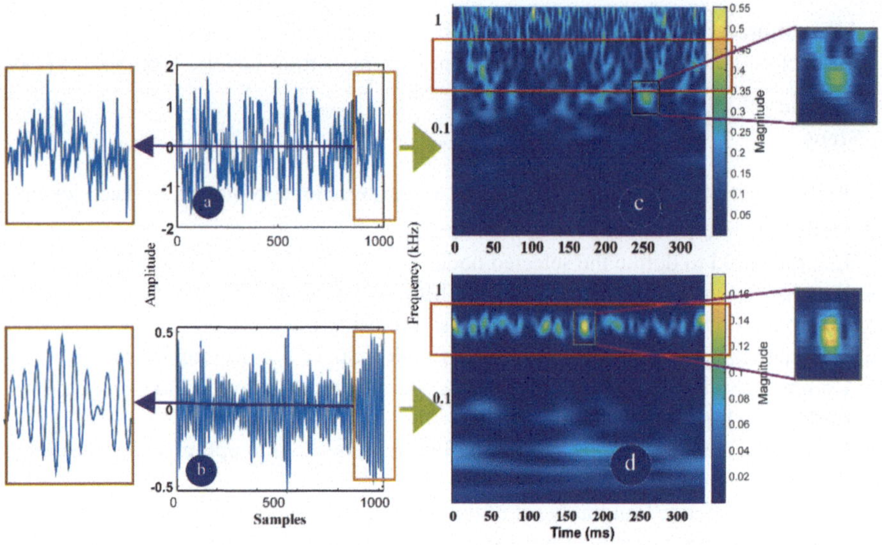

Fig. 3. (a) QPSK Signal (b) Refine Signal (c) WVD Signal (d) WPD Refine Signal.

We conducted additional tests further to evaluate the framework with different types of signals. The results for the four types of signals are shown in Table 3. The results demonstrate significant SNR improvements across various radar emitter signals using a specific enhancement framework. The BPSK signal improved by approximately 8.78 dB, while QPSK and 16-QAM saw increases of around 8.71 dB and 8.41 dB, respectively. Notably, the 8PSK signal exhibited the greatest improvement, increasing about 12.91 dB. These enhancements indicate the framework's strong capability in reducing noise and clarifying signals, which is crucial for radar systems. Future work could further optimize this framework to achieve even better SNR results and explore its efficiency and applicability in real-time scenarios.

Table 3. Radar Emitter Signal SNR Augmented.

Signal Emitter	SNR Initial	SNR Framework
BPSK	−11.8954	−3.1169
QPSK	−7.98260	0.7284
16-QAM	−10.8623	−2.4504
8PSK	−13.5919	−0.6794

4 Conclusion

The application of the APSO-MODWPT framework demonstrates a significant reduction in noise and an enhancement of SNR in RESC. The CWT analysis reveals that the refined signal maintains a stable amplitude within the same frequency range, making it more recognizable. Further evaluation with various signal types confirms that this framework consistently improves SNR quality across different scenarios. The results suggests that the framework is a robust tool for enhancing signal clarity, facilitating more accurate signal analysis, and suitable for machine learning classification.

References

1. Wang, S.Q., Wang, C.Y., Sun, Q., Gao, C.Y., Yang, S.G.: Radar emitter signal intra-pulse feature extraction based on empirical mode decomposition. Procedia Comput. Sci. **154**, 504–507 (2018). https://doi.org/10.1016/j.procs.2019.06.078
2. Wang, X.: Electronic radar signal recognition based on wavelet transform and convolution neural network. Alexandria Eng. J. **61**, 3559–3569 (2022). https://doi.org/10.1016/j.aej.2021.09.002
3. Singh, P., Shree, R., Diwakar, M.: A new SAR image despeckling using correlation based fusion and method noise thresholding. J. King Saud Univ. - Comput. Inf. Sci. **33**, 313–328 (2021). https://doi.org/10.1016/j.jksuci.2018.03.009
4. Cao, Y., Bai, J.L., Li, H., Zhao, J., Liu, S., Gao, Z.: Deep representation method for radar emitter signal using wavelet packets decomposition. J. Eng. **2019**, 6282–6286 (2019). https://doi.org/10.1049/joe.2019.0256
5. Jagannath, A., Jagannath, J., Kumar, P.S.P.V.: A comprehensive survey on radio frequency (RF) fingerprinting: traditional approaches, deep learning, and open challenges. Comput. Networks. **219**, 109455 (2022). https://doi.org/10.1016/j.comnet.2022.109455
6. Zhang, T., Chen, W., Li, M.: Classification of inter-ictal and ictal EEGs using multi-basis MODWPT, dimensionality reduction algorithms and LS-SVM: a comparative study. Biomed. Signal Process. Control **47**, 240–251 (2019). https://doi.org/10.1016/j.bspc.2018.08.038
7. Subramaniyan, N., Buch, J.J.U., Prince, A.A., Pathak, S.: De-noising of microwave reflectometry signal using maximal overlap discrete wavelet packet transform for plasma density measurement. Meas. J. Int. Meas. Confed. **222**, 113564 (2023). https://doi.org/10.1016/j.measurement.2023.113564
8. Zhang, W., Ma, D., Wei, J.J., Liang, H.F.: A parameter selection strategy for particle swarm optimization based on particle positions. Expert Syst. Appl. **41**, 3576–3584 (2014). https://doi.org/10.1016/j.eswa.2013.10.061
9. Shaik, M., Shaik, A.G., Yadav, S.K.: Hilbert–Huang transform and decision tree based islanding and fault recognition in renewable energy penetrated distribution system. Sustain. Energy, Grids Netw. **30**, 100606 (2022). https://doi.org/10.1016/j.segan.2022.100606
10. Tyagi, C.S.: 2008Comparative study of SVM classifiers and artificial neural networks application for rolling element bearing fault diagnosis using wavelet transform preprocessing Neuron (2008).https://doi.org/10.5281/zenodo.1072642
11. Toma, R.N., Kim, J.M.: Article bearing fault classification of induction motors using discrete wavelet transform and ensemble machine learning algorithms. Appl. Sci. **10** (2020). https://doi.org/10.3390/APP10155251
12. Zhang, T., Han, K., Li, H., Liu, S., Chen, F.: Two terminals protection method for HVDC transmission lines based on maximal overlap discrete wavelet transform and cross-correlation coefficient of traveling wave. Energy Rep. **9**, 1133–1148 (2023). https://doi.org/10.1016/j.egyr.2023.04.134

13. Shrifan, N.H.M.M., Akbar, M.F., Mat Isa, N.A.: Maximal overlap discrete wavelet-packet transform aided microwave nondestructive testing. NDT E Int. **119**, 102414 (2021)
14. Eberhart, R., Kennedy, J.: A new optimizer using particle swarm theory. In: MHS 1995. Proceedings of the Sixth International Symposium on Micro Machine and Human Science, pp. 39–43 (1995)
15. Qin, K., Zhang, Q., Liu, Y., Yang, B., Tang, X., Zhuang, N.: Dimensionality reduction combined with particle swarm optimization algorithm for solving high-dimensional nuclear data target accuracy evaluation problem. Ann. Nucl. Energy **198**, 110285 (2024). https://doi.org/10.1016/j.anucene.2023.110285
16. Shlens, J.: Notes on Kullback-Leibler Divergence and Likelihood. 1–4 (2014)

Author Index

E. Fitrianingsih et al. (Eds.): ISAST 2024, SPPHY 416, pp. 517–520, 2025.
https://doi.org/10.1007/978-981-96-1344-1

W

Y

The manufacturer's authorised representative in the EU is Springer Nature Customer Service Centre GmbH, Europaplatz 3, 69115 Heidelberg, Germany. If you have any concerns regarding our products, please contact ProductSafety@springernature.com

Printed and bound by CPI Group (UK) Ltd, Croydon, CR0 4YY
01/10/2026
02225931-0006